Developments in Geotechnical Engineering 14B

ROCKSLIDES AND AVALANCHES, 2

Further titles in this series:

1. *G. SANGLERAT*
THE PENETROMETER AND SOIL EXPLORATION

2. *Q. ZARUBA AND V. MENCL*
LANDSLIDES AND THEIR CONTROL

3. *E.E. WAHLSTROM*
TUNNELING IN ROCK

4A. *R. SILVESTER*
COASTAL ENGINEERING, I

Generation, Propagation and Influence of Waves

4B. *R. SILVESTER*
COASTAL ENGINEERING, II

Sedimentation, Estuaries, Tides, Effluents and Modelling

5. *R.N. YONG AND B.P. WARKENTIN*
SOIL PROPERTIES AND BEHAVIOUR

6. *E.E. WAHLSTROM*
DAMS, DAM FOUNDATIONS, AND RESERVOIR SITES

7. *W.F. CHEN*
LIMIT ANALYSIS AND SOIL PLASTICITY

8. *L.N. PERSEN*
ROCK DYNAMICS AND GEOPHYSICAL EXPLORATION

Introduction to Stress Waves in Rocks

9. *M.D. GIDIGASU*
LATERITE SOIL ENGINEERING

10. *Q. ZARUBA AND V. MENCL*
ENGINEERING GEOLOGY

11. *H.K. GUPTA AND B.K. RASTOGI*
DAMS AND EARTHQUAKES

12. *F.H. CHEN*
FOUNDATIONS ON EXPANSIVE SOILS

13. *L. HOBST AND J. ZAJIC*
ANCHORING IN ROCK

14A. *B. VOIGHT* (Editor)
ROCKSLIDES AND AVALANCHES, 1

Natural Phenomena

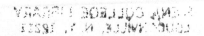

Developments in Geotechnical Engineering 14B

ROCKSLIDES AND AVALANCHES, 2

Engineering Sites

by

BARRY VOIGHT (Editor)

College of Earth and Mineral Sciences,
The Pennsylvania State University, University Park, Pennsylvania, U.S.A.

ELSEVIER SCIENTIFIC PUBLISHING COMPANY
Amsterdam — Oxford — New York 1979

ELSEVIER SCIENTIFIC PUBLISHING COMPANY
335 Jan van Galenstraat
P.O. Box 211, Amsterdam, The Netherlands

Distributors for the United States and Canada:

ELSEVIER NORTH-HOLLAND INC.
52, Vanderbilt Avenue
New York, N.Y. 10017

Library of Congress Cataloging in Publication Data   (Revised)

Main entry under title:

Rockslides and avalanches.

   (Developments in geochemical engineering ; 14A-14B)
   Includes bibliographies and indexes.
   CONTENTS: 1. Natural phenomena.--2. Engineering
sites.
   1. Mass-wasting. 2. Avalanches. 3. Soil
mechanics. I. Voight, Barry. II. Series.
QE598.2.R6   551.3   77-17810
ISBN 0-444-41507-6 (v. 1)

  ISBN 0-444-41507-6 (Vol. 14a)
  ISBN 0-444-41508-4 (Vol. 14b)
  ISBN 0-444-41662-5 (Series)

Printed in The Netherlands

"In the United States, the catastrophic descent of the slopes of the deepest cut on the Panama Canal issued a warning that we were overstepping the limits of our ability to predict the consequences of our actions . . ."

Karl Terzaghi
1936

PREFACE

Rockslides and Avalanches attempts to provide a foundation for studies of mass movement phenomena in the Western Hemisphere. The project began in 1973 during excursion preparations for the 3rd International Congress on Rock Mechanics, when it had become apparent that sufficient modern work had been accomplished at most important North and South American landslide localities to make possible the preparation of a comprehensive standard reference on the subject.

The work is divided into two volumes. The first emphasizes natural phenomena, and the second deals with mass movements as related to engineering projects. Unlike Albert Heim's classic summary of Alpine slide phenomena, *Bergsturz und Menschenleben*, our volumes are multiple-authored, containing 48 contributions from 70 authors. The overall effort is dedicated equally to four outstanding men, namely Albert Heim, Josef Stini, Karl Terzaghi, and Laurits Bjerrum. Introductory chapters in each of two volumes outline many of their important contributions. The dedication chapters for Stini, Terzaghi, and Bjerrum are given in Volume 2, because to a great extent the works of these men were carried out in conjunction with full-scale engineering projects, where precise field measurements could often be made. For them, the distinction between theory and practice had little meaning. The chapter on Heim is given in Volume 1 in accordance with the volume theme, viz. studies of natural phenomena. In retrospect, it may be appropriate to mention that one of these gentlemen had at one time recommended to me that an attempt be made

> "to collect details of slides and slips occurring on plane or partly plane surfaces. . . A collection of available data — evaluated critically — would be of great value even if it. . . should not be possible to end up with final solutions to the problem."

To a great extent this goal seems realized by completion of this work, certainly to a much fuller extent than any single person could manage by individual effort.

To list all those who have contributed to this work, directly or indirectly, is a difficult task; but the attempt must be made. Many are by preference anonymous, and for these and others who may be momentarily forgotten as this list is compiled, I express my deep appreciation.

The authors have obviously devoted considerable effort to chapter preparation; without exception they have shown extraordinary patience and courtesy to my editorial attempts, and they have my sincere gratitude. Next, for wise counsel and/or helpfulness in various matters, I am indebted to Kenneth S. Lane, L. "Spike" Underwood, W.R. Judd, G.B. Wallace, C. Fairhurst, G.F. Sowers, Richard E. Gray, Dewayne Misterek, R. Goodman, D.C. Banks, J.R. Lutton, W.G. Pariseau, F. Jørstad, D.T. Griggs, J.C. Sharp, B. Ladanyi, C.O. Brawner, H.W. Nasmith, R.W. Tabor, R. Stroud, J.F. Shroder, Jr., R.L. Slingerland, G. Eisbacher, R. Parizek and S. Thorarinsson.

Publication of the work was for a time scheduled by the Geological Society of America and numerous chapters were improved as a result of comments by the Society editorial staff and reviewers. Special mention is due in this regard to Bennie Troxel, D.J. Varnes, A. Wellck, C. Barteldes, D. Merrifield and I. Woodall.

All chapters have been subjected to technical review and all have been thereby improved. A reviewer's work too often remains unpraised, but as editor I have been impressed by the importance of their efforts. An important share of any credit for this project is theirs: J.F. Abel, Jr., J.T. Andrews, E.T. Cleaves, R. Colton, E. Dobrovolny, J.J. Emery, M.C. Everitt, R.S. Farrow, R.W. Fleming, R.M. Goodman, W.R. Hansen, R. Hardy, E.L. Hamilton, B.C. Hearne, F. Heuzé, L.B. James, A.M. Johnson, R.B. Johnson, T.C. Kenney, B. Ladanyi, F.T. Lee, P.J. Lorens, B. Mears, Jr., D.M. Morton, H.H. Neel, W.R. Normark, L.C. Pakiser, W.G. Pariseau, R.I. Perla, S.S. Philbrick, G. Plafker, W.G. Pierce, P.A. Schaerer, R. Scholten, R. Shreve, H.W. Shu, R.L. Shuster, R.L. Slingerland, D.L. Turcotte, D.J. Varnes, E.M. Winkler. My personal indebtedness to the following must also be recorded: R.C. Gutschick and E.M. Winkler, who first introduced me to research in slope mechanics at the University of Notre Dame; Archie MacAlpin, Leroy Graves, Ray Plummer, Steponas Kolupaila, Harry Saxe, Bill Fairley, and M.J. Murphy, also of Notre Dame; Fred Donath, Marshall Kay, Walter Bucher, and Rhodes Fairbridge at Columbia University; my Penn State colleagues, especially Rob Scholten, Gene Williams, the late Jon Weber, Dick Parizek, L.A. Wright, Rob Texter, D.P. Gold, and Barton Jenks; Guntram Innerhofer, of Tschagguns; O. Schmidegg, Innsbruck; Jacques Dozy, Delft; Andrej Werynski, Warsaw; Bengt Broms, Stockholm; Anders Rapp, Uppsala; Claire, Agnes, Rose and George Voight, Mary Raak, Arthur Cheesman and Frank Joseph of Yonkers; Tony Kamp, Miami; Nellie and George, Croton Dam and Shickshinny; fellow naturalists Chip Taylor, New York City, and J.P. Voight, Togwotee, Wyoming; Barbara and Elmer; and Lisa, Barb and MaryAnne.

Enormous assistance in preparation of this work was provided by Barb Dauria, Judy Bailey, and Dotty Duck of the Geosciences Department, and by Emilie McWilliams and coworkers of the Earth and Mineral Sciences Library at Penn State.

I thank the Albany Institute of History and Art, N.Y., for permission to include Thomas Cole's lithograph, "Distant view of the slides that destroyed the Willey family, White Mountains", as the frontispiece of Volume 1. Funds which permitted reproductions in color were generously supplied by Charles Hosler of the College of Earth and Mineral Sciences at Penn State.

Finally, I express my appreciation to F. van Eysinga and H. Frank of Elsevier Scientific Publishing Company, for pleasant collaboration, consideration in several matters and for a job well done.

BARRY VOIGHT
State College, Pennsylvania

LIST OF CONTRIBUTORS

D.C. BANKS — U.S. Army Engineer Waterways Experiment Station, P.O. Box 631, Vicksburg, Mississippi, U.S.A.

C.O. BRAWNER — Golder-Brawner & Associates, 224 West 8th Ave., Vancouver, British Columbia, Canada

B.R. CARTER — Law Engineering Testing Company, Atlanta, Georgia, U.S.A.

G.W. CLOUGH — Department of Civil Engineering, Stanford University, Stanford, California, U.S.A.

D.F. COATES — Director-General CANMET, Department of Energy, Mines and Resources, 555 Booth St., Ottawa, Ontario, Canada

A.M.M. COSTA COUTO e FONSECA — Secretaria de Obras e Serviços Públicos do Estado do Rio de Janeiro, Rio de Janeiro, Brazil

A.J. DA COSTA NUNES — Universidade Federal do Rio de Janeiro, Rio de Janeiro and Tecnosolo, S/A, Rua Pedro Alves 13, 15, Santo Cristo, Rio de Janeiro, Brazil

H.K. DUPREE — Geology Section, Upper Missouri River Division, U.S. Bureau of Reclamation, Billings, Montana, U.S.A.

L. ESPINOSA — Sociedad Mexicana de Mecanica de Rocas, A.C., Londres 44-2° piso, Coyoacán, México, D.F., Mexico

V.R. EYZAGUIRRE — Cerro de Pasco Corporation, Lima, Peru

H.F. FERGUSON — Foundation and Materials Branch, Pittsburgh District U.S. Army Corps of Engineers, Pittsburgh, Pennsylvania, U.S.A.

R.E. GRAY — General Analytics, Inc., 570 Beatty Red., Monroeville, Pennsylvania, U.S.A.

J.V. HAMEL — Hamel Geotechnical Consultants, 1992 Butler Drive, Monroeville, Pennsylvania, U.S.A.

R.E. HUNT — Tecnosolo, S/A, Rua Pedro Alves, 13, 15, Santo Cristo, Rio de Janeiro, Brazil

B.A. KENNEDY — Golder Associates, Inc., 10628 N.E. 38th Place, Kirkland (Seattle), Washington, U.S.A.

F.T. LEE — U.S. Geological Survey, Federal Center, Denver, Colorado, U.S.A.

B. LESTER — College of Earth and Mineral Sciences, Pennsylvania State University, University Park, Pennsylvania, U.S.A.

J.K. LOU — British Columbia Hydro and Power Authority, 700 Pender St., Vancouver, British Columbia, Canada

R.J. LUTTON — U.S. Army Engineer Waterways Experiment Station, P.O. Box 631, Vicksburg, Mississippi, U.S.A.

B.C. McLEOD — Geotechnical and Materials Testing Branch, British Columbia Department of Highways, Victoria, British Columbia, Canada

R.H. MERRILL	*Denver Mining Research Center, Bureau of Mines, U.S. Department of the Interior, Denver, Colorado, U.S.A.*
N.R. MORGENSTERN	*Department of Civil Engineering, University of Alberta, Edmonton, Alberta, Canada*
L. MÜLLER	*Paracelsusstrasse 2, Salzburg, Austria*
L.T. MURDOCK	*Dames & Moores Consulting Engineers, 445 S. Figueroa St., Los Angeles, California, U.S.A.*
W. MYSTKOWSKI	*Colorado Division of Highways, Denver, Colorado, U.S.A.*
W.G. PARISEAU	*Department of Mining, Metallurgy, and Fuels, University of Utah, Salt Lake City, Utah, U.S.A.*
D.R. PARKES	*Planning Branch, British Columbia Department of Highways, Victoria, British Columbia, Canada*
R.B. PECK	*1101 Warm Sands Drive, S.E. Albuquerque, New Mexico, U.S.A. and Professor Emeritus, University of Illinois, Urbana, Illinois, U.S.A.*
D.R. PITEAU	*D.R. Piteau and Associates, Ltd., Kapilano 100, West Vancouver, British Columbia, Canada*
G. PLAFKER	*U.S. Geological Survey, Menlo Park, California, U.S.A.*
R. SÁNCHEZ-TREJO	*Sociedad Mexicana de Mecanica de Rocas, A.C., Londres 44-2° piso, Coyoacán, México, D.F., Mexico*
B.L. SEEGMILLER	*Seegmiller Associates, 447 East 200 South, Salt Lake City, Utah, U.S.A.*
R.L. SLINGERLAND	*College of Earth and Mineral Sciences, Pennsylvania State University, University Park, Pennsylvania, U.S.A.*
G.F. SOWERS	*Chairman of the Board, Law Engineering Testing Company, P.O. Box 98008, Atlanta, Georgia, U.S.A. and Regents Professor of Civil Engineering, Georgia Institute of Technology, Atlanta, Georgia, U.S.A.*
P.F. STACEY	*Golder-Brawner & Associates, 224 West 8th Ave., Vancouver, British Columbia, Canada*
R.M. STATEHAM	*Denver Mining Research Center, Bureau of Mines, U.S. Department of the Interior, Denver, Colorado, U.S.A.*
W.E. STROHM, Jr.	*U.S. Army Engineer Waterways Experiment Station, P.O. Box 631, Vicksburg, Mississippi, U.S.A.*
R. SWEIGARD	*College of Earth and Mineral Sciences, Pennsylvania State University, University Park, Pennsylvania, U.S.A.*
G.J. TAUCHER	*Geology Section, Upper Missouri River Division, U.S. Bureau of Reclamation, Billings, Montana, U.S.A.*
R.D. TERZAGHI	*3 Robinson Circle, Winchester, Massachusetts, U.S.A.*
S. THOMSON	*Department of Civil Engineering, University of Alberta, Edmonton, Alberta, Canada*
B. VOIGHT	*College of Earth and Mineral Sciences, Pennsylvania State University, University Park, Pennsylvania, U.S.A.*
L.J. WEST	*Dames & Moore Consulting Engineers, 445 S. Figueroa St., Los Angeles, California, U.S.A.*
Y.S. YU	*Department of Energy, Mines, and Resources, 555 Booth St., Ottawa, Ontario, Canada*

SELECTED SI CONVERSION FACTORS

English unit	SI unit	Conversion factor F ($F \times$ English unit = SI unit)
inch (in)	metre (m)	0.02540
foot (ft)	metre (m)	0.3048
square inch (in^2)	square metre (m^2)	6.452×10^{-4}
square foot (ft^2)	square metre (m^2)	0.09290
cubic inch (in^3)	cubic metre (m^3)	1.639×10^{-5}
cubic foot (ft^3)	cubic metre (m^3)	0.02832
pound mass (lb)	kilogramme (kg)	0.4536
minute (min)	second (s)	60
degree (plane angle — $^{\circ}$, deg)	radian (rad)	1.745×10^{-2}
pound/cubic inch (lb/in^3)	kilogrammes/cubic metre (kg/m^3)	2.768×10^{4}
pound/cubic foot (lb/ft^3)	kilogrammes/cubic metre (kg/m^3)	16.02
pound force (lbf)	newton (N)	4.448
pound force/square inch (lbf/in^2)	newton/square metre (N/m^2)	6895
pound force/square inch (lbf/in^2)	bar (bar)	0.06895
foot pound (f) (ft-lbf)	joule (J)	1.356

CONTENTS

SLIDES NEAR CANALS, LAKES, RESERVOIRS, AND FJORDS

CONTENTS

ROCKSLIDES AND AVALANCHES: BASIC PRINCIPLES, AND PERSPECTIVES IN THE REALM OF CIVIL AND MINING OPERATIONS

W.G. PARISEAU and BARRY VOIGHT

Even a casual review of the papers contained in these volumes cannot but impress one with the great diversity of environments, the vast range of size and time scales, and the rich variety of rockslides, avalanches and kindred phenomena. At one extreme, there is possible gravitational sliding on a continental scale occurring at a rate of a few centimetres per year. At the other extreme are the relatively small, fast failures of benches that sometimes occur in highway cuts and open pit mines. All sizes of slides and rates of movements fill the spectrum between these end-members.

The reader may discover a second impression: the larger the mass movement (usually) the further back in time the event occurred and consequently the more descriptive and less quantitative is our knowledge of the specifics of the event. Natural rockslide and avalanche mechanics are almost necessarily vague. This should not be too surprising for almost all natural mass movements of geologic materials including the slides of the present day come to our attention after the fact. We often know little more than where the slide material was originally, when it moved, and where it came to rest. We are left to infer from often scanty evidence what "caused" the slide, how "fast" it moved, and so on. Originality on the part of the investigator may be a requisite factor in data gathering and even so the weight of acquired evidence may finally stand on fragile foundations. The lack of hard data presents difficulties in discriminating between theoretical models of rockslides and avalanches. Any model that allows the slide mass to move from its place of origin to its resting place in the time limits that bound the slide motion is likely to be consistent with the principal observable fact — that of the slide occurrence itself.

Much more data have been collected in the realm of civil and mining operations. Here the sure economic impact of slope failures justifies more detailed studies than those associated with the uncertain occurrences of natural rockslides and avalanches. Moreover, the volume of material moved in man-made slope failures is generally many times less than that moved in natural slides. Monitoring requirements are correspondingly less.

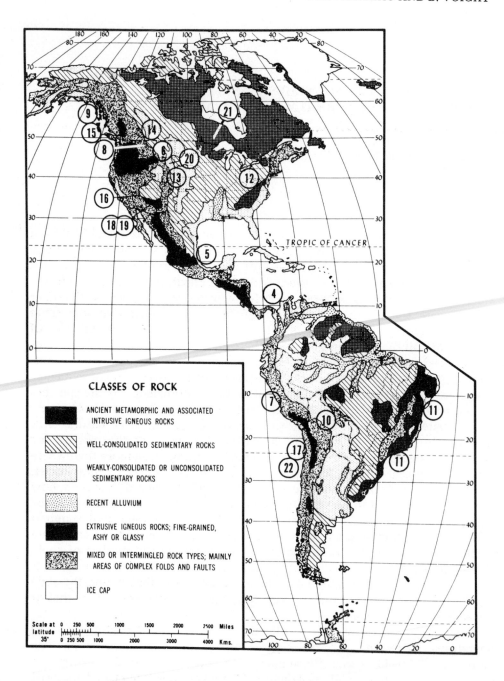

Fig. 1. Classes of rock in the Western Hemisphere (after Finch et al., 1957). Slide location numbers refer to chapters, this volume.

As a consequence, the intensity of instrumentation for monitoring mass movements is usually much greater in man-made or man-influenced slopes than in those where natural slopes have been monitored.

A third impression one gains is that slide dynamics seem commonly of more quantitative interest to investigators of natural movements, while slide initiation is of paramount concern in civil and mining works. In the present context, slide initiation refers to the onset of catastrophic motion, that is, to a sudden loss of equilibrium in a static or steadily creeping rock mass. Slide dynamics appear to be most often modeled by direct application of Newton's second law to the motion of simple blocks on an inclined plane, and slide initiation is also frequently modeled as a sliding block problem. Although this approach is almost certainly too elementary, the coupling of deformable body analyses to slide dynamics has not been much attempted, and probably will not receive widespread emphasis for some time because of computational difficulties.

The organization of *Rockslides and Avalanches* into two volumes reflects these associations. The present volume contains 22 contributions grouped into three main sections: Slopes near Canals, Lakes, Reservoirs, and Fjords; Slope Excavations for Transportation Routes; and Open Pit Mine Slopes (Fig. 1).

Despite diversity of detail, the phenomena described in this collection of papers have in common basic physical principles. No matter how large or small nor how fast or slow, nor whether of historic or prehistoric occurrence, *all* movements of masses of geologic materials must be consistent with basic physical principles. Mass must always be conserved; momentum and energy balances must always be maintained. There can be no serious debate concerning the applicability of these classical principles to the mechanics of rockslides, avalanches and kindred phenomena. Given a factual description of the geologic setting, they constitute the essential members of an interpretative framework for understanding the mass movements of geologic media. In this introduction we therefore first review the basic principles and modeling procedures in some detail; we then establish a perspective for the individual chapter contributions.

BASIC PHYSICAL PRINCIPLES

If a system is defined as a specified mass of material M, then the conservation of mass requires that:

$$\dot{M} = 0 \tag{1}$$

where the dot means time rate of change. The changes in linear momentum P and angular momentum H per unit of time must be balanced by the resultants F and L of the forces and torques acting on the system. Thus:

$$F = \dot{P} \text{ and } L = \dot{H} \tag{2a,b}$$

The balance of energy can be expressed as:

$$\dot{E} = W + Q \tag{3}$$

where E is the energy of the system, W is the work done on the system per unit of time, and Q is the heat added to the system per unit of time. Because the heat term Q appears in the energy balance, the second law of thermodynamics must also be considered in the list of basic laws.

Application of the principles expressed in equations [1], [2], and [3] occurs primarily in two distinct ways. The simplest way consists of treating the geologic mass as a single "particle" following Newton's second law of motion $F = Ma$ where a is the acceleration of the particle of mass M under the resultant force F. Another way is to view the geologic mass as a rigid body whose center of mass follows Newton's second law. If rotation of the body is neglected, this is equivalent to the first description. A more sophisticated elaboration of the basic principles views the geologic mass as a deformable body.

At each level of sophistication, the nature of required input information is the same. The geometry and properties of the slide mass, and the external forces acting on the mass must be known for the duration of the slide event.

In each case the question arises as to whether a simplified mathematical representation is adequate or not. The answer is seldom clear-cut; many subjective factors including the interests, skills, and experience of individual investigators are involved. Regardless, it seems reasonable to require mathematical representations of mass movements to be consistent with experimental observations, physical principles and the purpose of the analysis.

General purpose of analysis

The purpose of investigating the mass movements of geologic materials is to develop a predictive capability. Natural curiosity leads inevitably towards an understanding of the mechanics of mass movements and thus towards such a capability. The basic physical principles involved are known; the eventual obtainment of a detailed predictive capability is therefore theoretically possible. However, implementation of such a capability even if obtained would present formidable obstacles, especially in the realm of input information. Theoretical prediction schemes are nevertheless important because they serve as consistent interpretative frameworks within which approximations and empirical innovations can be evaluated.

Practical considerations and objectives particular to any given set of circumstances will naturally constrain the details of the analysis purpose and predictions. For example, the onset of a slide may be of primary interest. The investigator may wish to determine whether a rockslide would occur, a snow avalanche would be set in motion, a glacier would undergo a surge or rapid retreat, or that a creep rate of certain magnitude could be expected under a given set of conditions. However, it may be that the actual motion is of greatest interest. One may wish to be able to describe the motion of a mass movement as it occurs in order to predict how fast the material will move and how far it will travel. Or perhaps one would like to prescribe remedial measures for improving the stability of a slope in order to prevent failure.

There are, of course, circumstances where details of such predictions have serious consequences, e.g., in establishing zoning regulations for geologic hazards where the possibility of rockslides or snow avalanches pose direct threats to public and private safety in the operation of open pit mines and in the construction of highways, dams and canals. Rockslides and avalanches may also pose indirect threats to safety, e.g., through the generation of water waves in reservoirs that in turn threaten inhabited near-shore areas (Chapters 3, 7, 9) [1] and downstream dwellings in the event of dam rupture or over-topping (Chapters 3, 4, 9, Volume 1).

Specific analyses are diverse in purposes, but they have in common a predictive objective that is necessarily based on first principles. The inevitable idealizations and models that follow should never be so simplified or complicated as to preclude either obtainment of the analysis objective or a rational comparison with experimental observations. Balancing purpose, analytical tools and experimental observations will undoubtedly remain the perennial task of all investigators of mass movements of geologic materials.

Mass movement models

The macro-world is deterministic, so that the establishment of a predictive capability seems ultimately possible. To cite P.S. Laplace's famous declaration: "Given for one instant an intelligence which could comprehend all the forces by which nature is animated and the respective situation of the beings who compose it — it would embrace in the same formula the movements of the greatest bodies of the universe and those of the lightest atom; for it, nothing would be uncertain, and the future, as the past, would be present in its eyes" (Jaunzemis, 1967). The statement is one of great optimism. The difficulties are clear, that is, knowing what the present state of the universe is (or even what is meant by "state") and carrying out the calculations. The same difficulties are present in the much smaller "universe" of the mass movements of geologic media. Even if the type of information necessary to

---

[1] Chapter citations refer to this volume unless otherwise stated.

forecast the motion of all material particles in a geologic mass were known, the volume of information would be astronomical; calculations would be impossible with even the largest and fastest electronic computers. Of necessity, the elaboration of physical principles towards predictive schemes must involve a great sacrifice of detail. Predictive schemes are further constrained by practical considerations of purpose and availability of data. Consequently, mathematical representations of mass movements of geologic materials are largely conceptual and rather primitive, but occasionally useful.

Sliding rock block model

The most widely used and perhaps the earliest model proposed for the analysis of rockslides and similar phenomena is that of a rigid block resting on an inclined plane as shown in Fig. 2. The block represents the mass of the potential slide. Friction between the block and plane prevents sliding below some critical angle of inclination; above the critical angle the mass accelerates according to Newton's second law. Once the mass is in motion, deceleration occurs at angles of inclination below the critical angle.

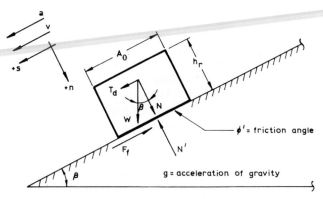

Mass: M Tangential driving component of weight: T_d
Volume: $V = A_0 h_r$ Frictional resisting force: F_f
Weight: $W = \gamma V$ Reaction normal to surface of sliding: N'

Equilibrium:

$\Sigma F_n \qquad = 0 \qquad\qquad \Sigma F_s \qquad\qquad\qquad = Ma \ (a \neq 0)$

$N - N' \qquad = 0 \qquad\qquad T_d - F_f \qquad\qquad = (W/g)a$

$W \cos \beta - N' = 0 \qquad\quad W \sin \beta - N' \tan \phi' = (W/g)a$

$\therefore a \qquad\qquad = g(\sin \beta - \tan \phi' \cos \beta)$

Note: If $T_d \leqq F_f (max) = N' \tan \phi'$ then $T_d = F_f$ and $a = 0$

Fig. 2. Sliding block model with notation.

The mass of the block remains constant, so that equation [1] is satisfied. According to equation [2a]:

$$(Wa/g) = W \sin \beta - W \cos \beta \tan \phi'$$

Thus:

$$a = g \sin(\beta - \phi')/\cos \phi' \qquad\qquad [4]$$

where g is the acceleration of gravity, $\tan \phi'$ is the coefficient of sliding friction between the block and the plane inclined at an angle β to the horizontal (positive counter-clockwise). Equation [4] shows that the critical angle is the friction angle ϕ'.

Equation [2b] is neglected (assumed satisfied).

The energy E in equation [3] is composed of kinetic energy K and internal energy. In the rock block model the internal energy is assumed to be constant, and Q is assumed to be zero. The kinetic energy is entirely translational. Under these conditions, the energy balance is purely mechanical; no new information is made available by equation [3]. In essence equation [4] contains the entire rock block story.

Over a section of slope having inclination β, the acceleration is constant. The velocity and distance traveled are then obtained by two elementary integrations. Let $G_f = g \sin(\beta - \phi')/\cos \phi'$. The symbolism G_f is intended as "downslope gravity $-$ dry friction".[2] The block acceleration is then just $a = G_f$, the velocity $v = G_f t + v_0$, and the slope distance traveled $s = s_0 + v_0 t + G_f t^2/2$ where v_0 and s_0 are velocity and distance obtained at time $t = 0$.

An interesting graphical representation of the rock block dynamics is readily developed in terms of an energy balance as shown in Fig. 3. From equation [3], $\dot{K} = W$ or $\dot{K} = F(ds/dt)$; hence $K = Fs$ or $(Wv^2/2g) = (W \sin \beta - W \cos \beta \tan \phi')s$ for the mass starting from rest and moving down a slope of constant β and ϕ'. Substituting from Fig. 3, $(Wv^2/2g) = Wh - Wd \tan \phi'$. The first term on the right of this last equation is just the work done [3] by the force of gravity in moving the block through the vertical distance h. The second term is the work done by the block against friction acting over the horizontal distance d; the difference is the kinetic energy increase in the system (from rest). Friction slows the block, but even in "free fall" the block velocity cannot exceed $\sqrt{2gh}$.

If the rock block dynamic analysis is carried out from slide initiation to

---

[2] "Dry" as used here indicates no fluid pressure, but not necessarily the absence of a fluid medium.

[3] The action of gravity could be viewed as a change in the "gravitational potential energy" defined as the work required to elevate the weight above some arbitrary datum. The result is the same.

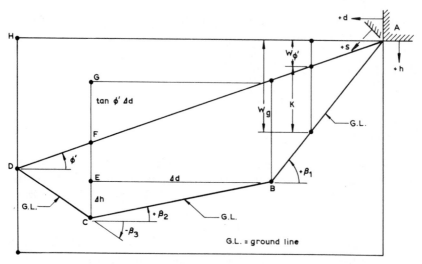

W_g = Wh = work done on slide mass by gravity
$W_{\phi'}$ = $Wd \tan \phi'$ = work done on slide mass against friction
K = $(W/2g)v^2$ = kinetic energy of slide mass
ΔK = $\Delta W_g - \Delta W_{\phi'}$ = change over any section
ΔK = $W\Delta h - W \tan \phi' \Delta d$ = $EC - GF$
K at C = K at B + ΔK = $EG + (EC - FG)$ = CF
Total vertical drop = HD
Total horizontal distance = AH
$\tan \phi'$ = (HD/AH)

Fig. 3. Graphical representation of kinetics of sliding rock block.

cessation of motion as shown in Fig. 3, then $h = d \tan \phi'$. Hence, the inclination of a line drawn between the starting and stopping points of the block is
ϕ', the friction angle. This is true regardless of the block trajectory (assumed
to be coincident with the ground surface) because $K = \int \dot{K}dt = \int Fds = (w/g)$
$\int (\sin \beta - \cos \beta \tan \phi')ds = (w/g)(\int dh - \tan \phi'dd) = (w/g)(h - d \tan \phi')$. If the
ground profile should project above this line at any point, the slide must
stop at the point of intersection between the line and ground profile. To be
consistent, the line should be drawn between the mass centers of the slide at
the beginning and end points of the slide. If this is done, velocities, distances
and times can be determined from a figure similar to Fig. 3 drawn to scale
(cf. Fig. 5, Chapter 1, Volume 1). Thus, the motion is completely determined within the confines of the model, that is, by Newton's second law
of motion.

Consistency of the sliding block model

An obvious test of the usefulness of the sliding block model is a comparison

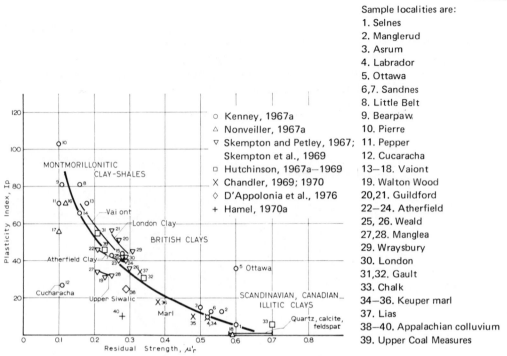

Fig. 4. Relation of residual friction coefficient to Atterberg plasticity index for clays and clay shales (Voight, 1973). Note that massive minerals, such as quartz, feldspar, calcite, ordinarily have friction coefficients greater than 0.6. Most rocks correspondingly have friction coefficients greater than 0.6.

of slip surface angles, friction angles determined from rockslide data, and experimental measurements of friction angles made in the laboratory. Apart from clay shales, colluvium, and polished or clay- or mica-coated fracture surfaces (Fig. 4) most laboratory measurements on a variety of rock types show coefficients of friction (tan ϕ') ranging between 0.6 and 1.0, or $\phi' = 30–45°$ (Jaeger and Cook, 1969, p. 59). Many natural slip surface angles are much less, and the question immediately arises as to how slides are initiated. Moreover, friction coefficients determined by back-calculation from slide data tend to be rather low, often in the range 0.1–0.3, even with neglect of cohesion. The basic rock model seems therefore in apparent contradiction to the facts.

There are at least two plausible explanations for resolving the apparent conflict between natural slope angles and sliding friction angles: either (1) the gravity forces that tend to cause sliding must be augmented in some way, or (2) the effectiveness of friction and other forces that resist sliding must be reduced.

Slide initiation

Earthquake forces are certainly capable over large regions of augmenting the gravity load acting on a potential slide mass (Fig. 5; Chapters 5, 16, 17; Chapters 4, 7, 8, 23, Volume 1). Blasting in open pit mines and even vibration from machinery may also augment gravity forces. Such transient forces are often sufficient to trigger a slide.

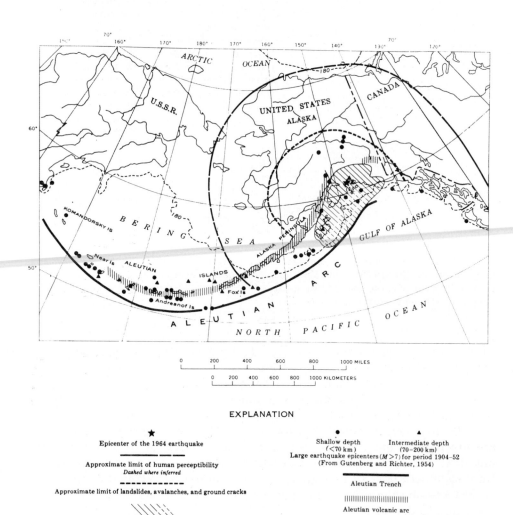

EXPLANATION

★
Epicenter of the 1964 earthquake

━━ ━━
Approximate limit of human perceptibility
Dashed where inferred

━ ━ ━ ━ ━ ━
Approximate limit of landslides, avalanches, and ground cracks

\\\\\:
Approximate area of major tectonic deformation
Dashed where inferred

●   ▲
Shallow depth   Intermediate depth
(<70 km)   (70–200 km)
Large earthquake epicenters (M >7) for period 1904–52
(From Gutenberg and Richter, 1954)

━━━━━
Aleutian Trench

||||||||||||||||||||||||||||
Aleutian volcanic arc

━ ━ ━ ━ 180 ━ ━ ━ ━
Approximate outer edge of continental shelf
Depth in meters

Fig. 5. Map of Alaska and adjacent areas showing the epicenter location of the 1964 earthquake, the area affected by it, and major geologic features. Note the large region of slides and avalanches associated with the 1964 events, and the concentration of epicenters of prior major earthquakes. (U.S. Geological Survey.)

Slide initiation and earthquake occurrence are not always simultaneous, but the association in some cases appears to be more than mere coincidence (Chapters 5, 17; Chapter 3, Volume 1). Inasmuch as slide development may progress for months and sometimes years before a final catastrophic movement, it may be that under such circumstances transient loads caused by earthquakes, blasting and vibrations trigger processes that eventually result in a catastrophic movement. The nature of such processes is admittedly obscure, but there must certainly be a reduction in the forces resisting sliding with the passage of time because the slide occurs well after the transient disturbance. In this regard, time dependency includes creep, consolidation and aging phenomena. Peak-residual phenomena involving displacement-dependent strength may also be involved, as may be diurnal and seasonal cycles of pore water pressure and heating and cooling.

Fluid pressure. A reduction of the slide resisting forces may occur through an actual reduction in strength or in the effectiveness of frictional resistance. The latter occurs in conjunction with the concept of effective pressure and the "wet" or "effective" friction model. If an average fluid pressure $\bar{p}$ acts over the area A_0 of the block shown in Fig. 6 then the net force trans-

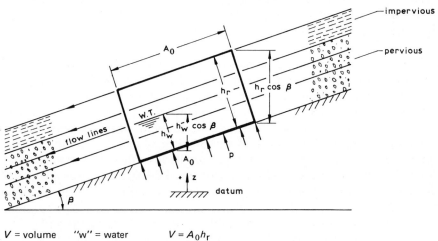

V = volume   "w" = water $V = A_0 h_r$
$W = \gamma_r V$ "r" = rock

1. Flow lines are parallel to slope surface
2. Datum is an arbitrary elevation
3. Piezometric head $h = (p/\gamma_w) + z$
4. W.T. = water table, $p = 0$
5. On slide surface $p = \Delta p = \gamma_w \Delta z = \gamma h_w \cos \beta$
6. $(W/A_0) = \gamma_r h_r$
7. $pA_0/W \cos \beta = \gamma_w h_w/\gamma_r h_r$

Fig. 6. Seepage and scale effects on sliding of rock blocks.

mitted across the interface is $N' = W \cos \beta - \bar{p}A_0$, and the frictional resistance is then $N' \tan \phi'$. The acceleration of the block is $a = g \sin \beta - g \cos \beta (1 - \bar{p}A_0/W \cos \beta) \tan \phi' = G_f'$. Comparison with equation [4] shows that the "effective" coefficient of sliding friction is $(1 - \bar{p}A_0/W \cos \beta) \tan \phi'$. Thus with sufficient fluid pressure, frictional resistance can be reduced to zero. If the potential surface of sliding lacks cohesion, then acceleration of the block is possible for all slope angles greater than zero. Because the acceleration is a constant G_f', velocity and distance are obtained by two elementary integrations, as before.

The term $\bar{p}A_0/W \cos \beta$ can be viewed as the ratio of water pressure to "rock pressure" normal to the slide plane if seepage is assumed to occur parallel to the slope, as shown in Fig. 6. In this case $\bar{p} = \gamma_w h_w \cos \beta$ where γ_w is the specific weight of water and h_w is the depth of water measured perpendicular to the slope so that $(\bar{p}A_0/W \cos \beta) = (\gamma_w h_w/\gamma_r h_r)$. Under the circumstances $h_w < h_r$ water pressure never fully overcomes the frictional resistance. Thus, a class of slopes can be determined such that artesian pressure $(h_w > h_r)$ is required to initiate sliding; this commonly involves a relatively impermeable layer over the slide surface.

Cohesion. Initiation of large slides even in the presence of high fluid pressures poses two related questions. The first concerns the existence of the sliding surface and the second the possibility of cohesion being present. Are cohesionless surfaces of great areal extent as common as rockslides? Do such surfaces exist before sliding or are they perhaps generated concurrently with sliding? The fact that slopes often remain relatively stable for many years before catastrophic sliding occurs is suggestive of progressive failure and therefore of slide surface generation concurrently with slide initiation. Regardless, slide initiation must overcome cohesion when present on the potential slide surface. Reduction of frictional resistance to sliding by high pore pressure on the slide plane may therefore be insufficient to explain slide initiation. If cohesion k acts over area A_0 of the slide surface the acceleration of the block is given by $a = G_k = g(\sin \beta - \cos \beta \tan \phi') - gkA_0/W$. If pore pressure $\bar{p}$ acts on the slide surface, then $a = G_k' = g[\sin \beta - \cos \beta(1 - \bar{p}A_0/W \cos \beta)\tan \phi'] - gkA_0/W$. The notation is intended to suggest downslope gravitational acceleration − "dry, with cohesion" and downslope gravitational acceleration − "wet, with cohesion". As used here "wet" refers to the presence of pore fluid pressure on the slide plane. In both cases the velocity and displacement of the slide mass are obtained by two elementary integrations.

Scale effect. The pore fluid pressure and cohesion terms in the expression for acceleration contain the ratio A_0/W. With reference to Fig. 6, this ratio has an obvious interpretation as the reciprocal of slide mass thickness or average depth to the slide plane measured perpendicular to the slope surface. As such it introduces a "scale" effect into the expressions for slide mass acceleration, velocity and displacement. A relatively thick slide mass will exhibit a lower total cohesive resistance to sliding and a reduced effect of pore fluid pressure on the slide plane compared to a thinner slide mass of the same properties and boundary forces. The combined effect on acceleration is $(\bar{p} \tan \phi' - k)/\gamma_r h_r$ where γ_r is the specific weight of rock and h_r is the thickness of the rock block. Other factors being equal, deep-seated slides will accelerate faster, achieve higher velocities, and move farther than slides on failure surfaces nearer ground level.

Progressive failure on a discontinuous sliding surface. Progressive failure can be incorporated into the sliding block model by introducing, in essence, a displacement-dependent cohesion. Momentary overloading of the slide mass by earthquake forces or by blasting during excavation and mining may cause a temporary failure of the slide mass. The motion results in shearing of intact rock bridges between joint surfaces that define a discontinuous failure plane for the slide mass. The total shearing resistance is composed of joint and intact resistances, thus:

$$A_0 \tau = A_j (\sigma \tan \phi' + k') + A_r (\sigma \tan \phi + k)$$

where A_0 = total area, $A_j + A_r$
A_j = joint plane area
A_r = rock bridge area
τ = average shear strength of failure surface
σ = average normal stress on failure surface
$k', \tan \phi'$ = joint strength parameters
$k, \tan \phi$ = intact rock strength parameters

Let $r_0 = A_r/A_0$, then:

$$\tau = (1 - r_0)(\sigma \tan \phi' + k') + r_0 (\sigma \tan \phi + k)$$

If the situation is simplified by assuming a cohesionless joint and a joint friction angle equal to the internal friction angle of the intact rock bridge, then:

$$\tau = \sigma \tan \phi' + r_0 k$$

Momentary overloading and shearing of rock bridges reduces the cohesive resistance to sliding by reducing the area of the bridges. The total original area remains the same. This is expressed by setting $r = r_0 - bs/s_0$ where b is a property of the material, s is the displacement, and s_0 is the slope length of the block. Initially $r = r_0$. The dependence of cohesive resistance on displacement is then $(r_0 - bs/s_0)kA_0$ and the acceleration of the block is given by:

$$a = g\left[\sin \beta - \cos \beta \left(1 - \frac{\bar{p}A_0}{W \cos \beta}\right) \tan \phi' - \frac{r_0 A_0 k}{W}\right] + \frac{gkA_0 bs}{Ws_0}$$

With the notation that
   s = displacement,
   $\dot{s} = v$ = velocity,
   $\ddot{s} = \dot{v} = a$ = acceleration,

$$G_k' = g\left[\sin \beta - \cos \beta \left(1 - \frac{\bar{p}A_0}{W \cos \beta}\right) \tan \phi' - \frac{r_0 A_0 k}{W}\right],$$

$$B^2 = \frac{kA_0 bg}{Ws_0},$$

the acceleration is given by:

$$\ddot{s} - B^2 s = G_k'$$

which has the general solution:

$$s = c_1 e^{Bt} + c_2 e^{-Bt} - G_k'/B^2$$

where t is time. If at $t = 0$, $v = 0$ and $s = 0$, then:

$$s = (G_k'/B^2)(\cosh Bt - 1)$$

and:

$$v = (G_k'/B) \sinh Bt.$$

In series form:

$$s = G_k' t^2/2 + 0(t^3) \quad G_k' = t/2 + 0(t^3)$$

and:

$$v = G_k' t + 0(t^2).$$

The higher-order terms are positive, so that the degradation of cohesive resistance with displacement leads to greater acceleration than otherwise. However, the block will not accelerate at all under static conditions as long as the resistance exceeds the downhill gravity load on the block. What is therefore postulated is a process of transient dynamic loads that augment the static slide load so as to cause a momentary slip of the slide mass. The process is cumulative; each small slip shears additional rock bridges, further degrading the cohesive resistance. Although the slide mass comes to rest after each momentary acceleration, ultimately catastrophic failure may occur. The danger will be especially high in the presence of pore fluid pressure on the slide plane.

Terzaghi (1962, p. 253; cf. Jaeger, 1971, p. 118) referred to r_0k as the "effective cohesion" of a discontinuously jointed rock mass. He described natural mechanisms that tended to reduce r_0 in the vicinity of a slope, and commented on the practical difficulty of determining r_0 in the field; he therefore recommended that the effective cohesion be neglected in order to be on the conservative side in estimating safety factors against sliding.

Such a degree of conservatism, perhaps tolerable in civil works, could make many open pit mines uneconomical. For example, a value of $r_0 = 0.1$ would seem to justify an order of magnitude reduction in the cohesion of intact rock as an estimate of the cohesion of the jointed rock mass. The resulting effective cohesion, even for weak intact rock, is by no means negligible. However, despite some progress in the matter, accurate field determination of r_0 remains an unresolved problem.

A displacement failure criterion. Displacement of the slide mass during the progressive failure stage prior to the obtainment of a sustained non-zero acceleration is of considerable interest because of its potential as a predictive index for the onset of catastrophic failure (e.g., Chapter 17). The loss of cohesion resulting from shearing the bridges of intact rock on a discontinuous failure plane is associated with peak-residual strength phenomena, as illustrated by the data of Lajtai (1969) obtained in laboratory direct shear tests. The phenomena relative to the simplified model discussed previously can be represented by failure criteria in the τ-σ plane and by shear stress/shear displacement curves. The shear displacement is normalized by division by s_0, the width of the shear test specimen. In this regard s/s_0 is a displacement per unit of original length, a "strain" of sorts. Both are shown in Figs. 7 and 8. The stress drop from peak to residual shear strength is:

$$\tau_{pk} - \tau_{res} = [\sigma \tan \phi' + r_0k] - [\sigma \tan \phi' + (r_0 - bs_c/s_0)k]$$

where s_c is the "critical" displacement associated with the peak-to-residual stress drop. As shown in Fig. 8, s_c' is the displacement associated with the stress increase to the peak condition, and s_e represents elastic distortion. The stress drop is simply kbs_c/s_0. With complete loss of cohesion $bs_c/s_0 = r_0$ and $\tau_{pk} - \tau_{res} = r_0k$. One interpretation of b can be obtained as follows. With designation M_τ as the slope or "modulus" of the post-peak shear stress-normalized shear displacement curve, one has $(1/b) = (k/M_\tau)$ and M_τ

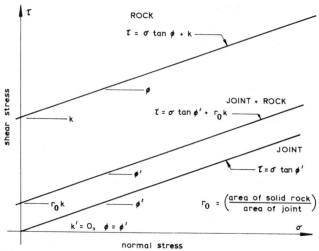

Fig. 7. Rock and joint failure criteria.

$(s_c/s_0) = r_0 k$, thus $(s_c/s_0) = r_0 k/M_\tau$. These last two relationships are reminiscent of one-dimensional stress-strain and strain to failure relationships in which the system cohesion is $r_0 k$ and the system shear modulus is M_τ. If $r_0 = 10^{-1}$, $k = 10^4$ kN/m^2 and $M_\tau = 10^5$ kN/m^2, then $(s_c/s_0) = 10^{-2}$. A slope stressed to approximately the peak condition and 300 m in length would show 3 m of additional displacement before losing cohesion along the entire failure surface. These numbers are probably near the correct order of magnitude for slopes and slides in open pit mines.

Clearly, progressive loss of cohesion may dispose the rock mass to catas-

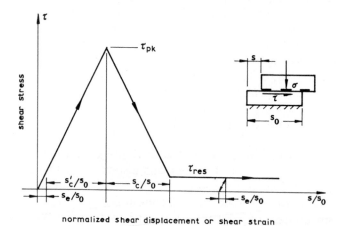

Fig. 8. Idealized shear stress-normalized shear displacement ("strain") test curve for a discontinuous joint.

trophic sliding induced by slight disturbances or "triggers" that would earlier in the development of the slide merely have induced a momentary acceleration of the mass. Actual displacement records made at several open pit mines show the phenomena involved quite clearly. It is encouraging to see that the very simple model presented here captures somewhat quantitatively the major features of the problem.

Blasting effect on progressive failure. If one knew the displacement Δs of the slide mass that occurs during each blasting event in an open pit mine, then an estimate of the critical number of blasts n_c leading to catastrophic failure would be approximately $n_c = s_c/\Delta s$. The assumption is that the peak condition has already been achieved in order for the slide to develop. If the blasting frequency is f times per day, then catastrophic failure occurs after n_c/f days. Blasting twice a day for a period of three years represents approximately 2000 blasts. If $s_c = 3$ m and $s_0 = 300$ m as before, then $\Delta s = 1.5$ mm per blast. Of course, this is only an order of magnitude estimate of the movement of the center of mass of the slide during a blast. In practice, the amount of explosive and the location of the charge with respect to the slide mass would significantly influence the acceleration of the mass which is actually accelerated differentially as the blast wave traverses the slide mass (Fig. 9). Moreover, the loss of cohesion following each blast would on the average lead to a larger Δs during each succeeding blast because of the reduction in cohesion. Progressive failure itself "accelerates".

Suppose the slide mass (rock block model) is statically stable, that is, at rest under existing loads. The forces resisting sliding exceed the forces driving the slide, so that G_k is negative, although the acceleration is zero. Blasting momentarily augments the driving forces accelerating the slide mass, although the mass comes to rest after that blast. If the blast transient is assumed to be a single rectangular pulse and the loss of cohesion occurs after the blast, then a simple graphical analysis illustrates the main features of blasting effects on progressive failure and the possibility of subsequent triggering of a catastrophic slide. Fig. 10 is a hypothetical analysis of such a sequence of events. The cumulative effect of dynamic forces associated with periodic earthquakes (Fig. 5) could be similarly evaluated.

Fluctuating fluid pressure. A fluctuating fluid pressure has the same effect on progressive failure as periodic blasting. However, instead of momentarily increasing the forces causing sliding, the forces resisting sliding are momentarily decreased. The result is the same; the slide mass begins to move and then comes to rest. After a sufficient number of events, the cumulative effect makes the slide mass vulnerable to triggering and catastrophic failure.

Suppose the fluid pressure fluctuates according to $p = p_0 + \Delta p \sin \omega t$ where p_0 is the mean pressure, Δp is the excursion above (and below) the mean, $\omega = 2\pi f$, f is the frequency and $T = (1/f)$ is the period. The differential

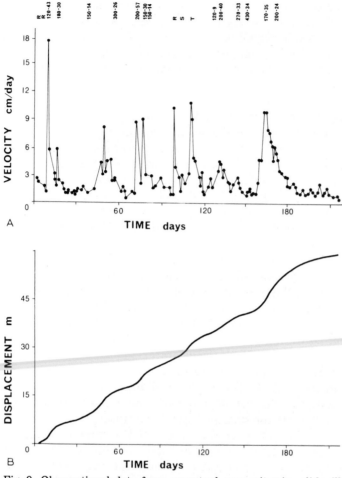

Fig. 9. Observational data from an actual open pit mine slide, illustrating (A) velocity of horizontal movement and (B) total displacement, as a function of time. Blasting occurred twice daily. Occasional remarks at top of figure refer to observed conditions possibly associated with velocity peaks. R = rainfall, S = snowfall, T = base of slope excavation. Numbers refer to approximate distance of blast in metres, and tonnes of explosive.

equation for the slide mass center acceleration is:

$$\ddot{s} - B^2 s = G'_k + P \sin \omega t$$

where $P = (\Delta p A_0 g \tan \phi')/W$.

If at $t = 0$, $v = 0$, and $s = 0$, then:

$$s = \left(\frac{G'_k}{B^2}\right)(\cosh Bt - 1) + \left(\frac{P_\omega}{B(B^2 + \omega^2)}\right)[\sinh Bt - (B/\omega)\sin \omega t]$$

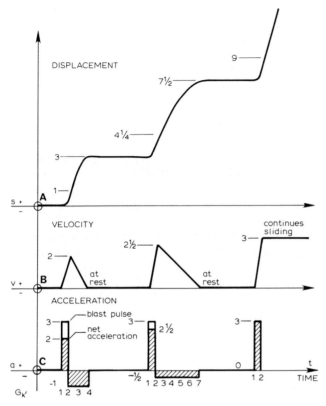

Fig. 10. Hypothetical slide history with sequence of blasting events leading to catastrophic failure. Factor of safety with respect to friction is unity. Loss of cohesion occurs between blasts. Scales are relative.

$$v = \left(\frac{G'_k}{B}\right) \sinh Bt + \left(\frac{P\omega}{B^2 + \omega^2}\right) [\cosh Bt - \cos \omega t]$$

$$a = (G'_k) \cosh Bt + \left(\frac{P}{B^2 + \omega^2}\right) \left[\left(\frac{B}{\omega}\right) \sinh Bt + \sin \omega t\right]$$

If $\Delta p = 0$, then $P = 0$ and the displacement-dependent cohesion case is recovered. If the cohesion is not displacement dependent, then $B = 0$ and $a = G'_k + P \sin \omega t$. Since the slide mass is initially at rest, values of acceleration less than zero do not actually occur for time $t < t_0$.

Fig. 11 shows a hypothetical acceleration history that varies periodically; every T days movement occurs when $P \sin \omega t$ overcomes the otherwise adequate slide resistance represented by a negative G'_k.

Fluctuating fluid pressure may reflect seasonal changes, so that T is measured in months (Fig. 12). Conceivably, dilatation and drainage created by shearing over the failure surface could cause a reduction in pore pressure bringing the slide mass to rest. Subsequent inflow and pressure build-up may again initiate motion. In this case, the period T would be coupled to the fluid flow aspects of the problem.

Fluid pressures are also enhanced by individual storm events. These more erratic fluctuations must be added to seasonal, annual, and other long-term variations. In practice T for some fluid pressure components may be measured in days, others in months and still others on longer-term periods.

Such fluid pressure fluctuations and blasting pulses of more complicated shape than sinusoidal could be described with the aid of a Fourier series expansion composed of sine (and cosine) terms. Alternatively, if the driving pulse form is known, then it could be directly inserted into the differential equation for the slide mass acceleration. The equation remains an ordinary linear differential equation with constant coefficients and can therefore be integrated. In this regard, the initial conditions for specifying the constants of integration could more generally be expressed as a time $t = t_0$, $v = v_0$ and $s = s_0$. A step by step graphical portrayal of the slide mass acceleration, velocity and displacement history facilitates the analysis.

Comparison with field measurements. Versions of the sliding block model that predict intermittent motion of the slide mass also reveal the need for

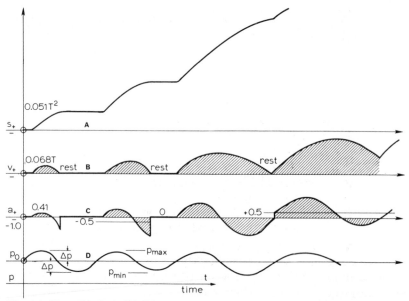

Fig. 11. Hypothetical slide history with varying pore fluid pressure.

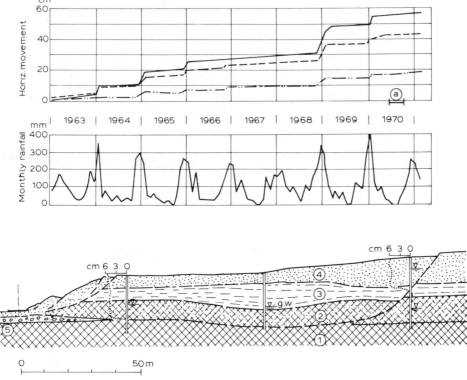

Fig. 12. Omsi-Zoo slide, Portland, Oregon, U.S.A. (after Radley-Squier and Versteeg, 1971; Záruba and Mencl, 1976). *1* = weathered basalt; *2* = decomposed basalt; *3* = stiff clay with sand interbeds; *4* = loess loam; *5* = silt with pebbles. Displacements at boreholes in centimetres for period April 1969 to February 1970. Observational data on monthly rainfall and associated movements given for period 1963—1970; note reactivation of movements when winter rainfall intensity exceeded 200 mm/month. The slide was stabilized by relief wells, drainage borings, and a toe buttress constructed in period "*a*" indicated for 1970.

observational detail in order to discriminate between theoretical models of rockslides. The frequency of displacement measurements should be high with respect to the frequency of motion events. Continuous displacement recording would reveal the entire history of the slide mass motion (see Chapter 18, Volume 1). Intermittent recording of displacements tends to obscure the step-like nature of the displacements. Very widely spaced displacement measurements would show a steadily increasing displacement with time thus obscuring the real nature of the slide motion. Under the latter circumstance, there is little hope of understanding the actual mechanics of the slide. Predictions of the time of the catastrophic sliding obtained by simply extrapolating trend divined from displacement-time plots that are based on very widely spaced displacement measurements may therefore be quite misleading. Slide monitoring programs should thus be designed with the idea of establishing the slide mechanism as well as for short time "warning" of slide movement. Fig. 9 shows observations made on an actual slide. It is of interest to note that despite about 7 m of slide displacement, motion ceased following a cessation of excavation activity. This development would not have been anticipated from an extrapolation of trend (cf. Chapter 17). Indeed, the opposite — catastrophic sliding — would have been expected.

Viscous resistance. Once a slide is initiated, a viscous type of resistance proportional to the velocity of the slide may develop. Let shearing resistance due to "viscosity" η be $2A_0\eta v$. The acceleration is:

$$\ddot{s} = G'_k + B^2 s - 2A\dot{s}$$

where $A = \eta A_0 g/W$. The general solution of this equation is:

$$s = c_1 e^{m_1 t} + c_2 e^{m_2 t} - G'_k/B^2$$

If at $t = 0$, $v = s = 0$, then:

$$s = e^{-At}(G'_k/B^2)\left[\cosh\sqrt{A^2 + B^2}t + \left(\frac{A}{\sqrt{A^2 + B^2}}\right)\sinh\sqrt{A^2 + B^2}t\right] - \frac{G'_k}{B^2}$$

$$v = e^{-At}G'_k[\sinh\sqrt{A^2 + B^2}t]/\sqrt{A^2 + B^2}$$

If the viscosity is zero, the $A = 0$ and the previous case of displacement-dependent cohesion is recovered. If $B = 0$, then:

$$s = (G'_k/4A^2)[2At - 1 + e^{-2At}]$$

and

$$v = (G'_k/2A)[1 - e^{-2At}] = (G'_k/A)(e^{-At}\sinh At)$$

The velocity in the latter case approaches in a steady state a value of $G'_k/2A$. Fig. 13 shows the slide history in this case. Thus, depending upon the relative magnitudes of A and B, the velocity behavior of a slide following initiation may range from an initially unlimited catastrophic increase to a very gradual increase and approach to a steady state value with time. A full spectrum of slide velocities is therefore possible according to this model.

Blasting effects and pore fluid pressure fluctuations could also be included with the viscous resistance model. Analytic integration is still possible, but the combined effects obscure the role of viscous resistance. However, even in the presence of a viscous resistance to slide motion, a cumulative loss of

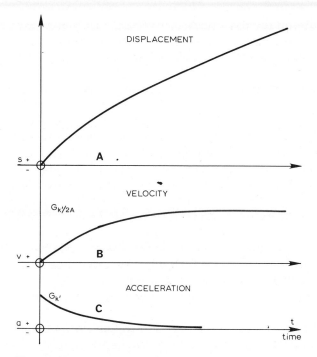

Fig. 13. Hypothetical slide history with viscous resistance.

cohesion due to intermittent movement may dispose the slide to triggering and catastrophic failure depending on the relative magnitudes of A and B. Viscous resistance does not automatically imply creep behavior to the exclusion of catastrophic sliding.

Sliding block model variations

Variations of the basic model exist. The main ones include considerations of different block shapes (e.g., wedge analysis) and non-planar surfaces of sliding (method of slices and modifications). These modifications are mainly associated with the prediction of incipient sliding, $a = 0$ in equation [4]. The latter modification (method of slices and variants) raises the question of the distribution of stress over the surface of incipient slip and is actually a venture into the realm of deformable body mechanics.

Slide continuation

The sliding block model is an adequate conceptual guide for slide initiation, but seems somewhat less satisfactory as a dynamic model for slide continuation although it is frequently used in this capacity. A basic contradiction involves low back-calculated "friction" values and high measured sliding friction angles. This contradiction has not been completely resolved by arguments involving fluid pressure on the basal surface. The basic problem seems to stem from the fact that the actual mechanisms of "slide" motions are poorly understood, and in this sense the term "sliding" may indeed prove to be a misnomer (see, e.g., Fig. 14).

An alternative model of a similar degree of simplification is a "rolling cylinder model".

Rolling cylinder model. The slide failure surface is actually a shear zone of finite thickness. With the formation of headwall cracks and the release of restraint at the top of a slide mass, shearing progresses over the developing slide surface. Rotations in the shear zone become large, and a progressive loss of cohesion occurs. The slide mass begins to accelerate. Acceleration of the mass continues with local disintegration and acceleration of the rotary motion in the shear zone. The portion of the slide mass in the shear zone is eroded by this action while the major portion of the mass moves relatively intact. The leading block segment must topple forward so that the "nose" of the slide is also eroded. The rolling cylinder model is shown in Fig. 15.

As in the sliding block model, the mass of the system is constant, so that equation [1] is satisfied.

Using equations [2a] and [2b] and the assumptions of rolling without any slip, the acceleration of the block riding on the rolling cylinders is simply:

$$a = fg \sin \beta \qquad\qquad\qquad [5]$$

where $f = (1 + w/2W)(1 + 3w/8W)$, w is the cylinder weight, and W is the weight of the block portion supported by a cylinder. The block acceleration

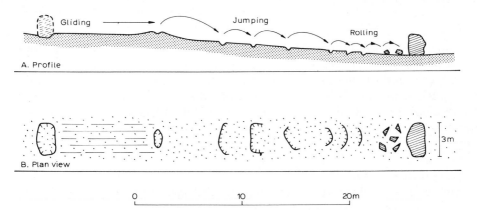

Fig. 14. Mechanism of transport of boulders by avalanches (Rapp, 1960; Fig. 40, avalanche V-45). Different types of erosion marks are typical of avalanche tracks. In this case one boulder (3 m × 1.5 m × 1.5 m) was moved by a dirty snow avalanche about 30 m over nearly horizontal ground.

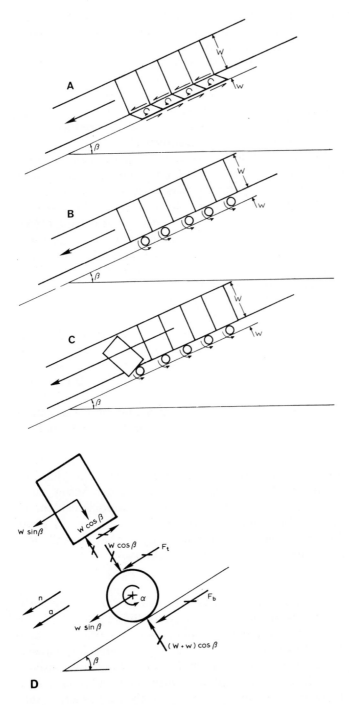

Fig. 15. Rolling cylinder model of rock slide. A. Initial shearing. B. Large rotation with continued shearing. C. Continued motion with front erosion. D. Cylinder and riding block.

may exceed the downhill gravity acceleration because f ranges from 1.0 to 1.33 as the ratio w/W ranges from zero to a large value ("small" to "large" cylinder relative to block size). The mass center of the cylinder accelerates at $a/2$, and moves with velocity $v/2$ where v is the velocity of mass center of the superincumbent block.

The energy balance is still purely mechanical. No new information is made available by equation [3], though it now includes rotational and translational kinetic energy terms. The energy balance reduces to $(v^2/2g)(W + 3w/8) = h(W + w/2)$, where h is the vertical distance that the mass center of the block moves; the left side of the equation is the total kinetic energy, and the right side is the total work done on the system (only gravity does work in this case; the cylinder mass center moves half the vertical distance that the block does). The kinetic energy of the block is slightly greater than the work done by gravity because of the driving effect of the cylinders on the block. Because as a practical matter $f \approx 1$, once rolling is initiated the acceleration will continue regardless of inclination. The slide accelerations and velocities are noticeably greater in the rolling cylinder model than in the sliding block model.

The assumption of no slip can be tested by comparing the ratio T/N of tangential to normal forces acting at the cylinder contacts with the coefficient of friction $\tan \phi'$. At the top of the cylinder $T/N = (\sin \beta/\cos \beta)(f - 1)$. At the bottom of the cylinder $T/N = (\sin \beta/\cos \beta)(1 - ff_1)$ where $f_1 = (W + w/2)/(W + w)$. Both expressions are positive and close to zero, so that even a very low friction angle would be compatible with the no slip assumption.

The rolling cylinder model could be viewed as a simple model for dynamic friction, that is, as an explanation for a reduction in the coefficient of "sliding" friction once motion is initiated.

Slide disintegration. Both the sliding block and rolling cylinder models are deficient because of the assumption of rigid body motions. Actual slide masses tend to disintegrate, some rather completely. The models can therefore have meaning only for the translation of the mass center. A great amount of interesting detail is then lost in the analysis that nevertheless seems worthwhile because of its consistency and simplicity.

Deformable body models

Relaxation of the rigid body assumption leads to various deformable body models of the mass movements of geologic materials. In such models, the explicit assumption is made that the geologic mass can be represented as a continuum. The meaning of this assumption is often misunderstood, particularly in rock mechanics where the assertion that rock masses are "discontinua" is frequently made. However, structural features such as faults, joints, bedding planes and so forth in a rock mass no more invalidate the applicability of continuum mechanics to the analysis of movements of geologic media than the presence of dislocations, grain boundaries, and crystal imperfections of all kinds, invalidates the usefulness of continuum mechanics in the analysis of steel structures. There is no question that rock masses have structural imperfections. So do intact laboratory-size test specimens, but of a somewhat different kind at a much smaller scale. The real difficulty lies in the extrapolation of material properties from the laboratory scale to the scale of rock masses in the field. [4] It is also fair to say that a "discontinuum" is not a well-defined or very useful entity. Most attempts to model geologic media as discontinua eventually result in field theories of sorts anyway as descriptions of the incipient motion of a single rigid rock block.

Basic approach and possibilities. Deformable body mechanics, continuum mechanics and field theories are rather synonymous and utilize concepts such as stress, strain and displacement fields (that may be discontinuous) for the description of the motion of the body. The basic equations expressing the physical principles outlined previously are first recast in integral form. Concepts of stress, strain and displacement are then introduced into the integral forms of the basic physical principles. Stress is the internal mechanical reaction of the deformable body to the externally applied loads. Strains characterize the deformation of the body, and displacements describe the movement of material points as before. All are usually functions of position, time, temperature and perhaps history of the material. The integral relationships expressing the conservation of mass and the balances of linear and angular momentum are manipulated to obtain the equations of motion expressed in terms of the stresses.

The stress equations of motion together with the geometry of strain or deformation and a relationship between stress and strain constitute the mathematical representation of the physical behavior of the deformable mass of material. If temperature is explicitly involved then some form of the Clausius-Duhem inequality should be introduced as it restricts the constitutive equations. A solution of the resulting system of equations under a specified set of applied loads (boundary conditions and initial conditions) enable determination of the stresses, strains and displacements throughout the mass during its motion. Much more detail concerning the mechanics of the body is thus obtained compared to the rigid body models. However, the mass center of a deformable body still translates according to Newton's second law ($F = Ma$). In this respect, the sliding block model is identical to

---

[4] In this regard, the extent of imperfections as indicated by the area ratio r_0 can be used as a rough guide for reducing laboratory strength properties for field application. The procedure is conceptually simple; multiply all dimensional material properties by r_0. Poisson's ratio and the angle of internal friction are then constant; cohesion, tensile strength and Young's modulus are reduced.

the deformable body model. (For a comparison with a segmented-block model, see Chapter 8.) However, the continuum model offers the possibility of calculating not only the acceleration, velocity and position of the center of mass but the distribution of these quantities throughout the slide mass for the duration of the slide. A significant amount of detail is thus added to the analysis, but not without a "cost".

Continuum model difficulties. The cost of the additional detail introduced into the mechanics of the mass movements of geologic materials through the deformable body assumption is represented by three major difficulties: (1) knowing how to characterize the geologic mass in terms of constitutive equations (stress-strain relationships, for example), (2) knowing the boundary and initial conditions, and (3) solving the governing system of equations (see e.g., Chapter 24, Volume 1). Much progress has been made in overcoming the latter difficulty through finite element techniques, although its indiscriminate use continues to be a source of misunderstanding. Of the former difficulties the first receives mainly theoretical attention, while the second is mainly the subject of field observation and measurements.

In present practice, only static or quasi-static analyses are generally pursued. Despite the power of the finite element techniques to cope with complicated material behavior, the vast majority of analyses are based on the assumption of an elastic material response. Plastic and viscous behavior are commonly assumed in quasi-static analyses. In these one recognizes the classical continuum models.

If all elements in a finite element mesh remain elastic, the model results indicate a "safe" excavation. However, if even a few elements exhibit safety factors less than one, the situation can be far more dangerous than it appears on the basis of a purely elastic analysis. Fig. 16A shows the results of an elastic analysis that indicate only a few elements have a safety factor below one. Accordingly the slope does not appear to be seriously threatened with a

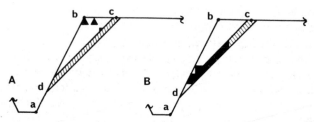

Fig. 16. Finite element representation of 300 m deep pit slope with fault zone indicated by points *d-c*. A. Results of elastic model; black triangles indicate elements in which stresses are higher than estimated material strength. Interpretation: safe slope. B. Elastic-plastic model; black zone indicates elements which have "failed". Interpretation: hazardous slope.

major slide. Fig. 16B shows the same slope analyzed as an elastic-plastic material. The slope seems seriously threatened after all. This example shows that the redistribution of stress associated with inelastic behavior cannot always be anticipated by a purely elastic analysis (Pariseau, 1972). If failure is cause for concern then an appropriate material model should be used in the analysis.

In this regard it may be added that there are many post-elastic approaches, and simply to specify "elastic-plastic" is not to guarantee success of a modeling venture. Such aspects as peak-residual behavior and path-dependency may be critical to simulation, but are commonly neglected (Pariseau et al., 1970; Voight, 1970; Duncan, 1972).

Scale of observation. A granitic test specimen in the laboratory that has a diameter of 3 cm may also contain grains several millimetres in mean diameter or even larger. Concrete test specimens sometimes contain particles as large as one-fourth the test specimen diameter. Homogeneity of stress is still assumed. The equivalent assumption in the field would be to neglect structural details, joint spacings, for example, approximately an order of magnitude less than the scale of observation. The latter is arbitrary but certainly conditioned by practicalities of purpose and measurement. In open pit mines, the analysis of slope stability is not noticeably affected by omitting bench geometry. The slope may be of the order of 300 m deep; benches are of the order of 15 m. It would be inconsistent to incorporate individually into the analysis joints shorter than the bench interval. Only the major structural features need to be modeled individually, either in terms of solid element assemblages or as discrete joints; still, the jointed rock mass properties must be defined in terms of an equivalent continuum. The real problem is always to make a wise choice and then to quantify the rock properties dictated by the model or constitutive equations selected. Unfortunately, such data are almost never known in the case of natural slides and only rarely in slides associated with civil and mining activity. Usually interest is aroused only after the appearance of headwall cracks when design opportunities are lost, and one can only witness the inevitable.

Fragmented rock block model
Dynamics of catastrophic mass movements are difficult to investigate from the continuum viewpoint. The disintegration of the mass cannot be adequately taken into account. It is not clear, for example, if disintegration is essentially complete just prior to the final catastrophic movement of the rock mass or if most of the disintegration occurs during the motion. There is, of course, the question as to the amount of disintegration that can be expected in the first place. If disintegration could be considered complete before acceleration of the rock mass, then some account of the slide dynamics could by made by assuming (1) that each fragment of the rock

mass is a rigid body of known geometry and (2) that the nature of the forces acting between the rock fragments is known (Cundall, 1974). If in two dimensions there are not too many fragments present and the inter-particle forces are not too complicated, then the computer can perform the calculations that give position and orientation of all fragments at any time (Fig. 17). The slide stops when all fragments come to rest.

Assumption (1) means that equations [2a] and [2b] apply. Although the actual distribution of stresses between blocks is not critical to satisfying [2a], it is to [2b]. Any distribution of inter-block stresses not identical to the actual distribution will give the block a spurious angular acceleration, even if the assumed forces are equivalent to the actual stresses. This is a serious defect of the model that does not appear to be generally recognized. Still, further development of this model appears to hold some promise for investigating slide dynamics involving disintegration and other problems as well. The alternative Lagrangian description of continuum models would require an enormous development effort, although there are computer codes for throwout calculations associated with cratering experiments using nuclear explosives that could conceivably be adapted for slide dynamic studies. It appears that the Lagrangian description and the fragmented rock

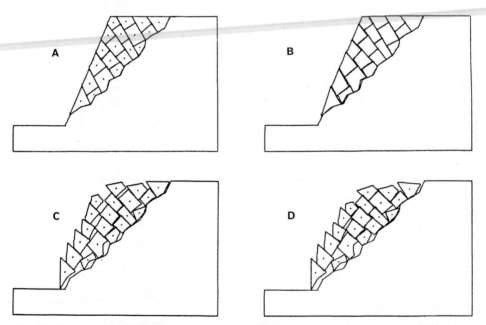

Fig. 17. Sequence of failure in hypothetical fragmented rock-block model (after Cundall, 1974). A. Friction angle set at 30°; the jointed mass is stable. B. Friction angle arbitrarily set at 25° for same slope, leading to movement of blocks. C. Motion continues. D. Equilibrium re-established.

block model would tend to converge as the former is refined and made more efficient while the latter is developed towards greater realism.

Groundwater flow models

Holes drilled in the upper portion of slopes normally display a downward pore fluid pressure gradient, whereas the reverse situation is encountered in lower portions of slopes (Fig. 18). Such a situation is contrary to that shown in Fig. 18A, as often illustrated in geotechnical literature, but recognition of such systems dates at least to the work of Hubbert (1940; cf. Tóth, 1963, 1972; Meyboom, 1966a, b, c; Freeze and Witherspoon, 1966, 1967, 1968). In areas of relatively low relief, recharge and discharge areas can be separated by distances of tens of kilometres.

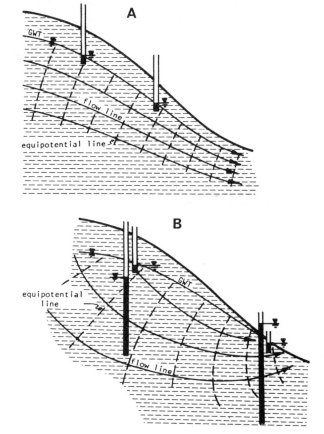

Fig. 18. Groundwater flow in slopes (after Patton and Hendron, 1974). A. Flow assumed parallel to groundwater table. Situation commonly assumed but seldom found in practice. B. More realistic flow pattern from groundwater recharge to discharge areas.

Slope failures can be anomalously concentrated within groundwater discharge areas, even where geologic conditions appear at first glance to favor
stability.

Significant variations occur in areas of regional groundwater recharge or in
zones of variable permeability, and here it is well to remember that
heterogeneity is the rule in rock masses, homogeneity the exception. Moreover, slope movements can themselves modify the "normal" pattern by
blocking normal slope discharge (Fig. 19). The effect of the emplacement of
slide debris is to increase the groundwater table level and to increase groundwater pressures in the areas of slide debris (Parizek, 1971; Patton and
Hendron, 1974). The overall result is an acceleration of continued slide
displacements. Surficial ice formation in winter can cause a similar effect in
discharge areas, leading to high fluid pressures and to the enhanced possibility of induced slope movements (Chapters 8, 12).

The normal presence of high fluid pressures in valleys and valley walls has
been documented by theoretical and field studies (e.g., Van Everdingen,

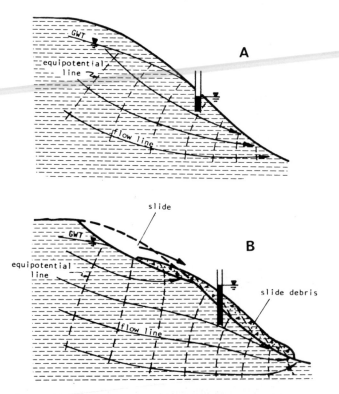

Fig. 19. Comparison of groundwater flow in slopes (after Patton and Hendron, 1974). A.
Before slide. B. With mantle of slide debris.

1972). Such discharge patterns could cause uplift, faulting, and clastic intrusions in the valley and along valley walls, and over relatively long time periods could have greatly influenced rock mass strength and rheology in the formation of valley anticlines (Chapter 14; Chapter 17, Volume 1). Induced high pressure can also act within pre-existing fracture systems and bedding planes to cause major slides (Fig. 20).

Zones of fracture concentration revealed by fracture traces and lineaments can be associated with slope failures (Parizek, 1976). Topographic depressions associated with such features help to concentrate recharge along them. Bedrock along zones of fracture can be three orders of magnitude more permeable than adjacent strata, with the result that groundwater flow to discharge areas consumes less hydraulic head in fracture zones than in adjacent bedrock. High fluid pressures therefore can easily develop in fracture zone discharge zones which are commonly associated with a thick, metastable colluvial cover.

Finally, changes in groundwater conditions caused by construction of water storage or tailings dams, water conduits, or excavations have in numerous cases led to slope failure (e.g., Chapters 5, 6, 21).

Terzaghi's 1929 concept of "minor geologic details" (Chapter 2) was illustrated by considering the possible variations of groundwater flow in heterogeneous ground, so that it is wrong to characterize the pre-computer era of hydrogeologic analysis as if limited to flow net construction involving homogeneous, isotropic media. Nevertheless, computer-based numerical methods have clearly and considerably improved the state of the art.

Numerical methods provide the most powerful tools for analysis of two- and three-dimensional flow patterns and pressure distributions in heteroge-

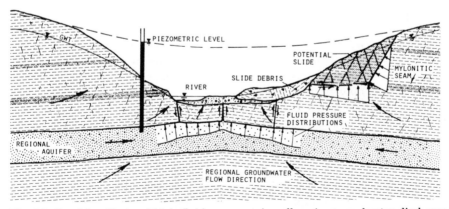

Fig. 20. Possible effects of high fluid pressures in valleys in groundwater discharge areas (after Patton and Hendron, 1974). Effects include increased deformation (e.g., bedding-plane slip) and slope movements in valley walls, and deformation of valley bottoms (e.g., uplift, faulting, clastic dike formation).

neous geological media. As with structural problems, uncertain or inadequate field data provide strong limitations. Nonetheless, given appropriate boundary conditions and the geometry of internal heterogeneous domains, our predictive capacity has advanced to a relatively sophisticated level (Sharp, 1970; Maini, 1971; Wittke et al., 1972; Cooley, 1974). The results of such flow analyses can be coupled to stability analyses or to deformable body models. With a knowledge of actual groundwater pressures at control points throughout a rock mass, it is possible to assess via numerical methods the distribution of fluid pressures throughout a slope (Sharp et al., 1972). These pressures can then be used to compute the effective normal stresses, and hence, the shear strength of critical points that may control stability.

Considered from a still more sophisticated viewpoint, the interplay between groundwater hydrology and rock mechanics has, as Snow (1969) put it, "scarcely been examined". For example, in considering hydraulic properties from one vantage point it has been customary to treat fractures as parallel-plate conduits, in rigid media. However, fractures are not only the main flow paths — they are also easily deformed. A stress-induced change in the aperture will change the flow characteristics of the fracture. This means that the interaction of mechanical and hydraulic effects must be considered (see, e.g., Witherspoon and Gale, 1977). Solutions to this problem are, however, not clear-cut. Both "continuum" and "discontinuum" formulations have been considered, but much basic research is required before the effects of this hydro-mechanical "interplay" can be properly examined, or predicted.

THE REALM OF CIVIL AND MINING OPERATIONS

Slides near canals, lakes, reservoirs, and fjords

Fluctuating water level

Significant stability problems can occur both under conditions of a rising reservoir and a rapid drawdown. Whereas the latter condition has received much emphasis, it may be that the rising condition is the most troublesome. Lane (1970) pointed out that about half (245) of 500 landslides occurring over a 12-year period in glacial lake sediments bordering Lake Roosevelt, developed during initial rise of the reservoir (Fig. 21). The next largest number (30%) occurred during two periods of major drawdown. Since these data were collected additional slides have formed, with rapid drawdown being a dominant factor, probably in association with displacement-dependent strength loss. In time a purely statistical argument may therefore shift emphasis toward drawdown as being the most unstable condition. Nevertheless, in terms of slide size and consequences, the rising reservoir may remain most significant.

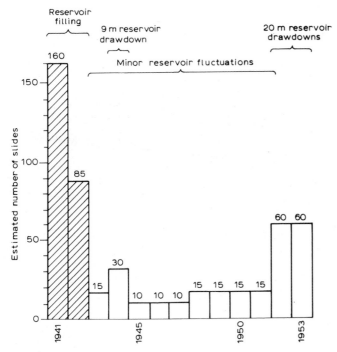

Fig. 21. Histogram of 500 slides along the reservoir shoreline behind Grand Coulee Dam (after Jones et al., 1961).

These generalizations may apply equally well to rock slopes. Indeed, it was the gigantic 1963 rockslide in the Vaiont reservoir in Italy that caused engineers to focus attention on the rising reservoir problem once again (Figs. 22—25). The Vaiont disaster has been widely studied and discussed in the engineering literature, and although the last word on the subject has not been written, it seems clear that the rising reservoir condition was a major factor in the ensuing catastrophe which claimed over 2000 lives and forced abandonment of dam and powerhouses. Slope movements had been observed several years prior to collapse; the 240×10^6 m^3 slide mass was also being observed at the time of catastrophic failure (Chapters 10, 17, Volume 1). In fact authorities had begun to lower the reservoir and were considering downstream evacuation. Neither action was implemented rapidly enough to avert disaster.

Observations of slope movement as a function of time suggest two causative factors of particular importance (Fig. 24). First, a correspondence existed between movement rate and reservoir stage; velocity typically increased with a rising stage, and diminished with drawdown. Second, periods of rapid movement seem associated with high precipitation. Even minor peaks in velocity, such as occurred in late 1961 and early 1962, can

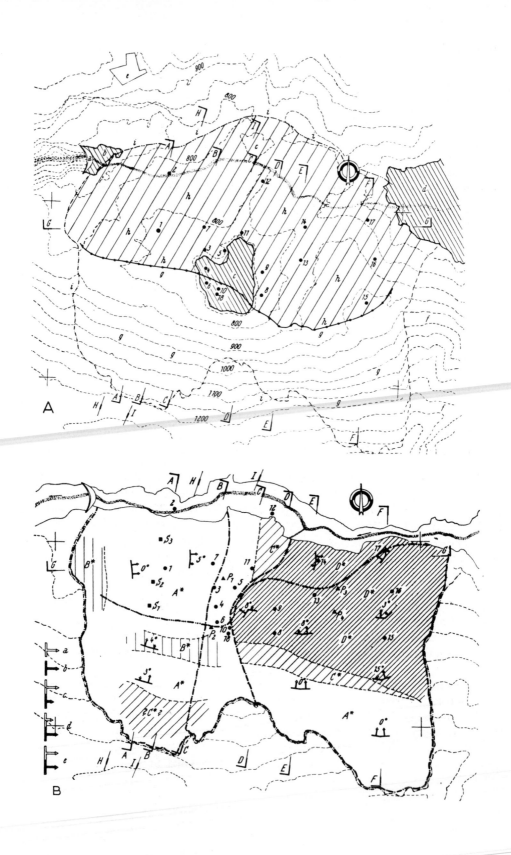

A

B

The generation of devastating water waves by landslides are of course known from many sites in addition to Vaiont. Classic occurrences are also known in Norway (Chapters 3, 9) and Japan, and in the Western Hemisphere, from Disenchantment and Lituya Bays, Alaska (Chapter 9), and Yanahuin Lake, Peru (Chapter 7). The combined death toll from these events exceeds 20,000. However, the important question of wave height and run-up prediction is at present difficult to resolve, mainly because of the non-linear theories needed to describe wave motion, complicated basin geometry, and uncertain knowledge of slide dimensions and emplacement history. The state of the art is given in Chapter 9, together with an examination of some better known field cases and an examination of wave hindcasts. Despite many difficulties, the hindcasts show that existing mathematical and experimental model approaches provide useful information upon which to base engineering decisions.

The plan of controlled slide displacement envisaged for Downie had previously been attempted with success in the Austrian Tyrol in the reservoir impounded by the 150-m Gepatsch Dam in Kaunertal (Fig. 27; Breth, 1967; Lauffer et al., 1967). It may also be said to have been attempted at Vaiont, with some initial success but ultimately disastrous consequences. During the filling of the Kaunertal Reservoir in August 1964, 20×10^6 m³ of rockslabs, waste and moraine material over a length of about one kilometre started to move (Fig. 28), reaching a maximum value of about 8 cm/day a month later. The reservoir level was held steady whereupon the slope velocity decreased and by January 1965 had practically ceased (Fig. 29). The reservoir was emptied and refilled, and only when the highest water level of the previous year had been reached did slope movements begin again. A maximum veloc-

Fig. 27. Gepatsch Dam and western slope of reservoir, Austria, illustrating landslide zone of Hochmaiss. Reservoir at full level, 1767 m.

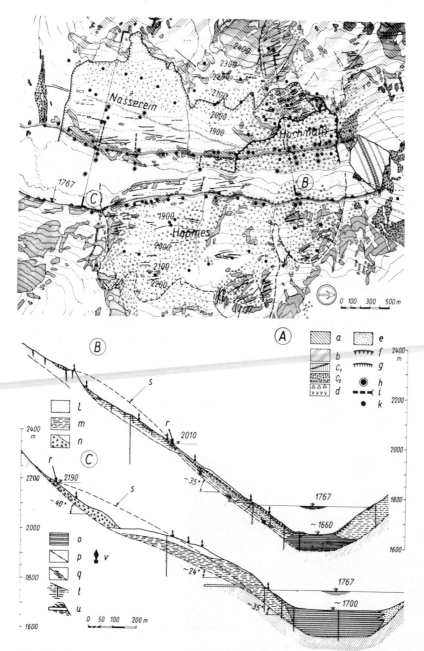

Fig. 28. Geology of Gepatsch Reservoir area (after O. Schmidegg, in Lauffer et al., 1967).
A. Geologic map on topographic base. B. Section in the Hochmaiss zone. C. Section in
the Nasserein zone. a = mass of post-glacial "sagging"; b = zone of loosening; c_1 = gneiss,
schist with cracks; c_2 = augen gneiss; d,n = moraine and talus material with blocks;
e,o = alluvium; $f,g,r,$ = recent fissures in movement zones; h,t = sounding boreholes; i,u =
exploratory gallery; k,v = monitor points; l = gneiss, schist (stable bedrock); m = slipped
rock with intense fissuration; p = possible slip zone; q = recognized slip zone; s = original
slope surface before post-glacial movements.

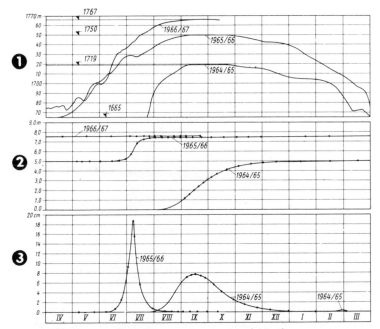

Fig. 29. History of Gepatsch slide area, 1964—1967 (Lauffer et al., 1967). 1. Reservoir stage. 2. Average total settlement (compare to filling curves). 3. Average daily settlement (almost zero for 1966—1967).

ity of 34 cm/day (15 cm/day vertical component) was achieved in July 1965, but slope motion had almost come to a halt a month later.

Analysis suggests that prior to the reservoir rise a stress ratio of $\tau/\sigma = 0.61$ was necessary for equilibrium. During the 1964 reservoir rise, toe buoyancy caused an increase of the stress ratio to 0.65, at which point slope movements commenced. Slope movement compensated for the shift in center of gravity caused by fluid uplift; to balance this effective weight loss a horizontal displacement of about 11 m total had been necessary (Fig. 30). The condition corresponds to an average required effective friction angle of 33°. Breth (1977) asked how a slight reduction in stress ratio contributed to stopping the slope movements, and conducted laboratory tests on the morainal material. These tests confirmed that the moraine quickly regains effective "viscosity" when the applied shear stress, after initially increasing, remains constant or slightly decreases. Thus the stabilization of the slope which occurred after every movement was traced back to the regeneration of the "viscosity" of the moraine after the shear stress ratio had been reduced by slope movement. Accordingly, "The rheological properties of the moraine make it possible to maintain the velocity of the slide displacement within controllable limits, when the dam filling is taken in stages, when the

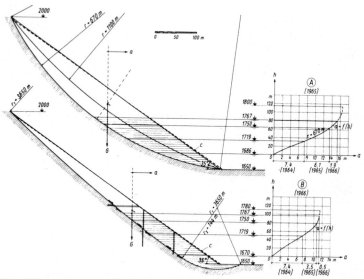

Fig. 30. Necessary displacement for compensating the uplift effect in the case of Hochmaiss (Lauffer et al., 1967). A. Forecast for slope movements in 1965 and 1966, assuming sliding mass as a circular segment with radius of curvature (r) of 670 m. B. Forecast for slope movements in 1966, assuming sliding mass boundary of compound curvature, with r_1 = 3650 m and r_2 = 144 m. G = weight of slide before reservoir filling; a = horizontal displacement of mass center required for maintaining the limiting position; h = relative reservoir level above initial limiting position; c = slope surface after movement.

filling velocity is limited, providing the filling is stopped as soon as the displacement of the slide has reached a critical velocity and not continued until the movement has ceased. This procedure is valid not merely for the special case but for every unstable slope, providing the movement takes place in a more or less homogeneous mass with properties similar to the moraine investigated and does not occur in a very thin layer or zone.

It is not valid for water saturated sands and silts with unstable structure, which can be fluidized by a light disturbance of the equilibrium within a short time" (Breth, 1967).

Since 1965, secondary creeping of the valley slopes has been observed in connection with reservoir fluctuations. Continuous observations show that the displacements have been progressively decreasing year by year (Figs. 31, 32). In connection with observations of slope creep, distinctly elastic rises along reservoir border roads could be observed during the filling periods between May and October of each year (Fig. 30). These rises have been explained as a reflection of expansion of a permeable rock mass due to uplift, which more than counterbalances the elastic compression of subjacent bedrock due to water load. Analogous movements have been recorded in the

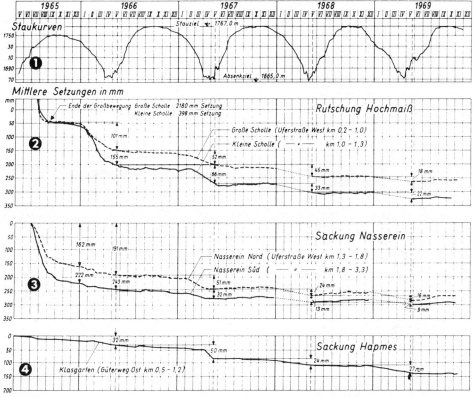

Fig. 31. History of the Gepatsch Reservoir slopes. 1965—1969 (after Neuhauser and Schober, 1970). Survey of vertical displacements of river-bank road as a function of reservoir stage (1), for three main movement zones as illstrated in Fig. 28, (2) Hochmaiss, (3) Nasserein, and (4) Hapmes. Note difference in displacement scale as compared to Fig. 29.

permeable upstream shell zone of the Gepatsch rockfill dam (Neuhauser and Schober, 1970).

The Kaunertal and Vaiont case studies represent a contrast of extremes. Given broadly similar slope conditions and reservoir fluctuations, displacements involved on one hand stable sliding, and on the other, complete collapse due to loss of strength.

The Santa Rosa spillway rockslide in Mexico is similar to the case at Kaunertal in several respects (Chapter 5). Development of slides occurred as a consequence of a rising reservoir, and gradual slope stabilization was achieved with approximate maintenance of the reservoir level. The Santa Rosa slide is not particularly large, roughly 10^6 m^3, but a careful instrumentation and investigation program makes the case history valuable.

An excellent North American example of a critical rising reservoir con-

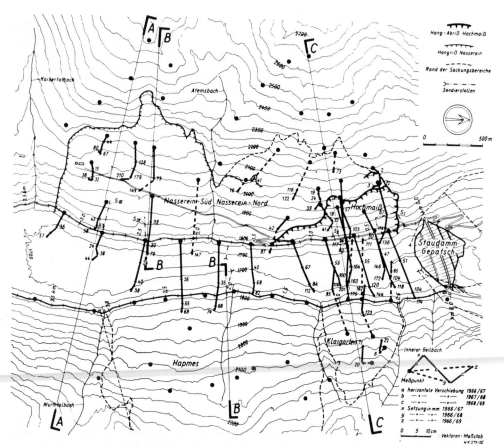

Fig. 32. Movement vectors for the Gepatsch Reservoir slopes, 1966—1969 (after
Neuhauser and Schober, 1970).

cerns slides in the Bighorn Reservoir, impounded behind Yellowtail Dam in
Montana (Chapter 6). Filling of the reservoir had started when the first slide,
involving $30-40 \times 10^6$ m^3, was detected by aerial reconnaissance in a poorly
accessible part of the reservoir. Subsequent slides have occurred with filling
periods in the late spring of each year. The slides are reactivated portions of
older slide masses; they have not interfered with reservoir operation, and
hence have been incompletely studied. It is possible that artesian and joint
water pressure may have played some role in slide movement, but the major
factor seems to have been the reduction in effective stress caused by the
rising water table.

The Panama Canal
Another example of a rising critical pool is that of the Panama Canal. Over

the first two years of operation, the canal was closed to traffic for about one-third of the time, while dredges excavated about 15×10^6 m^3 in order to reopen the waterway. While the Panama slides frequently involved an accelerated displacement rate during the rainy season, the evidence shows that slide activity also increased after the slope toe was submerged by the rising canal level.

The complete Panama Canal case history involves many complexities. The earliest slides along the route of the canal occurred in 1884, soon after the start of excavation by the French. The earliest slides — typically mudflows — occurred in the Cucaracha region, principally during the rainy season; for the most part these slides involved a 10 m thick surficial zone of residual impure red clay. A study of conditions associated with the slides was published in 1889 by Bertrand [5] and Zurcher, the geologists attached to the second French commission. It did not escape their attention (Cross, 1924, p. 25) that the underlying rock — the Cucaracha Formation — was particularly weak and yielded to pressure very easily when water-saturated. Further work on the canal seemed to verify this observation because the number and extent of slides increased following acquisition and resumption of the project by the United States in 1904. The most serious movements occurred in the deepest part of the cut, the Culebra district (see Frontispiece).

By 1915 the use of the canal was threatened by slides, and it became necessary to give careful attention to the slide causes. President Woodrow Wilson appointed a committee of the National Academy of Sciences to "consider and report upon the possibility of controlling the slides which are seriously interfering with the use of the Panama Canal". A preliminary report was published in the *Proceedings of the National Academy of Sciences* and in the *Annual Report of the Isthmian Canal Commission* in the following year. A somewhat more comprehensive report followed in 1924.

Experiences in Panama, in cuts of the Swedish State Railways, and in construction of the Kiel Canal in Germany were later cited by Karl Terzaghi (1936; Chapter 2) as events which led directly to the origin of soil mechanics as an engineering discipline. In retrospect it is therefore of interest to note that one of the authors of the 1924 National Academy report, H.F. Reid, had access to one of the early but obscure classics of geotechnical literature:

"Since the main report of the commission was drawn up I have been fortunate enough to secure a copy of a French Report, published in 1846 by M. Alexandre Collin, which described very fully slides occurring in argillaceous terranes in France... It is most interesting to note that these French slides, although very much smaller than those in the Canal Zone, present all the characteristics of the latter. The movement began with the opening of a crevasse at the time of the slide; there was then a drop of this part, followed by a slow movement and the thrusting

---

[5] Marcel Bertrand is a familiar name to tectonic geologists; he is best known for a successful reinterpretation of the Glarus *Doppelfalte* of Escher von der Linth and Albert Heim, which led to a revolution in Alpine geology.

forward of the lower part, and, in some instances, the elevation of the ground in front. In all of these cases, the movement took place along a surface of fracture which Collin takes great pains to point out was not a preexistant surface. . . He insisted that Coulomb's method of calculating the stability of slopes, by considering the forces acting along plane surfaces (a method which is still in vogue among engineers), is not applicable to these slides; and that slopes in argillaceous materials are far more unstable than Coulomb's formula would lead one to suppose.

Collin found that slides occurred at all seasons of the year, and some cases, years after the operation had been completed, and pointed out that the strength of the material may deteriorate in time under the forces acting on it, quite independently of any change in its character due to increase of water content. . . He experimented on the strength and friction of clays, and got some idea of the variation of these quantities with water content. All together, Collin's work is a model of its type, and it is very unfortunate that it has been entirely lost sight of by engineers."

Yet Collin's treatise continued to remain comparatively unknown to all but a few until relatively recently (Fig. 33). It may now be read in an English translation by W.R. Schriever, published in 1956. [6] Above all, Collin clearly

---

[6] With reference to Fig. 33, the material in the Hesse cut consisted of "colored marls between layers of calcareous shells (Muschelkalk) and layered sands. . . It is formed of successive and alternating layers of red, gray, blue and greenish marl in which one finds some kidney-shaped gypsum stones. Before this cut was completely excavated, cracks parallel to the crest, which came closer to the axis of the canal near their extremities, opened on the slope. This is one of the characteristics represented on the plans of the slides. . . At the point where the profile in plate X was taken, a crack A had opened to 2.1 m behind the edge of the slope; at other points the rupture A began up to 10 m behind the edge of slope B.

When the cut was completely excavated, the slide, the occurrence of which had been indicated in advance in the way we have just described, soon developed with a new intensity over a considerable length. . . The flat part AB, which is at the top of the slope, was lowered by 1.8 m, the slope CDL advanced towards the cut, and the bottom LI of the canal rose by 60 cm at FH. With the hypothesis of a spontaneous cycloidal movement, this disturbance is easy to explain. . . the line of sliding could only be the cycloid $ACSP$, whose generating circle has the diameter AGK of 12.8 m. The appearance of the movement and all its consequences would seem to justify this solution.

But direct observation has confirmed these hypotheses. The surface of sliding had to be exposed for the restoration of the slope, and the geometrical form of the surface was then recorded as the diagrams show. All uncertainty, therefore, disappears and one will conclude, no doubt, from the comparison of this surface with the cycloid, that it is impossible to ask for, or to obtain, a closer similarity.

It is one of the best examples we have been able to obtain. The failures of the slopes of this cut have been explained by saying that the sliding mass was softened by the infiltration of groundwater which is probably one of the principal causes of these disturbances; that numerous leaks had been observed in the slopes; and that an abundant spring, which existed 500 m away, had been dried up by the excavation of the cut. The underground derivation of nearby springs has no doubt had a more or less marked influence in producing this failure. This does not seem at all doubtful, but does it determine the failure completely or, in other words, would this slide have occurred without their influence? One can only reply that the slide might have occurred independently of the action of the springs, but that probably with this hypothesis the failure would only have occurred after a longer period of time"(Collin, 1846; see translation by W.R. Schriever, 1956, pp. 64—65).

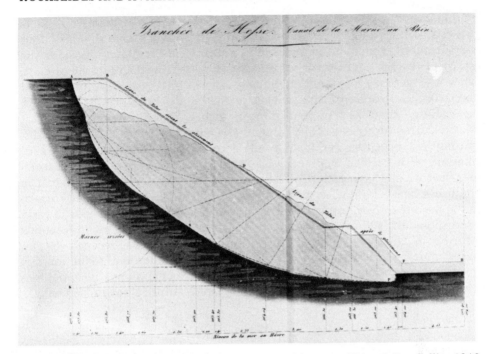

Fig. 33. Hesse cutting in Keuper Marl, canal from Marne to Rhine (after Collin, 1846, plate X). Slip occurred at completion of excavation, 1842. Height 12 m, slope 1.5 : 1. [6]

recognized that satisfactory solutions to the various problems of geotechnical engineering must "one day be the reward of those who, without separating mechanics from natural philosophy, are able to correlate the principles of the former with those facts which it is the purpose of the latter to discover and coordinate" (Skempton, 1956).

The National Academy report recognized that "A complete description of these earth and rock movements, with a discussion of attendant conditions and the character of the phenomena, would be of great scientific interest and of much practical value to the engineering profession" (Cross, 1924).

In 1924, this was not possible, for "the data for a satisfactory account of the slides do not exist. The movements have been observed principally by engineers, whose measure of their importance, under the dominating ambition to accomplish a gigantic task in an allotted time, has been the degree of inconvenience, delay, and expense caused by the slides. . ." (Cross, 1924, p. 24).

A few years later Karl Terzaghi was moved to comment on his deep regret that no lesson had been learned from the gigantic earthwork experiment — no serious attempt had been made to investigate the underlying physical causes for the benefit of future enterprises (Chapter 2). Indeed, until the comprehensive study presented in Chapter 4 of this volume, no single synthesis of the problem had been attempted despite a period of continuous observation of almost a century. The authors of Chapter 4 consider the history of the 64 significant slide and active areas from 1882 to the present, review the geology of the Canal Zone in relation to slide problems, and summarize the results of comprehensive field and laboratory investigations and stability analyses. An important finding is that the majority of past slides occurred along beds containing weak clay shale layers or structural discontinuities; effective residual strength parameters determined are zero cohesion, with friction angle of 7.5°. In addition to stratigraphic and structural factors and excavation of lateral support, many years of experience establishes heavy rainfall as an important factor in these slides. The correlation appeared so obvious that the topic was not usually mentioned in early descriptions of slides, except in those unusual cases where movements occurred in the dry season. Pore pressure measurements, recently begun, suggest much of the shale has not fully rebounded under load reduction due to excavation. Indeed, the piezometric head on canal slopes is sometimes below canal water level, and recovery may lead to future slope movement.

Fort Peck

The 1938 failures of the Fort Peck embankment illustrate analogous phenomena. The slide also furnished rarely available data for redesign; in this manner Fort Peck served as its own prototype (Middlebrooks, 1942). Other aspects of ground behavior at Fort Peck of more direct interest include movement of the powerhouse slope and creep-like bulging of the spillway.

The powerhouse slope movement was initiated in 1934 during excavation for outlet work construction. Movements have continued to the present time (Fig. 34). The 1934 slides were strongly influenced by discontinuities; the basal surface followed a bentonite bed, and the rear, a fault (Fig. 35). Hamel (1973) later suggested that the basal slip surfaces were located in gouge or fractured zones about 0.3 m thick near the base of weathered shale, which he interpreted as slip zones of ancient slides. Old slide masses are recognizable in 1934 topography (Fig. 36). During design studies for a second powerhouse, investigations were made to determine whether the slide was sufficiently deep-seated to affect the excavations. Slope studies began in 1953, exploratory holes were drilled, and piezometers were installed; additional holes and tiltmeters were installed in 1957 (Fleming et al., 1970b, pp. 189—209). Relatively minor excavation at the toe of slope in 1970 accelerated slide movement and initiated new slide areas nearby. The most recent synthesis of site investigations is due to Hamel (1973).

Reference points indicate displacement rates of 3—60 cm/yr for the slope. Movement of slide debris slowed significantly while the ground was frozen in the winter of 1970-71, accelerated during the 1971 spring thaw, then slowed again. Although the average piezometric head on failure surfaces is

Fig. 34. Slickensided scarp of slide 4, powerhouse slope, Fort Peck Dam, Montana (Corps of Engineers 34/1184, 2 October 1934; courtesy of G.S. Spencer). Location upslope of inclinometer casing WS-4, installed in 1957. Upper part of slide scarp is dry and cracked while lower part is smooth and bright below an undulating line probably marking the limit of dampness. This suggests that significant movement had occurred shortly before photo was taken (Hamel, 1973, p. 78). The slide scarp occurred along faults.

known to be moderate, movements appear to be quite sensitive to this parameter. Hamel predicted long term creep would continue at presently known average rates, with seasonal accelerations to short-term velocities of 0.7—3 m/yr during spring thaw or following heavy rains. The residual strength of the Bearpaw shale at the site is considered to be represented by zero cohesion, and a residual friction angle of $11.5°$, based on back-calculations. The 1934 slides mobilized a strength somewhat greater than this, suggesting partial "healing" of the shale along ancient landslide surfaces. The initial failure and reactivations are typically caused by excavation or rise in groundwater level. Any increase in water level causes an instantaneous response consisting of relatively slow downslope movement, much as observed at Kaunertal. The movement presumably opens cracks in the slide debris or foundation (Hamel, 1973, p. 34), permitting drainage and hence a decrease in driving force due to water; deceleration results.

Finally, movements had also been a problem in the Fort Peck spillway since the construction period (Fleming et al., 1970b, p. 203; Wilson, 1970, pp. 1530—1531). Vertical displacement contours indicate the general pattern

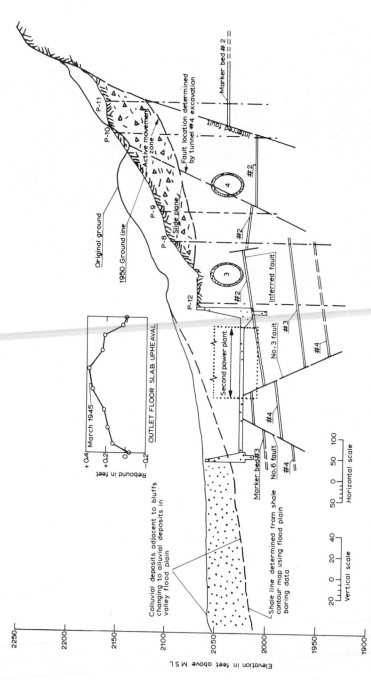

Fig. 35. Geologic profile through slide 4 and second powerhouse, Fort Peck Dam (after Fleming et al., 1970b). The slide dates at least to 21 September 1934.

Fig. 36. Oblique aerial view of slide 4 and powerhouse construction site (Corps of Engineers 34/1204, 3 October 1934; courtesy G.S. Spencer). Downstream to left. Water tank on right is landmark (cf. Fig. 34). The toe of slide 4 cuts across the slope between two haul roads. Powerhouses are now located at the toe of slope. The angular relations of sets of cracks and scarps is strikingly displayed and suggests the influence of pre-existing structures.

following three decades of movement. The movements are localized to specific areas, in part related to recognized fault zones and bentonite layers. Movement rate is reportedly steady and a function of slope height.

Slope excavations for transportation routes

Multiple slip planes
The Loveland Basin case history is important because movements of the slide mass were a continuing threat to important and costly tunnel construction (Chapter 13). The slide has been monitored and treated for over a decade. Triggered by road excavation, the slide occurred in a sheared granitic slope whose fabric had been long established by multiple episodes of tectonic deformation; slide motion was strongly influenced by structures.

A graben (about 300 m long, 50 m wide, and 10 m deep) and several small uphill-facing scarps lie at the continental divide summit 0.7 km due west of Loveland Basin. These structures have been interpreted as indications of large-scale gravitational spreading of steep-sided ridges (Radbruch-Hall et al.,

1976); the term "Sackung" (literally "sagging") of Zischinsky (1966, 1969) has been adopted in order to distinguish this kind of relatively large-scale movement in rock masses from other types of flow, particularly those involving unconsolidated material. The "expected" shear pattern in a two-layer slope undergoing gravitational spreading is illustrated by Fig. 37, in which uphill-facing scarps are clearly indicated. Similar features have been recognized at many localities in the Rocky Mountains (Fig. 38; see Chapter

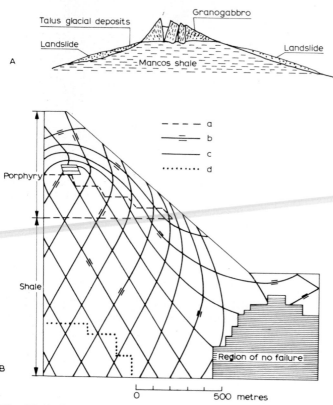

Fig. 37. A. Diagrammatic profile across Dolores Peak area, Colorado, showing extensional deformation of laccolithic ridge and landslides perhaps related to bulged slopes. See Fig. 38 for locations. The principal trench, about 60 m deep and 120 m wide, probably originated after removal of ice cover left the ridge unsupported; extrusion of Mancos Shale then subjected the laccolith body to tension. B. Two-dimensional finite element simulation of expected shear pattern in laccolith and underlying shale. a = boundary of region of potential tensile failure; b = shear failure trajectories, assuming conjugate failure on planes oriented at 30° to principal compression trajectories; c = boundaries of shear failure region assuming zero cohesion; d = boundaries of failed region assuming laboratory-determined cohesion values. Not drawn, but considered probable, are extension fractures parallel to maximum compression trajectories. (After Radbruch Hall et al., 1976.)

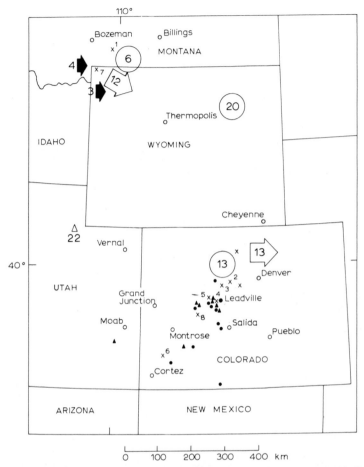

Fig. 38. Map of Rocky Mountain states, western U.S.A., indicating locations of rockslide and gravitational spreading sites. Gravitational spreading at steep-sided ridges denoted by × symbols; *1* = Stillwater Complex; *2* = Loveland Pass; *3* = Shrine Mountain; *4* = Bald Eagle Mountain; *5* = Mt. Nast and Surprise Ridge; *6* = Dolores Peak; *7* = Sepulcher Mountain; *8* = Crested Butte (Radbruch-Hall et al., 1976; cf. Chapter 17, Volume 1). Circled numbers indicate chapters of this volume: (6) Bighorn Reservoir; (13) Loveland Basin; (20) open pit sites in shale, Wyoming. Chapters in Volume 1 indicated by dark arrows: historic slides at (3) Gros Ventre and (4) Madison Canyon; light arrows: prehistoric slides of (12) Heart Mountain and (13) Northern Front Range; and triangle: (22) snow avalanche site at Alta, Utah.

17, Volume 1). Radbruch-Hall et al. (1976, pp. 25—27) suggest that the Loveland Basin slide may be related to bulging and consequent slope oversteepening due to gravitational spreading, with accompanying loosening of the rock mass.

Of particular interest at Loveland Basin is the evidence which suggests that deeper zones of movement were initiated and became more pronounced with time as the slide progressed; this is akin to the "multiple-storied" slides of the Black Sea coast in the Caucasus (Ter-Stepanian, 1967, pp. 38—39; Ter-Stepanian and Goldstein, 1969; cf. Hofmann, 1973). Movements were influenced by freeze and thaw cycles prior to emplacement of a massive buttress in 1971. Similar downward enlargement of the slide mass has been observed at Panama (Chapter 4); there deepening excavations were responsible, a cause not present at Loveland Basin.

Slip on numerous planes has also been noted in landslides at the Ventura Avenue oil field in California, where ten slide areas affected nearly 50% of the eastern producing area (Figs. 39, 40). The slides have been noticed for at least a century. In operations from 1924 to 1969, 61 wells were destroyed by slides and 53 were returned to production. A total of 132 wells were influenced by eleven major slides but have been kept in production. A brief summary of some aspects of the problem has been published by Kerr et al. (1971), but most of the detailed case history information still remains in corporate and private files.

The occurrence of wells in unstable areas has been helpful in understanding the slide problem. These provided a basis for repeated surface surveys, and in addition served as inclinometers in many instances thus providing important data on the geometry of the sliding surfaces. Excavations at damaged well sites also provided direct observation of slip planes. With some exception, slip surfaces tended to develop in thin clay seams

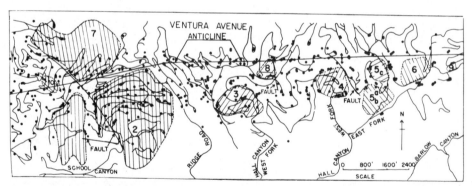

Fig. 39. Main slide areas in eastern part of Ventura Avenue oil field (after Kerr et al., 1971). Black dots indicate producing oil wells. Slide areas are numbered; parallel lines indicate direction of movement.

Fig. 40. A. Ventura Avenue oil field. Aerial view of upper part of slide 2 in 1969 (Mark Hurd Aerial Survey photo; Kerr et al., 1971). Eleven zones of movement have been identified, with movement ranging from 1 to 18 m. B. Downhole photograph illustrating multiple slip planes. 60 cm observation borehole. Slide 2.

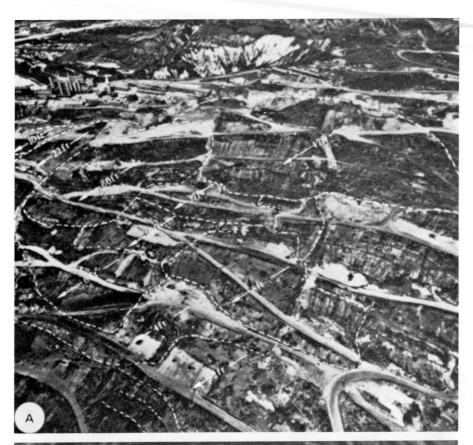

parallel to bedding within a tilted interbedded Pliocene sandstone-clay shale sequence (Fig. 40B). Pressure ridges typically developed at the toe, and peripheral cracks or graben formed at the rear. With a few exceptions, the larger slides occurred in years of heavy rainfall (Table I), although many

TABLE I

Rainfall and slide movements at Ventura Avenue oil field

Year	Rainfall (mm)	Rainfall histogram	1	2	3	4	5b	5c	6	7	8	9	10
Slide area ($10^4\,m^2$)			12	65	13	9	8	4	7	16	3	1	1
1937	651	*******	*										
1938	694	*******	*										
1939	358	****	*	*	*	*							
1940	380	****	*	*	*	*							
1941	932	*********	*	*	*	*							
1942	402	****	*	*	*	*							
1943	570	******	*	*	*	*	*						
1944	514	*****	*	*	*	*	*						
1945	340	***	*	*	*	*	*						
1946	300	***	*	*	*	*	*						
1947	271	***	*	*	*	*	*						
1948	201	**	*		*								
1949	206	**	*		*								
1950	345	***	*										
1951	267	***	*										
1952	762	*******	*	*				*					
1953	294	***	*					*					
1954	405	****	*			*		*					
1955	375	****	*					*					
1956	448	****	*	*							*		
1958	714	*******	*	*	*			*			*		
1959	166	**		*	*			*			*		
1960	319	***		*	*						*		
1961	157	**		*	*						*		
1962	573	******		*	*	*	*	*			*	*	
1963	290	***		*	*						*		
1964	238	**		*							*		
1965	420	****		*									
1966	335	***		*									
1967	574	******		*	*								*
1968	432	****		*	*				*				
1969	940	*********		*	*				*	*	*		
1970	659	*******		*									
1971	340	***		*									
1972	191	**		*									
1973	517	*****		*									

could also have been influenced by precipitation of preceding years. In 1941, a record year for rainfall, a major slide destroyed 27 wells overnight. Only one slide, in 1949, is known to have been triggered by an earthquake. Artesian pressures were associated with massive, permeable sandstones underlying some slide masses. Slip on multiple-planes as discussed in Chapters 13 and 18 has been well documented at Ventura; in some instances total displacements were distributed among perhaps 30 individual planes (H. Neel, personal communication).

The type of slide movement in granitic rock illustrated by Chapter 13 may seem unusual, because discussions of failures involving a granitic rock mass more commonly record the dominant influence of exfoliation joints. Excavation for the spillway of Surry Mountain Dam in New Hampshire provides an informative example of the effect of such joints. A steep cut was made in massive, fresh granite containing a few widely spaced exfoliation joints (Lane, 1970). Three years later, during a period of cyclic freeze and thaw, a small rock fall occurred; this event was followed by a major slide (Fig. 41). Joint water pressure behind an ice-blocked outlet was considered to be the major cause (Lane, 1970, p. 525; cf. Chapter 2, Fig. 5; Chapters 8, 12). After a period of 19 years, an adjacent rock slab resting on the same prominent discontinuity also began to move, requiring significant excavation and structural modification of a highway bridge attached to the unsteady rock slab. Rehabilitation costs totaled about 70% of the initial cost of the entire dam.

Weathered rock and colluvial slopes
In regions of more intense weathering, granitic rock can be altered to a

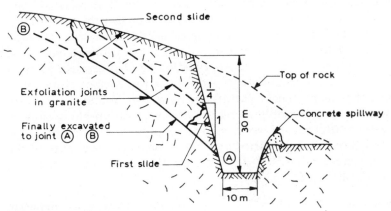

Fig. 41. Rockslides at Surry Mountain Dam, Keene, New Hampshire (after Lane, 1970). Side channel spillway cut originally at 1 : ¼, 30 m high, in granite of monument quality but containing widely-spaced exfoliation joints. Slides occurred three years after excavation, in the spring of 1943.

rather weak material. Discontinuities may nevertheless remain more or less intact as effective planes of weakness throughout the weathering process. The problems of slope stability in such material are many, and are nowhere dramatized better than in Brazil. Here strong topographic relief, rainfall of remarkable intensity (Fig. 42), and weathered rock all contribute to classic mass movement phenomena, reviewed and superbly illustrated in this volume in Chapter 11. The great variation in slide masses include exfoliation slabs, irregularly weathered complexes of residual soil and relatively intact crystalline blocks, and colluvial soils. Base-of-slope excavations for construction materials, together with hydrogeologic factors, seem to have triggered most slides. Where geologic conditions are favorable to slides, almost any construction activity that cuts into a slope may result in sliding at some future date unless protective measures are provided (cf. Japan Society of Landslide, 1972).

Many of the slope failures common in residual and colluvial soils and weathered rock are the result of special effects imposed on the slopes by the development of the weathering profile. These effects include a profound decrease in the shearing resistance of the rock mass, a fact well known, and a reduction of permeability of the upper region of the slope, a fact less appreciated (Parizek, 1971; Deere and Patton, 1971; Patton and Hendron, 1974). The latter condition effectively creates an inclined artesian system parallel to the ground surface and causes an increase of fluid pressures beneath discharge areas (Fig. 43). The more permeable portions of the weathering profile are especially susceptible to slope movements. A common mode of failure, even in soils, involves relic defects or layering inherited from the original fabric.

The influence of initial stratigraphy on weathering profiles is well illustrated by hill slopes on coal measures of eastern North America. In the Appalachian Plateau, repeated observations have shown that beds of coal, clay, and shale, instead of extending horizontally to intersect the hill surface, thin abruptly and bend downslope in a zone of near-surface creep (Sharpe and Dosch, 1942). Gradual thinning continues in the downslope extension of the beds, for a hundred metres or more, creating layered artesian systems and zones of weakness approximately parallel to the surface slope. It is hardly surprising that many recent earthflows have been found to occur on such slopes, and not uncommonly along a uniform elevation along slope — thus reflecting the influence of specific stratigraphic conditions on slope stability. Careful study of these earthflows has given a clearer picture of the influence of slope creep (Figs. 44, 45) although the phenomenon has long been recognized (Lesley, 1856, p. 35):

"... the whole surface of all hills have been in the slow but perpetual movement downward from the beginning, so that in the present day the soil or weathered broken edge of any stratum overlies the strata below it, while it is itself covered by the soil of some stratum above it. On slight slopes this transition of material has gone to no great distance, but on slopes of 20° or 30° the smut of a given coal

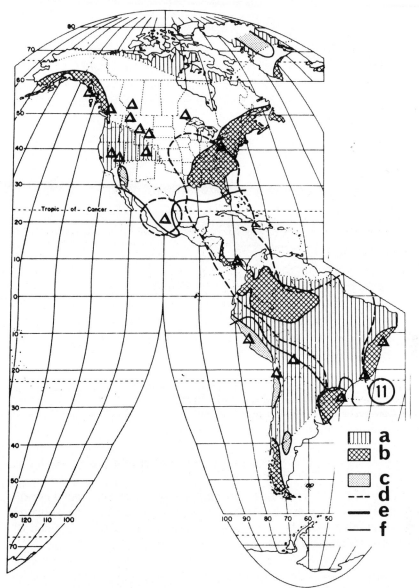

Fig. 42. Precipitation patterns of significance to slope movements (data from Common, 1966; after Snead, 1972, courtesy of J. Wiley and Sons). Locations associated with chapters of this volume indicated by triangles. a = areas essentially outside temperate and tropical storm tracks; b = areas with at least 5 cm of precipitation in January, April, July, and October; c = areas with precipitation less than 13 cm per annum; d = important thundershower zones; e = areas of monsoon climate; f = general equatorward snowfall limit; 11 = sites of "hydraulic excavation" as discussed in Chapter 11, this volume, with precipitation intensities as much as 10 cm/hr. Similar precipitation intensities extend into North America as indicated by curve d.

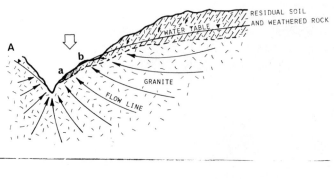

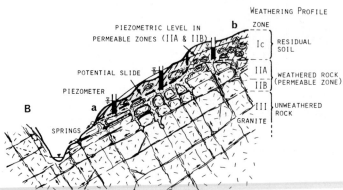

Fig. 43. Groundwater flow conditions in areas of thick residual and colluvial soils and deeply weathered rock (after Patton and Hendron, 1974). A. Position of portion of slope *a-b* in regional flow system; B. Detailed flow and fluid pressure conditions in vicinity of *a-b*.

> bed has probably been drawn out in a long knife-like wedge, the edge of which is to be seen many yards below its proper place."

These slope movements were most strongly active under periglacial conditions affecting much of eastern North America, but the present equilibrium of such slopes remains delicate and is easily upset by man's disturbance of natural conditions (Chapter 12).

Climatic factors

It may also be noted that debris slides have also been particularly common in the Appalachians, where the combination of slope parameters and precipitation regimes have produced about 3000 recognizable debris slides (Bogucki, 1977). More than 1800 slides have formed in Georgia, North Carolina, Tennessee, Kentucky, West Virginia, and Virginia in the present century alone, and as many as 200 deaths in the southern area may have been directly caused by such slides (Scott, 1972). Such slides also occur in the Katahdin Range in Maine, the Green Mountains of Vermont, and the

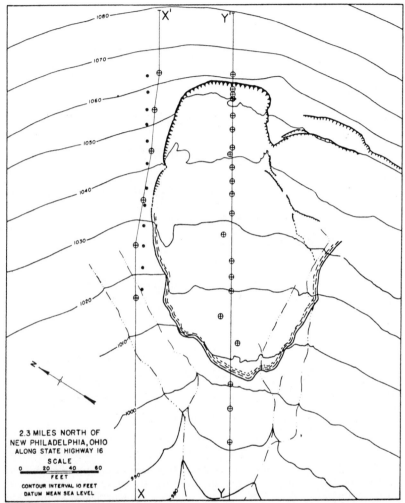

Fig. 44. Map of 1937 earthflow in colluvial slope, Ohio (after Sharpe and Dosch, 1942). Larger and deeper borings are indicated by crosses in circles; smaller or shallower borings are represented by smaller symbols (cf. Fig. 45). Precipitation of January 1937 was the heaviest recorded for any month except September 1866. A large number of earthflows in eastern Ohio reflected the high ground saturation of this period. The toe of the flow advanced 6 m in the year and a half following the major movement and completely blocked an old mine road.

Berkshires of Massachusetts. Flaccus (1958) identified almost 500 slides in the White Mountains of New Hampshire (see Frontispiece, Volume 1) and documented 11 slide-related deaths; over 400 slide scars have been recognized in the Adirondacks (Bogucki, 1977).

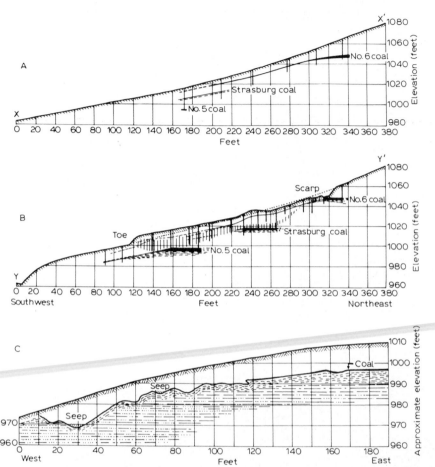

Fig. 45. Slope sections showing disturbance of coal and clay beds by creep and earthflow.
A. Profile X-X' adjacent to earthflow of Fig. 44, showing flexure and drawing out of
No. 6 coal, covering the lower beds. The coal bed becomes unrecognizable about 60 m
downslope. B. Profile Y-Y' on axis of earthflow. Three recognizable coal beds are present,
each of which bends and thins downslope in a zone of colluvial creep. The base of the
earthflow is a zone of intensified flow rather than a surface of slip. C. Comparative profile
of north wall of fresh road cut on U.S. 250, west of Tappan, Ohio. The coal and clay
layers thin westward and drop irregularly 9 m in a horizontal distance of 40 m. The sags
in the coal layer are troughs that reportedly formed by creep at right angles to section
illustrated.

The specific mechanism of failure in debris slide formation has not been
witnessed, but the slides are typically associated with steep, forested slopes
and conditions of either prolonged or high intensity precipitation. Investigators have shown that Appalachian slides originate on slopes of 17—44°, with
rainfall intensity as much as 10 cm/hr (or 56 cm over a two-day period)

(Moneymaker, 1939; Williams and Guy, 1971; Woodruff, 1971; Scott, 1972; Schneider, 1973; Bogucki, 1976, 1977). These precipitation intensities are equivalent to those associated with the "hydraulic excavation" phenomena of Brazil (Fig. 42; cf. Chapter 11).

Inasmuch as the potential for excavation-triggered catastrophic rock failure increases approximately with the sine of the slope angle, no area of the Western Hemisphere presents more acute debris slide and rockslide problems to the construction engineer than the Andes (Fig. 46). The Bolivian case

Fig. 46. Slide scar at Trans-Andean pipeline excavation near Túquerres, Colombia (McIntyre, 1970; courtesy of *National Geographic*). Five men were carried by debris slides to the Rio Sucio, 400 m below.

history of Chapter 10 illustrates the situation quite well. Excavation of a highway bench 18 m deep into the mountain face exposed several shale seams in sandstone; months later, sliding occurred suddenly along bedding planes dipping at about 45°. The slide volume was small, but seven men were killed, thus creating conditions that threatened progress on the highway. The timing of the slide is of interest. The trigger mechanism has not been unambiguously identified, but may have involved cyclic thermal expansion and contraction, causing progressive downslope movements on an undulating slip surface. Sudden failure may have occurred when the slide mass advanced past the summit of the dominant asperities on the slide plane.

Climatic factors also seem important at the Hell's Gate cut on the Trans-Canada Highway, where slope movement over a 6-year period could be correlated with precipitation, temperature, and cyclic freeze-thaw conditions (Chapter 15). Freezing conditions and snowfall appear to retard movement, because infiltration of precipitation is prevented or retarded. It seems conceivable that similar movements could have preceded and aided ultimate collapse of large-scale natural slope failures subjected to similar climatic conditions, e.g., at Hope and Frank localities.

Analogous observations of the influence of temperature and precipitation on rock movements have been recorded for the Lower Gros Ventre slide (Chapter 3, Volume 1) and for several locations in the Scandinavian Arctic (Fig. 47). In the Arctic mountains mass movements have two peak seasons (Rapp and Strömquist, 1976); one is in May and June, the period of rapid snowmelt and thawing. Rockfalls, snow avalanches and solifluction have their seasonal peak at this time. The other maximum period is in summer and autumn when debris slides and flows are triggered by heavy rainstorms with long intervals.

Clay shales

The influence of both artificial and natural excavations on stability of clay shales is considered in Chapter 14. This chapter provides a regional review of slides in the Interior Plains of western Canada. In the case of Gardiner Dam (formerly South Saskatchewan River Dam), both rebound and slides provided serious problems; neither swelling nor strength could be reliably determined from short-time laboratory experiments. Similarly, the Grierson Hill slide at Edmonton was also presumably influenced by excavation, in this case due to river erosion, although subsidence associated with subsurface coal mining may have permitted ingress of water to the slide mass and hence contributed to instability. The Devon slide was directly triggered by road excavation, although detailed investigation revealed that this "first time" slide was actually located within a massive prehistoric slide block.

Chapter 14 emphasizes the effects of channel erosion of low modulus sedimentary bedrock. Load removal was characteristically accompanied by a

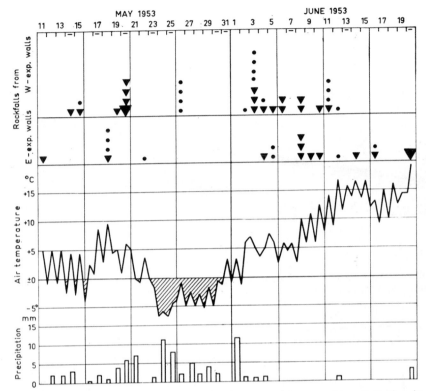

Fig. 47. Correlation between rockfalls, air temperature, and precipitation in May-June, 1953, at Kärkevagge, northern Scandinavia (after Rapp, 1960). Rockfalls recorded by inventories on snow all days except those marked with a minus (−) on top of graph. The highest frequency of rockfalls occurs on thawing. Each case of rockfall recorded is marked by a dot (pebble-fall), a small triangle (small boulder-fall) or big triangle (large boulder-fall). Air temperature at 01^h and 13^h from recordings in Kärkevagge at 820 m altitude. Precipitation at Riksgränsen weather station.

rebound that gave rise to upwarping and flexural slip of strata in valley walls and to a gentle anticlinal structure beneath the valley floor (Figs. 48, 49, 20). The displacement patterns are at least qualitatively predictable by simple elastic theory (Fig. 50, Table II), although the elastic idealization is clearly an oversimplification of rather complex material behavior. Rebound at the valley center ranges from about 3 to 10% of the valley depth for sites where Young's modulus is less than 350 MN/m². In contrast, artificial excavation associated with engineering sites in the same region have maximum rebounds of less than 1% of excavation depth, thus illustrating the importance of time-dependent rebound continuing over geologic periods of time (Matheson, 1972; Matheson and Thomson, 1973).

Engineering problems in clay shale have, of course, been encountered

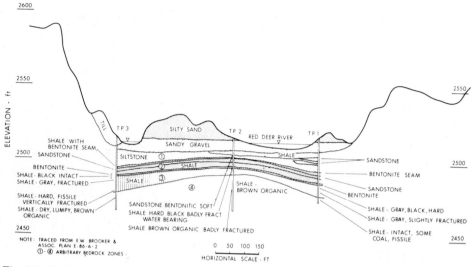

Fig. 48. Valley flexure as evidenced from test pits at Ardley damsite, Alberta, Canada (after Matheson and Thomson, 1973).

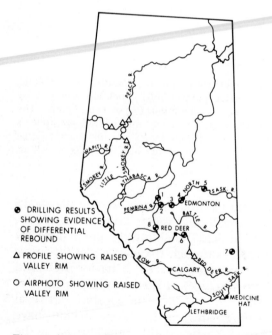

Fig. 49. Observed valley flexure locations in Province of Alberta, Canada (after Matheson and Thomson, 1973). *1* = Pembina River damsite; *2* = Tomahawk damsite; *3* = Carvel damsite; *4* = Edmonton bridges; *5* = Hairy Hill damsite; *6* = Ardley damsite; *7* = Sounding Creek damsite; *8* = Rocky site A.

TABLE II

Compilation of maximum values of rebound and reported values of E (Matheson and Thomson, 1973)

Site	Maximum rebound (m)	δ/H	E (MN/m^2)	
Ardley Damsite	3.7	0.08	58—83	
Boundary Dam	1.8	0.07	180	(shale)
Carvel Damsite	1.5	0.025	365	(sandstone)
Hairy Hill Damsite	1.8	0.033	43	
Rocky A Damsite	1.2	0.036	—	
Sounding Creek	1.5	0.125	—	
St. Mary Dam	0.5	0.01	6900	
Three Rivers Damsite	0.2	0.002	4100	
Tomahawk Damsite	2.1	0.035	100	
Pembina 3A Damsite	1.5	0.038	310—520	
James MacDonald Bridge	1.2	0.08	—	
Garrison Dam	6.1	0.09	97—280	
Gavins Point	2.1	0.05	140	
Bowman-Haley	6.1	0.12	11	
A.G.T. Excavation	0.1	0.007	41—140	
South Saskatchewan Dam	—	0.007	120	

for many years. Alexandre Collin worked with such materials in the 19th century (Fig. 33), and in the Western Hemisphere the classic problem has been the Panama Canal. But geotechnical studies associated with construction in the Great Plains of western United States (Fig. 51, Table III) have particularly added to our understanding of clay shale behavior, e.g., at Fort Peck (Middlebrooks, 1942; Hamel, 1973) as previously discussed, Garrison (Smith and Redlinger, 1953; Lane, 1961), Oahe (Knight, 1963; Underwood et al., 1964; Fleming et al., 1970b, pp. 223—242) and Waco Dams (Beene, 1967). Similar contributions have been made by Canadian investigators as outlined in Chapter 14. Related problems have been observed elsewhere in North and South America, most especially in areas of Tertiary deposition. In South America the problem has not yet been subjected to intensive study, although it has been recognized in the Recôncavo Basin and in southern areas of Brazil (Chapter 11) and in Peru (Fleming et al., 1970b, p. 288).

The two fundamental factors which influence geotechnical properties and mass behavior of clay shales are degree of consolidation and lithology, both of which reflect geologic history (Skempton, 1964; Bjerrum, 1967; Scott and Brooker, 1968; cf. Fleming et al., 1970a). Diagenetic bonds can be

TABLE III

Landslide-susceptible clay shales in United States

Stratigraphic unit	Description
Amsden Formation	Pennsylvanian; Wyoming; marine shale, limestone, sandstone; 450 m (*Chapter 3, Volume 1*)
Bearpaw Shale	Cretaceous; Montana, Wyoming, and Alberta, Canada; marine clay shale; 200 m; in Montana Group (*Chapter 11, Volume 1; Chapter 21, Volume 2*)
Carlile Shale	Cretaceous; Colorado and Wyoming, Nebraska, Kansas, South Dakota, Montana and New Mexico; shale; 60 m; in Colorado Group
Cherokee Shale	Pennsylvanian; Kansas, Nebraska, Missouri and Oklahoma; shale; 150 m; in Des Moines Group
Chugwater Formation	Triassic; Wyoming; siltstone and shale; 450 m
Claggett Formation	Cretaceous; Montana, Wyoming; marine clay shales and sandstone beds; 120 m; in Montana Group (*Chapter 11, Volume 1*)
Cody Shale	Cretaceous; Wyoming; marine shale, glauconitic sandstone, bentonite; 700 m
Colorado Shale	Cretaceous; Montana, Wyoming, Colorado; shale, bentonite; 450 m; in Montana Group (*Chapter 11, Volume 1*)
Conemaugh Group	Pennsylvanian; Pennsylvania, Ohio, West Virginia; cyclic sequences of claystone, shale, siltstone, sandstone, limestone and coal; 150—260 m (*Chapter 12, Volume 2*)
Dakota Group	Cretaceous; Colorado, Wyoming, North and South Dakota; shale, sandstone, bentonite; 200 m (*Chapter 13, Volume 1*)
Dawson Formation	Cretaceous/Lower Tertiary; Colorado; non-marine clay shales, siltstones and sandstone; 300 m
Del Rio Clay	Cretaceous; Texas; laminated clay with beds of limestone; in Washita Group
Eden Group	Ordovician; Ohio, Indiana, and Kentucky; shale with limestone; 75 m; in Cincinnati Group
Fort Union Group	Paleocene; Montana, Wyoming, North Dakota, South Dakota, and Colorado; massive sandstone and shale; 1200 m (*Chapter 21, Volume 2*)
Frontier Formation	Cretaceous; Wyoming and Montana; sandstone with beds of clay and shale; 700 m; in Colorado Group
Fruitland Formation	Cretaceous; Colorado and New Mexico; brackish and freshwater shales and sandstones; 160 m
Graneros Shale	Cretaceous; Colorado, Wyoming, Montana, South Dakota, Nebraska, Kansas and New Mexico; argillaceous or clayey shale; 60 m; in Colorado Group

TABLE III (continued)

Stratigraphic unit	Description
Gros Ventre Formation	Cambrian; Wyoming and Montana; shale with sandstone and limestone; 240 m (*Chapter 6, Volume 2*)
Landslide Creek Formation	Cretaceous; shale (*Chapter 17, Volume 1*)
Jackson Group	Eocene; Gulf Coastal Plain (Alabama to Texas); calcareous clay with sand, limestone, and marl beds
Mancos Shale	Cretaceous; Colorado, New Mexico, Utah, Wyoming; marine, carbonaceous clay shale with sand; 600 m
Merchantville Clay	Cretaceous; New Jersey; marly clay, 15 m; in Matawan Group
Modelo Formation	Miocene; California; clay, diatomaceous shale, sandstone, and cherty beds; 3000 m (*Chapter 18, Volume 1*)
Monongahela Group	Pennsylvanian; Pennsylvania, Ohio, West Virginia; non-marine cyclic sequence of claystone, shale, siltstone, sandstone, limestone, and coal; 150 m (*Chapter 12, Volume 2*)
Monterey Shale	Miocene; California; hard silica-cemented shale and soft shale; 300 m
Morrison Formation	Jurassic; Colorado and Wyoming, Montana, South Dakota, Kansas, Oklahoma, New Mexico, Arizona and Utah; shale with sandstone and limestone beds; 60 m (*Chapter 13, Volume 1*)
Mowry Shale	Cretaceous; Wyoming, Montana and South Dakota; hard shale; 45 m; in Colorado Group
Pepper Formation	Cretaceous; Texas; clay shale
Pierre Shale	Cretaceous; North Dakota, South Dakota, Nebraska, Minnesota, Montana, Wyoming, and Colorado; marine clay shale and sandy shale; 210 m; in Montana Group
Rincon Shale	Miocene; California; clay shale with limestone; 600 m
Sundance Formation	Jurassic; South Dakota, Wyoming, Montana, Nebraska, and Colorado; shale with sandstone; 200 m
Taylor Marl	Cretaceous; Texas; chalky clay; 360 m
Thermopolis Shale	Cretaceous; Wyoming, and Montana; shale with sandy bed near middle; 240 m; in Colorado Group
Trinity Group	Cretaceous; Texas, Oklahoma, Arkansas, Louisiana; shale, fine sand, gypsiferous marl and occasional limestone
Wasatch Formation	Eocene; Wyoming, Montana, North Dakota, Colorado, Utah, and New Mexico; sands and clay; 1500 m (*Chapter 11, Volume 1; Chapter 21, Volume 2*)
Wind River and Indian Meadows Formations	Eocene; Wyoming; alternating beds of shale and sandstone; 1000 m (*Chapter 21, Volume 2*)

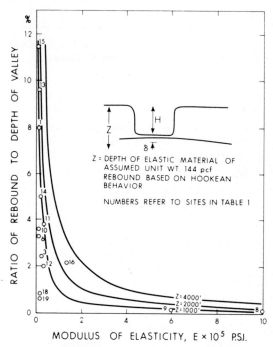

Fig. 50. Normalized maximum values of elastic rebound versus Young's modulus (after Matheson and Thomson, 1973).

Fig. 51. Location of major damsites sited on clay shales in western United States and Canada.

occurred in the spring "break-up" period, and yet the pit face may not have thawed. Two major shear zones had a major influence on stability. In addition the failed area consisted of a rock promontory (convex slope) and pit blasting had been increased. Remedial measures included waterproofing of the drainage duct and filling cracks to reduce infiltration of water, and installation of horizontal drains. Piezometers indicated major reduction in joint water pressures within a month after installment of drains.

Slope monitoring and failure prediction

Long-term stability of an excavation is the general goal in civil engineering projects, whereas in open pit mining projects extraction is the primary goal, and slope stability is necessary only for attainment of this primary goal. The economic necessity of minimizing waste rock removal means that steep slope angles are desirable. At today's costs, enormous savings accrue for each degree the slope can be steepened. The probability for slope failure is thus high, and as pits get deeper, failures become more frequent and costly. The optimum slope angle is not therefore the steepest. In some mine projects it is possible to "live" with a slide, and to eliminate certainly the possibility of all slides would require unacceptably conservative slope angles.

Because progressive slope failure is accompanied by rock fracture and energy release, the use of acoustic emission methods seems practical for purposes of monitoring and perhaps predicting slope distress or failure. The present state of the art is reviewed in Chapter 16 with case history data from pits in the American southwest. Chapter 16 shows that acoustic emission surveys can predict slope failure before visual evidence appears in the form of surficial tension cracks. In addition, these surveys can be used to monitor the effects of external events on exposed slopes, such as excavation by mining or stripping, blasting, or earthquakes.

During a period in which microseismic studies were being carried out at the Kimbley pit in Nevada, an interesting slope failure involving about 3000 m^3 of soft altered rhyolite occurred (Hamel, 1971). Indications of failure were first noted on September 20, with small rock fragments "popping" from the slope face; by 9:00 a.m. on September 22, a vertical peripheral crack was observed, and by the afternoon numerous cracks had developed in the mass. Ravelling of material at the edge of the mass was almost continuous by 5:20 p.m., increasing in intensity until 5:30 p.m. at which time the main failure occurred. The main mass appeared to rotate pitward about its toe as a quasi-rigid body; it then slumped downward, breaking up as vertical displacements became dominant (Fig. 56). Hamel suggested a block model for progressive failure, in which stress redistribution following toe failure caused a change in moment (from resisting to overturning) and hence to pitward rotation of the slope.

In Chapter 22, Canadian experiences in the application of finite element methods of slope analysis are presented. Parametric studies indicate that a knowledge of the pre-mining stress field (which is known to vary from point to point) is important for simulation of slope behavior. The effects of rock mass inhomogeneity are examined, and are shown to strongly influence behavior of the rock mass; in addition, several case histories are presented in which attempts were made to model field conditions.

An open pit mine is a dynamic system in contrast to an excavation for civil engineering purposes. Experience has shown that it is important to monitor slopes during a mining operation, for information helpful to design of subsequent mine modifications. The Chuquicamata pit failure is the classic case of slope monitoring (Chapter 17), although other slides and falls have also been predicted in time to cause minimal interruption to mining operations or to minimize accidents to mine personnel. It may not always be possible to accommodate a massive slide as "comfortably" as was accomplished at Chuquicamata, but as Hoek and Bray (1974) have suggested, "it is nice to know that one has this possibility in reserve. On the more positive side. . . the monitoring of surface displacements in slopes will become an important method of slope control since it has been shown to be possible to detect incipient instability a number of years before serious problems develop, perhaps early enough for effective remedial measures. . ."

A second example of a successful program of pit slope monitoring is afforded by the Atikokan, Ontario, case history of Chapter 20. Here, slope monitoring and warning systems permitted the safe removal of recoverable ore reserves for a ten-month period prior to a massive toppling failure. The slope was observed and photographed at the moment of collapse, providing rare documentation of an actual failure.

Toppling failures are complex, and much has yet to be learned about the associated slope displacements, particularly at the base of slope and within the rock mass itself (cf. Goodman and Bray, 1976). The study of Braced-up Cliff may be mentioned as an example of this complexity; this locality is at the same time of much historic significance, inasmuch as it represents the earliest known example of native American ingenuity in the art of geotechnical engineering (Keur, 1933).

Braced-up Cliff, or Threatening Rock as it was known to National Park Service personnel, toppled on January 22, 1941, damaging a portion of the ruins of Pueblo Bonito in Chaco Canyon National Monument, New Mexico. The huge sandstone monolith, about 50 m long, 30 m high, and 10 m thick, was completely detached from a nearby cliff. It rested completely on shale which had been partially undercut by erosion. An earth and masonry terrace had been constructed by inhabitants of the Pueblo Bonito in the 11th Century to prevent erosion at the base of the block and to support the block.

By 1933 the block was separated by a fissure about 4 m wide at the top and about 1 m wide at the base. The fissure was partly filled by loose rocks, and the difference in fissure width from bottom to top may in part have reflected weathering and rockfall. The block had not only moved out from the cliff, but had settled about 20 cm. The amount of undercutting varied from 1 to 5 m, so that at places the undercutting extended nearly to the axis of the center of gravity.

Measurements of progressive movement were recorded by T.C. Miller and L.T. McKinney, custodians of the National Monument. Two steel bars were set in concrete in 1935, one on the cliff and the other on the block, and additional points were established in 1936 and 1940. Between November 1935 and December 1940, the block moved outward 30 cm; in the final month, 23 December 1940 to 22 January 1941, the block moved outward 26 cm, westward 36 cm, and settled 10 cm. On the evening of 21 January a large rock fragment fell from the top of Braced-up Cliff. McKinney estimated that the rock moved outward 23 cm during the night of January 21. While making his regular monthly measurements of movement on the following day at 12:40 p.m., he could hear the rock popping and cracking and the block moved outward about 1 mm during his measurement. At 3:24 p.m. Braced-up Cliff fell.

Three Indians observed the fall from a distance of 100 m. According to their description as recorded by the Park Service (R.C. Heyder, cited by Schumm and Chorley, 1964, pp. 1049—1050):

> "The Indians on the woodpile heard the rock groan and looked up to see dust shooting out of the cracks in it. The slab leaned out about 30 or 40 feet from plumb, settled sharply, and when it hit solid bottom, rocks from the top of it were broken loose and propelled into the ruin. The lower two-thirds then pivoted on its outer edge and fell down the slope toward the ruin. The whole mass broke into many fragments and an avalanche of rocks catapulted down the slope into the walls of the back portion of Pueblo Bonito."

The westward movement component noted for the final month suggests sliding or tilting toward the southwest, the place of greatest undermining. The actual details of rock movement during the period of measurement are uncertain because measurements were made only at the top of the cliff; sliding, tilting, or both processes acting in combination would account for the observed displacements. The fall of the block, according to the eyewitness accounts, was a combination of tilting and sliding. In McKinney's opinion the earlier movements appeared to be a combination of settling and movement away from the cliff, rather than tilting; in final movements the block seems to have shifted its center of gravity beyond the zone of shale support, and toppling naturally occurred.

Displacement-time records (Fig. 57) indicate alternating periods of relative stability (in summer) and periods of relatively rapid movement (between dates of the first killing frost in the fall and the last killing frost in the spring). The higher precipitation of summer was less effective in promoting movement than the hydrologic and climatic conditions of the remainder of the year. A likely factor of importance was collection of snow in the gap between the monolith and the cliff face (Schumm and Chorley, 1964). Snow

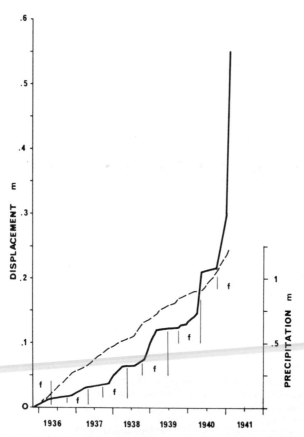

Fig. 57. Cumulative displacement of Braced-up Cliff, New Mexico, versus time (after Schumm and Chorley, 1964). Cumulative precipitation given by dashed line. Freeze-thaw periods designated *f*.

melt would likely have caused greater saturation of the shale at the base of the monolith than individual precipitation events; average annual precipitation at Chaco Canyon is only about 21 cm.

A plot of block displacement versus time, on logarithmic paper, is approximately linear (Schumm and Chorley, 1964). If the line linking displacement against time in Fig. 58 is extended to the total width of 180 cm to be moved, the data suggest that movement of the slab from the canyon wall began about 2500 years before its actual fall. The inferred rate of natural slope retreat by slab failure is thus about 1 m in 2500 years, a rapid rate by geologic standards. Many of the assumptions implied by this analysis are dubious, yet the estimate seems of the correct order of magnitude, for displacements were clearly sufficient long ago to cause concern to the inhabitants of the Pueblo Bonito; trees wedged beneath the monolith in their

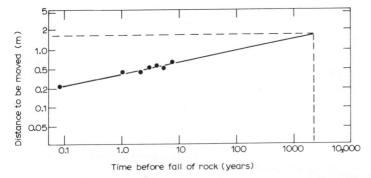

Fig. 58. Estimate of length of time involved in the failure of Braced-up Cliff, New Mexico, based on displacement measurements in 1935—1941, and known total displacement of 1.8 m (after Schumm and Chorley, 1964). Movements had to be significant enough to be of concern to inhabitants of Pueblo Bonito about 900 years B.P.

rockslope remedial engineering efforts have been dated by radiocarbon methods at 1057 and 1004 A.D. (Judd, 1959).

Prediction of catastrophic slope failure is thus hardly a new subject. A century ago Albert Heim well understood that slope collapse was preceded by subtle warning phenomena of various kinds. At Goldau, for example, Heim noted that the animals had, literally, "sense" enough to leave the slope prior to its collapse. At Elm the quarry operations were closed due to increasing frequency of stone falls, a prelude to the catastrophe which followed (Chapter 1, Volume 1). Tension cracks invariably preceded major movements of the rock mass.

In the opinion of most authors, high-frequency or continuous displacement monitoring remains the most effective means of failure prediction. If significant increases in slide velocity are detected, the indication is that the shear strength of some portion of the rock mass has been reduced; complete failure might be an ultimate consequence. Other methods, such as permanently installed water pressure transducers and acoustic emission monitoring, have also been employed, but mainly for research purposes rather than for practical engineering use. A useful summary of monitoring equipment is given by Franklin and Denton (1973). In all cases, however, fundamental rock mechanics problems are involved, not so much in the acquisition of data, but rather in establishing specific predictive criteria for catastrophic sliding. At Chuquicamata (Chapter 17), failure occurred on the *earliest* of a range of predicted dates inferred from displacement measurement. "Maximum displacement" was employed as the criterion for instability. Determination of the actual value of maximum displacement to be used in such a criterion is, of course, an open question, as is the method of extrapolation of displacement curves. Chuquicamata also provides an excellent example of acoustic emission in relation to progressive failure, and in addition apparently provides one of the few documented cases of delayed seismic effects, inasmuch as the onset of slide motion was triggered by an earthquake. However, the velocity of slope movement was also noticeably influenced by blasting events during mining of the pit; the effect seems much as described earlier in this introduction (Fig. 9).

REFERENCES

Barron, K., Coates, D.F. and Gyenge, M., 1970, Artificial support of rock slopes. *Can. Dep. Energy, Mines and Resources, Mines Branch R228.*

Beene, R.R.W., 1967. Waco Dam Slide. *Proc. Am. Soc. Civ. Eng., J. Soil Mech. Found. Div.*, 93(SM4) Paper 5306: 35—44.

Bjerrum, L., 1967. Progressive failure in slopes of overconsolidated clay and clay shales. *Proc. Am. Soc. Civ. Eng., J. Soil Mech. Found. Div.*, 93(SM5): 1—49.

Bogucki, D.J., 1976. Debris slides in the Mt. Le Conte area, Great Smoky Mountains National Park, U.S.A. *Geogr. Ann.*, 58(3) 179—191.

Bogucki, D.J., 1977. Debris slide hazards in the Adirondack Province of New York State. *Environm. Geol.*, 1(6): 317—328.

Brawner, C.O., 1971. Case studies of stability on mining projects. *Proc., 1st Conf. on Stability in Open Pit Mining, Vancouver, B.C.*, pp. 205—226.

Brawner, C.O., 1974. Rock mechanics in open pit mining. *Proc., 3rd Congr. Int. Soc. Rock Mech.*, 1-A: 755—773.

Breth, H., 1967. The dynamics of a landslide produced by filling a reservoir. *9th Int. Congr. on Large Dams, Istanbul*, Q.32, R.3, pp. 37—45.

Broili, L., 1967. New knowledge on the geomorphology of the Vaiont slide slip surface. *Rock Mech. Eng. Geol.*, 5: 38—88.

CANMET, 1976—1977. *Pit Slope Manual.* Canadian Centre for Mineral and Energy Technology, Ottawa, Ont. Report in 10 chapters (1976—1977).

Chandler, R.J., 1969. The effect of weathering on the shear strength properties of Keuper Marl. *Geotechnique*, 19: 321—334.

Chandler, R.J., 1970. A shallow slab slide in the Lias clay near Uppingham, Rutland. *Geotechnique*, 20: 253—260.

Coates, D.F. and Sage, R., 1973. *Can. Dep. Energy, Mines and Resources, Mines Branch TB1181.*

Collin, A., 1846. *Experimental Research on Landslides in Clay Strata, Accompanied by Considerations of Several Principles of Soil Mechanics.* Carilian-Goeury and Dalmont, Paris; translated by W.R. Schriever, 1956, Univ. of Toronto Press, Toronto, Ont., 160 pp.

Common, R., 1966. Slope failure and morphogenetic regions. In: G.H. Dury (Editor), *Essays in Geomorphology.* Heinemann, London, pp. 53—81.

Cooley, R.L., 1974. *Finite Element Solutions for the Equations of Ground-Water Flow. Hydrology and Water Resource Publication 18.* Center for Water Resources Research, Desert Research Inst., Univ. of Nevada, Reno, Nev., 134 pp.

Coulomb, C.A., 1776. Essai sur application des règles de maximis et minimis à quelque problèmes de statique, relatifs à l'architecture. *Mem. Acad. Sci. (Savants Etrang.)*, 7: 343—382.

Cross, W., 1924. Historical sketch of the landslides of the Gaillard Cut. In: National Academy of Sciences, *Report of the Committee of the National Academy of Sciences on Panama Canal Slides. U.S. Natl. Acad. Sci. Mem.*, 18: 23—43.

Cundall, P., 1974. Rational design of tunnel supports, a computer model for rock mass behavior using iterative graphics for the input and output of geometrical data. *U.S. Army Corps of Engineers, Tech Rep.*, MRD-2-74.

D'Appolonia, E., Alperstein, R. and D'Appolonia, D.J., 1967. Behavior of a colluvial slope. *Proc. Am. Soc. Civ. Eng., J. Soil. Mech. Found. Div.*, 93(SM4): 447–473.

Deere, D.U. and F.D. Patton, 1971. Stability of slopes in residual soils. *Proc., 4th Pan-Am. Conf. Soil Mech. Found. Eng., San Juan, Puerto Rico*, 1: 87–170.

Douglass, P.M., 1974. Slope stability investigations at the Centralia coal mine. In: B. Voight and M.A. Voight (Editors), *Rock Mechanics — The American Northwest. 3rd Congr. Exped. Guide, Int. Soc. Rock Mech., Spec. Publ.* Experiment Station, College of Earth and Mineral Sciences, Pennsylvania State Univ., University Park, Pa., pp. 283–285.

Duncan, J.M., 1972. Finite element analyses of stresses and movements in dams, excavations and slopes. In: C.S. Desai (Editor), *Application of the Finite Element Method in Geotechnical Engineering.* U.S. Army Waterways Experiment Station, Vicksburg, Miss., pp. 267–324.

Eigenbrod, K.D., 1975. Analysis of the pore pressure changes following the excavation of a slope. *Can. Geotech. J.*, 12: 429–440.

Finch, V.C., Trewartha, G.T., Robinson, A.H. and Hammond, E.H., 1957. *Elements of Geography.* McGraw-Hill, New York, N.Y., 4th ed., 693 pp.

Flaccus, E., 1958. *Landslides and Their Revegetation in the White Mountains of New Hampshire*, Ph.D. Thesis, Duke Univ., Durham, N.C., 187 pp.

Fleming, R.W., Spencer, G.S. and Banks, D.C., 1970a. *Empirical Study of the Behavior of Clay Shale Slopes. USAE NCG Tech. Rep.*, No. 15, Vol. 1, 93 pp.

Fleming, R.W., Spencer, G.S. and Banks, D.C., 1970b. *Empirical Study of the Behavior of Clay Shale Slopes. USAE NCG Tech. Rep.*, No. 15, Vol. 2, 304 pp.

Francais, J.F., 1820. Recherches sur la poussée des terres, sur la forme et les dimensions des murs de revêtment et sur les talus d'excavation. *Mém. de l'Office du Génie*, 4: 157–193.

Franklin, J.A. and P.E. Denton, 1973. The monitoring of rock slopes. *Q. J. Eng. Geol., Geol. Soc. London*, 6: 259–283.

Freeze, R.A. and P.A. Witherspoon, 1966. Theoretical analysis of regional groundwater flow, 1. Analytical and numerical solution to the mathematical model. *Water Resour. Res.*, 2(4): 641–656.

Freeze, R.A. and P.A. Witherspoon, 1967. Theoretical analysis of regional groundwater flow, 2. Effect of water table configuration and subsurface permeability variations. *Water Resour. Res.*, 3(2): 623–634.

Freeze, R.A. and P.A. Witherspoon, 1968. Theoretical analysis of regional ground water flow, 3. Quantitative interpretations. *Water Resour. Res.*, 4(3): 581–590.

Giudici, F. and Semenza, E., 1960. *Studio Geologico del Serbatoio del Vaiont.* Relazione inedite S.A.D.E., 21 pp.

Goodman, R.E. and Bray, J.W., 1976. Toppling of rock slopes. *Proc. Am. Soc. Civ. Eng. Spec. Conf. on Rock Engineering for Foundations and Slopes, Boulder, Colo.*, pp. 201–234.

Gregory, C.H., 1844. On railway cuttings and embankments with an account of some slips in the London Clay. *Minutes Proc. Inst. Civ. Eng.*, 3: 135–145; Discussion, 3: 145–173.

Gutenberg, B., and Richter, C.F., 1954. *Seismicity of the Earth and Related Phenomena.* Princeton Univ. Press, Princeton, N.J., 310 pp.

Hamel, J.V., 1970a. *Stability of Slopes in Soft, Altered Rocks.* Ph.D. Thesis, Univ. of Pittsburgh, Pittsburgh, Pa.

Hamel, J.V., 1970b. The Pima Mine slide, Pima County, Arizona. *Geol. Soc. Am., Abstracts with Programs*, 2(2): 335.

Hamel, J.V., 1971. Kimbley pit slope failure. *Proc., 4th Pan-Am Conf. Soil. Mech. Found. Eng., San Juan, Puerto Rico*, 2: 117–127.

Hamel, J.V., 1973. Rock strength from failure cases, powerhouse slope stability study, Fort Peck Dam, Montana. *U.S. Army Corps of Engineers, Tech. Rep.*, MRD-1-73, 159 pp.

Hoek, E., 1970. Conference summary. *Proc., 1st Int. Conf. on Stability in Open Pit Mining, Vancouver, B.C.*, pp. 239—242.

Hoek, E. and Bray, J.W., 1974. *Rock Slope Engineering*. Institution of Mining and Metallurgy, London, 309 pp.

Hofmann, H., 1973. Modellversuche zur Hangtektonik. *Geol. Rundsch.*, 62: 16—29.

Hubbert, M.K., 1940. The theory of groundwater motion. *J. Geol.*, 48: 785—944.

Hutchinson, J.N., 1967a. Written discussion. *Proc. Geotech. Conf., Oslo*, 2: 183—184.

Hutchinson, J.N., 1967b. The free degradation of London Clay Cliffs. *Proc., Geotech. Conf. on Shear Strength of Natural Soils and Rocks, Oslo*, 1: 113.

Hutchinson, J.N., 1969. A reconsideration of the coastal landslides at Folkestone Warren, Kent. *Geotechnique*, 19: 6—38.

Jaeger, J.C., 1971. Friction of rocks and stability of rock slopes. *Geotechnique*, 21: 97—134.

Jaeger, J.C. and Cook, N.G.W., 1969. *Fundamentals of Rock Mechanics*. Methuen, London, 513 pp.

Japan Society of Landslide, 1972. *Landslides in Japan*. Ekoda 2-21-2, Nakano-ku, Tokyo, 41 pp.

Jaunzemis, W., 1967. *Continuum Mechanics*. Macmillan, New York, N.Y.

Jones, F.O., Embody, D.R. and Peterson, W.L. 1961. Landslides along the Columbia River Valley, northeastern Washington. *U.S. Geol. Surv., Prof. Paper*, 367, 73 pp.

Judd, N.M., 1959. The Braced-up Cliff at Pueblo Bonito. *Smithsonian Inst. Annu. Rep. 1958*, pp. 501—511.

Kankare, E., 1969. Geotechnical properties of clays at the Kimola Canal area with special reference to the slope stability. *State Inst. Tech. Res., Helsinki, Publ.*, 152.

Kenney, T.C., 1967a. Influence of mineral composition on the residual strength of natural soils. *Proc. Geotech. Conf., Oslo*, 1: 123—129.

Kenney, T.C., 1967b. Stability of the Vaiont valley slope. *Rock Mech. Eng. Geol.*, 5: 10—16.

Kerr, P.F., Stroud, R.A. and Drew, I.M., 1971. Clay mobility in landslides, Ventura, California. *Am. Assoc. Pet. Geol. Bull.*, 55(2): 267—291.

Keur, J.Y., 1933. *A Study of Primitive Indian Engineering Methods Pertaining to Threatening Rock*. Chaco Canyon National Monument, unpublished report.

Knight, D.E., 1963. Oahe Dam: geology, embankment, and cut slopes. *Proc. Am. Soc. Civ. Eng., J. Soil Mech. Found. Div.*, 89(SM2), Paper 3466: 99.

Kwan, D., 1971. Observations of the failure of a vertical cut at Welland, Ontario. *Can. Geotech. J.*, 9: 283—298.

Lajtai, E.Z., 1969. Strength of weakness planes in rock. *Int. J. Rock Mech. Min. Sci.*, 6(5): 499—575.

Lane, K.S., 1961. Field slope charts for stability studies. *Proc., 5th Int. Conf. Soil Mech. Found. Eng.*, Paris, 2: 651.

Lane, K.S., 1970. Engineering problems due to fluid pressure in rock. *Proc., 11th Symp. on Rock Mechanics*, pp. 501—540.

Lauffer, H., Neuhauser, E. and Schober, W., 1967. Uplift responsible for slope movements during the filling of the Gepatsch Reservoir. *9th Int. Congr. on Large Dams, Istanbul*, Q.32, R.41, pp. 669—693.

Lesley, J.P., 1856. *A manual of Coal and its Topography*. Lippincott, Philadelphia, Pa., 224 pp.

Lo, K.Y., Lee, C.F. and Gelinas, P., 1972. An alternative interpretation of the Vaiont slide. *Proc., 13th Symp. on Rock Mechanics*, pp. 595—623.

Maini, Y.N.T., 1971. *In Situ Hydraulic Parameters in Jointed Rock, Their Measurement and Interpretation.* Ph.D. Thesis, Univ. of London, London.

Matheson, D.S., 1972. *Geotechnical Implications of Valley Rebound.* Ph.D. Thesis, Univ. of Alberta, Edmonton, Alta., 424 pp.

Matheson, D.S. and Thomson, S., 1973. Geologic implications of valley rebound. *Can. J. Earth Sci.,* 10: 961—978.

McIntyre, L., 1970. Columbia. *Natl. Geogr.* 138(2): 234—273.

Mencl, V., 1966. Mechanics of landslides with non-circular surfaces with special reference to the Vaiont slide. *Geotechnique,* 16: 329—337.

Meyboom, P., 1966a. Unsteady groundwater flow near a willow ring in hummocky moraine, *J. Hydrol.,* 4: 38—62.

Meyboom, P., 1966b. Groundwater studies in the Assiniboin River drainage basin, I. The evaluation of a flow system in south central Saskatchewan. II. Hydrologic characteristics of phreatophytic vegetation in south-central Saskatchewan. *Geol. Surv. Can. Bull.,* 139, Part I, 65 pp., Part II, 64 pp.

Meyboom, P., 1966c. Current trends in hydrogeology. *Earth-Sci. Rev.,* 2: 345—364.

Middlebrooks, T.A., 1942. Fort Peck slide. *Trans. Am. Soc. Civ. Eng.,* 107: 723—742.

Moneymaker, B.C., 1939. Erosional effects of the Webb Mountain (Tennessee) cloudburst of August 5, 1938. *J. Tenn. Acad. Sci.,* 14: 190—196.

Müller, L., 1964. The rock slide in the Vaiont Valley. *Rock Mech. Eng. Geol.,* 2: 148—212.

Müller, L., 1968. New considerations of the Vaiont slide. *Rock Mech. Eng. Geol.,* 6: 1—91.

Neuhauser, E., and Schober, W., 1970. Das Kriechen der Talhänge und elastische Hebungen beim Speicher Gepatsch. *Proc., 2nd Congr. Int. Soc. Rock Mech., Belgrade,* Paper 8-7.

Nonveiller, E., 1967a. Shear strength of bedded and jointed rock as determined from the Zalesina and Vaiont slides. *Proc. Geotech. Conf., Oslo,* 1: 289—294.

Nonveiller, E., 1967b. Mechanics of landslides with non-circular surfaces with special reference to the Vaiont slide. *Geotechnique,* 17: 170—171.

Pariseau, W.G., 1972. Elastic-plastic analysis of pit slope stability. In: C.S. Desai (Editor), *Application of the Finite Element Method in Geotechnical Engineering.* U.S. Army Waterways Experiment Station, Vicksburg, Miss.

Pariseau, W.G., Voight, B. and Dahl, H.D., 1970. Finite element analyses of elastic-plastic problems in the mechanics of geologic media, an overview. *Proc., 2nd Congr. Int. Soc. Rock Mech., Belgrade,* Paper 3-45.

Parizek, R.R., 1971. Impact of highways on hydrogeological environment. In: D.R. Coates (Editor), *Environmental Geomorphology. Proc., 1st Annu. Geomorphol. Symp., Binghamton, N.Y.* State Univ. of New York, Binghamton, N.Y., pp. 151—200.

Parizek, R.R., 1976. On the nature and significance of fracture traces and lineaments in carbonate and other terranes (and discussion). In: *Karst Hydrology and Water Resources.* Water Resource Publications, Fort Collins, Colo., 1: 47—108.

Patton, F.D. and Hendron, A.J., 1974. General report on mass movements. *Proc., 2nd Int. Congr. Int. Assoc. Eng. Geol.,* Paper V-GR, 57 pp.

Radbruch-Hall, D.H., Varnes, D.J. and Savage, W.Z., 1976. Gravitational spreading of steep-sided ridges ("Sackung") in western United States. *Bull. Int. Assoc. Eng. Geol.,* 14: 23—35.

Radley-Squier, L. and Versteeg, J.H., 1971. The history and correction of the Omsi-Zoo landslide. *Proc., 9th Annu. Eng. Geol. Soils Eng. Symp., Boise, Idaho,* pp. 237—256.

Rapp, A., 1960. Recent developments of mountain slopes in Kärkevagge and surroundings, northern Scandinavia. *Geogr. Ann.,* 42(2/3): 71—200.

Rapp, A. and Strömquist, L., 1976. Slope erosion due to extreme rainfall in the Scandinavian Mountains. *Geogr. Ann.,* 58A(3): 193—200.

Reid, H.F., 1924. The movement in the slides. In: National Academy of Sciences, *Report of the Committee of the National Academy of Sciences on Panama Canal Slides. U.S. Natl. Acad. Sci., Mem.*, 18: 79—84.

Schneider, R.H., 1973. *Debris Slides and Related Flood Damage Resulting from Hurricane Camile, 19—20 August, and Subsequent Storm, 5—6 September, 1969 in the Spring Creek Drainage Basin, Greenbrier County, West Virginia*, Ph.D. Thesis, Univ. of Tennessee, Knoxville, Tenn., 131 pp.

Schumm, S.A. and Chorley, R.J., 1964. The fall of threatening rock. *Am. J. Sci.*, 262: 1041—1054.

Scott, J.S. and Brooker, E.W., 1968. Geological and engineering aspects of upper Cretaceous shales in western Canada. *Geol. Surv. Can., Paper*, 66-37.

Scott, R.C., 1972. *The Geomorphic Significance of Debris Avalanching in the Appalachian Blue Ridge Mountains*. Ph.D. Thesis, Univ. of Georgia, Athens, Ga. 185 p.

Seegmiller, B., 1974. Cable bolt design for the East N-2 pit slope, Nacimiento Mine, Cuba, N. Mexico, *Consulting Eng. Rep.* (May).

Selli, R. and Trevisan, L., 1964. Caratteri e interpretazione della frana del Vaiont. *Ann. Mus. Geol. Bologna, Ser. 2*, 32: 1—68.

Sharp, J.C., 1970. *Fluid Flow through Fissured Media*. Ph.D. Thesis, Univ. of London, London.

Sharp, J.C., Maini, Y.N.T. and Harper, T.R., 1972. Influence of groundwater on the stability of rock masses, 1. Hydraulics within the rock mass. *Trans. Inst. Min. Metall.*, A13—A20.

Sharpe, C.F.S. and Dosch, E.F., 1942. Relation of soil-creep to earthflow in the Appalachian Plateaus. *J. Geomorphol.*, 5: 312—324.

Skempton, A.W., 1956. Alexandre Collin (1808—1890) and his pioneer work in soil mechanics. In: W.R. Schriever, *Translation of Landslides in Clay Strata by A. Collin (1846)*. Univ. of Toronto Press. Toronto, Ont., pp. xi—xxxiv.

Skempton, A.W., 1964. Long-term stability of clay slopes. *Fourth Rankine Lecture, Geotechnique*, 44: 77.

Skempton, A.W. and Petley, D.J., 1967. The strength along structural discontinuities in stiff clay. *Proc. Geotech. Conf., Oslo*, 2: 29—46.

Skempton, A.W., Schuster, R.L. and Petley, D.J., 1969. Joints and fissures in the London Clay at Wraysbury and Edgware. *Geotechnique*, 19: 205—217.

Smith, C.K. and Redlinger, J.F., 1953. Soil properties of Fort Union Clay Shale. *Proc., 3rd Int. Conf. Soil Mech. Found. Eng., Zürich*, 1: 62.

Snead, R.E., 1972. *Atlas of World Physical Features*. Wiley, New York, N.Y., 158 pp.

Snow, D.T., 1969. Discussion of reservoir leakage. *Bull. Assoc. Eng. Geol.*, 6: 83—88.

Tamburi, A.S., 1974. Creep of single rocks on bedrock. *Geol. Soc. Am. Bull.*, 85: 351—356.

Ter-Stepanian, G., 1967. The use of observations of slope deformation for analysis of mechanism of landslides. *Probl. Geomech., Acad. Sci. Armenian SSR*, 1: 32—51.

Ter-Stepanian, G.I. and Goldstein, M.N., 1969. Multi-storied landslides and strength of soft clays. *Proc., 7th Int. Conf. Soil Mech. Found. Eng., Mexico, 1969*, 2: 693—700.

Terzaghi, K., 1929. Effect of minor geologic details on the safety of dams. *Am. Inst. Min. Metall. Eng., Tech. Publ.*, 215: 31—44.

Terzaghi, K., 1936. Presidential Address. *Proc., 1st Int. Conf. Soil Mech. Found. Eng.*, 3: 22—23.

Terzaghi, K., 1962. Stability of steep slopes in hard unweathered rock. *Geotechnique*, 12: 251—270.

Tóth, J., 1963. A theoretical analysis of groundwater flow in small drainage basins. *J. Geophys. Res.* 68: 4795—4812.

Tóth, J. 1972. Properties and manifestations of regional groundwater movement. *Int. Geol. Congr. 24th Sess., Sect. II, Hydrogeology*, pp. 153—163.

Underwood, L.B., Thorfinnson, S.T. and Black, W.T., 1964. Rebound in redesign of Oahe Dam hydraulic structures. *Proc. Am. Soc. Civ. Eng., J. Soil Mech. Found. Div.,-* 90(SM2), Paper 3830: 65.

Van Everdingen, R.O., 1972. Observed changes in groundwater regime caused by the creation of Lake Diefenbaker, Saskatchewan. *Inland Waters Branch, Dep. of the Environment, Ottawa, Ont., Tech. Bull.,* 59, 65 p.

Voight, B., 1970. Idealization error, applied mechanics, and the art of engineering. *Proc., 1st Int. Congr. Int. Assoc. Eng. Geol.,* 2: 1352—1357.

Voight, B., 1973. Correlation between Atterberg plasticity limits and residual shear strength of natural soils. *Geotechnique,* 23: 265—267.

Williams, G.P. and Guy, H.P., 1971. Debris avalanches — a geomorphic hazard. In: D.R. Coates (Editor), *Environmental Geomorphology. Proc., 1st Annu. Geomorphol. Symp., Binghamton, N.Y.* State Univ. of New York, Binghamton, N.Y., pp. 25—46.

Wilson, S.D., 1970. Observational data on ground movements related to slope stability. *Proc. Am. Soc. Civ. Eng., J. Soil Mech. Found. Div.,* 96(SM5): 1521—1544.

Witherspoon, P.A. and Gale, J.E., 1977. Mechanical and hydraulic properties of rocks related to induced seismicity. *Eng. Geol.,* 11: 23—55.

Wittke, W., Rissler, P. and Semprich, S., 1972. Three-dimensional laminar and turbulent flow model through fissured rock according to discontinuous and continuous models. *Proc. Symp. on Percolation through Fissured Rock.* Deutsche Gesellschaft für Erd und Grundbau, Stuttgart, T1-H: 1—18.

Wood, A.M., 1971. Engineering aspects of coastal landslides. *Proc., Inst. Civ. Eng.,* 50: 157—276.

Woodruff, J.F., 1971. Debris avalanches as an erosional agent in the Appalachian Mountains. *J. Geogr.,* 70: 399—406.

Záruba, Q. and Mencl, V., 1976. *Engineering Geology.* Elsevier, Amsterdam, 504 pp.

Zischinsky, V., 1966. On the deformation of high slopes. *1st Congr. Int. Soc. Rock Mech., Lisbon,* 2: 179—185.

Zischinsky, U., 1969. Über Sackungen. *Rock Mech.,* 1: 30—52.

NOTES ADDED IN PROOF

[a] (p. 39) L. Broili argued that the sequence contained a high proportion of carbonate (marly limestones, marls) and thus could not be considered clay-like in regard to mechanical behavior. But recent work by F.D. Patton and A.G. Hendron suggests, to the contrary, that montmorillonite clay-rich layers were in fact involved along the Vaiont slip zone (F.D. Patton, oral communication, 1977). The hindcast friction angle cited above is therefore not necessarily anomalous. Indeed, laboratory measurements of slip zone material by Patton and Hendron, and Kenney, suggest residual friction angles appreciably *less* than the $20°$ figure cited above. The difference between laboratory and hindcast field values may be accounted for by such considerations as undulations (folds) and other forms of slip surface roughness, heterogeneity, local cementation, mixing of lithologies in brecciated zones, stiffness of the sliding mass, etc. The reader should note that although the problem of static friction has been emphasized in the above discussion, there is in addition a problem of dynamic friction. A drastic loss in dynamic frictional resistance is implied by the inferred emplacement velocity of the Vaiont slide mass. Vaporization of pore fluid due to frictional heating on the slip surface has been proposed as a possible mechanism for reduction of frictional resistance (see Chapter 20, Volume 1). More recent unpublished work along similar lines by C. Faust and B. Voight indicates that frictional heat-induced volume changes in pore water is sufficient as a friction-reducing mechanism; heating to the point of vaporization does not seem necessary for loss of strength.

[b] (p. 41) The recent field observations by F.D. Patton and A.G. Hendron support this opinion (F.D. Patton, oral communication, 1977), viz., that much larger displacements of the Vaiont slide mass had occurred prior to reservoir filling.

[c] (p. 41) The Downie slide is at present thought to be about 300 m deep, with a volume of perhaps 2×10^9 m³. Surface surveys from 1965 to 1976 imply a relatively slow, uniform rate of about 2 cm/yr. The average laboratory residual friction angle for failure zone gouge samples is 21°. Under present conditions, computed factors of safety for most profiles (high groundwater level assumed) vary between 0.9 and 1.1, confirming the marginal stability implied by the movements. Submergence of the slide toe by a reservoir would decrease the safety factor by perhaps 10%. Reduction of piezometric pressures along the failure surface by drainage, on the other hand, would increase the factor of safety by a significant amount (F.D. Patton, oral discussion, 1977, Geological Society of America—American Society of Civil Engineers Penrose Conference on Landslides). Dynamic analyses suggest the possibility of rapid slide movements only if rapid strength loss occurs (cf. Chapter 20, Volume 1).

Dedication

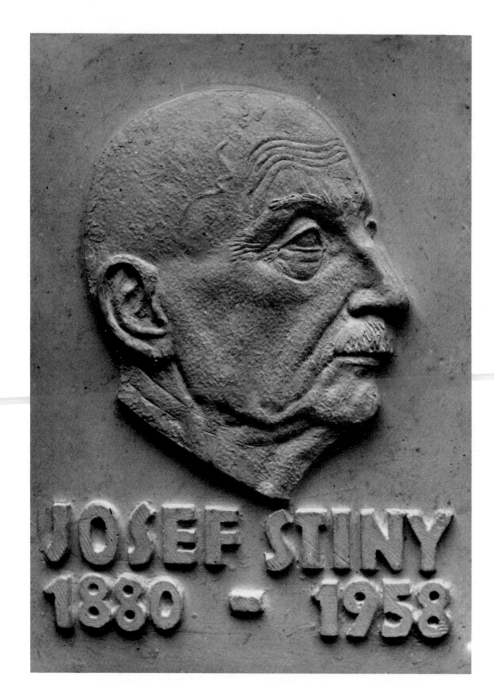

JOSEF STINY
1880 – 1958

Chapter 1

JOSEF STINI: CONTRIBUTIONS TO ENGINEERING GEOLOGY AND SLOPE MOVEMENT INVESTIGATIONS

LEOPOLD MÜLLER

INTRODUCTION

Men who are ahead of their time and thereby inspire future developments are often not appreciated during their lifetime, least of all in their home country, where (according to an old proverb) the prophet is not praised. Josef Stini was in some ways an exception; his influence on contemporary engineering science and practice was great, especially in Austria, his native country. Abroad, however, his name was not well known. Stini's ideas were, in spite of their relevance, not immediately spread over the world, in contrast to the work of his countryman Karl Terzaghi.

Gradually, however, an international collection of engineers and geologists have become aware of Stini's vital work; it now seems most appropriate to make the man and his life's work better known. As one of Stini's students, I appreciate the request made by the Editor of this volume, in conjunction with the activities of the Third International Congress on Rock Mechanics, for this presentation. It is an opportunity for those of us who have profited from Stini's powerful personality to express our gratitude to him.

TEACHING, ENGINEERING PRACTICE, RESEARCH

We owe it to the initiative of Professor Oerley, the designer of two railway lines in the Alps, that by the beginning of the 1920's lectures on technical geology had replaced those of general geology at Vienna's Technical University. This was a field almost entirely new to science at that time, and Josef Stini was called to teach and to direct research at the new institute. Already author of more than two dozen papers, in 1919 he had published a textbook on *Technische Gesteinskunde* (Engineering Petrology) which is valuable even today (cf., 1929g[1]). With his textbook on *Technische Geologie* (Technical

---

[1] Reference citations without author's names refer to publications by Stini as listed in the bibliography.

Geology), published in 1922, he took an advanced position compared with the contemporary literature of those days.

But even in his own opinion he had not yet executed the decisive philosophical step, from "geology-applied-to-engineering", to an independent "engineering geology", which Stini designated more correctly as "Baugeologie" (Construction Geology), and which, in 1929, was fundamentally presented by Redlich, Terzaghi and Kampe (Redlich et al., 1929). Stini's lectures at that time, nevertheless, encompassed all presently included as fundamental topics of this science. These lectures were aimed entirely at the engineering project and related activity, and all problems were presented from the engineering viewpoint instead of that of basic geology. Stini consciously avoided burdening engineering students with historical geology or paleontology. Instead, he made them acquainted especially with the facts of fundamental geology and rock properties. Endeavoring to supply the engineer not so much with historical facts but with more easily grasped numbers, he developed in those early times a technique of quantitative geological field analysis. He was anxious to train the eyes of the young engineer to see

Fig. 1. In the field with students.

connections between structure and strength of rock masses, between mor-
phology of a landscape and related mechanical parameters (Fig. 1).

It can be stated, now, that thorough information concerning the importance and applicability of engineering geology given to each civil engineering
student is *more important* than the education of excellent engineering
geologists. It is mainly due to Stini's recommendations that today, in Austria, every public office in civil engineering and reclamation as well as most
of the construction firms employ one or more "construction geologists".

Stini's 333 publications are an impressive endowment. Many of these were
published in *Geologie und Bauwesen*, a journal founded by Stini and characterized by Terzaghi as the world's first periodical on this subject (Fig. 2);
it was the predecessor of today's *Rock Mechanics* journal, the official
publication of the International Society for Rock Mechanics. His textbooks
on *Die Quellen* (Springs; 1933) and *Technische Gesteinskunde* (1919) are
useful even today, and *Tunnelbaugeologie* (Tunneling Geology; 1950) remains one of the classics of its field. It has often been regretted that his consultant reports have not been published, since they conveyed an incredible
wealth of knowledge and scientific findings, and were of high didactic quality. An extensive series of excerpts from Stini's works have recently been
translated into English (Austrian Society of Geomechanics, 1974).

One of Stini's greatest capabilities was his mediatorship between an enormous range of technical fields, such as geology, engineering, mining, soil
mechanics, geophysics, biology, hydrology, and mechanics. As one example
of his versatility, Stini derived much profit from his education as a technical
biologist; his lectures contained valuable data concerning the significant
indications plants can give with regard to the rock mass and to the associated
groundwater conditions. The basis for this versatility involved his multi-dis-
ciplinary education in both agricultural engineering and technical geology,
eleven years of concentrated engineering practice with the highly esteemed
Austrian Torrent Regulation Authority, and last but perhaps not least, his
activity as an engineering officer in World War I. During his 22 years as Professor for Technical Geology at Vienna he continued to work in connection
with engineering practice. As one of the best-known experts of his country,
his consultation was requested for almost all of the important engineering
projects of that period. His lectures were filled with references to practical
engineering problems, and he strongly believed that the *Baugeologie* lecture
series for civil engineers ought to be presented by an engineering geologist
rather than a general geologist. His deep understanding of engineering geology made it possible again and again for him to point out those subtle indications of nature which make it possible to foresee and explain engineering
problems. The necessary intuitive grasp was characteristic for Stini. This,
and the fact that the problems encountered were completely different from
those typical for ordinary geological studies, led to the demand for education of specialists in engineering geology.

Geologie und Bauwesen

Zeitschrift für die Pflege der Wechselbeziehungen
zwischen Geologie, Gesteinkunde, Bodenkunde usw.
und sämtlichen Zweigen des Bauwesens

Herausgegeben von J. Stiny

Jahrg. 14, Heft 3

(Abgeschlossen im Juli 1943)

Inhalt:

Springer - Verlag in Wien

Fig. 2. *Geologie und Bauwesen* in 1943.

THE CONCEPT OF "ROCK MASS" AND THE BIRTH OF "ROCK MECHANICS"

Josef Stini led the way towards quantitative instead of descriptive statements of geologic observations. At a time when joints were scarcely considered important, he was — with Hans Cloos and Salomon — among the first to introduce and systematically apply "statistical joint measurement". He originated the "Kluftkörper" (joint body), the conceptual model of a rock block bounded by joints and containing the important parameters "joint separation" and "degree of rock separation". This model deserves to be widely used because it well illustrates the properties of jointed rock masses and the characteristic difference between intact rock and the rock

Fig. 3. At the outcrop — the jointed rock mass.

mass, a concept apparently first recognized by Albert Heim in 1905.

He defined an imaginary rock body, limited by pairs of joints of mean orientation at a mean joint separation; the model provided information on the degree and kind of rock jointing and fragmentation, on the quality of a rock mass with regard to excavation and its capacity to bear heavy loads, difficulties to be expected during the engineering of underground excavations, and certainly not least, the anisotropy of water percolation or "Wasserwegigkeit" — a term also introduced by Josef Stini. From these considerations he was led to practical suggestions, e.g., that the forces in arch dams should be introduced into the abutment rock in a suitable direction in regard to the existing joint systems.

With emphasis and personal engagement he convinced his students and engineering colleagues about the fundamental importance of planar structural elements such as rock joints, bedding surfaces, and foliation (Fig. 3). In 1943, involved in the execution of large in-situ tests, Stini compared a regularly jointed rock mass with ancient dry-laid masonry, and thus gave a most illustrative explanation of the mechanical behavior of a rock mass. This simple illustration was a basic concept in the discovery of the essential axiom of rock mechanics.

From this fundamental idea, rock mechanics could be established as the mechanics of discontinua, and with subsequent work it became possible to derive statements about the anisotropy of rock masses, the favoured directions of greatest and least load-bearing capacity, and the important theorem that the mechanical properties of a rock mass are determined primarily by the nature of its discontinuities and only in a lesser way by the strength of the intact rock itself (1930a). Therefore Stini can be regarded as one of the important originators of modern Rock Mechanics. One of the founders of the International Circle for Geomechanics, he inspired this group with his ideas; subsequently this group expanded to form the International Society for Rock Mechanics.

SLOPE CREEP, ROCKSLIDES AND AVALANCHES

Stini's interest in gravitational movements of slopes began early in his career, when as a young engineer in his mid-twenties he had to deal extensively with problems of torrents and mud, water and gravel avalanches (Muren). His first scientific report in 1907 dealt briefly with Muren and related phenomena, and, following descriptions of local disasters (1908a-d, 1909a,b), he published an important monograph on Muren, particularly considering conditions in the Tirol; this book (1910) was regarded as a classic example of an engineering geological presentation which fitted the experience of engineering practice into the larger overview of geologic processes (Kieslinger, 1958).

I. unmittel= bar auf	Fels	festes Material stark über= wiegend	Fels= } sturz schlipf
	Schutt		Erd= } sturz schlipf Gekriech
II. mittelbar auf	Schutt mit Hilfe	Flüssiger Körper (Wasser)	1. Die Masse festen Ma= teriales über= wiegt noch	Eis= Aschen= Moor= Geröll= } Mure
			2. Die Masse des Wassers überwiegt	Niederwasser Mittelwasser Hochwasser } Flie= ßender Ge= wässer
		Fester Körper	Masse des während kur= zer Zeitab- schnitte beweg= ten Materiales meist gering	Schneebewegungen (Lawinen ꝛc. ꝛc.) Eisströme (Gletscher ꝛc.)

Fig. 4. Stini's classification of slope phenomena (from *Die Muren*, 1910).

In this early work, Stini presented a classification of slope phenomena (Fig. 4) which Sharpe (1938) subsequently called "one of the most inclusive and satisfactory" yet developed:

"Stini divided movements into those in which the force of gravity acts directly and those in which its action is indirect. Under direct action he placed falls and slips of rock, and falls, slips, and creeps of earth. Under indirect action he described movements aided by water, ice, or snow. These included ice, ash, bog, or rubble flows or avalanches (Muren); flows in which water is predominant; snow avalanches, and ice streams or glaciers. Few classifications have shown so well the relations of landslides and other mass-movement to stream and glacial action."

Stini's work in this general field has been reviewed by Kieslinger (1958) in his *Opus Impressum*:

"Technical measures against mudflows remained one of Stini's favorite themes. Beside a few later papers on sources of fluvial gravel deposits (Geschiebeherde) and their control (1921, 1930c) and on the local distribution of surfaces of initiation of slides (Rutschungsanbrüche), he summarized his experience in the book 'Geological Principles in Controlling the Source Area of Fluvial Gravels' (1931c). This book includes a fine classification on the systematics of such rupture initiations. All of these investigations pursue the aim of 'Near-natural torrent control' (1939c)...

The question of... the connections between forestry and geology (1932a, 1942a,c) regularly received his attention, along with reports of experience and recommendations on difficult turfing in the high altitude barrenland (1934c, 1935a). The characteristics of plants for stabilization of natural slopes in rock debris were also observed, from the viewpoint of mechanics, as in a paper on the tensile strength of plant roots (1947). The execution of technical measures for controlling avalanches, e.g., the construction of snow-fencing (1932d), also belongs to this complex problem; it is very extensive, because it not only concerns the forest engineer but also the civil engineer engaged in the construction of roads in high mountains and at hydropower construction sites in the Alps."

Gradually Stini's attention shifted from agricultural and forest soil science to technical soil mechanics.

"Indicative of Stini's method of work there appeared at first a systematic review and a clarification of nomenclature, an 'Attempted classification of soils in a technical sense' (1929b). Then followed developments of testing methods, partly new (1946) and partly improvements of existing methods. Beside publications on individual testing procedures (1918, 1927a, 1929a, 1930f, 1932b) a comprehensive review of the then-existing soil testing methods was presented in a paper on a special (anniversary) occasion (1929f).

Another group of papers deals with important question of imperviousness of a soil or ground and its improvement... In addition to permeability tests (1936a), investigations were made on volume of voids in different spherical packings in order to understand the behavior of soil fabric (Gerüstboden) better (1932c). Further, tests on the actual flow in soil capillaries were performed (1932f). All these tests were performed with an underlying desire to verify the different mathematical formulae through tests on real objects and to define the range of their applicability.

The common use of clayey impervious materials motivated the publication of a review of the level of knowledge on clay minerals and clayey rocks (1938a). In a critical review he put forward his views on the 'Fundamentals of Foundation Soils' (1930d). Another paper points to anisotropy in rocks and soils (1929c).

The significance of these works of Stini can only be appreciated when one keeps in mind the period of their publication, viz. the years in which the testing methods of soil mechanics were only in their initial development. Many of his basic ideas were later developed by others (e.g. A. Kiener in *Geologie und Bauwesen*, 3, Wien, 1931). Again and again the generalised results of research flow into practical engineering application, e.g. the findings of soil science research in road construction (1927b)."

Especially in high mountains, the engineer must assess the likely influences of natural mass movements on proposed construction in order to avoid initiating or facilitating such movements, often caused by a careless interference with the natural ground conditions. Instead, of course, a method of construction should be selected which causes the least possible disturbance to existing conditions. Stini repeatedly expressed himself on these problems

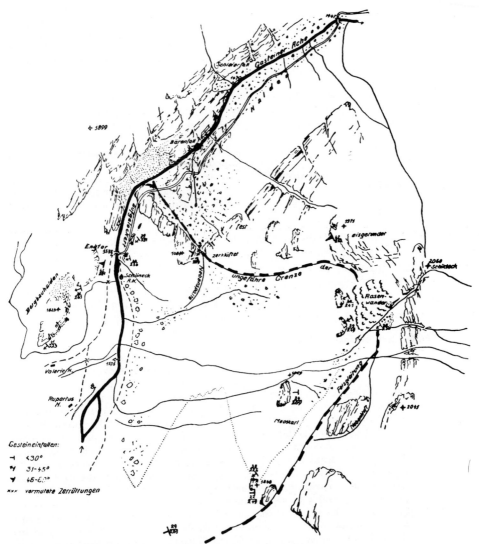

Fig. 5. "Die Felsgleitung am Nordostrande des Nassfeldes bei Böckstein" (from *Nochmals der Talzuschub*, 1942c).

(1939a,b). On each category of mass movement Stini has expressed himself, partly through accurate individual description and measurement (Fig. 5), and partly through summarizing the state of knowledge (1926f, 1930b, 1934b), thus presenting general concepts useful to the engineer for practical applications (1928, 1952a).

"Slow *block movement* was followed through repeated accurate measurements of blocks at altitudes higher than 2000 m (1935b, 1936b); movements of a few milli-

metres, rarely of a few centimetres/year, were observed. On steep slopes this movement could lead to rock bursts, which then formed heaps of detrital material
(Schutthalden) which endanger roads and other constructions (1934a, 1952d). A
cut in steep slopes in highly jointed rock, especially when the jointing is unfavour-
ably inclined, could lead to a gradual loosening of the rock mass and ultimately
cause a rockfall (1934a).

The very slow downward movement of steep slopes, frequently observed in alpine
valleys, is of special technical significance. These movements go very deep into the
rock (probably up to about 200 m) and are thus of a different order of magnitude
than near surface movements of detrital material. Under certain conditions this
movement could lead to heavy damage to pressure tunnels of a power house... in
this zone. Stini called this movement *slope creep* (Talzuschub). Stini's descriptions
(1941, 1942a,c, 1952b,c) stress the importance of this phenomenon for civil engineering constructions. Some cases of slope creep (leading finally to large rockfalls)
were thoroughly investigated (1926e, 1931a) and one of them was even simulated
in model tests."

Stini emphasized the temporal nature of rock slopes, realizing that continuous shape change was occurring due to ordinary geologic processes. These
changes sometimes occur quite imperceptibly, sometimes catastrophically;
but the change in shape is inexorable, one state being replaced by another,
according to the ancient law that "everything flows...". This view is well por-
trayed in Stini's 1941 paper on *Unsere Täler wachsen zu* (Valley Accretion):

"The modern engineer often carries out his constructions in the high mountain regions, shying neither snow nor ice. But the starting point of his activity and his
foothold remains the valley; starting from there, he extends his roads like grasping
arms up to the building sites high at the mountain slopes. The construction expert
working in such regions has to recognize the masses sliding down from the gaping
summits, closing the space of the valley where he is building. For him the disintegration of the mountains means the pushing together of the valley-slopes; the valleys, carved out by erosion, thus tend to close again. Some examples may demonstrate how important sometimes this process, which may occur slowly or in 'jerks',
can be for the professional activity of the engineer. This phenomenon is new to
most engineers, contrary to the well known, violent, sudden rock falls, or comparatively rapid, frequent rockslides. The natural phenomenon that I want to deal with
hides itself carefully and anxiously from the eyes of the multitude. Only he who is
lovingly engaged in the study of the small forms of the terrain and with their gene-
sis is able to recognize the phenomenon. His deep contact with nature and his
manifold experience render him sage. To the construction expert who pays attention only to the obviously noticeable slides and rock falls, this natural phenomenon
generally remains misleading and deceptive due to its extraordinarily well disguised
small-scale form. An example can take the place of many words.

In a time in which, continuing old traditions, tunnels are constructed again and
long galleries are driven, it is of course very important to pay proper attention to the
accretion of valleys. Where the scarcely-noticeable settlements accompanying the
phenomenon of the 'Talzuschub' have resulted in a disintegration of the strained
rock mass, it may be inevitable to drive the tunnel or gallery right through this
zone.

But it would already be of great value if the dangers were correctly and timely recognized. In this case the engineer can take precautions against the heavier rock
pressure to be expected within the zone of disintegration and against the expected

high water pressure. If the tunnel axis is nearly parallel to the zone of disintegration, one prefers when practicable to have the tunnel located deeper within the mountain where the rock mass is not yet disintegrated so much. If this is not done, one runs the risk of the tunnel being steadily menaced by the heavy pressure of the valley accretion and the dangerous displacements which accompany it. The person concerned with the construction of pressure galleries will in such cases tend to construct his tunnel deeper within the mountain mass, since this assures the tightness of the tube against water loss and brings further rewards through an increased stability of roof and sidewalls of the gallery."

These findings, published in 1941 and 1942, have recently gained attention when similar observations were gained in other countries (see, e.g., Terzaghi, 1962; Zischinsky, 1966; Chapter 17, Volume 1).

Other types of slope movements were also of concern to Stini; to return to Kieslinger's (1958) *Opus Impressum:*

"The presence of joint surfaces facilitates loosening and ultimate sliding of large rock masses (Bergschlipf, Felschlipf). Stini has described such *rockslides* (1932e, 1942b), thereby pointing out the extremely low rock mass strength caused by jointing and thus making a strong recommendation for using the least-disturbing kinds of construction methods in such rocks (1939a,b, 1940).

Loose rock masses, along with highly jointed and weathered rocks, are highly susceptible to mass movements. In high mountains this tendency is further accelerated by frost action. Stini had followed these movements over the years through observations and measurements. This *surface creep* (Rasenwandern), whose velocity was about 15 mm/year (1934a, 1935b, 1936b), is of significance especially for road construction because of the pressure exerted by it on retaining walls.

All rapid movements of loose masses could conveniently be classified as *landslides* (Rutschungen). Stini has repeatedly described individual noteworthy landslides (1908b,d, 1926c,d,g, 1930e, 1938b), and often discussed the nature of such slides and the possibilities of their control (1928, 1929d, 1934a, 1938b-d). Tests on the stability of sandy and clayey rocks were started at an early date (1918). The consideration of shear strength of such loose rocks led to a uniform evaluation of mass movements, earth pressure and mountain pressure (1928)... The dependence of permissible inclination of slope in such soils on height of the slope was pointed out (1929d), and formulae for determining earth pressure on retaining walls developed. The sliding surfaces (1935c) often start along water filled, mostly tectonic, joints. Stini's long dealing (30 years) with problems of such slides led him to the view that some large tectonic processes such as *gravity faulting* (Gleitfaltung) and overthrusting (Gleitüberschiebung) are in principle nothing but large-scale slides (1929e)."

This latter view again reflects the breadth of Stini's grasp — rockslides on all scales are of interest, from movement of individual fragments to displacements and distortion of an entire mountain range. The mechanical-physical deductions which he seeks to employ seem valid in nearly the same way for both rock masses on an ordinary scale and gigantic overthrust sheets (1929e):

"Folding and overthrusting by sliding are possible processes, withstanding mechanical-technological verification. To indicate this was the only aim of this paper. It is worthwhile to demonstrate this in a rough first approximation...

Results of field surveys unravelling the structure of mountain ranges will decide if and to what extent sliding processes have actually contributed to the process of mountain building. At present there is every reason to believe that their role was not insignificant. It remains to be seen whether the whole process of folding and overthrusting can be explained by sliding due to differential uplift alone. Future research, possibly carried out by future generations, will judge thereupon. Moreover, sliding can very well exist as an important secondary process apart from lateral compression and convection currents. Nature is acting with a plurality of means, many a cog-wheel hooks on to the other, many a lever acts together with the other. Why should mountain building be a simple process, for ever and solely governed by one driving mechanism?''

CONCLUSIONS

Stini's way of working was characteristic for his personality: while in the field he always proceeded from unprejudiced observation to interpretation, from the single, isolated case to the generalized statement. His research activity was invariably inspired by practical problems. He immediately and most accurately pursued every observation and experience, trying to understand it theoretically, and compared it self-critically with experiences of his own and others. Hypotheses were formulated only after profound comparison with other opinions, and in this his imposing knowledge of the technical literature was employed to great advantage.

That his life's work was an avocation seems obvious; indeed, as Alois Kieslinger has pointed out, even while working during vacations he contributed so much as a field geologist that this work alone can be favorably compared with that of many having field geology as their main profession.

All who had the opportunity to collaborate with Josef Stini were enriched by the experience. Absolute thoroughness, austere logic and integrity of reasoning, righteousness, and self-criticism were characteristics of his personality, combined with an unusual modesty, shyness, and sensitivity. He was able to combine his heavy pessimism with a benevolent and dry humor. This pessimism seems unfortunately confirmed by keen experiences of ingratitude at the end of his years of fruitful activity. On the other hand, Stini was capable of much gratitude toward all those who provided foundations for his efforts. In particular, his wife, Leopoldine, was actively engaged in nearly all of his works, theoretical as well as practical, and inspired him in no small degree by her collaboration.

Josef Stini remains a giant in his field; indeed, a paper such as this can at best give but an incomplete impression of the work of a lifetime. His influence remains alive in his work and in the work of his students, and his original papers can still be studied with considerable profit to the reader.

REFERENCES

Austrian Society for Geomechanics, 1974. *Josef Stini: Excerpts from Publications.* Salzburg (Franz-Josef-Strasse 3/111), Translation No. 18, 102 pp.

Kieslinger, A., 1958. Josef Stini: Opus Impressum. *Mitt. Geol. Ges. Wien,* 50. English version by B. Sharma in Translation No. 18, Austrian Society for Geomechanics, Salzburg.

Redlich, K.A., Terzaghi, K. and Kampe, R., 1929. *Ingenieurgeologie.* Springer, Vienna, 708 pp.

Sharpe, C.F.S., 1938. *Landslides and Related Phenomena.* Columbia Univ. Press, New York, N.Y., 136 pp.

Stini, J., 1907. Das Murenphänomen. *Mitt. Dtsch. Naturwiss. Ver. Hochschulen,* 1: 7—22, Graz.

Stini, J., 1908a. *Die Berasung und Bebuschung des Ödlandes im Gebirge als wichtige Ergänzung getroffener technischer Massnahmen und für sich betrachtet.* Selbstverlag, Graz, 155 pp.

Stini, J., 1908b. Der Erdschlipf im Schmaleckerwalde (Zillertal). *Mitt. Geol. Ges. Wien,* 1: 408—412.

Stini, J., 1908c. Über Bergstürze im Bereich des Kartenblattes Rovereto-Riva. *Verh. Geol. Reichsanst. Wien,* pp. 320—326.

Stini, J., 1908d. Die Erdschlipfe und Murbrüche bei Kammern. *Mitt. Naturwiss. Ver. Steiermark,* 45: 264—273.

Stini, J., 1909a. Die Ursachen der vorjährigen Vermurung im Zillertale. *Mitt. Geol. Ges. Wien,* 2: 213—226.

Stini, J., 1909b. Die jüngsten Hochwässer und Murbrüche im Zillertale. *Österr. Wochenschr. Öffentl. Baudienst (Wien),* 7, 4 pp.

Stini, J., 1910. *Die Muren. Versuch einer Monographie mit besonderer Berücksichtigung der Verhältnisse in den Tiroler Alpen.* Wagner, Innsbruck, 139 pp.

Stini, J., 1918. Einige Bezichungen zwischen Kolloidchemie, Geologie und Technik. *Jahrb. Geol. Bundesanst. Wien,* 68: 259—284.

Stini, J., 1919. *Technische Gesteinskunde.* Waldheim-Eberle, Vienna, 1st ed., 335 pp.

Stini, J., 1921. Die Geschieberde der Wildbäche. *Mitt. Geol. Ges. Wien,* 14: 275 (Vortragstitel).

Stini, J., 1922. *Technische Geologie.* F. Encke, Stuttgart, 789 pp.

Stini, J., 1925. Die Ausführung der Kluftmessung. *Geologie,* 38: 873—877.

Stini, J., 1926a. Kluftmessung und Quellenkunde. *Int. Z. Bohrtech., Erdölbau Geol.,* 34: 97—100.

Stini, J., 1926b. Kluftmessung und Erdölgeologie. *Int. Z. Bohrtech., Erdölbau Geol.,* 34: 137—138.

Stini, J., 1926c. Die Erdrutschungen des Jahres 1924 bei Monachil in Spanien. *Z. Geomorphol.,* 1: 54—58.

Stini, J., 1926d. Die Erdbewegungen bei Amalfi. *Z. Geomorphol.,* 1: 58.

Stini, J., 1926e. Bersturz bei Gnigl. *Z. Geomorphol.,* 1: 60.

Stini, J., 1926f. Massenbewegungen in den Alpen. *Z. Geomorphol.,* 1: 156.

Stini, J., 1926g. Unwetterwirkungen in Österreich während des Sommers 1925. *Z. Geomorphol.,* 1: 296—298.

Stini, J., 1927a. Eine Abänderung des Wiegnersch'n Schlämmverfahrens. *Fortschr. Landwirtsch.,* 2: 810—811.

Stini, J., 1927b. Strassenwesen und Baugrundgeologie. *Strassenbau,* 18: 354—356.

Stini, J., 1928. Rutschungen, Gebirgsdruck. Bergbauschäden und Baugrundbelastung. *Int. Z. Bohrtech., Erdölbau Geol.,* 36: 66—72.

Stini, J., 1929a. Zerrüttungsstreifen und Steinbruchbetrieb. *Geol. Bauwes.,* 1: 51—59.

Stini, J., 1929b. Versuch einer Einteilung der Böden im technischen Sinne. *Geol. Bauwes.*, 1: 67—69.

Stini, J., 1929c. Richtungsbedingtheit der Gesteinfestigkeit und der Bodeneigenschaften. *Geol. Bauwes.*, 1: 120—123.

Stini, J., 1929d. Zur Kenntnis und Abwehr der Rutschungen. *Geol. Bauwes.*, 1: 190—202.

Stini, J., 1929e. Faltungen und Überschiebungen durch Gleitung (Rutschungen grössten Massstabes). *Zentralbl. Mineral., Geol. Palaeontol, Teil B*, pp. 116—125.

Stini, J., 1929f. Neuzeitliche Untersuchung des Bodens für Gründungen. In: *Festschrift zum 50jährig. Bestand der Städt. Prüfanstalt für Baustoffe 1879—1929*, Vienna, pp. 112—117.

Stini, J., 1929g. *Technische Gesteinskunde*. Springer, Vienna, 2nd ed., 550 pp.

Stini, J., 1930a. Zum Begriff "Festigkeit" bei natürlichen Gesteinen. *Z. Prakt. Geol.*, 38: 59—60.

Stini, J., 1930b. Schäden durch Naturgewalten in Österreich im Jahre 1929. *Geol. Bauwes.*, 2: 134—136.

Stini, J., 1930c. Zur Verbauung der Feilenanbrüche in den Wildbacheinzugsgebieten. *Geol. Bauwes.*, 2: 208—217.

Stini, J., 1930d. Grundsätzliches über den Baugrund. *Geol. Bauwes.*, 2: 217—228.

Stini, J., 1930e. Der jüngste Erdrutsch in Lyon. *Geol. Bauwes.*, 2: 237—239.

Stini, J., 1930f. Ein für bodentechnische Zwecke geeignetes Schlämmverfahren. *Geol. Bauwes.*, 2: 233—236.

Stini, J., 1931a. Ein nicht gewöhnlicher Felssturz bei Langen am Arlberg. *Geol. Bauwes.*, 3: 148—150.

Stini, J., 1931b. Zur örtlichen Verteilung von Rutschungsanbrüchen auf Steilhängen. *Geol. Bauwes.*, 3: 143—148.

Stini, J., 1931c. *Die geologischen Grundlagen der Verbauung der Geschieberde in Gewässern*. Springer, Vienna, 121 pp.

Stini, J., 1932a. Forstwirtschaft und geologischer Aufbau von Niederösterreich. *Österr. Vierteljahrsschr. Forstwes.*, 3, 16 pp.

Stini, J., 1932b. Das Absaugeverfahren in der technischen Bodenkunde. *Geol. Bauwes.*, 4: 283—284.

Stini, J., 1932c. Der Hohlrauminhalt tatsächlicher Bodengerüste. *Geol. Bauwes.*, 4: 145—148.

Stini, J., 1932d. Die Bewährung von Schneezäunen bei der Lahnenverbauung. *Geol. Bauwes.*, 4: 105—106.

Stini, J., 1932e. Ein Felsschlipf an der Vintschgauer Bundesstrasse. *Geol. Bauwes.*, 4: 101—104.

Stini, J., 1932f. Zur Wasserbewegung in Haarröhrehen. *Geol. Bauwes.*, 4: 149—154.

Stini, J., 1933. *Die Quellen. Die geologischen Grundlagen der Quellenkunde für Ingenieure aller Fachrichtungen sowie für Studierende der Naturwissenschaften*. Springer, Vienna, 255 pp.

Stini, J., 1934a. Geologie und Bauen im Hochgebirge. *Geol. Bauwes.*, 6: 24—30; 33—65.

Stini, J., 1934b. Schäden durch Naturgewalten in Österreich in den Jahren 1932 und 1933. *Matér. Etude Calamités*, 33, 74 pp.

Stini, J., 1934c. Die Begrünung von Böschungen und anderen technischen Ödflächen im Hochgebirge. *Geol. Bauwes.*, 6: 134—140.

Stini, J., 1935a. Die Begrünung von sehr hoch gelegenen Aubrüchen in Wildbacheinzugsgebieten. *Wiener Allgem. Forst. Jagdz.*, 53(12/14).

Stini, J., 1935b. Zur Kenntnis der Geschwindigkeit langsamer Bodenbewegungen im Hochgebirge. *Geol. Bauwes.*, 7: 111.

Stini, J., 1935c. Zur Kenntnis der Rutschflächen. *Geol. Bauwes.*, 7: 120—121.

Stini, J., 1936a. Uber einen Durchlässigkeitsversuch. *Geol. Bauwes.*, 8: 63f.

Stini, J., 1936b. Die Geschwindigkeit des Rasenwanderns im Hochgebirge. *Geol. Bauwes.*, 8: 96.

Stini, J., 1938a. Sammelbericht: Neuzeitliche Erkenntnisse betraffend die Natur und die technischen Eigenschaften der Tonmineralien und Tongesteine. *Geol. Bauwes.*, 10: 59—62.

Stini, J., 1938b. Die heurigen Rutschungen im Grüntale (Gau Wien). *Geol. Bauwes.*, 10: 104—109.

Stini, J., 1938c. Die Rutschgefährlichkeit des Baugeländes und seine Untersuchung. *Geol. Bauwes.*, 10: 113—123.

Stini, J. 1938d. Die Beschürfung von Zerrüttungsstreifen beim Talsperrenbau. *Geol. Bauwes.*, 10: 140—145.

Stini, J., 1939a. Formschonendes Bauen. *Geol. Bauwes.*, 11: 94—95.

Stini, J., 1939b. Formschonendes Bauen. *Strasse*, 6: 333—335.

Stini, J., 1939c. Naturnahe Wildbachverbauung. *Geol. Bauwes.*, 11: 140—151.

Stini, J., 1940. Naturnahes Bauen im Fels. *Geol. Bauwes.*, 12: 15—29.

Stini, J., 1941. Unsere Täler wachsen zu. *Geol. Bauwes.*, 13: 71—79.

Stini, J., 1942a. Talzuschub und Bauwesen. *Bautechnik*, 1942.

Stini, J., 1942b. Abbrüche von Felskeilen. *Geol. Bauwes.*, 13: 107—110.

Stini, J., 1942c. Nochmals der Talzuschub. *Geol. Bauwes.*, 14: 10—14.

Stini, J., 1943. Die Geologie als eine der Grundlagen der Wasserwirtschaft, der Wildbachverbauung und des Wasserbaues überhaupt. *Geol. Bauwes.*, 14: 73—82.

Stini, J., 1946. Baugeologisches vom Loibl-Tunnel. *Österr. Bauzeitschr.*, 1: 7—10.

Stini, J., 1947. Die Zugfestigkeit von Pflanzenwurzeln. *Geol. Bauwes.*, 16: 70—75.

Stini, J., 1950. *Tunnelbaugeologie*. Springer, Vienna, 366 pp.

Stini, J., 1952a. Neuere Ansichten über "Bodenbewegungen" und über ihre Beherrschung durch den Ingenieur. *Geol. Bauwes.*, 19: 31—54.

Stini, J., 1952b. Ein "Talzuschub" im Burgenlande. *Geol. Bauwes.*, 19: 137.

Stini, J., 1952c. Talzuschub und Wildbachverbauung. *Geol. Bauwes.*, 19: 135—136.

Stini, J., 1952d. Massnahmen zur Sicherung von Bauarbeiten und Baustellen gegen Steinschlaggefahr. *Geol. Bauwes.*, 19: 231—232.

Terzaghi, K., 1962. Stability of steep slopes on hard unweathered rock. *Geotechnique*, 12: 251—270.

Zischinsky, U., 1966. On the deformation of high slopes. *Proc. 1st Int. Congr. on Rock Mechanics, Lisbon, 1966*, 2: 175—185.

KARL TERZAGHI (1883—1963)

Chapter 2

KARL TERZAGHI ON ROCKSLIDES: THE PERSPECTIVE OF A HALF-CENTURY

RUTH D. TERZAGHI and BARRY VOIGHT

> 1883—1963: 25 years of groping in the dark a. 55 years of strenuous efforts. One veil torn down, but many other ones, concealed behind it, encountered.
>
> K.T.: 1963, first entry in diary

INTRODUCTION

"An engineer can hardly look at personally conducted tours or at tourists, plain and simple, without wonder and pity because he is spoiled by his profession. He craves for insight... The ordinary travelers merely skim over the surfaces, and what they are shown is practically worn out by thousands of preceding glances..."

Thus did Karl Terzaghi introduce his lively 1928 travelogue on *Landslides in Central America*. Up to that time, Terzaghi's experience had been limited to earthwork operations in temperate climates. Therefore, in the spring of 1928 he asked the United Fruit Company to give him an opportunity to study problems involving geological materials in their Latin American holdings. In Costa Rica, for example, he wanted to get acquainted with the classical landslides along the company-owned railroad from Port Limon to San Jose; it was his intention to combine the study with a side trip with "mules and men" to one of the active volcanoes, for Terzaghi never really lost contact with his favorite science, geology, and his favorite sport of mountaineering.

Nonetheless this trip was, with tongue-in-cheek and without mention of the various tropical ailments acquired, considered a partial failure, "because even the steepest slopes refused to move while I was there..." Terzaghi's claim was that "the slopes owe their temporary stability to a strange act of Providence rather than the shearing strength of the material"; this conclusion was strongly supported by events of the next rainy season, when, according to Casagrande (1960), more than a mile of the line was wrecked by landslides, and communications between the coast and the capital (San Jose) had to be maintained by plane.

The previous stop of this trip was Panama, where Terzaghi prepared a memorandum for the U.S. Army Engineers on the hydrogeology of a reservoir site in karst of the Chagres River basin. Landslides were, however, not merely of incidental interest:

"I wished, first of all, to establish personal contact with the Panama Canal slides. Since I failed to see them in 1912 when they were alive, I wanted to pay tribute at least to the corpses..."

Panama was of special significance to Terzaghi because of the role it played concerning the origins of soil mechanics. As expressed in his Presidental Address at the First International Conference on Soil Mechanics and Foundation Engineering in 1936,

"Ten years ago the investigations which led to this conference still had the character of a professional adventure with rather uncertain prospects for success. This adventure began a short time before the war, simultaneously in the U.S.A., in Sweden, and in Germany. It was forced upon us by the rapid widening of the gap between the requirements of canal and foundation design and our inadequate mental grasp of the essentials involved.
    In the United States, the catastrophic descent of the slopes of the deepest cut on the Panama Canal issued a warning that we were overstepping the limits of our ability to predict the consequences of our actions..."

Arrangements in Panama had been made by a former student, Lt. Vandevoort, and within two days Terzaghi

...."knew already how beastly the tropical sun can burn when one climbs over a slide from the tongue towards the rear. The slides were rather disappointing except for three beautiful specimens (East-, West-Culebra and Cucaracha slides) located on both sides of rugged, rocky hills called Gold, Zion, and Contractors Hills. Here the movements spread over a broad belt, far beyond the neat lines of excavation. After a couple of days of strenuous physical exercise between huge chunks of broken earth I was compensated for my efforts by looking down on the scene of destruction from the cockpit of a gorgeous 900-horsepower bombing plane. The view was most impressive. The black basalt hills stand like huge ruins between trough-like, yawning gaps produced by the greenish shales of the soft Cucaracha formation flowing towards the canal over a width of several thousand feet. One cannot see these gaps without deeply regretting that no lesson was learned from the gigantic earth work experiment. Subduing the slides was a triumph of steam shovels, persistence and nerves, but no serious attempt was made to investigate the underlying physical causes for the benefit of future enterprises of a similar nature. Even the very modest research program proposed by a committee of the National Academy of Sciences in 1924 remained on paper."

This early impression of the brute-force approach (Fig. 1) to geological engineering problems, whereby "the spirit of analysis kept modestly in the background," was not in Terzaghi's style. No lessons had been learned. (See, however, Chapter 4, this volume.) As expressed in his impressive paper on *Mechanisms of Landslides*, published in a volume honoring the engineering geologist Charles Berkey (1950, p. 121),

Fig. 1. Cucaracha slide adjacent to Gold Hill, looking north from a point on the west bank of the canal south of Contractors Hill, October 16, 1913. This was one of the blasts by which it was hoped to facilitate the clearance of a channel past the toe of the slide. These efforts were unavailing because the plastic clay of the slide quickly closed the gaps made by the explosions (National Academy of Sciences, 1924).

> "Every landslide or slope failure is the large-scale experiment which enables competent investigators to draw reliable conclusions regarding the shearing resistance of the materials involved in the slide. Once a slide has occurred on a construction job, the data derived from the failure may permit reliable computation of the factor of safety of proposed slopes on the same job..."

WATER PRESSURE IN PORES AND FRACTURES

The development and application of the paradigm of effective stress was perhaps foremost among Terzaghi's major contributions to the new sciences dealing with the mechanics of geological materials (1923, 1925, 1936a, 1943b; Redlich et al., 1929). The early history of the evolution of this principle has been recorded with characteristic attention to detail by Skempton (1960) and need not be repeated here. Terzaghi's appreciation of the important effect of pore- and cleft-water pressure resulted in his rejection of the old tenaciously popular myth concerning the alleged "lubricating" effect of water and led to a quantitative explanation of the true effect of ground water on slope stability (1950, pp. 91—92):

> "If a slide takes place during a rainstorm at unaltered external stability conditions, most geologists and many engineers are inclined to ascribe it to a decrease of the shearing resistance of the ground due to the "lubricating action" of the water which seeped into the ground. This explanation is unacceptable for two reasons.
>
> First of all, water in contact with many common minerals, such as quartz, acts as an anti-lubricant and not as a lubricant. Thus, for instance, the coefficient of static friction between smooth, dry quartz surfaces is 0.17 to 0.20 against 0.36 to 0.41 for wet ones (Terzaghi, 1925, pp. 42—64).
>
> Second, only an extremely thin film of any lubricant is required to produce the full static lubricating effect characteristic of the lubricant. Any further amount of lubricant has no additional effect on the coefficient of static friction... since practically all the sediments located beneath slopes are permanently "lubricated" with water, a rainstorm cannot possibly start a slide by lubricating the soil or boundaries between soil strata...
>
> Last — but not least — water which enters the ground beneath a slope always causes a rise of the piezometric surface, which, in turn, involves an increase of the pore-water pressure and a decrease of the shearing resistance of the soil...
>
> Throughout a saturated mass of jointed rock, soil, or sediment, the water which occupies the voids is under pressure. Let
>
> p = pressure per unit of area at a given point P of a potential surface of sliding, due to the weight of the solids and the water located above the surface,
> h = the piezometric head at that point,
> w = the unit weight of the water, and
> ϕ = the angle of sliding friction for the surface of sliding.
>
> ...If the potential surface of sliding is located in a layer of sand or silt, the shearing resistance s per unit of area at the observation point is equal to
>
> $$s = (p - hw) \tan \phi \qquad\qquad [1]$$
>
> Hence, if the piezometric surface rises, h increases, and the shearing resistance s decreases. It can even become equal to zero. The action of the water pressure hw can be compared to that of a hydraulic jack. The greater hw, the greater is the part of the total weight of the overburden which is carried by the water, and as soon as hw becomes equal to p the overburden "floats". If a material has cohesion, c per unit of area, its shearing resistance is equal to the sum of s, equation [1], and the cohesion value c, whence
>
> $$s = c + (p - hw) \tan \phi ..."\qquad\qquad [2]$$

Terzaghi recognized quite early that the effective stress concept applied to crystalline rock quite as well as to clays and sands. This conclusion was based in part on his own experiments, for Terzaghi was the first to show that variation of pore pressure alone, in "unjacketed" triaxial experiments, had a negligible influence on the compressive strength of concrete (Terzaghi and Rendulic, 1934). Later, he interpreted the experiments by Griggs on Solnhofen limestone and marble in a similar fashion, in an important paper on *Stress Conditions for the Failure of Saturated Concrete and Rock* (Terzaghi, 1945).

Thus studies by Terzaghi were in advance of their time. In tectonic geol-

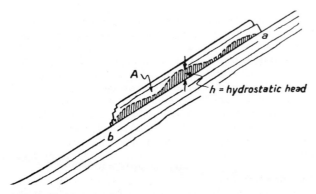

Fig. 2. Diagrammatic section through site of rock slide of Goldau (1806) prior to slide. Slab *A* was separated from its base by a thin layer of weathered rock. The dotted line represents the piezometric surface in this layer during a heavy rainstorm (Terzaghi, 1950).

ogy, they laid the foundation for the important treatises by M.K. Hubbert and W.W. Rubey in 1959 on the *Role of Fluid Pressure in Mechanics of Overthrust Faulting*, perhaps the most influential papers on tectonic geology thus far written in this century. The Hubbert and Rubey papers were concerned in large part with the question of gravity-driven overthrusts — rockslides on the enormous scale of geosynclines. Terzaghi's influence on this subject was considerable; a perspective on his contribution has been discussed in a volume on *Thrust Faults and Décollement* (Voight, 1976). Geophysicists too have become increasingly aware of the effective stress principle, and scientists are now examining its many implications on the important subjects of earthquake prediction and control.

However, the modified Coulomb relation expressed by equation [2] was already in use by Terzaghi for rock masses, over a quarter of a century ago (1950, pp. 93—94):

> "To illustrate its bearing on rockslides, the classical slide of Goldau in Switzerland will be discussed. This slide has always been ascribed to the "lubricating action" of the rain- and meltwater. Fig. 2[1] is a diagrammatic section through the slide area. It shows a slope oriented parallel to the bedding planes of a stratified mass of Tertiary Nagelflue (conglomerate with calcareous binder) which rises at an angle of 30° to the horizontal. On this slope rested a slab of Nagelflue 5000 feet long, 1000 feet wide, and about 100 feet thick. It was separated from its base by a porous layer of weathered rock.
>
> The fact that the slab had occupied its position since prehistoric times indicates that the shearing force, which tended to displace the slab, never exceeded the shearing strength, in spite of the effects of whatever hydrostatic pressures, hw, in equation [1], may have temporarily acted on the base of the slab in the course of its existence.
>
> On September 2, 1806, during heavy rainstorms, the slab moved down the slope, wiped out a village located in its path, and killed 457 people (Heim, 1882). This

---

[1] Figure and equation numbers have been changed to fit into this text.

catastrophe can be explained in at least three ways. One explanation is that the angle of inclination of the slope had gradually increased on account of tectonic movements, until the driving force which acted on the slab became equal to the resistance against sliding. A second explanation is based on the assumption that the resistance of the slab against sliding was due not only to friction, but also to a cohesive bond between the mineral constituents of the contact layer. The total shearing resistance due to the bond was gradually reduced by progressive weathering, or by the gradual removal of cementing material, either in solution or by the erosive action of water veins. The third explanation is that h in equation [1] or [2] assumed an unprecedented value during the rainstorm, whereas the cohesion c, in equation [2], remained unchanged, provided cohesion existed. In Fig. 2 the value h is equal to the vertical distance between the potential surface of sliding, ab, and the dash line interconnecting ab which represents the piezometric line. During dry spells h is equal to zero. In other words, the piezometric surface is located at the slope. During rainstorms the rain water enters the porous layer located between slab and slope at a and leaves it at b. Since the permeability of this layer is variable, the piezometric line descends from a to b in steps, and the value h in equations [1] and [2] is equal to the average vertical distance between the piezometric line and the slope.

The maximum value of h changes from year to year, and if the exits of the water veins at b are temporarily closed by ice formation while rain- or meltwater enters at a, h assumes exceptionally high values. However, the seasonal variations of h, the corresponding variations of s, equations [1] and [2], and the occasional obstruction of the exits at b have occurred in rhythmic sequence for thousands of years, without catastrophic effects. It is very unlikely that h assumed a record value in 1806, in spite of unaltered external conditions. Therefore, it is more plausible to assume that the slide was caused by a process which worked only in one direction, such as a gradual increase of the slope angle or the gradual decrease of the strength of the bond between slab and base. In no event can the slide be explained by the "lubricating effect" of the rain water".

Awareness of the effective stress principle and knowledge of the effect of water pressure in open joints contributed to Terzaghi's recognition that landslides in some areas seem to exhibit a *periodicity*, apparently related to an (approximately) twenty-year cycle of rainfall maxima. The existence of some law governing the periodicity of slide events had seemed intuitively evident to other workers, such as the Austrian Josef Stini (Kieslinger, 1958), but no specific hypothesis had been presented. Terzaghi's attention was drawn to the subject for the first time in connection with one of his professional assignments in England, an investigation of the causes of the intermittent landslides in the Folkestone Warren on the north channel coast (1950, pp. 102—104; cf. Hutchinson, 1969):

"The Warren can be described as a giant niche, about 10,000 feet long and 1000 feet wide, located between the channel coast and a steep cliff with a height of about 400 feet. The upper, vertical part of the cliff consists of chalk. The bottom of the niche consists of Gault clay buried beneath a chaotic accumulation of large fragments of chalk (Fig. 3a). The Southern Railway enters the niche from the west through the Martello Tunnel and leaves it at the east end through a tunnel leading toward Dover.

The railway was constructed in the middle of the nineteenth century. Prior to the

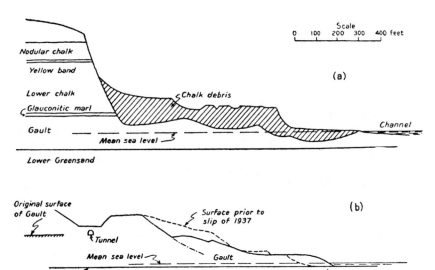

Fig. 3. Folkestone Warren, Channel Coast, England. (a) Diagrammatic cross section. (b) Displacement which occurred during the slide of 1937. The outer part of the slice of Gault clay involved in the slide advanced over a distance of about 70 feet toward the Channel by sliding along the boundary between Gault clay and Lower Greensand. Yet the surface topography of this part of the slice remained almost unchanged (Terzaghi, 1950).

construction of the railway, huge slides occurred at different points of the Warren in 1765, 1800, and 1839. After the railroad was built the periodic recurrence of slides continued. The first recorded slide affecting the railroad took place in 1877. In 1896 large movements occurred in the western part of the Warren. In December 1915 a slip affected nearly the whole of the Warren and effectively blocked the railway until the end of the first World War. It involved the movement of several millions of cubic yards. In the spring of 1937, slides took place over an area of 35 acres in the western part of the Warren and caused the formation of a crack across the lining of the Martello Tunnel.

Fig. 3b is a diagrammatic section through the area affected by the slide of 1937. The major part of the sliding mass advanced along the almost horizontal boundary between the Gault, which is a very stiff clay, and the Lower Greensand, which is a soft sandstone. A detailed account of the physical characteristics of the materials involved in the slide was published by Toms (1946). In Fig. 3b the original position of the ground surface is indicated by a dash line, and the final one by a plain line. It should be noted that the front part of the sliding mass moved bodily, without undergoing more than a slight deformation, over a distance of about 70 feet toward the channel. Such a movement would not be conceivable unless the resistance against sliding along the base of the moving section of ground was very low. Furthermore, the slide was not preceded by a change of the external conditions for the equilibrium of the slope. Hence we are compelled to assume that the resistance against sliding has decreased, which can be accounted for only by an increase of the hydrostatic pressure (hw in equation [1]) on the base of the sliding body.

The borings, made in 1939, showed that the piezometric surface for the Lower

Greensand was located at elevations up to 27 feet above sea level. These elevations are subject to seasonal variations and to variations within a longer period, caused by variations in the average rainfall. However, in the Warren, the rainwater has no opportunity to get into the Lower Greensand from above, because the Greensand is covered with a blanket of Gault clay having a very low permeability. The variations of the piezometric heads can be caused only by similar variations in the elevations of the water table in that region where the Greensand emerges at the surface. This region is located many miles north of the Warren.

The fact that the major slides in the Warren occurred once every 19 or 20 years suggests that the movements were due to corresponding maxima in the amount of rainfall in those regions where the aquifer located beneath the slide area reaches the surface. In agreement with this assumption the Warren slide of 1937 was preceded by abnormally heavy rainfalls, 'between 15 and 16 inches of rain falling during the first three months' (Seaton, 1938, discussion by Ellson on p. 438).

Considering the mechanics of the Warren slide one might expect a similar periodicity in connection with landslides on all those slopes whose factor of safety, with respect to sliding, varies with the elevation of the water table in distant aquifers".

In the succeeding decade, Terzaghi became gradually more immersed in questions concerning the mechanical behavior of rock masses (Fig. 4). In his 1961 diary (summary of the year, following the entry of December 31), he emphasized that

"My attention is now entirely concentrated on rock mechanics, where my experience in applied soil mechanics can render useful services..."

Fig. 4. Terzaghi in the field (Norway, 1957).

In the next year his influential paper on *Stability of Steep Slopes on Hard Unweathered Rock* was published; in it Terzaghi returned to the question of fluid pressure effects and to the question of slide periodicity, influenced by the recent work of Bjerrum and Jørstad in Norway (1962c, pp. 261–263; cf. Chapter 3, this volume):

"A portion of the precipitation on the higher parts of a topographically dissected area enters the joints in the underlying rock and emerges in the proximity of the foot of the slopes in the form of springs. If the secondary permeability of the rock were uniform the water-table would assume the shape indicated in Fig. 5a and vary between a lowest position (dash line) at the end of the dry season and a highest position (unbroken line) at the end of the wet or melting season. The volume of the continuous open joints in which the water travels through the rock is very small compared to the volume of the rock located between these joints.

Therefore, the vertical distance between the lowest and the highest position of the water-table is measured in tens of feet and not in feet, as in unconsolidated pervious sediments.

In reality the water-table is not well defined as shown in Fig. 5a because the secondary permeability of jointed rock commonly varies more or less erratically from

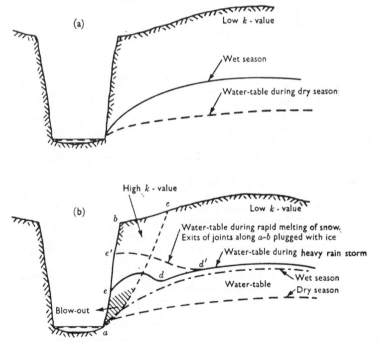

Fig. 5. (a) Location of lowest and highest water-table in jointed rock with uniform but low coefficient of permeability. (b) Water-table in the same rock during heavy rainstorms or snow melt if permeability in space *abc* is much higher than elsewhere. Line *ed* represents the water-table at times when the joints drain freely, and *e'd'*, when they are plugged with ice (Terzaghi, 1962c).

place to place and the water may rise in two adjacent observation wells to very different levels. Yet if the water-table has at least the general characteristics of that shown in Fig. 5a the seepage exerted by the flowing ground-water could not possibly be important enough to have a significant influence on the stability of steep slopes.

On account of the low secondary permeability of the jointed rock only a small portion of the rainfall descends through the joints. Part of it is temporarily retained in the voids of the weathered top layer and the remainder flows as surface run-off toward the edge of the slope. However, before it reaches the edge it crosses the upper surface bc of the wedge-shaped body of rock abc in Fig. 5b. On account of the shearing stresses prevailing in this body, the joints may be much wider and more numerous than those in the rock located farther from the slope. Because of this circumstance, the quantity of water which can enter the joint system in this body, per unit of area of its top surface bc, is much greater than the corresponding quantity which enters the rock elsewhere. Consequently the water-table in that body may rise temporarily to a position such as that indicated by the unbroken line de in Fig. 5b.

The water which occupies the joints exerts onto the walls of the joints a pressure u equal to the unit weight of the water times the height to which it would rise in an observation well terminating in the joint. This pressure corresponds to the pore-water pressure in soil mechanics and will be called the *cleft-water pressure*. Like the pore-water pressure, it reduces the frictional resistance along the walls of the joints (see equation [2]) and if a joint is very steep it tends to displace the rock between joint and slope towards the slope.

The cleft-water pressure is zero at the water-table and it increases in a downward direction. Hence if a slope fails on account of cleft-water pressures, the failure will start at the foot of the slope within the shaded portion of the area abc in Fig. 5, whereby the rock will be displaced by the water pressure in a horizontal direction. As a result of the initial failure, the rock located above the seat of the failure is deprived of support and it will descend owing to its own weight.

The influence of the cleft-water pressures on the stability of steep rock slopes is well illustrated by the rock slide statistics prepared by the Norwegian Geotechnical Institute (Bjerrum and Jørstad, 1957). Within the area covered by the observations, the winters are severe, the snowfall abundant, and the heaviest rainfalls occur during the autumn months.

...Slide frequency was greatest in April, during the time of the snow melt, and in October within the period of greatest rainfall. However, most of the major slides have taken place in April, because at that time of the year the exits of the joints are still plugged with ice while the snowmelt is feeding large quantities of water into the joints of the rock within the wedge-shaped body of rock abc in Fig. 5b. Owing to this condition, the water-table adjacent to the slope is raised from position ed into position e'd'...

If, during a series of exceptionally wet years, the least stable slopes are "cleaned off" by rock falls and slides, many decades pass before the deterioration of the remaining slopes has advanced far enough to cause a slope failure".

Terzaghi was, of course, well aware that natural precipitation was not the only cause of an increase of pore- or cleft-water pressure (1950, pp. 100—102; *Engineering News-Record*, 1953). A section of special interest in connection with hydroelectric power development was included in his 1962 paper; it dealt with the potentially disastrous effects of high cleft-water pressure associated with leakage from reservoirs and pressure tunnels, particularly in the proximity of portals. For the latter, Terzaghi cited an example from a paper by Stini (1956) on the blasting effects of water. With regard to reservoirs, the Malpasset disaster is mentioned (1962c, pp. 263—264):

"If water leaks out of a reservoir formed by a concrete dam, the greatest cleft-water pressure develops in the joints of the rock at the foot of the slope downstream from the toe of the dam. If a slope failure should occur as a result of cleft-water pressures it would start at the foot of the slope and proceed as indicated in Fig. 5b. The effect of such a slide on the stability of the dam depends on the type of dam...

If the dam is a thin arch dam, the cleft-water pressures are also greatest in the proximity of the toe of the dam at the foot of the slope. However, they are very much greater than the corresponding pressures near the toe of a concrete gravity dam of equal height, because the base of an arch dam is much narrower. Furthermore, the downstream toe of an arch dam rises along the slopes of the valley in a downstream direction and enters the area occupied by a potential slide scar. Finally, also in the proximity of the foot of the slope, the effects of the cleft-water pressures combine with those produced by the thrust of the arch which has a component in the downstream direction and thus tends to push the rock out of the slope.

The possible consequences of these conditions are illustrated by Fig. 6a, showing the plan of a fictitious arch dam. The cross-section of the dam is shown in Fig. 6b. The area in which a slope failure would start is indicated by shading. The slide scar would be located uphill from this area. It can be seen that a slide initiated by a blow-out at the foot of the slope would deprive the upper portion of the base of the dam of its support. Hence if a blow-out occurs in the rock supporting an arch dam the consequences are likely to be catastrophic. The failure of Malpasset Dam was probably started by such a blow-out. Fortunately the development of cleft-water pressures within the rock downstream from arch dams can be avoided by adequate drainage (Terzaghi, 1962a)".

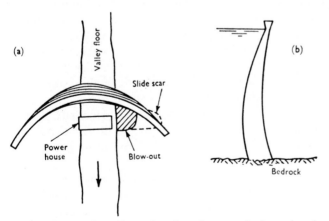

Fig. 6. Plan and section of an imaginary arch dam, showing slide scar produced by blow-out in area of maximum cleft-water pressures (Terzaghi, 1962c).

RELATION BETWEEN SLIDE PROCESS AND REMEDIAL TREATMENT

As an engineer, Terzaghi was frequently called upon to cope with slope movements which were already underway. His method was to adapt the means for stopping the movement to the specific processes which started the slide and kept it in motion (1950, table 1, column C). Because of this fact, Terzaghi was less interested in classifying kinds of landslides than in discriminating between the processes which conceivably could lead to slides, and to analyze each one of these processes. His subdivision of causes into two fundamental groups proved to be a convenient aid to analysis, and is commonly employed in present day practice (1950, p. 88):

> "The causes of landslides can be divided into external and internal ones. External causes are those which produce an increase of the shearing stresses at unaltered shearing resistance of the material adjoining the slope. They include a steepening or heightening of the slope by river erosion or man-made excavation. They also include the deposition of material along the upper edge of slopes and earthquake shocks. If an external cause leads to a landslide, we can conclude that it increased the shearing stresses along the potential surface of sliding to the point of failure.
>
> Internal causes are those which lead to a slide without change in surface conditions and without the assistance of an earthquake shock. Unaltered surface conditions involve unaltered shearing stresses in the slope material. If a slope fails in spite of the absence of an external cause, we must assume that the shearing resistance of the material has decreased. The most common causes of such a decrease are an increase of the pore-water pressure, and progressive decrease of the cohesion of the material adjoining the slope. Intermediate between the landslides due to external and internal causes are those due to rapid drawdown, to subsurface erosion, and to spontaneous liquefaction".

Despite the variety of causes thus far identified, it is hardly an exaggeration to say that most slides are due to an abnormal increase of pore water pressure in the slope-forming material or in a part of its base (Terzaghi, 1943a; 1950; 1953; 1962c). In such cases treatment invariably included the installation of a system of drainage designed to decrease the pore- or cleft-water pressure at the surface of sliding. Radical drainage was also the recommended precaution whenever it could be anticipated that leakage from a reservoir might result in sliding at the toe of a dam.

His method of dealing with landslide problems evolved, in the course of his practice, into three successive steps. The first was to determine the mechanical causes. The second was to install observational devices sufficient to verify or modify his concepts. The final step, in light of knowledge concerning the causative mechanisms, involved working out the most economical method for stopping the movement.

Terzaghi's approach is well illustrated by the manner in which he dealt with a landslide on a slope adjacent to penstocks of the Serra hydroelectric power station on the east coast of South America. Some observations made during installation of the drainage system were described as follows (1950, p. 120):

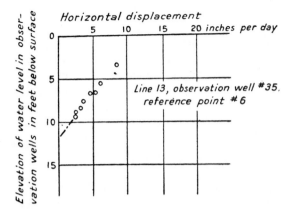

Fig. 7. Diagram showing the relation between the position of the water table with reference to a slope after failure (ordinates) and the horizontal component of the corresponding downhill movement of the surface of the slope (Terzaghi, 1950).

"The extraordinary efficacy of drainage has recently been demonstrated by the following observation. During a tropical cloudburst, involving a precipitation of 9 inches in 24 hours, a slide occurred on a slope rising at an average angle of 30°. The slope is located on deeply weathered metamorphic rocks, and the deepest part of the surface of sliding was about 130 feet below the surface. The slide area was about 500 feet wide, 1000 feet long, and the quantity of material involved in the slide exceeded half a million cubic yards.

Since the slide occurred in the close proximity of a hydro-electric power station, immediate action was indicated. In order to get quantitative information concerning the ground movements and the factors which determine the rate of movement, reference points were established on several horizontal lines across the slide area, and observation wells were drilled in the proximity of the reference points. By plotting the vertical distance between slope and the water level in the wells as ordinates, and the corresponding rate of movement of the adjoining reference points as abscissas, diagrams like Fig. 7 were obtained. Although the moving mass had a depth up to 130 feet, the diagrams showed that the lowering of the water table by not more than about 15 feet would suffice to stop the movement.

Drainage was accomplished by means of toe trenches, drainage galleries, and horizontal drill holes extending from the headings into water-bearing zones of the jointed rock. The movements ceased while the drainage was still in an initial state. The following rainy season brought record rainfalls; yet the ground movements in the slide area remained imperceptible".

Memoranda by Terzaghi concerning this project, written in 1947 and 1948, are of exceptional interest; excerpts have been published in the 1960 Festschrift volume, *From Theory to Practice in Soil Mechanics*. A case history report on the slide was presented by Fox (1957).

INFLUENCE OF GEOLOGICAL DETAILS

In both of his major articles dealing with landslides (1950, 1962c), Terzaghi emphasized the influence of more or less readily ascertainable geological details on slope stability. These included the presence of relatively weak layers subject to slow deformation under stresses below the *fundamental strength* (the creep threshold stress of Griggs, 1936, p. 564) of the material, the nature and orientation of joints, and the presence of extensive planes of weakness such as faults and deep-seated surfaces of prehistoric sliding.

Terzaghi recognized that

> "A civil engineer, trained in soil mechanics, may have a better grasp of the physical processes leading to slides. However, he may have a very inadequate conception of the geologic structure of the ground beneath the slopes, and he may not even suspect that the stability of the slope may depend on the hydrologic conditions in a region at a distance of more than a mile from the slope."

Because of this painstaking attention to significant details, many examples involving fundamentally different geologic conditions were cited in Terzaghi's papers; of particular interest is the account of the 1903 Frank slide (1950, pp. 95—96):

> "In hard, jointed rocks, resting on softer rocks, a decrease of the cohesion of the rock adjoining a slab may occur on account of creep of the softer rocks forming their base. The great Turtle Mountain slide of 1903 near Frank, Alberta (Fig. 8) seems to belong to this category. Percolating waters and frost action have contributed to the breakdown (Sharpe, 1938, p. 79). They always do, but they have done it for many thousands of years. Percolating waters cannot move blocks located between joints at great depth, and the frost action is only skin deep. Hence neither water nor frost could have altered the stability conditions in the rock adjoining the slope beyond a distance of a few feet from the slope. However, the limestones, forming the bulk of the peak, rested on weaker strata which certainly 'crept' under the influence of the unbalanced pressure produced by the weight of the limestone, and the rate of creep was accelerated by coal-mining operations in the weaker strata. The total cohesion along the potential surface of sliding in a jointed rock is equal to the combined shearing strength of all those blocks of rock which interfere, like dowels, with the sliding movement. The yield of the base of the limestone caused an increase of the shearing stresses; the increases of the stresses caused one dowel after another to "snap" and the slope failed when it was ripe for failure, at a time when the factor of safety assumed one of its periodic minimum values. Fig. 8b is a graph illustrating the process which led to the slide."

Terzaghi's logic in this classic case still seems unassailable; but the geologic cross-section, after McConnell and Brock (1904), has now been updated by recent studies, and the results are of significance in relation to inferred slide mechanisms. The reader will thus find the paper by Cruden and Krahn in the first volume of particular interest.

Terzaghi was also long familiar with, and in fact introduced, the concept of "minor geologic details", i.e., features that can be predicted with assurance neither from the results of careful investigations nor by means of a rea-

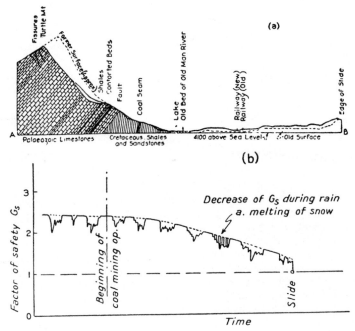

Fig. 8. Turtle Mountain after the great Frank, Alberta, slide in 1903 (after McConnell and Brock, 1904). (a) Cross-section. (b) Diagram illustrating concept of the changes of the safety factor of the slope prior to the slide (Terzaghi, 1950; cf. Chapter 2, Volume 1).

sonable amount of test borings (Terzaghi, 1929). He was to emphasize this concept and variations on the theme throughout his career, as he recognized that the degree of reliability assigned to rational design methods was in most cases exaggerated, and not at all commensurate with the amount of hard geological knowledge commonly on hand.

Recognizing that joints are among the minor geological features that commonly exert an important influence on the stability of slopes on hard rock, Terzaghi (1962c) analyzed in detail the relation between joint orientation and slope stability. An appreciation of the theoretical aspects of this relationship is, however, useless unless valid data concerning joint orientation are available. As Ruth Terzaghi (1965) pointed out in an article dealing with principles developed in the course of collaboration with Karl Terzaghi on a consulting assignment [2], the results of a joint survey may be intolerably misleading unless observations are made on exposed surfaces and/or boreholes and tunnels with an adequate variety of orientations, and the reported frequency of joints of any given set is adjusted for the angle between that set and the drill holes or exposed rock faces.

---

[2] See editorial inclusion on next page.

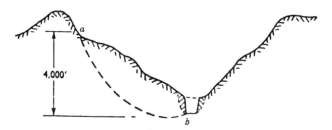

Fig. 9. Deep-seated rock slide beneath slope of deep valley between high mountains (Terzaghi, 1962c).

Regarding the question of deep-seated rock slides, Terzaghi was strongly influenced by the Alpine discoveries of the Austrian geologists Stini (1941, 1942a,b) and Ampferer (1939, 1940):

> "Practically nothing is known concerning the mechanics of these deep-seated, large-scale rock slides. It is not known whether the slides took place rapidly or slowly, and it is doubtful whether they are preceded by important creep deformation of the rocks located within the shear zone. However, it is known that the rock located above the surface of sliding has been damaged at least to a moderate extent. Existing joints have opened and new ones have been formed. Hence the compressibility and secondary permeability of the rocks has increased. Furthermore, in the immediate proximity of the surface of sliding, the rock is completely broken or crushed. Hence a site for a high concrete dam should not be considered suitable unless there is positive evidence that the underlying rock has never been subject to displacement by a deep-seated rock slide."

In connection with these features, Terzaghi published a sketch (Fig. 9) designed to illustrate the danger inherent in the presence of a zone of movement, associated with a prehistoric slide, beneath the slopes of a reservoir in high mountains. Although the article was published almost a year before the Vaiont disaster, the sketch could have served to illustrate the conditions existing on the left bank of the reservoir before the great slide took place.

---

[2] Editoral inclusion:

The story of the origin of this paper is of much interest, inasmuch as it illustrates the effective working relationship of Karl and Ruth Terzaghi (letter R.T. to B.V., 1975):

A geological report submitted to him in connection with a consulting assignment landed promptly on my desk for study and comment. It contained the results of a joint survey presented in a series of polar diagrams which indicated that the orientation of joints at the surface was strikingly different from that of joints in the rock at depth, observed in drill-holes. As I started to puzzle over this curious circumstance and to try to find an explanation for it, I suddenly realized that the difference in recorded orientation did not correspond to reality but was a result of what I later learned (from Bruno Sander) to call "Schnitteffect". When I informed Karl of my conclusions, he was at first incredulous ... a geologist, he thought, could not be so stupid. I finally convinced him, and the article resulted.

EARLY WARNING OF AN IMPENDING SLIDE

Of major practical importance was Terzaghi's recognition that landslides, with the exception of those triggered by earthquakes or due to spontaneous liquefaction, are preceded by slight movements associated with a gradual decrease in the ratio between the average shearing resistance of the ground and the average shearing stress on the potential surface of sliding (1950, p. 110):

"It has often been stated that certain slides occurred without warning. Yet no slide can take place unless the ratio between the average shearing resistance of the ground and the average shearing stresses on the potential surface of sliding has previously decreased from an initial value greater than one to unity at the instant of the slide. The only landslides which are preceded by an almost instantaneous decrease of this ratio are those due to earthquakes (table 1, column D, action 3) and to spontaneous liquefaction (action 13). All the others are preceded by a gradual decrease of the ratio which, in turn, involves a progressive deformation of the slice of material located above the potential surface of sliding and a downward movement of all points located on the surface of the slide. Hence if a landslide comes as a surprise to the eyewitnesses, it would be more accurate to say that the observers failed to detect the phenomena which preceded the slide. The slide of Goldau (Fig. 2) took the villagers by surprise, but the horses and cattle became restless several hours before the slide, and the bees deserted their hives (Heim, 1882)".

In his own engineering practice, wherever movement of rock or earth could have unacceptable consequences, he insisted on the installation of means for observation which were sufficiently extensive and sensitive to detect the slight displacements which commonly precede major movement. On several occasions, he noted with sorrow that even a well-designed monitoring system could not prevent disaster if those in charge failed to pay adequate attention to the observational data.

A PERSPECTIVE ON ROCK MECHANICS IN RELATION TO SLOPE DESIGN

It was not by oversight that the two principal articles (1950, 1962c) dealing with slope stability did not include a discussion of possible applications of the techniques of rock mechanics for determining in-situ properties of rock masses. Instead, Terzaghi recognized that (1962c, pp. 252, 255—256):

"Natural conditions may preclude the possibility of securing all the data required for predicting the performance of the real foundation material by analytical or any other methods. If a stability computation is required under these conditions, it is necessarily based on assumptions which have little in common with reality. Such computations do more harm than good because they divert the designer's attention from the inevitable but important gaps in his knowledge of the factors which determine the stability of slopes on hard, unweathered rock..."

Elaborating on this theme (1962c, pp. 255—256),

"Let s be the shearing resistance at a given point P of a potential surface of sliding in a porous and saturated material,

c_i = its cohesion,
ϕ = its angle of shearing resistance,
ϕ_f = angle of friction along the walls of a joint, [3]
ϕ_c' = critical slope angle of jointed rock with effective cohesion, equal to the slope angle of a plane through the foot and the upper edge of the steepest stable slope which can be produced by excavation, [3]
ϕ_c = critical slope angle of jointed rock without cohesion, equal to the slope angle of the steepest stable slope on such rock, [3]
p = the unit pressure at point P,
u = the hydrostatic pressure in the water located next to point P.

According to a well-established empirical law:

$$s = c_i + (p - u) \tan \phi \qquad [3]$$

All intact as well as jointed rocks with effective cohesion have the mechanical properties of brittle materials. Failure of slopes on brittle materials starts at the point where the shearing stress becomes equal to s (equation [3]). As soon as failure occurs at that point, the cohesion of the rock at that point becomes equal to zero whereupon the stresses in the surrounding rock increase and the rock fails. Thus the failure spreads by chain action and the process is known as *progressive failure*. In order to apply equation [3] to problems of rock mechanics, the influence of the joint pattern on the shearing resistance of jointed rock must be considered. If the rock has a random pattern of jointing, equation [3] is valid for any section through the rock. Therefore the rock performs like a stiff clay without joints, or an impure sand with considerable cohesion. If a slope on such material is undercut, the slope fails progressively by shear along a roughly concave surface of failure through the foot of the slope...

The only essential difference between a slope on brittle cohesive soil and a rock weakened by joints resides in the means at our disposal for determining the values c_i and ϕ in equation [3]. For brittle soils both values can be obtained by means of simple laboratory tests. On the other hand, the c_i-value of jointed rock cannot be determined by any of the presently available methods for rock investigations. The value of ϕ for rocks with random pattern of jointing can only be estimated on the basis of what we know about the ϕ-value of cohesionless aggregates in general and what we learned from case records as shown under the following heading. The value of ϕ of rocks with a well-defined joint pattern is a function of the orientation of the potential surface of sliding with reference to the joint system. Hence for any one type of rock it can have very different values."

Although keenly interested in the development of measurement techniques in rock mechanics (e.g., 1962b) and in the exploration of the geological significance of their results, Terzaghi became convinced that the variability of rock masses, combined with the great expense of available testing procedures, precluded the possibility of obtaining sufficient information for a reliable computation of slope stability. His views on the subject were expressed in one of his last publications (1963; cf. 1962d; Bjerrum and Jørstad,

[3] For a specific discussion of the role of ϕ_f, ϕ_c', ϕ_c, refer to Terzaghi (1962c, pp. 256–260).

1963), a discussion of an article by Klaus John (1962) in the Proceedings of the American Society of Civil Engineers:

"..analysis is based on the resistance quotient, R_q. In the paper, this quotient is defined 'as the ratio between the resistance of R of a jointed rock mass along any section against tensile, shear, or frictional failure, and the stressing S along the same section by tensile and shear stresses'. At a given tensile strength of the intact rock, the tensile strength of the jointed rock mass is determined by the average extent of the joints; however, the data required for computing the extent cannot be secured by any practicable means. The failure of jointed rock commonly occurs along uneven surfaces, formed by the walls of joints commonly belonging to several different sets. The shearing resistance along such surfaces depends primarily on (1) the extent of the joints, (2) the hydrostatic pressure exerted by the water occupying the joints, and (3) the degree of interlock between the rock surfaces on either side of the surface of sliding. An attempt to evaluate the resistance against failure by shear led the writer to the following conclusions. The resistance component (1) and the resistance-reducing hydrostatic pressure (2) cannot be determined by any practicable means, and the influence of interlock on the shearing resistance, item (3), can be defined by an angle of shearing resistance, which corresponds to the true angle of the shearing resistance in soil mechanics. This value may range for potential surfaces of sliding behind slopes on jointed rock between 30° and 70°, depending on the joint pattern and the orientation of the potential surface of sliding with reference to this pattern. It also depends, to a lesser extent, on the value of the coefficient of friction along the walls of the joints. However, an accurate evaluation of the angle of shearing resistance is commonly impracticable...

...Considering the uncertainties involved in estimating the value of R_q even under simple conditions similar to those prevailing along a potential surface of sliding in the rock behind a slope located above the water table, it is difficult to ascertain how a theoretical procedure... could furnish trustworthy results.

...Theoretical stability analysis requires, first of all, reliable information concerning the numerical value of the resistance quotient, R_q. For the time being (in 1962), it is still impossible to secure this information and it is still an open question whether or not adequate procedures for determining the value of R_q can be developed at all. Photoelastic methods and model studies are merely expedients to be used as substitutes for cumbersome computations. The results furnished by these investigations cannot be more reliable than those of the equivalent computations.

If the jointed rock underlying a slope is acted on by hydrostatic pressures exerted by water percolating through the joints toward the slope, the stability of the slope depends, to a large extent, on the pattern of seepage. This condition prevails on the slopes in many deep open-pit mines and on the slopes downstream from the abutments of arch dams (Terzaghi, 1962c). The permeability of most jointed rock varies greatly over short distances in a random fashion. This can be seen in every wet rock tunnel. Therefore, a reliable evaluation of the forces exerted by percolating water onto the rock is impossible."

Terzaghi was also skeptical regarding the usefulness of statistical methods in rock mechanics. The following paragraphs are quoted from a letter dated August 23, 1963:

"The general principles of the statistical method are known to every civil engineer. Yet the legitimate application of this method to problems of rock mechanics are extremely limited because the methods require (a) that the cost of performing the required tests should not be prohibitive, and (b) that the body subject to statistical investigation should be statistically homogeneous. These two conditions are rarely satisfied...

Failure in rock commonly occurs along a surface or in a narrow zone of exceptional weakness. Therefore the application of the statistical method requires first of all the discovery and thorough investigation of all localized defects... The risk of a failure of a structure supported by the rock depends primarily on the results of this investigation, yet reliable means for discovering all the significant localized defects of a large body of rock are not yet available. The risk of failure due to scattered defects such as joints is commonly compensated by a generous factor of safety. Yet the determination of the frequency of these defects requires a vast amount of time and labor. This statement applies particularly to the methods of *in situ* rock testing."

CONCLUSIONS

In 1963, Terzaghi summarized his views of the mutually supporting roles of geological observations, physical testing, and theory in the solution of problems of both soil and rock mechanics:

"Within the last few decades, in both soil and rock mechanics, a great number of new techniques for testing and field observations have been developed. As a consequence, knowledge of the engineering properties of the materials encountered in subsurface engineering has rapidly increased. In both domains, the successful application of theoretical procedures to the solution of engineering problems is limited to those instances in which the following two conditions are satisfied:
1. Within the range of influence of the proposed engineering operations, the boundaries between materials with significantly different engineering properties should be reasonably well defined. The boundaries divide the subsoil or bedrock into zones. Within each zone, the soil or rock should be sufficiently homogeneous to permit its replacement without serious error by an ideal, perfectly homogeneous substitute, whose engineering properties can be defined by a few numerical values.
2. The number of tests performed should be great enough to permit reliable evaluation of the statistical average of the significant engineering properties of the real material located within each zone.

If these fundamental requirements are not satisfied, the results of performance forecasts can be misleading in every field of subsurface engineering. In the field of soil mechanics, a moderate amount of subsoil exploration is commonly sufficient to determine whether or not the first condition is satisfied. If the investigation shows that it is satisfied, considerable sampling and testing is required to determine the statistical average of the engineering properties of the soils located within the seats of potential trouble. However, the individual tests are relatively inexpensive and, as a consequence, the second requirement can commonly be satisfied at a reasonable expense. The validity of the theoretical concepts on which the performance forecasts are based has already been demonstrated by numerous, well documented case records. These records have also disclosed the type and magnitude of the errors resulting from the differences between real soils and their ideal substitutes...

At an early stage of the development of soil mechanics, it was also discovered that the detrimental effect of seepage on the stability of slopes is caused by the seepage pressures exerted by the percolating water and not by the mere presence of the water. It was mentioned previously herein that the forecast of the intensity and distribution of the seepage pressure exerted by water percolating through natural soil deposits is commonly impracticable. Nevertheless, the newly acquired insight into the mechanics of interaction between soil and water had important and far reaching practical consequences. It led to the development of new means for observation, such as pore-pressure gages and multiple observation wells. On many projects, such means for observation permit determination of the intensity of the forces exerted by the percolating water onto the permeable medium before they reach a critical value. They also make it possible to prevent a failure by adequate provisions. These may consist in reducing the rate at which impervious dam construction materials such as clay are placed, in a modification of the cross section of the dam, or in the installation of supplementary drains.

In a general way, experience has shown that the principal benefits derived from soil mechanics reside in the newly acquired capacity to recognize and locate potential sources of trouble well in advance of construction and to eliminate the risk of an anticipated damage by adequate design. The possibility of an accurate prediction of the performance of soil-supported structures on a theoretical basis at sites satisfying the first requirement is merely a welcome by-product. At present, if a foundation or an earth dam fails unexpectedly despite the designer being reasonably familiar with the principles of soil mechanics, the failure is commonly due to reliance on theoretical procedures without conclusive evidence that the first requirement is satisfied. Unfortunately, such instances are by no means rare, but they are not inevitable.

If, in the realm of soil mechanics, the first requirement is satisfied, there is no serious obstacle against compliance with the second requirement, because sampling and soil testing are relatively inexpensive operations. On the other hand, in the field of rock mechanics, the cost of compliance with the second requirement may be, and often is, prohibitive, provided the requirement can be satisfied at all. For instance, the cost of securing reliable average values of the deformation characteristics of jointed rock masses is commonly prohibitive and the determination of the resistance of such masses against failure by shear is still impossible.

Because of the high cost of in-place rock testing, the temptation is great to underestimate the number of tests that are required to secure reliable numerical values for the constants that appear in the basic equations of theoretical rock mechanics. Hence, the influence of the future developments in rock mechanics on rock engineering will depend to a large extent on the degree to which the practitioners of rock mechanics are able to resist this temptation..."

A decade of work since the above was written has not suggested a need to modify these conclusions. We recall Terzaghi's notes written long ago on the retirement of von Mises, a former colleague in a research and design unit of the Austrian Air Force:

"In his field, theory reigns supreme. In my field it can be disastrous, unless it is kept on a shelf in a bottle with the label: *add not more than five drops to each gallon of experience.*"

REFERENCES

Ampferer, O., 1939. Über einige Formen der Bergzerreissung (Some types of mountain splitting). *Sitzungsber. Akad. Wiss. Wien, Math.-Naturwiss. Kl., Abt. 1*, 148.

Ampferer, O., 1940. Zum weiteren Ausbau der Lehre von den Bergzerreissungen (Further contributions to our knowledge of mountain splitting). *Sitzungsber. Akad. Wiss. Wien, Math.-Naturwiss., Abt. 1*, 149: 51—70.

Bjerrum, L. and Jørstad, F., 1957. Rockfalls in Norway. *Norw. Geotech. Inst. Internal Rep.*, F-230.

Bjerrum, L. and Jørstad, F., 1963. Discussion of "An Approach to Rock Mechanics" by K. John, 1962. *Proc. Am. Soc. Civ. Eng., J. Soil Mech. Found. Div.*, 89 (SM 1): 300—302.

Casagrande, A., 1960. Karl Terzaghi — his life and achievements. In: L. Bjerrum et al. (Editors), *From Theory to Practice in Soil Mechanics.* Wiley, New York, N.Y., pp. 3—21.

Engineering News-Record, 1953. Landslide destruction of power plant blamed on leaks in penstock tunnel. November 12.

Fox, P.P., 1957. Geology exploration and drainage of the Serra slide, Santos, Brazil. In: P.D. Trask (Editor), *Engineering Geology Case Histories, 1.* Geological Society of America, Washington, D.C., 7 pp.

Griggs, D.T., 1936. Deformation of rocks under confining pressures. *J. Geol.*, 44: 541—577.

Heim, A., 1882. Über Bergsturze. *Naturforsch. Ges. Zürich, Neujahrsbl.*, 84.

Hubbert, M.K. and Rubey, W.W., 1959. Role of fluid pressure in mechanics of overthrust faulting, 1. Mechanics of fluid-filled porous solids and its application to overthrust faulting. *Geol. Soc. Am. Bull.*, 70: 115—166.

Hutchinson, J.N., 1969. A reconsideration of the coastal landslides at Folkestone Warren, Kent. *Geotechnique*, 19: 6—38.

John, K.W., 1962. An approach to rock mechanics. *Proc. Am. Soc. Civ. Eng., J. Soil Mech. Found. Div.*, 88 (SM 4), Paper 3223.

Kieslinger, A., 1958. Josef Stini: Opus Impressum. *Mitt. Geol. Ges. Wien*, 50.

MacDonald, D.F., 1913. Some engineering problems of the Panama Canal in their relation to geology and topography. *U.S. Dep. Inter., Bur. Mines, Bull.*, 86, 88 pp.

McConnell, R.G. and Brock, R.W., 1904. Report on the great landslide at Frank, Alberta. *Can. Dep. Inter., Annu. Rep. 1902—1903*, Part 8, 17 pp.

National Academy of Sciences, 1924. *Report of the Committee of the National Academy of Sciences on Panama Canal Slides. U.S. Natl. Acad. Sci., Mem.*, 18, 84 pp.

Redlich, K.A., Terzaghi, K. and Kampe, R., 1929. *Ingenieurgeologie.* Springer, Vienna, 708 pp.

Rubey, W.W. and Hubbert, M.K., 1959. Role of fluid pressure in mechanics of overthrust faulting, 2. Overthrust belt in geosynclinal area of western Wyoming in light of fluid pressure hypothesis. *Geol. Soc. Am. Bull.*, 70: 167—206.

Seaton, T.H., 1938. Engineering problems associated with clay, with special reference to clay slips. *Inst. Civ. Eng. J. (London)*, Paper 5170, pp. 457—498.

Sharpe, C.F.S., 1938. *Landslides and Related Phenomena.* Columbia Univ. Press, New York, N.Y., 136 pp.

Skempton, A.W., 1960. Significance of Terzaghi's concept of effective stress. In: L. Bjerrum et al. (Editors), *From Theory to Practice in Soil Mechanics.* Wiley, New York, N.Y., pp. 42—53. Reprinted in: B. Voight (Editor), 1976. *Mechanics of Thrust Faults and Décollement.* Benchmark Press, Stroudsburg, Pa., pp. 122—132.

Stini, J., 1941. Unsere Talen wachsen zu (Our valleys close up). *Geol. Bauwes.*, 13: 71—79.

Stini, J., 1942a. Nochmal der Talzuschub (Some more about the closing up of our valleys). *Geol. Bauwes.*, 14: 10—14.

Stini, J., 1942b. Talzuschub und Bauwesen (Engineering consequences of the closing-up of our valleys). *Bautechnik*, 20: 80.

Stini, J., 1952. Neuere Ansichten über Bodenbewegungen und ihre Beherrschung durch den Ingenieur (New conceptions concerning ground movement and its control by the engineer). Geol. Bauwes., 19: 31—54.

Stini, J., 1956. Wassersprengung und Sprengwasser (Blasting effects of water). Geol. Bauwes., 22: 141—169.

Terzaghi, K., 1923. Die Berechnung der Durchlässigkeitsziffer des Tones aus dem Verlauf der hydrodynamischen Spannungserscheinung. Sitzungsber. Akad. Wiss. Wien, Math.-Naturwiss. Kl., Abt. 2A, 132: 125—138.

Terzaghi, K., 1925. Erdbaumechanik. Franz Deuticke, Wien, 399 pp.

Terzaghi, K., 1928. Landslides in Central America. Technol. Rev., 31: 12—16.

Terzaghi, K., 1929. Effect of minor geologic details on the safety of dams. Am. Inst. Min. Metall., Tech. Publ., 215: 31—44.

Terzaghi, K., 1936a. Stability of slopes on natural clay. Proc. 1st Int. Conf. Soil Mech. Found. Eng., Cambridge, Mass., 1: 161—165.

Terzaghi, K., 1936b. Presidential address. Proc. 1st Int. Conf. Soil Mech. Found. Eng., Cambridge, Mass., 1: 22—23.

Terzaghi, K., 1943a. Measurements of pore-water pressure in silt and clay. Civ. Eng., 13: 33—36.

Terzaghi, K., 1943b. Theoretical Soil Mechanics. Wiley, New York, N.Y., 510 pp.

Terzaghi, K., 1945. Stress conditions for the failure of saturated concrete and rock. Am. Soc. Test. Mater., Proc., 45: 777—801.

Terzaghi, K., 1950. Mechanism of landslides. In: S. Paige (Editor), Application of Geology to Engineering Practice (Berkey Volume). Geological Society of America, Washington, D.C., pp. 83—123.

Terzaghi, K., 1953. Discussion on stability and deformations of slopes and earth dams, research on pore-pressure measurements, groundwater problems. Proc. 3rd Int. Conf. Soil Mech. Found. Eng., 3: 217—218.

Terzaghi, K., 1960. Memoranda concerning landslides on slope adjacent to power plant, South America. In: L. Bjerrum et al. (Editors), From Theory to Practice in Soil Mechanics. Wiley, New York, N.Y., pp. 409—415.

Terzaghi, K., 1961. Past and future of applied soil mechanics. J. Boston Soc. Civ. Eng., 48: 110—139; Discussions and closure, 49: 96—110 (1962).

Terzaghi, K., 1962a. Discussion of "Control of seepage through foundations and abutments of dams" by A. Casagrande. Geotechnique, 12: 67—71.

Terzaghi, K., 1962b. Measurement of stresses in rock. Geotechnique, 12: 105—124.

Terzaghi, K., 1962c. Stability of steep slopes on hard unweathered rock. Geotechnique, 12: 251—270; also Norw. Geotech. Inst. Publ., 50.

Terzaghi, K., 1962d. Does foundation technology really lag? Eng. News-Rec., February 15, pp. 58—59.

Terzaghi, K., 1963. Discussion of "An approach to rock mechanics" by K. John, 1962. Proc. Am. Soc. Civ. Eng., J. Soil Mech. Found. Div., 89 (SM 1): 295—300.

Terzaghi, K. and Peck, R.B., 1948. Soil Mechanics in Engineering Practice. Wiley, New York, N.Y., 566 pp.; 2nd ed., 1967.

Terzaghi, K. and Rendulic, L., 1934. Die Wirksame Flächenporosität des Betons. Österr Ing.-Archit.-Verein, Z., 86: 1—9, 30—32, 44—45.

Terzaghi, R.D., 1965. Sources of error in joint surveys. Geotechnique, 15: 287—304.

Toms, A.H., 1946. Folkestone Warren landslips. Research carried out in 1939 by the Southern Railway Company. Inst. Civ. Eng., Rail Road Eng. Div., Railway Paper, 19.

Voight, B. (Editor), 1976. Mechanics of Thrust Faults and Décollement. Benchmark Press, Stroudsburg, Pa., 471 pp.

LAURITS BJERRUM (1918—1973)

Chapter 3

LAURITS BJERRUM: CONTRIBUTIONS TO MECHANICS OF ROCKSLIDES

RALPH B. PECK

INTRODUCTION

The intimate relation between Laurits Bjerrum and the Norwegian Geo-technical Institute makes impossible the clear separation of the contributions made by Bjerrum as an individual and by his colleagues. Indeed, the marked success of the Institute in dealing with problems of geotechnics has been due

Fig. 1. In Bjerrum's office at the Norwegian Geotechnical Institute, 1966; from left to right, O. Eide, B. Kjaernsli, Bjerrum.

in no small measure to the close rapport among its members. The reader should appreciate that Bjerrum was not only an outstanding scientist and engineer in his own right, but the active head of a most productive team. Hence, it is inevitable that some of the original ideas of other members of the Institute should, in this paper, be attributed to Bjerrum. I do not believe his colleagues will be offended.

Laurits Bjerrum became head of the newly formed Norwegian Geotechnical Institute in 1951. In accordance with the purposes of the Institute, he set about to investigate those geotechnical problems of peculiar significance to Norway (Fig. 1). He often commented that the main problems were associated with the presence of extremely soft soil or extremely hard rock. The economic importance of problems associated with extremely soft soil justified concentration of the Institute's work on the prediction and control of settlements of structures on such soils and on the characteristic quick-clay landslides to which they are often susceptible. There is little question that the contribution of Laurits Bjerrum to our understanding of the origin and physical properties of quick clays, to the mechanism of quick-clay slides, and to methods for preventing such slides is his outstanding achievement with respect to the stability of slopes. The story of this investigation, of how the fundamental phenomena were identified and evaluated, and of how,

Fig. 2. Tafjord. Slide of 1934 on right (photo Widerøe).

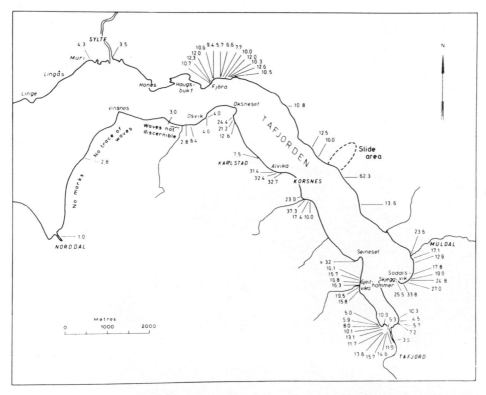

Fig. 3. Tafjord. Wave height in metres after rockslide of April 7, 1934 (after Jørstad, 1968).

finally, simple practical procedures were developed for preventing most such slides, deserves to be included among Norse saga. It is not, however, part of the story of rock mechanics.

To a smaller degree, Bjerrum and the Institute also made contributions to our knowledge of submarine slides. This aspect of the stability of slopes, of obvious interest to Norway, also belongs in the realm of soil mechanics rather than rock mechanics.

To a third aspect of stability of slopes, however, Bjerrum and the Institute gave continued attention: that of rockfalls and rockslides. Although the economic consequences of these occurrences are comparatively minor in Norway, their catastrophic nature, the occasional high loss of life, and the ever present possibility of danger carried social implications demanding that the Institute turn its attention to them. The deep fjords with their bordering steep cliffs, with all their beauty, nevertheless engendered a sense of foreboding that a fall might occur and that it might set up a destructive wave capable

of overwhelming villages, boats and any persons along the shoreline (Figs. 2—
5). Undoubtedly, the Norwegian government and the Norwegian people
expected their new Institute to make efforts to reduce these dangers; they
were not disappointed. By 1952 the Institute was acting as expert consultant
on rockfalls and rockslides to the Natural Catastrophe Fund of Norway.

Fig. 4. Ravnefjell and Lake Loen, showing scar of 1936 slide.

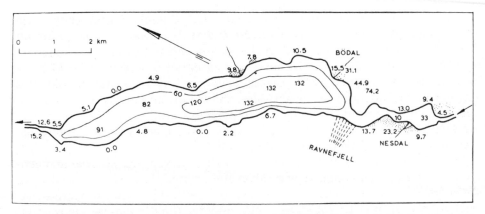

Fig. 5. Lake Loen. Wave height in metres due to 1936 slide at Ravnefjell (after Th.J. Selmer and Gustav Saetre).

STABILITY OF HARD ROCK SLOPES

The first step in Bjerrum's approach to the problems of rock slope stability was to become intimately acquainted with those slides and falls that had
already occurred, in order to define the processes leading to their occurrence
and the circumstances under which instability developed. The study included
a search of archives for historical accounts of earlier catastrophes (Fig. 6).
An example of these findings was published by Finn Jørstad in *Naturen*

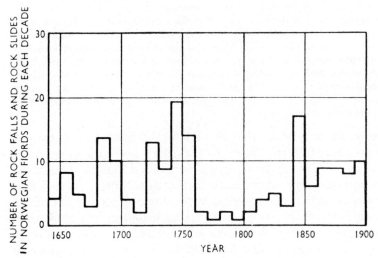

Fig. 6. Distribution of rockfalls and rockslides in Norwegian fjords for periods 1640—
1900.

magazine 1956 and later reprinted as an N.G.I. Publication under the title "Rock Slide at Tjelle" (*Fjellskredet ved Tjelle*). It contains a description by Gerhard Schøning, who traveled in the area between 1773 and 1775, some twenty years after the slide. By investigating detailed accounts of old rockfalls and rockslides such as the one at Tjelle, and by revisiting the old sites and examining many more recent ones, the N.G.I. investigators soon developed a comprehension of the factors of greatest significance. Moreover, they came to the realization that no such slide takes place without observable warnings over an appreciable period of time.

By 1954 the general pattern of the means for preventing fatalities and economic losses had emerged. It was taken for granted that many rockfalls and slides could not be prevented. By cooperation with the authorities, areas particularly susceptible to rockfalls could be defined and land-use maps could be prepared to permit location of residences and industries in comparatively safe places. Important installations such as industrial plants and power stations in localities not previously occupied could be established in the least dangerous places. Where important and costly installations might have to be located in potentially dangerous areas, attempts might be justifiable to improve the stability of the slopes by anchorages or by removal of materials likely to fall and cause damage.

Over the next few years additional investigations and experience led to greater understanding and a greater refinement of the approach. The results of the studies may be crystallized into the following quotation (Bjerrum and Jørstad, 1968):

> "Until the time when a mountain slope reaches its final state of equilibrium a slow, cumulative process of rock destruction proceeds in the slope, and such phenomena as progressive failure, the occurrence of stress concentrations, and successive denudation of the surface play a decisive role. This time dependent behaviour cannot at the present time be included in theoretical calculations of the type which come under the heading of 'rock mechanics'. Since the theoretical approach would only give an approximate picture of the actual conditions, the authors have been compelled to abandon the hope of finding a solution based on theoretical calculations. We have decided instead to study the phenomena in the field and to try to understand the processes in action in a mountain slope, and then to use this knowledge to make an appraisal of individual cases."

It became clear at an early stage that two distinctly different occurrences had to be recognized; rockfalls (by far in the majority), and rockslides. After an examination of some 300 locations, it was concluded that most rockfalls were the result of climate-induced processes acting at and near the rock surface (Fig. 7). Frost-shattering was observed to be the most significant of the causes. Statistics gathered from many locations indicated clearly that rockfalls occurred most frequently in the spring and in the late autumn (Fig. 8; cf. Fig. 7) when the average temperature fluctuates around the freezing point. However, although rockfalls were found to be fairly frequent, they produced few fatalities.

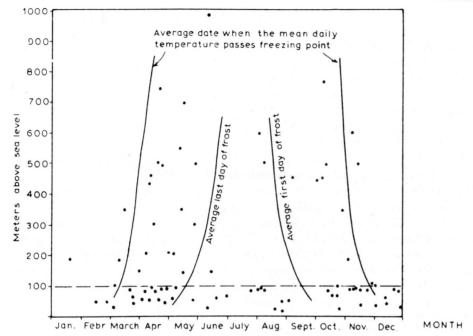

Fig. 7. Rockfalls in eastern Norway in relation to altitude, time of fall, and temperature. Dots indicate rockfalls as a function of elevation and date (after Bjerrum and Jørstad, 1968).

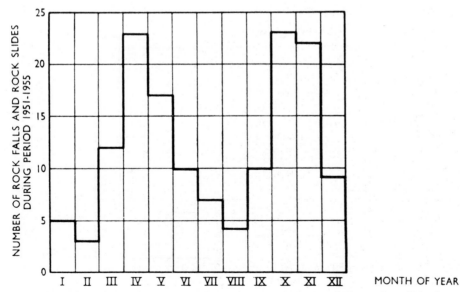

Fig. 8. Distribution of Norwegian rockfalls in period 1951—1955 as a function of month of year.

On the other hand, three rockslides in the period 1873—1940 accounted for more than half of the total loss of life by rockfalls and rockslides (Fig. 9). The three slides were of a deep-seated nature, and the rock in each instance descended into a body of water where it created a destructive wave (Jørstad, 1967, 1968; Bjerrum and Jørstad, 1968). Such large deep-seated rock slides were found to be a consequence of four factors: the jointing, the residual stresses found at depth within the rock mass, the action of water filling the open joints in the rock, and the variations in water pressure in the joint system. The variations in water pressure in particular were judged to be of significance because much of the deformation associated with the pressure fluctuations is irreversible. Consequently, cracks widen, the joints enlarge progressively at the expense of intact rock, and the internal stresses within the intact rock increase.

Although the mechanism of slope deterioration was clear, there appeared to be no possibility of accurately evaluating the factor of safety with respect to the degree of stability of a steep mountain slope. There appeared to be no equipment adequate to survey the distribution or extent of the joints, particularly the sheeting joints, nor to assess the stress concentrations in the intact rock. Therefore, it was concluded that no basis could be found for evaluating quantitatively the degree to which the processes of disintegration

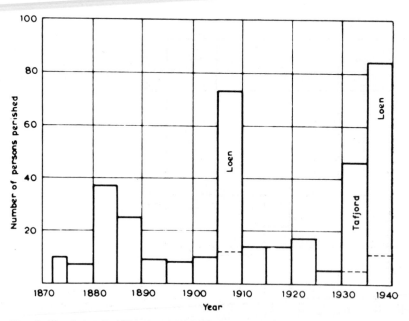

Fig. 9. Number of persons perished due to rockfalls and rockslides in Norway in period 1873—1940, from official sources (after Bjerrum and Jørstad, 1968).

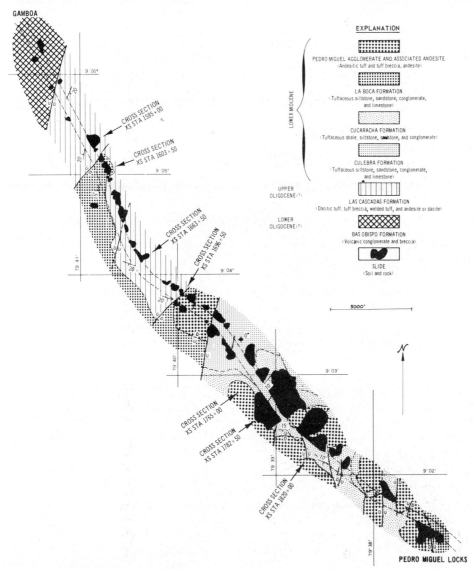

Fig. 8. General geology of Gaillard Cut and locations of Figs. 9—16 (cf. Fig. 1). Minute grid lines approximately 1830 m apart.

Las Cascadas Formation. The Las Cascadas Formation outcrops in the banks of the canal in Bas Obispo, Las Cascadas, Cunette, and Empire Reaches (Fig. 8). The age has generally been regarded as Oligocene but without verification. Four different rock types are recognized — dacitic tuff breccia, tuff, welded tuff, and andesite.

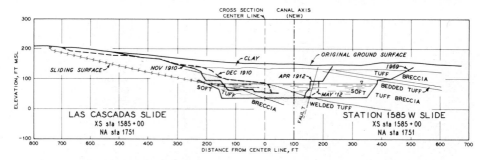

Fig. 9. Geology and construction-slide sequence at Las Cascadas and Station 1585W slides.

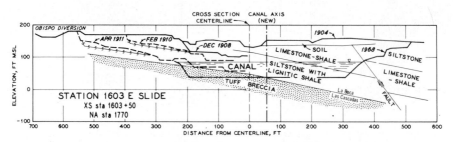

Fig. 10. Geology and construction-slide sequence at Station 1603E slide.

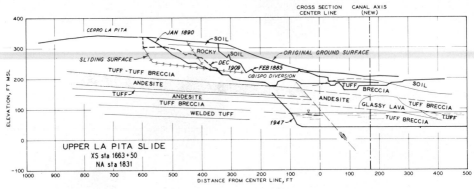

Fig. 11. Geology and construction-slide sequence at Upper La Pita slide.

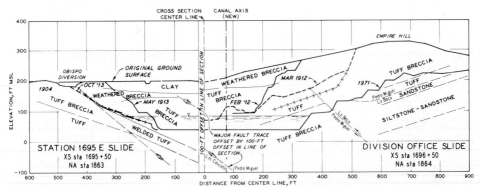

Fig. 12. Geology and construction-slide sequence at Station 1695E and Division Office slides.

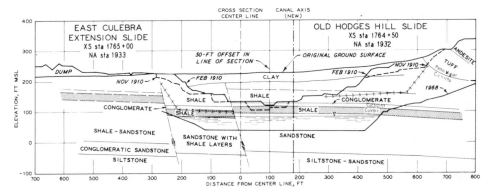

Fig. 13. Geology and construction-slide sequence at East Culebra Extension and Old Hodges Hill slides.

The dacitic tuff breccia and associated welded tuff and the andesite are massive rocks that characteristically occur in beds 9—18 m thick. Bedding is usually not obvious in borings although faint structure may be visible in out-crops. Joints are widely spaced in most of these massive rocks though the andesite contains some closely spaced platy joints. The tuff breccia, welded tuff, and andesite resist instability rather well in comparison to the finer-grained units of the Las Cascadas. Columnar structure is characteristic of welded tuff and seems to have made these beds relatively permeable for transmission of groundwater and therefore locally susceptible to sliding.

Tuff in the Las Cascadas Formation has considerable significance from an

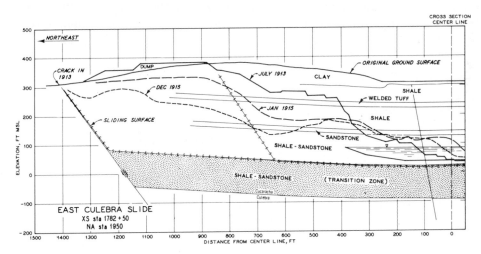

Fig. 14. Geology and construction-slide sequence at East Culebra slide.

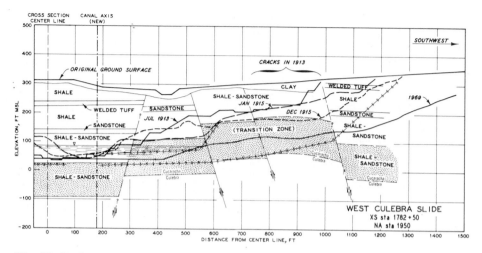

Fig. 15. Geology and construction-slide sequence at West Culebra slide.

engineering point of view; it is essentially a clay shale in its finer types. In contrast to the other rocks, the tuff layers have bedding, and this bedding has commonly functioned as parts of the sliding surfaces. The typical tuff layer is thin, on the order of 3 m in thickness, and soft. The bedding and jointing may vary from moderate to well developed. Slickensides have been reported in borings so consistently that they appear to be characteristic of the tuff. Most slides in the Las Cascadas appear to be related to critically located tuff beds.

Gatuncillo Formation. Borings have passed below the siltstone of the Culebra Formation in Culebra Reach and entered a similar siltstone identified as

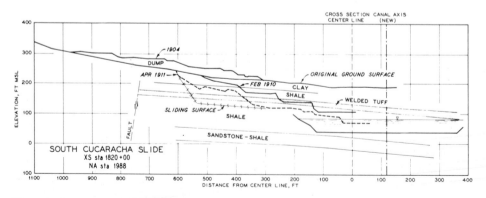

Fig. 16. Geology and construction-slide sequence at South Cucaracha slide.

Gatuncillo Formation. The Gatuncillo Formation is regarded as Eocene in age and consists of medium-hard, sandy siltstone grading downward to a finer siltstone. There is a high carbonaceous content. Bedding is thin to moderate. The locations of these beds are apparently below the slide activity; however, additional borings may eventually modify this conclusion.

Culebra Formation. The Culebra Formation forms both canal banks along the northwest half of Culebra Reach and locally occurs in Cucaracha Reach. Other beds previously classified as Culebra Formation have been reassigned to the La Boca Formation. The formation consists of soft, well-laminated, carbonaceous, shaly siltstone with subordinate pebbly or tuffaceous layers and a concentration of limy sandstone beds in the upper 30 m (Fig. 17). The sandstone beds are commonly 1—3 m thick and separated by thin beds of

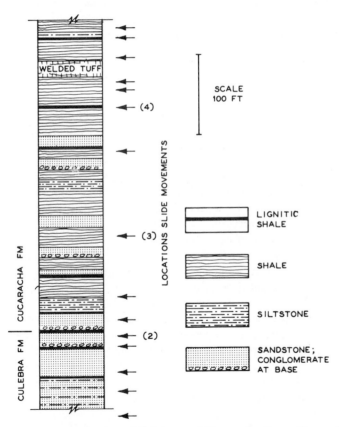

Fig. 17. Stratigraphy of Culebra and Cucaracha Formations with locations of sliding surfaces deduced from records. Parentheses indicate occurrences (more than one).

soft carbonaceous shale and laminated tuffaceous siltstone. Shale and silt-stone usually crumble after exposure to the atmosphere. Fossils of oysters, crabs, and other marine animals indicate that the environment of deposi-tion was near-shore marine and the age is Early Miocene (Woodring, 1957). Interbedded dark layers rich in plant fossils suggest local coastal swamps formed intermittently. During canal construction the freshly exposed Cule-bra Formation commonly emitted an odor of natural gas or petroleum. Locally these rocks oxidized so rapidly when exposed to the air that they became heated and much of the carbonaceous material which they contained was burned off. The known thickness of the Culebra Formation is about 107 m. The upper half is much more sandy than the lower half.

The Culebra Formation is second only to the Cucaracha in the seriousness of its slides. Although the geographic positions of the two formations near the Isthmian divide, i.e., the area of deepest cutting, help explain their shared importance, it must still be concluded that the Cucaracha Formation was intrinsically more troublesome. Some sections of the banks in Culebra Reach, composed of the Culebra Formation, have remained largely intact long after only slightly higher adjacent sections in the Cucaracha Formation failed on a massive scale.

Cucaracha Formation. The Cucaracha Formation overlies the Culebra For-mation and consists of shale and subordinate layers of sandstone and con-glomerate. The unit is tuffaceous in nature and its high montmorillonite con-tent (see "Mineralogy", p. 171) results from reconstitution of volcanic glass particles that once were a major component. Horizons of black, carbonace-ous to lignitic shale (Fig. 17) indicate intermittent swampy conditions of origin. The Cucaracha Formation is divided in two portions by a distinct marker bed of welded tuff occurring about 90 m above its base. The upper portion consists of shale with at least one lignitic bed and discontinuous beds of coarser-grained rock. The lower portion consists of shale, sometimes car-bonaceous, with seven or eight layers of sandstone and/or conglomerate (Fig. 17). These coarser-grained layers average 5 m in thickness. They usually exhibit graded bedding, with a sharp break at the bottom between conglom-erate and underlying shale, a gradual decrease in grain size upward through sand, and a more or less gradational contact with the overlying shale.

Microscopic studies (Mead and MacDonald, 1924) have revealed a relict granular texture as coarse as sand even in montmorillonite-rich shale. Pale to colorless rims of a scaly mineral, probably a mica-like clay, appear around certain grains of minerals or of altered rock fragments. The rock tends to crack along these rims when it dries. There are also films or irregular veins of the scaly mineral in parts of the rock that are not distinctly granular. The clay rims suggested that shrinkage had occurred in situ, with certain grains shrinking more than the matrix and forming minute concentric openings. The openings subsequently became filled with the micaceous (or clayey) mineral substance. It was suggested that these natural microfractures facilitated the development of slickensides.

Several geologists, e.g., Thompson (1947), have studied the Cucaracha Formation and concurred in observing that certain horizons are characterized by the presence of intricate systems of slickensides. The present studies have concluded that lignitic shale layers are the most consistently slickensided, and this, along with the continuity of these beds, helps to explain their high susceptibility to development of sliding surfaces. Otherwise the abundance of slickensides in the Cucaracha Formation seems to have been exaggerated commonly, and as often as not, slickensided portions are concentrated along faults that cut across all beds. In such cases slickensiding decreases to a point of non-existence as the distance between faults increases.

The Cucaracha Formation is mostly terrestrial in origin. A 27- to 45-m transition zone can be delineated within the Cucaracha to include characteristics of both the more usual, terrestrial Cucaracha and the underlying marine Culebra Formations. Both formations are Early Miocene in age. Limy concretions and joint fillings are widely scattered in the transition zone, suggesting that the shale may locally be cemented and slightly stronger than shale above. However, our tests for slaking, shrinkage, and insoluble residue and petrographic examinations have found no conclusive support for the presence of cementation, and no strength difference has been established.

La Boca Formation. The La Boca Formation overlies the Las Cascadas Formation along Las Cascadas Reach and is said to overlap the Culebra and Cucaracha Formations elsewhere away from the canal banks. The La Boca along with the Culebra and Cucaracha Formations formed early in the Miocene. Upward and laterally in the section the La Boca apparently passes into the Pedro Miguel Agglomerate. The La Boca is dominated by siltstone and sandstone, but there are also appreciable amounts of conglomerate, shale, and limestone. The lower 30 m of the formation in Las Cascadas Reach is mostly siltstone. Two or three thin zones of lignitic shale are included, and they apparently have facilitated the development of sliding surfaces that formed in the La Boca.

Pedro Miguel Agglomerate. Coarse and fine andesitic tuff breccia and conglomerate are present in the Pedro Miguel Agglomerate, and these are interbedded with subordinate tuff and even shale. The finer-grained layers seem to be more prevalent in the lower part of the unit, and the contact with siltstone in the underlying La Boca Formation is inconspicuous. The age is Miocene. Mild alteration has brought about reconstitution of glassy material to clays and related minerals and a development of zeolites in openings. The Pedro Miguel is a strong formation that only gives slope stability problems where bedding, faults, or joints provide ready-made sliding surfaces.

Igneous formations. Irregular bodies of basalt, andesite, and dacite are widely distributed in the region. Most of the high hills in the southeastern part of the Canal Zone are resistant masses of intrusive or extrusive basalt of Miocene or possibly Pliocene age. The igneous rocks are among the strongest, and they generally have given stable slopes.

Geological structure

Faults. The dominant geological structure is a regional system of faults trending generally north-northeast. The westernmost major fault of this system in Gaillard Cut (Fig. 8) crosses near NA sta 1711 (XS sta 1544). This nearly vertical fault separates the Bas Obispo Formation on the west from a block of Las Cascadas Formation on the east. The Haut Obispo slide is localized along the fault.

The next known major fault strikes N40°E and dips 50°SE in crossing the canal near NA sta 1763 (XS sta 1596), and it brings marine beds of the La Boca Formation down relatively, against the older Las Cascadas beds. This fault is probably not of the same magnitude as that at NA sta 1711 but may involve considerable strike-slip displacement. The old West Whitehouse slide formed along this fault.

A major fault, locally limiting exposures of the Las Cascadas Formation on the southeast, strikes N50°E across the canal at NA sta 1863 (XS sta 1696). The dip is steeply to the southeast. The block to the south is composed of Pedro Miguel Agglomerate for the most part, but near the fault a sizable block of the underlying marine La Boca beds has been dragged down. Small slides are localized along the fault on both banks.

A conspicuous major fault crosses the canal near NA sta 1887 (XS sta 1720) along a strike of N15°E. Although the fault surface is poorly exposed, it is marked by a contrast in rock type. Pedro Miguel Agglomerate occurs to the west, and tentatively Culebra Formation to the east. The fault functioned in the development of large slides on both banks of the canal. Numerous faults with northward trends are present southeast of NA sta 1887, but exposures are not adequate to confirm geological importance or function in sliding.

Bedding. Bedding generally dips to the southwest in Gaillard Cut from a geologically elevated region to the northeast. Many slides moved down bedding so that there is a strong and important tendency for instability to be most common on the updip or east bank. Bedding was the dominant geological factor in development of sliding, and its role is discussed further in the section "Generalized slide mechanics".

Little or no bedding was recognized in the Bas Obispo Formation. Bedding within the Las Cascadas and La Boca Formations appears to change gradually from a dip of about 20°W at the north end of exposures near NA sta 1796 (XS sta 1630) to a dip of about 15°S at the southeastern end of the exposures near NA sta 1863 (XS sta 1696), as though the bedding wraps around the nose of a broad anticline plunging gently to the south-south-west.

Blocks of Pedro Miguel Agglomerate and associated basalt in the southern half of Gaillard Cut appear to have the form of depressed blocks or graben with bedding inclined steeply along the boundary faults. Most of the cut south of NA sta 1887 passes through nearly horizontal beds of the Culebra and Cucaracha Formations. However, along faults the beds are inclined as steeply as 60°. In the vicinity of Cerro Nitro in Cucaracha Reach the Culebra, Cucaracha, and Pedro Miguel beds dip off the west flank of the hill as steeply as 40°. These local, steep dips are a result of laccolithic intrusion of basalt from below.

Property uniformity and differences among clay shales

Several material properties quantitatively characterize the important clay shales and show their similarity from one formation to another: water content, grain size gradation, Atterberg limits, specific gravity of solids, and mineralogy.

Classification indexes. The laboratory classification of shale samples (Table I) from the Cucaracha, Culebra, and La Boca Formations indicated that in a remolded state, most are clays of high plasticity, but with several classifying as silts of high plasticity. Liquid limits and plasticity indexes vary widely although there is little difference in average values. A wider range of values is evident within a formation than between formations. A wide variation also occurs with depth and within individual samples.

Water content. Natural water contents of the clay shales (Table I) range from 10 to 30% and are 5—20 percentage points below the plastic limit, indicating overconsolidation. Samples from known and possible slide zones showed a significantly greater water content. Water contents about twice the average were determined for slickensided zones near the current depth of sliding in East Culebra and West Culebra slides, indicating wetter and softer clay shales are localized there. Generally a noticeable variation exists between siltstones and clay shales, with the latter having higher water contents.

Specific gravity and density. Specific gravity of solids of clay shales (Table I) range from 2.70 to 2.85, while average values consistently fall near 2.76. Dry densities on laboratory specimens (Table I) also show a wide range of 1.2—2.1 Mg/m^3 (75—131 pcf), indicating a more or less random occurrence of low-density strata with depth and in areal extent. Rather consistently higher dry densi-

TABLE I

Summary of laboratory index test results on clay shale

WES boring no. (except as noted)	Atterberg limits		Specific gravity of solids	Dry density (pcf)
	LL	PI		
Cucaracha Formation				
WEC-1 ⎫	47—116 [1]	19—78 [1]	2.73—2.85	90.3—131.4
WWC-1 ⎬	(65)	(36)	(2.75)	(112.0)
WMS-1 ⎭				
WCSE-1	59—114 [2]	32—88 [2]	2.74—2.81	98.6—121.1
			(2.77)	(110.6)
Recommended average	*86*	*54*	*2.76*	*110.6*
Culebra Formation				
WEC-1 ⎫	67—83 [1]	33—48 [1]	2.83	95.5—102.4
WWC-1 ⎭	(75)	(41)	—	(99.0)
WEPA-1	71—104 [2]	48—74 [2]	2.72—2.80	112.6—124.1
			(2.75)	(117.2)
Recommended average	*82*	*56*	*2.78*	*114.2*
La Boca Formation				
WLWS-1	67—111 [2]	34—86 [2]	2.72—2.76	92.2—106.8
	(82)	(52)	(2.75)	(100.2)
WEWS-1	61—113 [2]	35—78 [2]	2.66—2.78	80.6—114.7
	(81)	(48)	(2.72)	(97.0)
WLCH-1	69—115 [2]	33—84 [2]	2.62—2.85	74.9—111.0
	(97)	(65)	(2.75)	(99.6)
Recommended average	*88*	*56*	*2.74*	*98.8*
Las Cascadas Formation				
D1 [3] (SAD, 1968)	93 [4,5]	58 [4,5]	2.80	110—117
Pedro Miguel Agglomerate (shaly graywacke)				
CRW-15 [5] (PCC, 1969)	42 [6]	17 [6]	2.77	118

Values in parentheses are averages for each boring or set of borings. Recommended averages were determined using the blenderizing procedure.

[1] Grated, powdered, air-dried, slaked for 48 hours, dried back, and worked with spatula.
[2] Grated, air-dried, slaked for 48 hours, and blenderized for 10 minutes.
[3] Two samples.
[4] 19 cycles of slaking and blenderized for 4 minutes.
[5] One sample.
[6] Air-dried, slaked, and worked with a spatula.
To convert pcf to Mg/m^3, multiply by 0.016.

TABLE II

Semi-quantitative mineralogical composition of clay shale

Material	La Boca Formation		Culebra Formation	Cucaracha Formation
	WLCH-1	WEWS-1	WEPA-1	WCSE-1
Clays (%)				
montmorillonite	55	55	50	60
kaolinite	15	15	20	30
Non-clays (%)				
quartz	5	13	15	8
plagioclase feldspar	5	10	5	n.d.
calcite				n.d.
gypsum			n.d.	n.d.
natrojarosite	13	n.d.	n.d.	n.d.
siderite	n.d.	n.d.	4	n.d.
pyrite	4	5	4	n.d.
Organic (%)	assumed to be about 3			
Total	~100	~100	~100	~100

n.d. = not detected.

ties were found in Cucaracha and Culebra clay shale than in La Boca clay shale. Samples from the Culebra Formation tended to have a slightly higher dry density than those from the overlying Cucaracha Formation.

Mineralogy. Mineralogical examinations by X-ray diffraction techniques indicate a notable consistency for 36 clay shale samples from the Cucaracha, Culebra, and La Boca Formations. All samples contained montmorillonitic and kaolinitic clays in abundances of about 55 ± 5% and about 22 ± 8% respectively, with small amounts of other minerals as shown in Table II. Organic content was found to be about 3—5%. The high montmorillonite content in these slide-prone shales supports past experience indicating that troublesome shales usually contain a significant amount of montmorillonite. Chemical studies (Attewell and Taylor, 1973) indicate that the particular variety present is the somewhat less notorious Ca-montmorillonite with only subsidiary interstratification of sodium ions.

Consolidation characteristics and preconsolidation of clay shale

Results of consolidation tests on relatively large specimens of intact clay shale using special equipment capable of consolidation pressures up to 47,900 kN/m^2 (500 tsf), indicate preconsolidation pressures (Table III) as

TABLE III

Summary of consolidation data

Boring no.	Initial void ratio e_0	Estimated overburden pressure p_0 (tsf)	Estimated preconsolidation pressure p_c (tsf)	Expansion index * C_e	Coefficient of consolidation * c_v (10^{-4} cm²/s)	Coefficient of permeability * k (10^{-10} cm/s)
Cucaracha Formation						
WEC-1 } WWC-1 } WMS-1 }	0.657–0.482	2–8	>25–60	0.018–0.060	0.060–1.22	0.3–3.2
WCSE-1	0.679–0.489	3–11	11–98	0.05 –0.09	0.04 –0.06	0.05–0.07
Average	—	—	—	0.05	0.15 **	0.7 **
Culebra Formation						
WEC-1	0.725	4	>30	0.048	0.21	1.0
WEPA-1	0.506, 0.425 .	8, 15	84, 140	0.07, 0.02	0.08, 0.24	0.10, 0.07
Average	—	—	—	0.05	0.18	0.4
La Boca Formation						
WLWS-1	0.818, 0.631	2, 3	5, 31	0.10, 0.07	0.51, 0.10	5.8, 0.16
WEWS-1	0.469, 0.575	7, 11	70, 11	0.04, 0.06	0.13, 0.10	0.09, 0.13
WLCH-1	0.521, 0.713	2, 7	150, 100	0.05, 0.11	0.14, 0.44	0.22, 3.5
Average	—	—	—	0.07	0.24	0.8 **

* Calculated for low rebound loads to bracket estimated load decreases caused by excavation of canal.
** Excluding highest value.
To convert tsf to kN/m², multiply by 95.8.

high as 14,400 kN/m² (150 tsf). Preconsolidation pressure p_c was estimated from geological evidence to be as high as 12,500 kN/m² (130 tsf). However, consolidation testing also produced p_c values as low as 480 kN/m² (5 tsf) for specimens having low dry densities; p_c generally increased with increase in dry density. No consistent trend was indicated with depth of sample; low p_c was found for samples from as deep as 53 m.

Swell pressures measured on Cucaracha and Culebra clay shales at the East Culebra—West Culebra—Model Slope area range from 190 to 1240 kN/m² (2 to 13 tsf) and increase with depth of the sample. These pressures are 0.8— 1.3 times the computed effective overburden pressure assuming hydrostatic pore pressures, as though in-situ pressures in some strata have rebounded to equilibrium with existing overburden pressures.

Rebound of the consolidated specimens occurred during unloading. Free swell under zero load increased the specimen height by as much as 8.5% of the initial height. The expansion index C_e which is useful in estimating rebound under a given decrease in overburden load is consistently near 0.05. The coefficient of consolidation c_v for unloading which is useful in estimating the time required for rebound under a given decrease in overburden load ranges widely, but averages about 0.2×10^{-4} cm²/s. Computed vertical coefficient of permeability k values averaging 10^{-10} to 10^{-11} cm/s compare closely with the horizontal k of 10^{-9} to 10^{-10} cm/s determined from falling head tests in field piezometers placed in clay shale strata.

The consolidation characteristics appear to vary more within a formation than between formations and do not show a consistent relationship with depth, especially in the case of estimated p_c values. This implies that non-uniform rebounding has occurred. The consequences of preconsolidation and rebound to strength and slope stability are considered further under "Seepage and pore pressure", p. 209).

Groundwater conditions

Hundreds of borings by PCC have previously confirmed that the groundwater table or top of permanent saturation lies generally within 8 m of the ground surface. More sophisticated study of water levels and their fluctuation by open system Casagrande-type piezometers were carried forth by U.S. Army Engineers Waterways Experiment Station (WES) after a few initial installations in 1968 for the evaluation of Hodges Hill active area. A few WES piezometers remain in service after several years and continue to supply data.

In-situ permeabilities. Falling head or rising head tests indicated horizontal field permeability k varying from 10^{-10} to 10^{-6} cm/s. The lower field k values are associated with clay shale layers and agree with k values calculated from laboratory consolidation tests (Table III). The higher field k values are associated with sandstones and conglomerates. Several piezometer tips were placed in strata described as containing numerous slickensides; evidence of increased permeability there was not conclusive.

Piezometric levels. The information gained from the piezometers indicates several important groundwater conditions. Generally at older slides such as East Culebra and East Culebra Extension where considerable displacement has presumably caused internal fracturing, piezometer levels coincide closely with Panama Canal water level. At locations of less displacement such as South Cucaracha slide and West Empire active area, piezometer levels are substantially higher than the canal water level.

At relatively stable locations such as on the Model Slope, Hodges Hill, and Las Cascadas Hill, piezometers positioned near the canal reflect the canal wa-

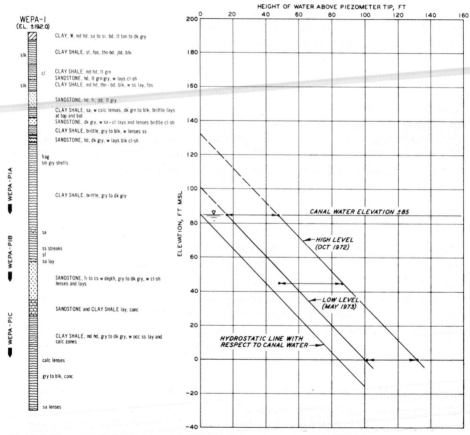

Fig. 18. Piezometric heads, Culebra 1925E active area piezometers.

ter level. At these locations the piezometer level is sensitive to rainfall with higher levels during the rainy seasons, e.g., in October (Fig. 18).

Anomalous pore pressures. Seven piezometers (with tips at 0 to −43 m below mean sea level) in the transition zone of the Cucaracha Formation or in the underlying Culebra Formation at East Culebra, West Culebra, and East Culebra Extension slides and the Model Slope indicated piezometric levels below the canal water level.

The reduced piezometric levels are believed to reflect incomplete rebound of deep strata due to canal and slide excavation. The reduced pore pressures are approximately 85% of those that would eventually develop under the canal water level. It was not possible to delineate the thickness of strata exhibiting reduced pore pressures. The reduced pore pressures seem to characterize only certain strata. The significance in regard to strength loss and future stability is discussed under "Seepage and pore pressure", p. 209).

MEASURED SHEAR STRENGTH OF CLAY SHALE

A comprehensive series of laboratory shear strength tests has been conducted on small specimens including samples prepared to simulate the effects of slickensides, joints, bedding planes, faults, and existing sliding surfaces.

Field shear strength

During foundation investigations for the Third Locks Project (PCC, 1942), six field direct shear tests (reviewed by Smith and Lutton, 1974) were made on 0.093-m² (1 sq.ft) sections of slickensided clay shale considered to represent weak Cucaracha shale. The tests were conducted in a drift located 11 m below ground surface. Two tests were made under a normal (vertical) load of 380 kN/m² (4 tsf) and four tests were made under a normal load of approximately 720 kN/m² (7.5 tsf). Failure in all tests was governed by the position of the slickensides below the top of the test block. Considerable variation [260—460 kN/m² (2.7—4.8 tsf)] occurred in the measured shear stress at failure under the higher normal load. The peak shear strength envelope considered to fit the data best was defined by undrained cohesion c and undrained friction angle ϕ of 62 kN/m² (0.65 tsf) and 21.6°, respectively.

Laboratory shear strengths

Because of its prominent role in the largest slides, clay shale of the Cucaracha Formation has been tested previously more extensively than shale from the other formations. During our study, tests were not only concentrated on the Cucaracha again but also extended to the shales of the Culebra

TABLE IV

Summary of tests used to explore the various aspects of strength of clay shale

Test	Shale material	Objective
Unconfined compression	Intact	Undrained strength and general variability
Q-triaxial	Intact	Undrained strength and general variability
Drained direct shear	Intact	Effective shear strength intact
Drained direct shear	Slickensided	Influence of slickensides
Drained direct shear (repeated)	Precut	Residual strength
Drained direct shear	Slurry-consolidated	Effective shear strength after rebound and softening
R̄-triaxial	Slurry-consolidated	Effective shear strength after rebound and softening

and La Boca (Table V). The laboratory tests used to explore the various aspects of the strength of clay shale are given in Table IV. For further discussion of the rationale for choosing the tests see section "Back analyses of slope stability".

Undrained strength. Results of a large number of unconfined compression (UC) and unconsolidated undrained triaxial (Q) tests from previous studies of Cucaracha clay shales (Table V) show a wide range in undrained strengths, i.e. 29—3260 kN/m² (0.3—34 tsf). Generally, strengths greater than 290 kN/m² (3 tsf) were measured with an average near 1340 kN/m² (14 tsf) from Q tests. In two cases (Table V), Q-strength envelopes were interpreted (PCC, 1942; SAD, 1968) which indicated that clay shale from the Cucaracha and La Boca Formations had low ϕ of 13 and 18.5°, respectively, but high c of 930 and 1100 kN/m² (9.7 and 11.5 tsf), respectively.

Peak effective strength, intact. The peak effective shear strength of intact specimens vary widely and only upper and lower limits were estimated on the basis of WES tests. Effective strengths near the upper limit (Fig. 19) correspond to specimens exhibiting brittle failure, i.e., with large drop in shear stress after the peak. Effective strengths near the lower limit correspond to specimens exhibiting plastic failure, i.e., with little or no drop in shear stress after the peak. The relative effective strength ranges for formations are compared in Fig. 19.

    Laboratory direct shear tests on 232-cm² (6 sq. in) pedestals cut in the top of undisturbed cubes (PCC, 1947) indicated a shear strength envelope near that for the upper limit from WES tests. Tests on silty clay shale from the

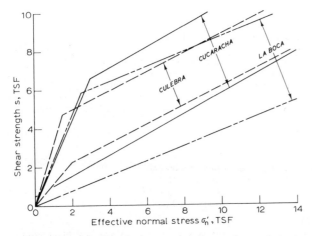

Fig. 19. Effective peak shear strength envelopes, intact specimens.

La Boca Formation (SAD, 1968) indicated a peak effective shear strength higher than the upper limit on the figure. Direct shear tests on specimens of clay shale from the Las Cascadas Formation (SAD, 1968) indicated a higher peak effective shear strength than for clay shales from the other formations.

Peak effective strength, slickensided. A special series of tests on highly slickensided specimens of Cucaracha clay shale indicated a strongly curved effective strength envelope (Fig. 20) for effective normal stresses less than about 385 kN/m² (4 tsf). At higher normal stresses, an approximately linear envelope was defined by the strength parameters, effective peak cohesion $c_p' = 150$ kN/m² (1.6 tsf), and effective peak friction angle $\phi_p' = 24°$. This envelope lies above the lower limiting effective strength envelop for intact specimens.

"Ultimate" effective strength, intact. Effective shear stresses bearing on intact specimens at displacements of about 1.3 cm (near the end of the test) were used to define an "ultimate" strength range (Fig. 20). Since the effective shear stress was still decreasing, a true ultimate strength had not been reached. These "ultimate" strengths indicate little or no c' and a slight reduction in ϕ' in comparison to peak strengths. For the Cucaracha clay shale specimens, the "ultimate" strengths are similar to the peak effective strength obtained on precut specimens (Table V). This is not true for the Las Cascadas clay shale specimens.

Peak effective strength, slurry consolidated. Clay shale samples from the Cucaracha, Culebra, and La Boca Formations were used to prepare one clay

TABLE V

Summary of shear strength parameters from laboratory tests on clay shale

Source of information	UC at natural water content (tsf)	Q-triaxial at natural water content, peak		Direct shear, intact specimens			
				peak		"ultimate" *	
		c (tsf)	ϕ (deg.)	c'_p (tsf)	ϕ'_p	c' (tsf)	ϕ' (deg.)
Cucaracha Formation							
Third Locks Project (PCC, 1942)		9.7 (composite of UC, Q, and S tests)	13	0.65 (field test on undisturbed SLK blocks at natural water content)	21.6		
Isthmian Canal Studies (PCC, 1947)		3.5—27.9 (14.7 avg.)	—	6.0 (at natural water content)	20	1.8 (at natural water content; termed "residual")	20
Hermann and Wolfskill (1966)							
Hodges Hill (PCC, 1969)	2.9—41 (generally <20)						
SAD (1969)	10.5—33.5	5.5—26.2 (12.9 avg.)	—				
WES, Report 1 (Lutton and Banks, 1970)	0.5— 8.1						
WES, Report 3 (Banks et al., 1975)	0.3			1.6 (very SLK specimen, linear envelope, $\sigma'_n > 4$ tsf) 0.5—5.0 ($\sigma'_n \geqslant 3$ tsf)	24 28 **	0.3—0.4	12—22 **
LaGatta (1970)							
Culebra Formation							
WES, Report 1 (Lutton and Banks, 1970)							
WES, Report 3 (Banks et al., 1975)				1.3—4.0 ($\sigma'_n \geqslant 1.5$ tsf)	26 **	0.2	18—24 **

Repeated direct shear, precut specimens				Slurry-consolidated or resedimented specimen, peak			
peak		residual		direct shear		R-triaxial	
c' (tsf)	ϕ' (deg.)	c'_r (tsf)	ϕ'_r (deg.)	c' (tsf)	ϕ' (deg.)	c' (tsf)	ϕ' (deg.)
		0	10				
		(at natural water content on polished surface)					
		0.11	5.7	0.23	17.5		
		0.08	5.6	0.6	17		
		(resedimented)					
		~0	8				
		(average)					
0—1.0	14—28 **	0	4.5—9.5				
($\sigma'_n > 4$ tsf)		(0 avg.)	(7.5 avg.)				
(0.2 avg.)	(19 avg.)						
		0	5.8—10.5				
		(0 avg.)	(7.5 avg.)				
		0.3	5.7	0	17.5— 7	0	20—18.5
						(bilinear envelopes)	
				0	15		
				(specimen trimmed vertically)			
		0	6.4	0	15 —10		
		(annular shear)		(curved envelope)			
		0	6				
		0	10.8	0	21	0	22—19
		0.06	4.6	($\sigma'_n \leqslant 8$ tsf)		($\sigma'_n \geqslant 2$ tsf, curved envelope)	

TABLE V (continued)

Source of information	UC at natural water content (tsf)	Q-triaxial at natural water content, peak		Direct shear, intact specimens			
				peak		"ultimate" *	
		c (tsf)	ϕ (deg.)	c'_p (tsf)	ϕ'_p	c' (tsf)	ϕ' (deg.)
Hermann and Wolfskill (1966)							
La Boca Formation							
SAD (1968), silty clay shale (MH and ML)		11.5	18.5	6	22.5		
WES (1968), silty shale (ML)				1.4—2.8 ($\sigma'_n \geqslant$ 5 tsf)	25 **		
WES Report 3 (Banks et al., 1975)				0 —4.9 ($\sigma'_n \geqslant$ 2.5 tsf)	21 **	0 —0.6	17—24 **
Las Cascadas Formation SAD (1968)				7.9	24.5	0	34
Pedro Miguel Agglomerate Hodges Hill (PCC, 1969), shaly graywacke							

UC = unconfined compression test, Q = unconsolidated-undrained test, $\bar{R}$ = consolidated-undrained test with pore pressure measured, S = consolidated-drained test, SLK = slickensided, c = cohesion intercept (total stress), ϕ = angle of internal friction (total stress), c' = cohesion intercept (effective stress), ϕ' = angle of internal friction (effective stress). c'_p = effective peak cohesion intercept, ϕ'_p = effective peak angle of internal friction, c'_r = residual cohesion intercept, ϕ'_r = residual angle of internal friction, and σ'_n = average effective normal stress.

slurry for each formation. Each slurry was then loaded vertically in a consolidometer to form a normally consolidated sample approximately 15 cm high by 20 cm in diameter. Direct shear tests and triaxial tests with pore pressure measurements ($\bar{R}$ tests) were performed on specimens from the slurry-consolidated samples. The effective shear strengths (Table V) indicate little or no c' and a further reduction in ϕ' as compared with the peak and "ultimate" effective shear strength parameters for intact specimens. Strengths tend to be lowest for Cucaracha specimens and highest for La Boca specimens.

Repeated direct shear, precut specimens				Slurry-consolidated or resedimented specimen, peak			
peak		residual		direct shear		$\bar{R}$-triaxial	
c' (tsf)	ϕ' (deg.)	c'_r (tsf)	ϕ'_r (deg.)	c' (tsf)	$\cdot\phi'$ (deg.)	c' (tsf)	ϕ' (deg.)
		0.1	7.5	0.3	22		
		(annular shear)		0.7	18		
		0.1	6.5				
		0.2	4				
		0.07	3.9	0	25 –13	0	25–21
		0.06	3.6	(curved envelope)		($\sigma'_n \geqslant 2$ tsf, curved envelope)	
1.8	15.5	0.2	4.5				
(0.5	(11.5	(remolded to in-situ					
minimum)	minimum)	density)					
		0	16				

* "Ultimate" is at a shear stress less than peak at a displacement of about 1.3 cm. It is not a true ultimate in some tests since the shear stress was still decreasing at the end of the test.
** Upper and lower limiting strength envelopes.
To convert tsf to kN/m², multiply by 95.8.

Residual effective strength, precut. The large number of residual effective strength determinations by repeated direct shear on precut samples of Cucaracha clay shale (see Table V), indicate little or no residual effective cohesion c'_r and a range in residual effective friction angle ϕ'_r of 4—11°. The range of values shown on a distorted scale in Fig. 21 clusters about an average ϕ'_r of 7.5°. This average also applies to clay shale from the Culebra Formation, but may be somewhat high for clay shale from the La Boca and Las Cascadas Formations. The one test on the Pedro Miguel Agglomerate (on shaly graywacke), indicated a relatively high ϕ'_r as might have been expected since this material is sandy.

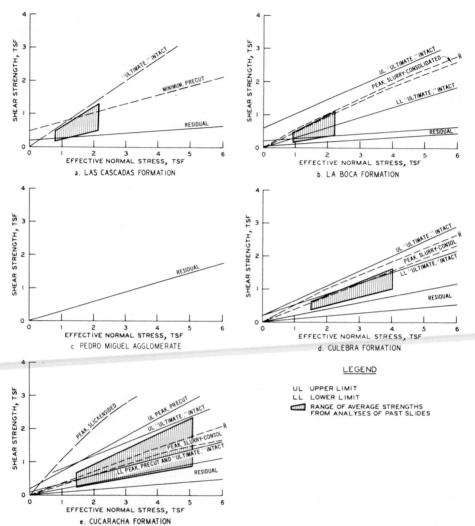

Fig. 20. Comparison of laboratory effective shear strength and range of average effective shear strengths from stability analysis for initial slides along bedding (see Table V for shear strength parameters from laboratory tests and Table VII for shear strength parameters from analyses of past slides).

Shear strength reduction

The laboratory effective shear strengths reviewed above show an important, orderly reduction from peak intact to residual conditions. A significant decrease in effective strength occurred for intact specimens sheared to displacements of about 1.3 cm, for precut specimens, and for slurry-consolidated specimens. The relative order of the effective strength ranges is shown

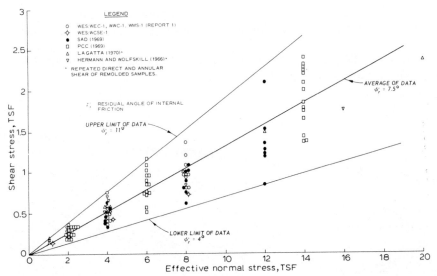

Fig. 21. Drained residual shear strength, Cucaracha clay shale. Note exaggeration of vertical scale.

in Fig. 20 and suggests the following conclusions:

(1) Slickensides apparently do not reduce the effective peak strength of intact shale as much as would be expected, although the reduction is significant.

(2) Similar large reduction in effective peak strength of intact shale generally occurs: (a) at relatively small displacements (as indicated by effective "ultimate" strength envelopes), (b) along preexistent surfaces such as joints (as indicated by effective peak strength envelopes for precut specimens), and (c) after complete softening (as indicated by effective peak strength envelopes for slurry consolidated specimens).

(3) The largest strength reduction occurred at large displacements, where the residual or minimum effective strength was reached.

SPECIFIC SLIDE EXAMPLES

In this section several specific examples of slides are presented to illustrate some of the more important features of the slides of Gaillard Cut. The slides selected are as follows (see Fig. 1 for locations identified 47 and 48, 7, 11, and 42, respectively):

(1) East Culebra and West Culebra slides — illustrating the largest slides through a sequence of progressively more serious movements, all within the Cucaracha shale.

(2) Las Cascadas slide — an old slide illustrating the same general development but in the different, predominantly pyroclastic Las Cascadas Formation.

(3) Station 1603E slide — illustrating slide development strictly governed by weak bedding surfaces in the La Boca Formation.

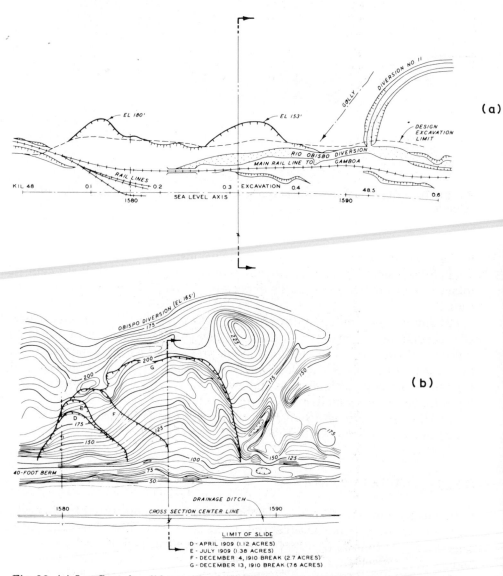

Fig. 22. (a) Las Cascadas slide area in about 1895 showing French cut, diversions, tracks, and slide limits. (b) Las Cascadas slide as of September 1912 with earlier limits superimposed.

(4) Culebra 1925E slide — illustrating a large slide recently developed in a section of bank that had previously remained largely intact for 60 years.

Las Cascadas slide

The east bank of the excavation opposite Las Cascadas is known to have given the French companies problems of instability. Fig. 22 shows the principal features in the vicinity in about 1895, most prominent of which was the end of Rio Obispo diversion ditch and a subsidiary channel draining nearer regions of the east side. By 1888 two slides had broken beyond the intended limit of excavation. These two unstable areas at XS sta 1579 and 1585 were near the starting points of later enlargements.

Between the years 1895 and 1907, rather little excavation was accomplished on the east side in this vicinity, i.e., approximately 6—16 m below natural ground surface. Apparently, as a result of this inactivity no problems of instability developed either. The slide reactivated in about September 1907 when a routine cut along the east side near XS sta 1579 effectively steepened the overall inclination of the bank by several degrees and presumably was followed within a matter of days by a slight subsidence of the bank edge 62 m east of the center line. By April 1908 the point of rupture most distant from the excavation was within 24 m of the crest of the hill, a line that was expected from experience to mark the limit of all slides of this character.

The slide increased in area in July 1909 and broke still farther to the south on 4 December 1910. Before the material from the movement of 4 December had been removed, a larger movement occurred on 13 December 1910. The moving material consisting of some hard rock, but mainly a layer of red clay, carried away all tracks in the east half of the excavation and extended slightly beyond the axis, blocking the pioneer drainage cut. Prior to the movement of 13 December, 137,000 m³ of material had been removed from the slide. Material in motion on 14 December 1910, due to the movements of 4 and 13 December, was estimated at 238,000 m³. For awhile in 1910, rock was dumped on the slide apparently as a buttress. Sheet piling was considered for holding the bank temporarily against future trouble.

Mud flows containing only a few thousand cubic metres of material developed as a result of rains in the spring of 1911. On 10 July 1911, the rain-saturated mud slide moved into the prism blocking all tracks east of the central drainage ditch and interrupting work. Similar minor activity continued intermittently for many years.

The mechanics of this slide was dominated by the geological structure in which beds dipped about 10° toward the excavation, especially for the largest movement. As the cut was deepened in 1910 and largely undercut the strong, relatively rigid welded tuff layer, the strength of underlying soft tuff

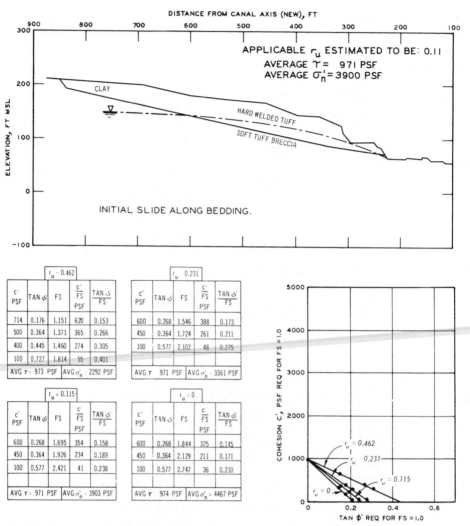

Fig. 23. Results of stability analyses, Las Cascadas slide. Date of section analyzed 1910; dates of active sliding: pre-1888—1942; present stations of slide limits: 1746—1755; present station of section: 1751.

breccia or bedded tuff was mobilized and quickly exceeded. The sliding surface is approximated in Fig. 9 accordingly. The preslide ground profile is dated November 1910. The groundwater surface was considered to have been about 15 m below the ground surface to its intersection with the toe of the slope. The r_u value was estimated to have been 0.11 (see subsection "Method of Analysis" for outline of procedure for back analyzing the strength). Highlights of the analyses are presented in Fig. 23.

Station 1603E slide

The slide centered at XS sta 1603 on the east bank constitutes the main one of the group known as Whitehouse slide. Its earliest movement took place in 1907 after little deepening below the French cut. Fig. 10 shows that the cut had reached a depth of only a few metres by July 1907. A cut of about 3 m caused a small slump of the bank by September 1907, but by October the slide had enlarged considerably by breaking back 30 m from the toe of the slope. The overall inclination was noticeably low. The slide was then inactive for several months.

Station 1603E slide soon became, however, one of the most annoying and troublesome of the year; in November 1908 the area was 9300 m² and calculated volume in motion was 81,300 m³. Seepage water from Obispo diversion was ponded in a graben-like depression at the rear. This depression was inclined down slightly toward both ends of the mass, and ponded water drained laterally to the excavation. The slide broke back in October 1910 to Obispo diversion dike, necessitating excavation of a new channel that carried diversion water about 300 m farther from the cut at this point. Fig. 24 shows slide limits in 1909 through 1911. The limit in April 1911 resulted from the enlargement that took place in October 1910, when the slide roughly doubled in area. Relatively minor movements occurred in 1937, 1938, 1952, and 1958 that required dredging in the canal prism.

The similarity in configuration and the well-established geological struc-

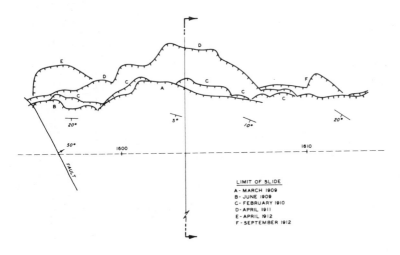

Fig. 24. Growth of Whitehouse slide during dry excavation showing limits of slides. Station 1603E slide constitutes the central part.

ture clearly indicate that slippage along bedding surfaces inclined into the canal was the dominant mechanism for the three more-or-less separate stages of the slide.

A tuff breccia is found at the top of the Las Cascadas Formation. It was logged as a volcanic conglomerate of moderate hardness in an old French

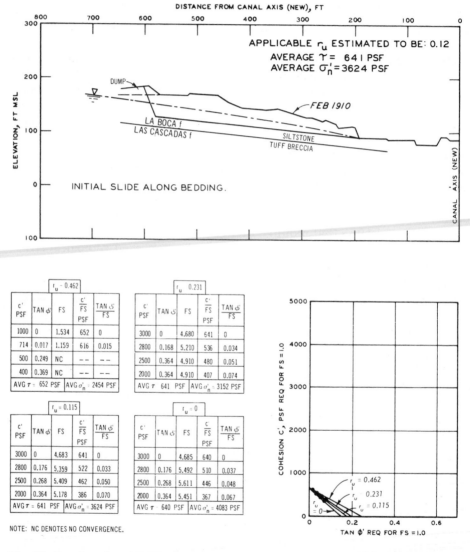

NOTE: NC DENOTES NO CONVERGENCE.

Fig. 25. Results of stability analyses, Station 1603E slide. Date of section analyzed: 1910; dates of active sliding: 1907—1958; present stations of slide limits: 1766—1776; present station of section: 1770.

boring. Original constituents were partly altered to clay. This apparently non-marine pyroclastic unit gives way upward in the section to interbedded shale and siltstone at the base of the La Boca (Fig. 10). Thick, lignitic shale beds were identified in French borings at the top and near the bottom of this 18 m thick unit. The La Boca Formation changes upward to limestone interbedded with subordinate shale layers.

Profiles after each of the three stages had long, linear portions inclined about 10° toward the canal from points at the base of the scarps. This uniform angle provides a first approximation of the residual friction angle of weak beds that functioned as the sliding surfaces. A lignitic shale bed at the base of hard limestone seems to have been the first sliding horizon. This bed was undoubtedly affected by weathering to the east where it lay near the surface. Sliding surfaces for stages 2 and 3 were localized in interbedded shale and siltstone below the limestone.

Back analysis of the 1910 movement is summarized in Fig. 25. The groundwater surface descended from the level of Obispo diversion to the toe of the slope so that $r_u = 0.12$.

Culebra 1925E active area and slide

The slope at NA sta 1925 experienced no serious problems during dry excavation. A considerable amount of terracing to unload the top in 1912 and 1913 was related to the large slide problems on either side along the bank. A crack extended back of East Culebra Extension slide in March 1913 and probably into the immediate vicinity from the southeast, but the slope remained intact for 15 years.

Minor movements occurred at adjacent Hagan's slide to the northwest in fiscal year (FY) 1922 and again in FY 1924 and in East Culebra Extension slide, intermittently in FY 1923—1927. The area was more directly affected by sliding in June 1929 between XS sta 1755 and 1760. A total of about 130,000 m³ of slide material was removed by dredges.

Rather serious movements of East Culebra Extension slide in 1931-1932 brought about a program of remedial dredging and unloading of that slide and adjacent marginally affected areas. A slide basin was dredged 57—76 m east of the prism line between XS sta 1756 and 1770 and a great volume of loose rock and earth was sluiced from the top of the bank.

Interest was focused in 1969 on this slope as one of several priority areas for stability surveillance. A crack had been discovered across the topographic nose (Fig. 26) and opened joints were noted farther back in the slope. The location between two serious old slides made the area particularly suspicious.

By early 1973, cracks formed a crescent-shaped pattern reaching up to 300 m from the canal axis. Fig. 26 shows the arrangement of cracks surveyed

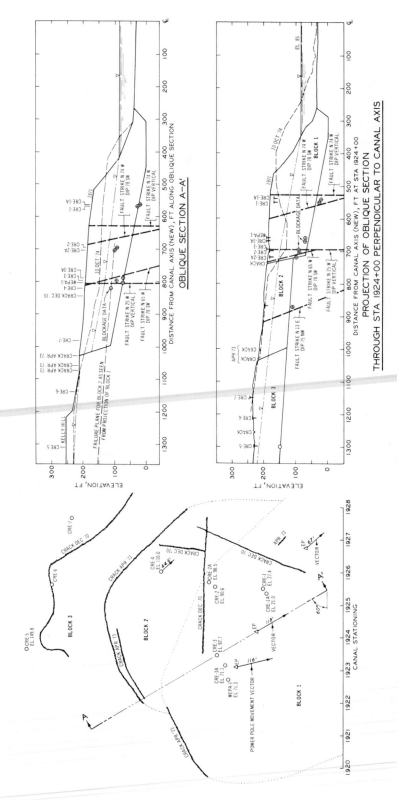

Fig. 26. Boring locations, crack data, movement vectors, and projected sections used in the back analysis of the Culebra 1925E active area, 10 October 1974 slide. Section *A-A'* parallels general movement direction.

in 1970 and 1973. Movements continued to accumulate, particularly during the wet season, and on October 10, 1974, the nose slid obliquely toward and partially into the canal (Fig. 7). A frontal block (block 1) moved approximately 30 m while a second (block 2) moved about 13 m. A graben formed between blocks.

Up to 20 m of Cucaracha Formation was originally present in this bank, but by 1968 this material had largely been excavated. A prominent conglomerate remaining at the top of the bank was considered to mark the base of the Cucaracha Formation. A sequence of alternating limy sandstone and thinner black shale or siltstone, constitutes the Culebra Formation below. A fault identified at East Culebra Extension slide may enter the back of the area on a strike trend of about N5°W, with a dip of 70°W.

Back analysis of this slide (Fig. 26) using blockage and groundwater data from borings indicated strength parameters $c' = 0$ and $\phi' = 13.0°$ for block 1 moving alone. A slightly smaller required strength was obtained assuming that block 2 provided a driving force to block 1.

East Culebra and West Culebra slides

The first indication of a slide near Culebra appeared on the east bank by September 1885 but seems to have been followed by a period of inactivity extending into the spring of 1886. An enlargement apparently occurred in about May since modest slope breakdowns were evident across a 355-m front (Fig. 27).

More serious movement was clearly in progress in June 1886, affecting the natural slope and three working benches. The slide continued to enlarge during the summer and by November extended as far as 300 m from the center line. Excavation continued in the vicinity, although a much greater volume of material now had to be removed. By 1890 the design of the canal had been changed from sea level to lock, and the slope inclination had been reduced. The seriousness of this slide seems to have been even greater than generally recognized. A "limite des terrains en mouvement" recognized at 350—400 m from the center line apparently bounded ground in motion. This limit was near the position of the eventual slide scarp 25 years later.

The French endeavored to overcome the slide problem by digging drainage tunnels, but the clay and shale were too dense and fine to permit free passage of water and the program was discontinued. A period of little or no excavation activity ensued in the interval 1890—1895 (see subsection "General history"). The second French company began operations with a concentrated effort in the divide cut. Between 1895 and 1903, the cut at XS sta 1790 was deepened by 30 m and widened considerably. The first indication of a slide on the west bank appeared in the records of the second French company during this interval. The bank (Fig. 27) apparently moved out

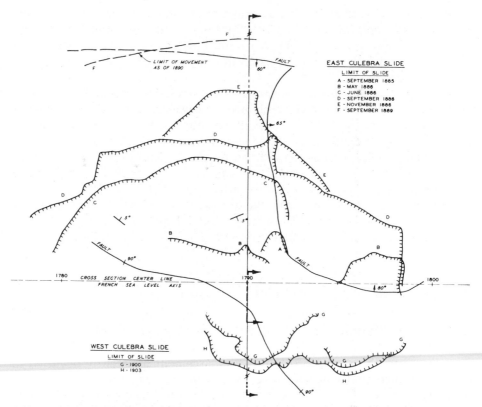

Fig. 27. Initiation and growth of East Culebra and West Culebra slides during operations by French companies, 1884—1903 (cf. Figs. 28—30).

about 3 m almost intact with benches preserved.

The first problem for the United States developed as cracking on the west bank near XS sta 1778 (location not well known) in the summer of 1906, after only nominal new excavation (Cross, 1924). Then in January 1907 a slide was reported on the east side (Goethals, 1916). This was followed in succession by more cracking and movement of both banks in 1907, 1908, and 1909.

West Culebra slide was recognized as a more serious problem in October 1909 when larger movements of the slope and disruption of railroad tracks occurred. For about two years there had been a crack in the west bank about 600 m long and 180—275 m from the center line. About 1.5×10^6 m^3 were estimated within this limit. The condition of the west bank at a later stage of development in 1911 is shown in Fig. 28.

At the same time as the foregoing enlargements were taking place in West Culebra slide, East Culebra slide was also growing in volume. More than

115,000 m^3 of slide material was removed by the United States from East Culebra slide by July 1909, but slide area and volume increased markedly in 1910 (Cross, 1924). Movements became more and more clearly characterized by rupture and crushing of rock at depth as witnessed by the common occurrence of upheaval in the bottom of the cut and subsidence on the bank.

A major development of East Culebra slide occurred in February 1911 (Fig. 28) when a section of bank 335 m long suddenly broke away. The section settled almost vertically about 9 m, moved laterally displacing a wide bench (elevation 135 ft) and then all the ground west of it, and closed the pioneer drainage ditch. About 420,000 m^3 were included in this mass. The movement came in the dry season, without warning, and no blasting had been done in the vicinity for 10 months.

Subsequent, increasingly more serious activity of the two slides is summarized dramatically in the topographic maps of Figs. 29 and 30. A large amount of material was unloaded from the top of the banks (particularly the west bank) immediately behind the slide areas during 1911—1913. This tended to decrease slide movement and fissuring, but favorable effects were cancelled by continual deepening of the excavation. At the beginning of 1914, no notable movement of West Culebra slide was in progress. Nevertheless, several new cracks had appeared on the upper benches of Office Hill, and in June 1914 a crack was found even higher to the south.

A significant development of East Culebra slide came in March 1913 when a major crack was discovered at a position 425 m from the center line between XS sta 1778 and 1788. The crack eventually extended southward across XS sta 1790 and is shown accordingly in Fig. 29; note the coincidence with the surface trace of a fault (Fig. 14). The culminating sequence of movements in the Culebra vicinity started with a massive movement of the east bank on 14 October that closed the canal temporarily.

No noteworthy sliding of the west bank took place during 1914, but reexamination of the west bank following the major movement of East Culebra slide in October revealed the persistence of several long and well-developed cracks (Fig. 29). The cracks were not only extensive, but they also curved to enclose partially large blocks of ground and thus to isolate them as unstable masses. The cracks began to join and divide the mass into long blocks apparently parallel to and governed by north-trending faults and joints. Finally, on 8 August 1915 the West Culebra slide moved into the canal and closed it temporarily. Intermittent serious movements continued (Figs. 6, 30) but gradually tapered off in 1916. Subsequent activity was occasionally serious but generally subordinate.

Fig. 31 presents results of an analysis of a progressive sequence of failures of East Culebra slide. Conventional wedge analyses were employed for simplicity rather than using a more refined method such as by Morgenstern and Price (1965) or Bishop (1955).

GENERALIZED SLIDE MECHANICS

In this part the mechanics of slide development, in some cases through a series of stages, is generalized into several different models. These models represent past slide behavior and provide useful general understanding of possible future sliding.

General types of slides

The classification of the Gaillard Cut slides developed in this study departs from the previous classification of MacDonald (1915) by emphasizing the most important aspect in slide mechanics, the geological structure. In a general way and for purposes of stability analysis, six types of movements are proposed:
(1) Movements with only minor influence of bedding and faults.
(2) Movements in soil, dump fill, and highly weathered rock.
(3) Movements mainly along bedding.
(4) Movements mainly along faults.
(5) Movements in rock weakened by faulting.
(6) Reactivated movements.
The sixth type for reactivated movements is added to distinguish renewed motion of a slide for which a significant portion of the sliding surface was formed during a previous movement of one of the types 1 through 5. Table VI shows the frequency of the various slide movement types 1—5 and relations to specific formations. The accurate distinction of reactivated movements became important in the analytical portion of the study (see "Reactivated slides", p. 217).

TABLE VI

Frequency of slide and active area movements in relation to geological setting

Principal formation	Dominant structural condition					
	mainly soil	no gross structure	weakened by faulting	along fault	along bedding	total
Pedro Miguel	1	2	3	3	1	10
La Boca	0	0	0	5	9	14
Cucaracha	4	2	3	1	20	30
Culebra	0	2	2	3	9	16
Las Cascadas	0	5	5	5	18	33
Bas Obispo	0	0	0	1	0	1
Total	5	11	13	18	57	104

Soil slides. Many of the early shallow slides were confined to soil, dump, and weathered rock materials and thus were distinct and relatively unimportant in regard to the serious slides that formed in bedrock. The tropical weathering has produced a clay-rich soil, and on the steep slopes around Gold Hill, Cerro Paraiso and other hills natural landslides have tended to form. One of these old natural slides, the Cartagena mudflow, is still active as a consequence of disturbance by road construction.

Other soil slides seldom present major problems any more because the long history of excavation and slope disturbance has brought down most masses with a tendency for instability. Soil slides are highly susceptible to excessive rainfalls, either in the form of excessively wet seasons or individual heavy falls. Such slides are also of much lesser seriousness in the context of present operations than those that might affect bedrock.

Many of the old soil slides would have been avoided in the better practice observed in present-day excavation. Thus, the old practice of dumping excavation debris near the edge of the cut, on what are now recognized as unconservative, steep slopes, would not be allowed in present practice. Both French and United States engineers appreciated the importance of adequate drainage. At the present time such drainage is given a higher priority than could be allowed in the busy days of canal excavation. An added factor brought out during this study was the apparent characteristic of porous excavation rock and talus debris to entrap rainwater and aggravate the conditions of instability.

Slides predominantly along bedding. The most important conclusion from this study regarding mechanics of development of slides has been that geological structure and particularly bedding have functioned in the mechanics of practically all slides in bedrock. Faults commonly participated in a subordinate role and often formed lateral boundaries of unstable ground. It has

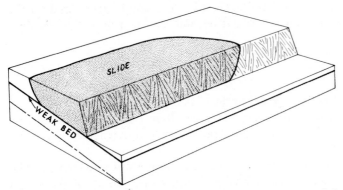

Fig. 32. Development of large slide along gently dipping weak bed.

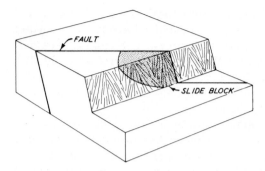

Fig. 33. Development and limitation of small slide dominated by a fault.

been necessary in reconstructing slide development for stability analyses to assume that the sliding surface departed from parallelism with bedding at the rear scarp and in some cases at the toe and cut diagonally across bedding.

One of the main reasons why slides dominated by movement along bedding were so serious was that they progressed to immense sizes once a sliding surface began to develop. The bedding commonly dips gently so that a large area became involved in sliding before the weak bedding surface emerged on the bank, intersected a fault, or in another way was removed from its critical spatial position. This concept is illustrated in Fig. 32 when compared with Fig. 33 for a slide dominated by a fault.

Slides predominantly along faults. Slides that moved predominantly along a fault or faults were fundamentally different from those that moved along bedding by virtue of the facts that faults were consistently steep in dip (Fig. 33) and they were ready-made sliding surfaces with low strength (see "Initial slides along bedding", p. 216). As a consequence of the steep dip of the fault-determined sliding surface, these slides were prevented from growing to very large dimensions as in size Classes 6 and 7; i.e., the fault crossed only a small area or width of the bank before it became too deep to influence slide development (Fig. 33).

Slide movements and stages

Sixty-one slides and active areas are described in WES Report 2 (Lutton et al., 1974—75), and most of these had more than one movement during periods of months or years; more than 100 separate movements have been recognized. In a general way the sequence of movements may be described as slow-intermittent, slow-continuous, and individual-separate movements. The distinction in this manner is somewhat misleading since some of the larger slides showed a more or less continuous movement of reference marks or base lines, but these reflected only surficial activity, and the slide was relatively stationary at depth. This sort of activity is highly influenced by rainfall. The more serious separate movements of the slides are of principal concern and were emphasized in our study. In some cases the separate movements were almost separate slides.

Some examples illustrating the various sequences of movements found in the slides of Gaillard Cut are noted here. The spectacular Hagan's slide of 1913 was one of the last serious slides to develop during dry excavation, and at the time of admission of water the slide protruded into the canal. The slide started in February 1913 and had reached essentially its ultimate development in four related large movements by June 1913. Each of these large movements appears to be an increment of one major stage extending over a three-month interval, and once this stage had come to an end, no further activity occurred, even after many years.

Perhaps more usual is the case of slow-intermittent activity or movements separated by periods of quiescence. Many slides at Gaillard Cut have a history extending through dry excavation and dredging and well along in the operation of the canal. The Las Cascadas, East Culebra, West Culebra, and Cartagena slides were all intermittently active for over 90 years. Documentation of continuing intermittent movements confirms independently that activity such as presently occurs is normal and could have been expected. The overall seriousness of slide movements has tapered off with time, but any major change in conditions such as future widening or deepening of the canal would tend to rejuvenate some slides and inevitably require at least remedial and/or clean-up work.

Delayed slide movements

The anomalous initiation of major slides more than ten years after opening of the canal warrants special concern. Station 1735W slide formed in a section of Culebra Reach previously unaffected, suggesting that a deterioration of slope strength does occur over a long period of time. The slide was Class 6 according to volume, and any reoccurrence on this scale in the future would constitute a major problem (see also subsection "Culebra 1925E active area and slide").

A similar occurrence on a smaller scale recently occurred in Las Cascadas Reach. Station 1619E slide broke into new ground in 1972 after more than 60 years of inactivity in this area. This stage proceeded mechanically in the same manner as for previous stages, and the instance illustrated the value of previous detailed history in rapid evaluation or even prevention of delayed slide activity. Neither Station 1735W slide nor the 1972 movement of Station 1619E slide followed a major change in bank conditions.

Slide enlargement

Enlargement outward. The usual manner of slide enlargement at Gaillard Cut and elsewhere is by incorporation of new ground laterally outward from that active at first. The sliding surface tends to extend outward along the same weak bedding surface that functioned in stage 1, except near the scarp where it crosses the bedding. Apparently, however, the peripheral area is marginally stable or in a condition of incipient failure, and subsequent stages of enlargement often follow.

The slide can thus progress in stages as shown in Fig. 34. The sequence of stages may be either in increments of one process or in relatively separate stages requiring some sort of additional aggravation to continue. Thus, it may be necessary in progressing from stage 1 to stage 2 that early slide debris be removed to open the toe area for additional movement.

Enlargement downward. The present study has uncovered a characteristic of many Gaillard Cut slides that has not been identified previously, a tendency to grow not only in lateral extent but also in depth, particularly as the cut was deepened. Fig. 35 illustrates stages of this slide process in relation to three progressively deeper stages of excavation. Stage 1 of sliding is related to sliding along a weak bedding surface daylighted at stage 1 of excavation. Stage 2 of sliding is similarly related to stage 2 of excavation and so forth. There are sufficient weak beds in the strata along Gaillard Cut to allow this sequence of geometrically independent failures to proceed locally.

Despite this separation of movements in the third dimension, the geographic position of the slide tends to remain localized. The Cucaracha slide is one of the better examples showing this superimposition of relatively independent movements. It seems that some sort of deep deformation, short of instability, started at the Cucaracha slide locality at an early date and eventually led to the deep slide by way of progressively deeper movements. Apparently, the gross geological structure at some localities, combined perhaps with some subtle, unrecognized, aggravating conditions, prepares certain sections of the bank for eventual, repeated instability.

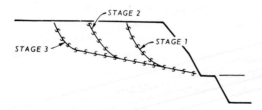

Fig. 34. Slide enlargement in stages outward from excavation cut.

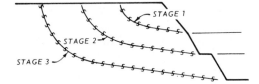

Fig. 35. Slide enlargement in stages downward as excavation is deepened.

Enlargement along the bank. A common manner of slide growth in Gaillard Cut proceeded by progressive enlargement parallel to the bank. Fig. 36 illustrates the process somewhat schematically for an actual example. Development required a weak bed inclined at a gentle angle. Sliding began in the bank near the point where the weak bed was first exposed in the course of excavation. The first stage of activity involved sliding of the bank down the weak bedding to as much as the full depth of excavation, ideally between the point on the top of the bank where the weak bed disappeared below.

A second stage followed more or less independently at a later time after the cut had been deepened appreciably. In the intervening period the same

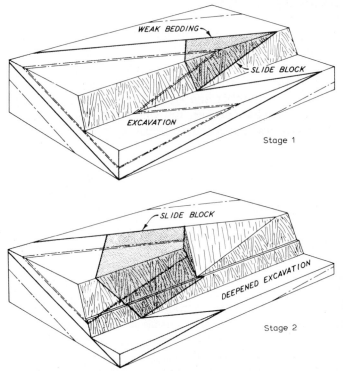

Fig. 36. Slide enlargement in stages 1 and 2 as weak bed is undercut at lower and lower levels.

weak bed had again been daylighted down dip, i.e., exposed with an adverse dip into the cut, and a new precarious condition formed. Stage 2 involved a sliding of the new, unstable section of the bank down dip from stage 1. This sequence could continue to stage 3 and more as the same process of under-cutting of the weak bed continued to greater depth and the excavation approached completion.

Importance of lithological types.

A composite stratigraphic column from boring data in Culebra Reach is presented in Fig. 17 to illustrate the gross characteristics and details of 185 m of Cucaracha and Culebra strata. The column indicates the gross alternation of sandstone and shale beds. Grain size increases more or less gradually downward in graded-bed cycles from lignitic shale and gray shale through siltstone, sandstone, and finally conglomerate. A marked discontin-uity in grain size lies at the base of the conglomerate, and the whole cycle is repeated. Below relatively massive and hard sandstone at the top of the Culebra Formation alternating thin beds of sandstone and siltstone grade from one to the other. The slide-prone horizons have been marked at the ap-propriate stratigraphic positions on the figure.

Sliding surfaces appear to correlate with shale horizons. Fig. 17 goes well beyond this in pinpointing certain horizons as the loci of several movements. These horizons tend to lie high in each graded-bed cycle, with the worst con-dition in lignitic shale. Elsewhere in Gaillard Cut appreciable weakness char-acterizes shale, siltstone, and tuff in other formations, and lignitic shale is again the weakest of all, based on number and size of cases. For all Gaillard Cut including Culebra and Cucaracha Reaches, the well-defined slide move-ments along bedding group as follows according to lithology:

lignitic shale	17
massive shale	8
siltstone	2
tuff	5
other rocks	5

Other conditions contributing to instability

Correlation of major sliding with rainfall. The correlation of heavy rainfall with slide activity (e.g., see Fig. 37) is a fact established by many years of experience here and elsewhere, and was so normal as to be commonly omit-ted from descriptions of the slides except in those unusual cases where move-ment came in the dry season. To illustrate the correlation, the dates for the fourteen largest movements of the West Culebra, East Culebra, and East Culebra Extension slides have been assembled and plotted in Fig. 38 with respect to the annual distribution of rainfall. It can be seen that the rainy

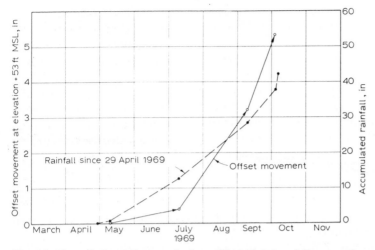

Fig. 37. Cumulative movement along West Culebra failure surface at elevation +53 as related to rainfall.

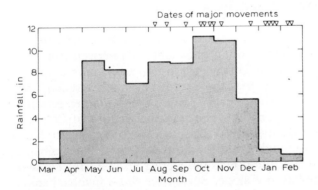

Fig. 38. Average monthly rainfall at Gaillard Cut and vicinity and temporal relationship to fourteen major slide movements.

season usually starts in April and ends in December. The peak in October and November falls in the center of a broad interval of slide activity from August through February.

Seepage and pore pressure. In the relatively tight shale, the groundwater table may have changed to a stable configuration more slowly than the rate at which the excavation was deepened, i.e., over a period of a few years, and pore pressure, though decreased from pre-excavation, may have remained relatively high, deep in the slopes. High groundwater would have been accentuated during construction where ponds and swamps had formed on the bank over the protracted period of French excavation. High pore pressures have been measured at numerous places during this study (see "Piezometric levels", p. 174). High groundwater is believed to have contributed to instability in at least twenty-two slides.

The low pore pressures measured in certain shale strata suggest that the shale has not fully rebounded under the reduction in load caused by canal excavation. The pore pressure u should continue to increase as the shale continues to recover, and in turn, the shear strength τ should decrease according to the relation:

$$\tau = c' + (\sigma_n - u) \tan \phi'$$

thus contributing to further slope instability such as described in subsection "Delayed slide movements".

A seepage force has also developed locally, most significantly as a consequence of groundwater moving along the more permeable layers or fault zones. Leakage of water from diversions at the top of the banks is known to have occurred in places, e.g., Lower La Pita slides, and seepage force may have been a significant aggravating condition at nine slides.

Surcharging of the excavation bank. It was common practice, particularly during French company operations, to dump rock and soil removed from the cut on the nearest available ground. This nearby dump area commonly lay just beyond the edge of the cut so that the dumps became a surcharge on the bank and also impeded surface drainage. Some of the most regrettable cases of such dumping occurred on both banks in the vicinity of Culebra, i.e., Culebra Reach. The serious Culebra Village slide (Fig. 4) appears to have begun as a result of such surcharging and the attendant impeded drainage. Eventually large volumes of old dump material were shoveled, sluiced, and dredged in attempts to stabilize the slope. Seventeen slides involved some surcharging.

Earthquakes as factors in sliding. Earthquakes felt in the Canal Zone suggest a concentration of shallow seismic activity about 160 km southwest of Gaillard Cut. The suspicion that such earthquakes affect canal slope stability arose during formulation of canal design and was revived at the time excavation was peaking. Although there is occasionally a general correspondence between earthquakes and slide movements, it is not at all convincing. A same-day relationship was found in this study only in the instance of the Lower La Pita slide. The bank is known to have been deforming in the previous month, but an earthquake may have triggered the rapid movement.

Disturbance by blasting. A concern with the disturbing effect of construction blasting on slope stability was expressed early after problems had developed from excessive charges and blastholes beyond design lines. Instability at West Culebra and Lower La Pita slides was thought to have been aggravated by excessive blasting of rock outside the design line. Damage of a somewhat less profound nature resulted elsewhere along the cut where the regular large rounds were used in close proximity to the prism line.

The slide problem soon became so well advanced that the importance of blasting as an aggravating condition for the large slides was dwarfed by other considerations. After a while in the Culebra district deep, heavily loaded, toe holes were routinely shot in the steep faces of slides for the purpose of bringing down as much material as possible to a gentler slope.

Softening of clay shale. According to Terzaghi (1936) lateral stress release resulting from an excavation should open joints or slickensides in stiff or over-consolidated clay like the shales in Gaillard Cut. By access of water softening should then proceed inward from the face of each of these fissures. The end product of such softening must be a clay weakened essentially to its normally consolidated condition (Skempton, 1970). Clay shale samples from WES borings commonly contained slickensides, but these slickensides were tight and remained obscure until exposed to the atmosphere. No softened material was noted along the slickensides. On the other hand, drilling fluid was frequently lost as though open fissures and joints are common at depth. No instances of failure by the softening process have been established in Gaillard Cut; but the general history of sliding, in some cases in delayed stages, is consistent with this process and makes it suspect as a factor in instability.

BACK ANALYSES OF SLOPE STABILITY

Stability analyses of 98 prior slide movements at 61 localities (e.g., Figs. 23 and 25), assuming factor of safety $FS = 1$ at failure, have led to back calculations of average shear and effective normal stresses acting along the failure surface. From these data for closely similar slides have come estimates of the shear strength envelopes or strength parameters operative in the field at the time of failure. It is shown below that these strength parameters from slide analyses compare closely with parameters obtained independently from laboratory testing.

General strength characteristics of clay shale

The conventional use of peak undrained laboratory strength for end-of-construction conditions and drained strengths for long-term conditions in limiting equilibrium stability analyses has proven satisfactory for soft clays and less frequently encountered stiff clays. However, many investigators have concluded that the usual methods of testing are not suitable for analyses of stability of slopes in stiff fissured clays and clay shales (Duncan and Dunlop, 1969). Detailed studies of factors affecting the strength of clay shales have shown that instability stems from their distinctive stress-strain and strength behavior, resulting in turn from the presence of slickensides, and other defects and the initial high horizontal stress conditions.

Peak versus residual strength. When a previously unsheared specimen of clay shale is caused to fail in drained direct shear, the resistance of the applied shear force exhibits two extreme values. Initial small displacements raise the shearing resistance toward a large value, the peak shearing resistance. With increasing displacements beyond the peak, the resistance decreases and eventually reaches a smaller limiting value, termed the residual shearing resistance. The peak shearing resistance can be visualized in concept but is rather difficult to determine in practice. Such factors as test specimen size and disturbance, and presence of slickensides, affect the peak shear strength parameters (Skempton and Hutchinson, 1969).

The residual strength, on the other hand, is much easier to determine. The initial condition of the specimen is unimportant (Skempton, 1964), and the only requirement is that the specimen be sheared sufficiently that resistance reaches a minimum.

Detailed evaluation of strength of bedded deposits containing slickensides, faults, etc., such as in Gaillard Cut is obviously difficult. Since these abundant discontinuities in situ tend to control the overall strength, they cannot be ignored. Instead a simplification must be invoked, such as offered in subsection "Laboratory shear strength", as a basis for choosing laboratory strength tests. For example, Calabresi and Manfredini (1973) in studying an overconsolidated fissured clay formation with bedding, joints, and faults showed that (1) along joints and bedding planes the drained peak friction angle ϕ_p' is the same as for intact clay, but c_p' is practically negligible, (2) the residual strength along discontinuities is reached after a small displacement, and (3) along faults the shear strength has already reached its residual value.

Total versus effective stress. Total stress stability analysis using undrained shear strength parameters is generally unattractive because of the difficulties of simulating in-situ stresses in the accompanying undrained triaxial testing. Drained strengths are relatively insensitive to normal stress so that the associated effective stress analysis appears to be the better approach for studying slides. This approach is particularly inviting because of the past research on the decrease from peak to residual strengths for drained conditions as discussed above. Thus our analyses of past slides were based on effective stress in which the pore pressures could be studied as a single unknown. Because of the limited information, a steady-state seepage condition was assumed and only average effective strength ranges were deduced for the different slide groups given in subsection "General types of slides".

Methods of analysis

A description of the techniques used to analyze the slides is contained i.. WES Report 3 (Banks et al., 1975). Usually a two-dimensional limiting equilibrium analysis was made according to the Morgenstern and Price (1965) technique, but where conditions warranted and sufficient data were available, a three-dimensional vectorial technique was used. Groundwater profiles inferred from geology and terrain were modified as appropriate with the piezometric levels from field measurements. The pore pressures were included in the analyses by casting estimated positions of the groundwater table in the form of r_u as defined by Bishop (1955).

Arbitrary effective shear strength parameters, c' and ϕ', were assumed and the resulting FS determined. Values of c'/FS were plotted against $\tan \phi'/FS$ to produce a straight-line plot for each of four pore pressure conditions. The resulting lines represented the relationship between c' and $\tan \phi'$ required for $FS = 1.0$. Two sets of values of c' and ϕ', representing $FS = 1.0$, can be substituted into the Coulomb strength relationship:

$$\tau = c' + \sigma'_n \tan \phi'$$

to determine the average shear stress τ and the average effective normal stress σ'_n acting along the failure surface. A plot of the average τ versus the average σ'_n for slides of a particular type provided sufficient points to define effective strength envelopes.

Results of back analyses for strength

Effective strength parameters from back analysis of slides and marginally active areas are compared in Table VII with corresponding parameters from laboratory tests (from Table V). Those slides and active areas identified specifically in the table were analyzed in greater detail than the "past slides" by using movement and piezometric data and laboratory test results generated during this study.

Initial slides in soil, dump fill, and highly weathered rock. The applicable strengths in soil or highly weathered rock are quite variable, considering the variety of ways in which these materials are developed. The range in effective shear strength deduced from past slides varied from an upper limit defined by $c' = 19$ kN/m² (0.2 tsf) and $\phi' = 23°$ to a lower limit of $c' = 0$, $\phi' = 7.5°$. Weathered materials near the ground surface probably are soft and contain mass defects such as open fractures and joints, and it appears reasonable that

TABLE VII

Summary of effective shear strength parameters from stability analyses and applicable laboratory she

Data source	Las Cascadas Formation		La Boca Formation		Pedro Miguel Agglomerate	
	c' (tsf)	ϕ' (deg)	c' (tsf)	ϕ' (deg)	c' (tsf)	ϕ' (de
Soil, dump fill, and highly weathered rock						
Analyses of past slides	0.2	23	0.2 *	23 *	0	28
	0	7.5	0 **	7.5 **	0	15
Applicable laboratory strength	0.5*(4)	15.5*	0 (5)	25	—	—
	0*(7)	7.5*	0*(7)	7.5 *	0 * (6)	16 *
Movement mainly along bedding						
Analyses of past slides	0	32	0	26	—	—
	0	14	0	10	—	—
Analyses of East and West Culebra slides and model slope	—	—	—	—	—	—
Analysis of Culebra 1925E active area	—	—	—	—	—	—
Applicable laboratory strength	0 *(1)	34 *	0(2)	24	—	—
	← ins. →		← ins. →		0 *(6)	16 *
Movement mainly along faults or in material weakened by faulting						
Analysis of past slides	0.3 *	20 *	0.3 *	18 *	0.65	18
	0 *	20 *	(average)		0.45	18
Analysis of West Empire active area	—	—	0	11	—	—
Applicable laboratory strength	0.5 (4)	15.5	0 (5)	25	—	—
	0.2 (8)	5	0.2 (8)	4	← ins. →	
Minor influence of bedding and faults (along and across bedding but mainly across bedding)						
Analysis of past slides	0.3	24	0.35 *	17 *	0.7 *	20 *
	0	24	0 *	17 *	0.2 *	20 *
Analysis of East Culebra Extension slide (back slope only)	—	—	—	—	—	—
Applicable laboratory strength	0 * (1)	34 *	0.6 (1)	24	—	—
	—	—	0 (3)	17	—	—
Reactivated slides						
Analysis of past slides	0 *	21 *	0 *	22 *	—	—
	—	—	—	—	—	—
Analysis of East Culebra Extension slide (entire sliding surface)	—	—	—	—	—	—
Analysis of South Cucaracha slide	—	—	—	—	—	—
Applicable laboratory strength	← ins. →		← ins. →		0 * (6)	16 *
	0.2 (8)	5	0.1 * (8)	4	← ins. →	

ins. = insufficient data to establish limiting values. (1) Upper limit of effective "ultimate" strength, intact specimens; (2) same as (1) neglecting cohesion; (3) lower limit of effective "ultimate" strength, intact specimens; (4) effective minimum strength, precut specimen; (5) effective peak strength, slurry-consolidated specimens (i.e., normally consolidated effective peak strength); (6) upper limit of residual strength; (7) average residual strength tests

Culebra Formation		Cucaracha Formation		Remarks
c' (tsf)	ϕ' (deg)	c' (tsf)	ϕ' (deg)	
0 **	20 **	0	20	upper limit
0	7.5	0	7.5	lower limit
0 (5)	21	0 (5)	17.5	upper limit
0 (7)	7.5	0 (7)	7.5	lower limit
0	22	0	25	upper limit
0	14	0 *	10 *	lower limit
—	—	0	19	average (force equilibrium stability analysis method)
0	13	—	—	average (complete equilibrium stability analysis method)
0 (2)	24	0 (2)	22	upper limit
0 (6)	11	0 (6)	11	lower limit
—	—	—	—	upper limit
—	—	—	—	lower limit
—	—	—	—	upper limit
0 (5)	21	0 (5)	17.5	upper limit
0 (8)	5	0 (8)	5	lower limit
0.75 *	16 *	0.65 *	18 *	upper limit
0 *	16 *	0 *	18 *	lower limit
0	16	—	—	across bedding
0.2 * (1)	24 *	0.4 (1)	22	upper limit
0.2 * (3)	18 *	0.3 (3)	12	lower limit
0 *	15 *	0	18	upper limit
(average)		0	7.5	lower limit
0	14	—	—	average
—	—	0 *	11 *	average
0 (6)	11	0 (6)	11	upper limit
0 (8)	5	0 (8)	5	lower limit

(for tests on Culebra and Cucaracha clay shale); (8) lower limit of residual strength.
  * Based on limited data.
** No data; applicable values inferred from values for companion formation (Las Cascadas for La Boca and Cucaracha for Culebra).
To convert tsf to kN/m^2, multiply by 95.8.

such material should exhibit strengths varying from normally consolidated effective peak strength to residual strength. Laboratory test data indicate that the effective shear strength of slurry-consolidated specimens and the residual strength generally bracket the range deduced from analyses of past slides.

Initial slides along bedding. Sufficient numbers of past slides fell into the slides-along-bedding group (Table VI) to establish meaningful strength parameters. Fig. 20 shows results for individual formations superimposed on laboratory results. The ranges of deduced strength, shown in Table VII, indicate $c' = 0$ but with ϕ' varying 14—22° for the Culebra, 10—25° for the Cucaracha and 14—32° for the Las Cascadas Formation. It is interesting to note that in the La Boca, Culebra, and Cucaracha the shales are comparable as would be inferred from similarity in indexes of plasticity (Table I). The detailed analysis of Culebra 1925E active area provided strength results along bedding in the Culebra Formation. The average effective strength parameters, $c' = 0$, $\phi' = 13°$ are comparable to those determined from the analyses of past slides. The effective strength parameters from less detailed stability analyses of the East Culebra and West Culebra slides and the Model Slope showed approximate agreement with those from past slides for the Cucaracha Formation. However, the values $c' = 0$ and $\phi' = 19°$ (Table VII) may be fortuitous since the ϕ' value is estimated to be about 3° higher than would be obtained from a more rigorous analysis.

The laboratory effective strength comparable to the upper limit of back-analyzed average effective strength appears to be the effective "ultimate" strength for intact specimens, neglecting cohesion. The "ultimate" strength range in the La Boca and Culebra Formations includes the effective peak strength for slurry-consolidated specimens which also corresponds to the upper limit of strengths for past slides (Fig. 20). It appears reasonable that the upper limit of the average effective strength prior to initial failure along bedding is much lower than the effective peak strength of intact shale.

The lower limit of back-calculated effective strengths along bedding in the Cucaracha Formation (Fig. 20) corresponds to the upper limit of residual effective strengths from laboratory tests and is somewhat higher than the upper limit of residual effective strengths for the Culebra Formation. For the Las Cascadas and La Boca Formations the data were insufficient to determine an upper limit for residual effective strength. Small displacements as may be expected to occur along bedding will reduce the effective strength to residual values. Thus the upper limit of residual effective strengths is not unreasonably low as a lower limit for initial sliding along bedding, considering that the numerous joints, slickensides, and surface cracks can reduce the average strength significantly.

Initial slides involving faulting. Effective shear strength parameters of $c' = 0$—62 kN/m² (0—0.65 tsf) and $\phi' = 18$—20° were deduced from relatively few analyses of past slides which have occurred along faults or in rock affected by faults. Only one slope analyzed in detail (West Empire active area) falls into this group. There, it was concluded that strength along any well-defined fault could be approximated by residual effective parameters. An upper limit to the required strength along the back one of three inferred sliding planes was calculated as $c' = 0$, and $\phi' = 11°$.

Large differences (Table VII) in strength parameters between past slide analyses and detailed analysis resulted largely from the crudeness of the past slide analyses which did not partition the effect of faulting. Since some structural details were lacking, the analyses were simplified to two-dimensional which usually only approximated the actual geological setting. The residual strength parameters appear to approximate a lower limit of strength along faults. An upper limit, especially where structural features control only a portion of the sliding surface, would be the effective peak strength for slurry-consolidated specimens.

Initial slides with minor influence of bedding and faulting. This group of slides, involving only insignificant shearing along bedding and faults is concentrated in the Las Cascadas Formation. The effective strength parameters at the upper limit were generally higher than those associated with initial slides along bedding and indicated a commonly important c' component. Effective strengths at the lower limit were also higher than those along bedding but generally showed $c' = 0$. In a detailed analysis of East Culebra Extension slide, the effective strength parameters operative along the back sliding plane for a high groundwater level were $c' = 0$ and $\phi' = 16°$ in proximity to the lower limit of average effective strengths calculated from past slides.

The comparison with laboratory shear strength (Table VII) indicates that the upper and lower limits for effective "ultimate" strength for intact specimens tend to correspond with the parameters from analyses of past slides. However, c' for the past slide strength limits has a wide range. Little guidance can be offered for the selection of the appropriate c' for analyses of presently stable slopes in which the critical failure surface would shear across bedding.

Reactivated slides. A sufficient number of reactivated slides occurred in the Cucaracha Formation to deduce average effective strength parameters. The available effective strength along a previously sheared surface of a reactivated slide is generally regarded as being at or very close to residual values. Actually the analyses of past reactivated slides gave effective strength parameters (Table VII) higher than residual values with few exceptions. Strength parameters for two slides in the Culebra Formation and the upper range for slides in the Cucaracha Formation approached values close to or equal to the effective peak strength for slurry-consolidated specimens. See Fig. 39 (cf.

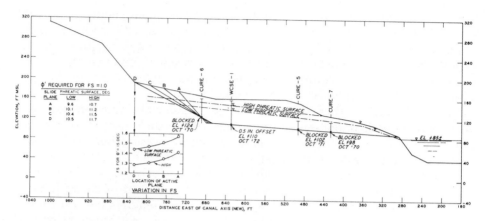

Fig. 39. Section and results of stability analyses, South Cucaracha slide. Profile taken from contours along section.

Fig. 16) for a section through South Cucaracha slide and for summary results of the detailed stability analysis.

Detailed stability analyses of East Culebra Extension slide helped explain the high estimates of effective strength, in this case in the Culebra Formation. Effective strength along the critical surface when partitioned into (1) residual along the previously failed portion, (2) peak normally consolidated on a new extension along bedding, and (3) peak normally consolidated across bedding at the back, gave a refined and more satisfactory approximation of the actual conditions in the slide. It was noted that if such a combination of strengths had not been used, overall strength parameters of $c' = 0$, and $\phi' = 14°$ would have been required for stability. These high strength parameters are close to values from analyses of past slides in the Culebra Formation, indicating that average strengths for reactivated slides are commonly overestimated. To properly assess strengths for such slides, careful attention must be devoted to partitioning the strength along the failure surface in a manner truly representative of the actual slide conditions.

CONCLUSIONS

General design strengths in shale

The effective shear strength parameters determined in the laboratory and substantiated by analyses of past slides are generally applicable and highly useful for the Gaillard Cut slopes and presumably also for similar shale slopes in other tropical to temperate regions. Slope stability can be assessed in a quantitative manner provided adequate documentation of geological struc-

TABLE VIII

Strength for use in analysis of future slope stability along Gaillard Cut

Geologic feature or soil condition	Effective shear strength parameters	Remarks
A. Bedding planes: (1) initial failure	$c' \approx 0 \quad \phi' = \phi'_p$	ϕ'_p varies from normally consolidated peak strength (or lower range of peak precut or "ultimate" intact) to upper range of residual strength (especially if small displacements have occurred in the field)
(2) reactivation along previous sliding surface	$c' \approx 0 \quad \phi' = \phi'_r$	ϕ'_r for residual strength
B. Fault planes:	$c' \approx 0 \quad \phi' = \phi'_r$	ϕ'_r for residual strength
C. Cross-bed shear planes: (1) initial failure	$c' \geqslant 0 \quad \phi' = \phi'_p$	c' and ϕ'_p from "ultimate" intact strength; c' generally greater than zero
(2) reactivation along previous sliding surface	$c' \approx 0 \quad \phi' = \phi'_r$	ϕ'_r for residual strength
D. Weathered overburden materials *	$c' \approx 0 \quad \phi' \geqslant \phi'_r$	ϕ'_r from residual strength to ϕ'_p for normally consolidated strength

Pore water pressures are required from piezometer measurements close to potential slide planes. Where jointing and faulting (or developed slide planes) indicate general seepage water communication, appropriate groundwater surface can be used as in steady seepage stability analyses.
* Variable strength; laboratory shear strength tests are required to determine applicable effective strength parameters.

ture, groundwater conditions, and movement data from field instrumentation is available. General guidance is given in Table VIII on proper strength parameters to use for analyses. Guidance on actual strength values is given in Table VII.

The WES studies have established that geological structure often controls slide mechanics and therefore forms a basis for the stability analyses. An important finding was that many past slides moved generally along bedding and specifically on weak shale layers. Other slides involved sliding along faults or through material weakened by faulting. Others occurred in weathered overburden materials. As a result, the effective shear strength parameters suggested for design use (Table VIII) should be related to those geologic features which control the overall stability.

The study has involved a concerted effort to distinguish initial slides from reactivated slides both in the field, and in turn, for analyzing stability. As a result it has been concluded that an average effective strength for a reactivated sliding surface is not as representative as a combination of strengths partitioned according to portions of the surface that are reactivated and portions that are through newly involved masses at the margin of the slide.

The laboratory tests on intact, slickensided, precut, and slurry-consolidated shale specimens (Table VII and Fig. 20) show a consistent decrease in effective strength with degree of defect and reworking. Intact specimens have the highest strength range followed in decreasing order by slickensided, precut, and slurry-consolidated specimens. It was found that intact specimens sheared to 1.3-cm displacement have a strength range which bracketed that for slurry consolidated specimens. The lowest (minimum) strength is the residual, $c'_r = 0$, and $\phi'_r = 7.5°$ as averaged from a large number of tests on Cucaracha and Culebra clay shales.

Design provisions for specific factors and conditions

The results of the stability analyses of this study have shown ways that available techniques can be successfully applied to slides in shales along Gaillard Cut and elsewhere. Major requisites are accurate assessment of potential sliding surfaces, corresponding strength parameters, and several other factors reviewed below.

Clay shale variability among formations. No conspicuous material variation was found among clay shales from the different formations. This apparent uniformity is illustrated generally by similarities in mineralogy and Atterberg limits. General similarities in strength averages of shale strata among the La Boca, Culebra, and Cucaracha sequences stand out despite wide strength variations that apparently occur from bed to bed within individual formations.

On the other hand, the formations seem in detail to have subtle differences in strength ranges as indicated in Table VII. Accordingly the best procedure in future analyses for stability appears to be on the basis of effective strengths determined by laboratory testing undertaken specifically for the analyses. In lieu of such laboratory test strengths the strength results of this study can be used for approximations provided good engineering judgment is exercised. For example, strength parameters near most of the lower limits in Fig. 20 would be appropriate in cases where a particularly conservative design stance has been taken.

Clay shale softening with time. Reduction of back-calculated strength has been documented for East Culebra and West Culebra slides from data calcu-

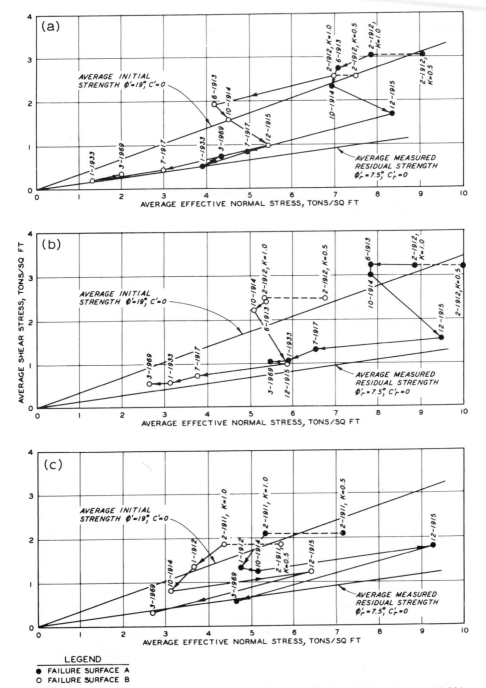

Fig. 40. Changes in average stresses from analyses of East Culebra slide at (a) NA sta 1950 + 10 (XS sta 1782 + 50); (b) NA sta 1955 + 10 (XS sta 1787 + 50); (c) NA sta 1943 + 60 (XS sta 1776 + 00). Factor K = 1.0 indicates pore pressures assumed hydrostatic with respect to phreatic surface (for other than "initial" movements). Factor K = 0.5 indicates assumed reduction of pore pressures by 50% ("initial" movements only).

lated by Binger (1948) and on the basis of test data for this study (Fig. 40; cf. Fig. 31). These data show a fairly rapid reduction in strength with time, but since they involved strength loss after the initial slides and thus the additional effect of large displacements, they cannot be used to estimate strength loss from softening without displacement.

The concept of strength softening with time implies that the cohesive strength component has been lost and that the limiting strength parameters can be estimated from tests on normally consolidated materials of the same composition. Accordingly the effective peak strength of slurry-consolidated specimens is recommended here as a measure of the reduced, long-term strength where no significant movement has occurred.

Strength decrease with movement. Although the softening process reduces or even eliminates the cohesive component of shear strength, only a small reduction in the frictional component occurs. Physical shearing is required to reduce ϕ' from the upper limit $\phi' = \phi'_p$ toward the lower limit $\phi' = \phi'_r$. Repeated direct shear tests showed that displacements of 0.2—0.3 m are required to reduce ϕ' to the residual value when testing a specimen having a precut plane. The detailed analysis of Culebra 1925E active area indicated that the strength was higher than the residual value along bedding, although displacements of surface markers were as large as 0.36 m. This displacement was determined from surface monuments installed after cracks were noted so the total displacement was even more.

It is concluded tentatively that some movement, perhaps as much as 0.6 m, can occur in areas in which newly formed surface cracks are being observed before large reductions in the average ϕ' are produced. This conclusion is restricted to situations where the geologic data clearly indicate that the slide involves previously unsheared clay shales. Prompt and adequate steps taken to arrest the motion will most likely stop further reduction in frictional resistance and prevent a large and massive, sudden failure.

Pore pressure increase with time. Study findings in WES Reports 1 and 2 have generally indicated that the depth of sliding is near the elevation of the canal bottom. This interpretation should not be taken as precluding the activation of deeper sliding surfaces in future movements. In this regard, the observation of reduced pore pressures may be particularly significant. The reduced pore pressures generally occur in the deeper clay shale layers, with the shallowest observed about 12 m below the canal bottom in the Model Slope. Study calculations suggest that the pore pressures are equalizing by increasing to values commensurate with the canal water elevations. Pore pressure changes, i.e., initially reduced in response to excavation, then increased toward some equilibrium state, may be used as an explanation of long-term failure of slopes in stiff and overconsolidated clays. Eigenbrod (1972) has specifically associated cracking instability at the Model Slope with the equalization of such "negative excess pore pressures".

REFERENCES

Attewell, P.B. and Taylor, R.K., 1973. *Clay Shale and Discontinuous Rock Mass Studies.* Final report to the U.S. Army European Research Office, London.

Banks, D.C., Strohm, W.E., Jr., De Angulo, M. and Lutton, R.J., 1975. Study of clay shale slopes along the Panama Canal — engineering analyses of slides and strength properties of clay shales along the Gaillard Cut. *U.S. Army Eng. Waterw. Exp. Stn., Tech. Rep.,* No. S-70-9, Rep. 3 (Corps of Engineers, Vicksburg, Miss.). WES Report 3.

Binger, W.V., 1948. Analytical studies of the Panama Canal slides. *Proc., 2nd Int. Conf. Soil Mech. Found. Eng.,* 2: 54—60.

Bishop, A.W., 1955. The use of the slip circle in the stability analysis of slopes. *Geotechnique,* 5: 7—17.

Calabresi, G. and Manfredini, G., 1973. Shear strength characteristics of the jointed clay of S. Barbara. *Geotechnique,* 23: 233—244.

Cross, W., 1924. Historical sketch of the landslides of Gaillard Cut. In: *Report of the Committee of the National Academy of Sciences on Panama Canal Slides. U.S. Natl. Acad. Sci. Mem.,* 18: 22—43.

Duncan, J.M, and Dunlop, P., 1969. Slopes in the stiff-fissured clays and shales. *Proc. Am. Soc. Civ. Eng., J. Soil Mech. Found. Div.,* 95 (SM2): 467—492.

Eigenbrod, K.D., 1972. *Progressive Failure in Overconsolidated Clays and Mudstones.* Ph.D. Thesis, Dep. of Civil Engineering, Univ. of Alberta, Edmonton, Alta.

Goethals, G.W., 1916. The dry excavation of the Panama Canal. *Trans. Int. Eng. Congr.,* 1: 335—385.

Herrmann, H.G. and Wolfskill, L.A., 1966. Engineering properties of nuclear craters — residual shear strength of weak shales. *U.S. Army Eng. Exp. Stn., Tech. Rep.,* No. 3-699, Rep. 5 (Corps of Engineers, Vicksburg, Miss.; prepared by Massachusetts Institute of Technology).

LaGatta, D.P., 1970. Residual strength of clay and clay shales by rotation shear tests. *U.S. Army Eng. Waterw. Exp. Stn., Contract Rep.,* No. S-70-5 (Corps of Engineers, Vicksburg, Miss.; prepared by Harvard University).

Lutton, R.J. and Banks, D.C., 1970. Study of clay shale slopes along the Panama Canal — East Culebra and West Culebra slides and the Model Slope. *U.S. Army Eng. Waterw. Exp. Stn., Tech. Rep.,* No. S-70-9, Rep. 1 (Corps of Engineers, Vicksburg, Miss.). WES Report 1.

Lutton, R.J., Hunt, R.W., Murphy, W.L. and Stewart, R.H., 1974-75. Study of clay shale slopes along the Panama Canal — history, geology, and mechanics of development of slides of Gaillard Cut. *U.S. Army Eng. Waterw. Exp. Stn., Tech. Rep.,* No. S-70-9, Rep. 2, Vol. I (1974), Vol. II (1975). (Corps of Engineers, Vicksburg, Miss.). WES Report 2.

MacDonald, D.F., 1915. Some engineering problems of the Panama Canal in their relation to geology and topography. *U.S. Bur. Mines, Bull.,* 86.

Mead, W.J. and MacDonald, D.F., 1924. Chemical and physical condition of the Cucaracha, the chief sliding formation. In: *Report of the Committee of the National Academy of Sciences on Panama Canal Slides. U.S. Natl. Acad. Sci. Mem.,* 18: 53—67.

Morgenstern, N.R. and Price, V.E., 1965. The analysis of the stability of general slip surfaces. *Geotechnique,* 15: 79—93.

PCC (The Panama Canal Company), 1942. *Report of Tests in Pedro Miguel Test Pit No. 3.* Diablo Heights, Canal Zone.

PCC (The Panama Canal Company), 1947. Slopes and foundations. *Isthmian Canal Stud.,* Appendix 12. Diablo Heights, Canal Zone.

PCC (The Panama Canal Company), 1969. *Hodges Hill Study, Report of Laboratory Tests on Rock Samples.* Balboa Heights, Canal Zone (unpublished).

SAD (U.S. Army Engineer District and Division, Jacksonville and South Atlantic, Corps of Engineers), 1968. Route 14, Panama subsurface data collection, raw data report, laboratory tests results. *Eng. Feasibility Stud., Atl.-Pac. Interocean. Canal, IOCS Memo.*, JAX-41 (Jacksonville, Fla.).

SAD (U.S. Army Engineer District and Division, Jacksonville and South Atlantic, Corps of Engineers), 1969. Laboratory results, East Culebra slide area, boring 14-D-37. *Eng. Feasibility Stud., Atl.-Pac. Interocean. Canal. IOCS Memo.*, JAX-80 (Jacksonville, Fla.).

Skempton, A.W., 1964. Long-term stability of clay slopes. *Geotechnique*, 14: 77—101.

Skempton, A.W., 1970. First-time slides in over-consolidated clay. *Geotechnique*, 20: 320—324.

Skempton, A.W. and Hutchinson, J., 1969. Stability of natural slopes and embankment foundations. *Proc., 7th Int. Conf. Soil Mech. Found. Eng.*, State of the Art, pp. 291—340.

Smith, C.K. and Lutton, R.J., 1974. Field tests of the Cucaracha Formation, Panama Canal, 1942—1946. *U.S. Army Eng. Waterw. Exp. Stn., Misc. Paper*, No. S-74-16 (Corps of Engineers, Vicksburg, Miss.)

Terzaghi, K., 1936. Stability of slopes of natural clay. *Proc., 1st Int. Conf. Soil Mech. Found. Eng.*, 1: 161—165.

Thompson, T.F., 1947. Origin, nature, and engineering significance of the slickensides in the Cucaracha clay shales. *Isthmian Canal Stud., Memo.*, 245 (The Panama Canal Company, Diablo Heights, Canal Zone).

WES (U.S. Army Engineer Waterways Experiment Station, Corps of Engineers), 1968. *Test Data on Silty Shale from Pedro Miguel Lock Site Investigations.* Vicksburg, Miss. (unpublished).

Woodring, W.R., 1957. Geology and paleontology of Canal Zone and adjoining parts of Panama. *U.S. Geol. Surv., Prof. Paper*, 306-A.

Chapter 5

ROCK SLOPE MOVEMENTS WITH HYDROELECTRIC POWER PROJECTS, MEXICO

ROBERTO SÁNCHEZ-TREJO and LEOPOLDO ESPINOSA

ABSTRACT

This paper summarizes the results of geotechnical investigations associated with rock slope problems at three hydro-power projects in Mexico. In each case sufficient data has been collected, primarily from the field, to permit identification of the causative mechanisms associated with slope movements. In two cases pore pressure conditions were recognized, and remedial measures involved the design of adequate drainage systems; in one of these cases, at the Chilapan project, associated activating factors were complex, involving both seismic activity and excavation at the slide toe. In a third case, at the Santa Rosa project, excellent documentation is given to the concept of a rising reservoir pool as a critical case in slope stability.

INTRODUCTION

Three geotechnical case history investigations of unstable rock slopes in Mexico are discussed in this paper. All three involve the construction of recent hydroelectric plants currently in operation, built by the Comision Federal de Electricidad (CFE) (see Fig. 1).

The purpose of this paper is to present some details of recent rock slope problems covering the period 1955 to the present. In each case, sufficient geotechnical data and field instrumentation has been acquired in order to permit discussions based on more than merely informed conjecture.

Different geologic and hydrologic mechanisms appear to be associated with the slope problems illustrated herein. At Santa Rosa, a causative relationship between slope movements and a rising reservoir pool was carefully documented. At Chilapan, slope movements were rather complexly associated with two earthquakes, and at Ixtapantongo, artesian fluid pressures were identified.

Sources of information include individuals in the technical departments of CFE and unpublished internal reports. The paper was prepared by the staff of Sociedad Mexicana de Mecánica de Rocas (SMMR) in collaboration with

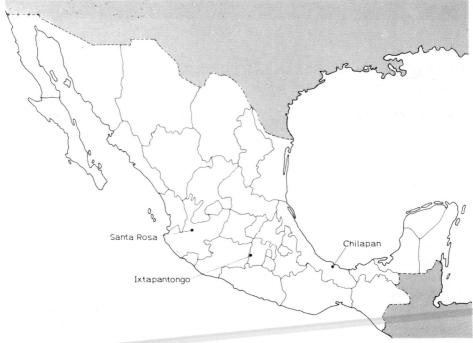

Fig. 1. Index map of Mexico showing location of projects.

engineers of CFE, and compiled by Roberto Sánchez-Trejo and Leopoldo
Espinosa.

SANTA ROSA SPILLWAY ROCKSLIDE

Introduction

Santa Rosa dam, across the Santiago River in Jalisco Province, is a 114 m
high double-curvature arch dam, 12 m thick at the base (Fig. 2). The dam
was constructed between 1959 and 1964. While the spillway access channel
was in the process of being excavated on the right bank, a rockslide occurred [1] which had all the appearances of being merely a local phenomenon.
This caused the spillway structure to be relocated. The observed movements
increased during the rainy season, then ceased. However, after the reservoir
outlet was closed in 1962 and the water level rose, the slide became significantly active once again, and the volume of active slide material extended

---

[1] Fractures were observed for the first time in the spring of 1961.

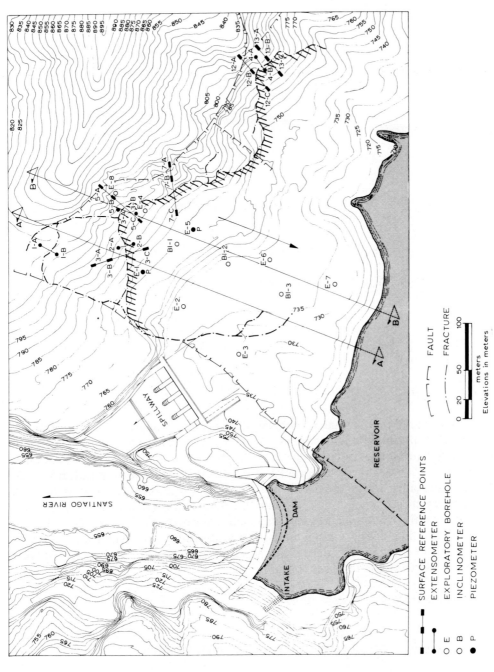

Fig. 2. Santa Rosa Dam. Location of sliding zone and instruments. Hachured area denotes boundary of slide.

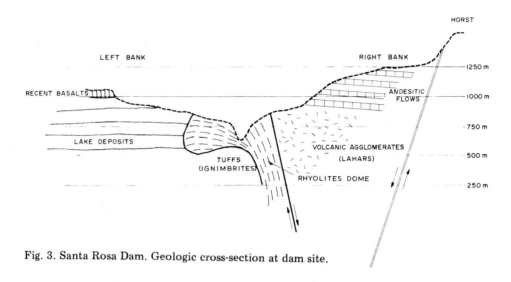

Fig. 3. Santa Rosa Dam. Geologic cross-section at dam site.

further upstream. The approximate volume of the slide is estimated at 8 × 10^5 m^3.

A more detailed geotechnical evaluation of the problem was therefore initiated (Mooser, 1965). This study included more detailed geological surface studies, subsurface borehole investigations, field instrumentation, laboratory material property studies, and stability analyses. Fig. 2 shows the zone affected by the slide, as well as the adjacent position of the sections under observation.

Geologic setting

According to regional geological studies, the topographic basin extends along deep canyons and valleys cut in a rather complex volcanic series of Middle and Upper Tertiary age (Fig. 3). It is assumed that Cretaceous limestones underly the volcanic rocks at depth; this assumption is not, however, of importance to the rockslide problem. The Santiago River runs along a fault which crosses the dam site on the right bank. Here volcanic agglomerates, mainly of the lahar type, stand on the northern side of the fault; the southern side of the fault bounds a rhyolite dome into which the river has eroded its canyon. The dam is founded upon the dome, whose banded lavas rest at depth upon compact tuffs.

Instrumentation

Open-hole piezometers, inclinometers, and surface displacement reference points were installed early in the geotechnical program. Subsequently, elec-

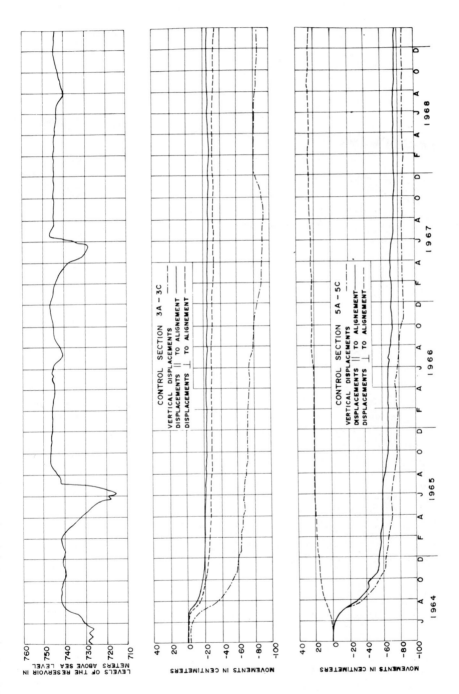

Fig. 4. Santa Rosa Dam. Slide displacements at two control sections, and reservoir pool level, for the period 1964–1968.

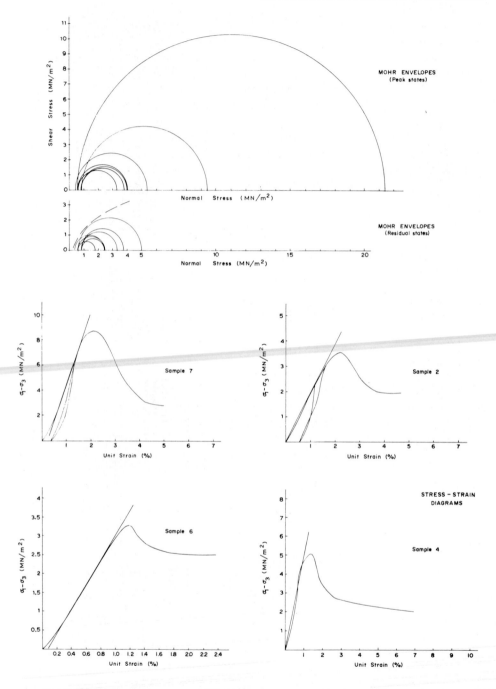

Fig. 5. Santa Rosa Dam. Summary of triaxial tests on samples from sliding zone.

trical extensometers were set up in order to permit detection and frequent automatic displacement data collection from a central instrumentation station. Four sections were equipped with electrical extensometers and the rest with a set of three surface benchmarks, originally set out in line. Fig. 4 shows the evolution of vertical and horizontal components of the displacements recorded at two representative sections in the area, together with fluctuations in the reservoir pool level.

Of most significance is the direct correlation of displacements with reservoir pool level. Displacements significantly increased as the reservoir pool rose above 730 m; then, as the pool was maintained at approximately the 741-m level, the displacements appeared to stabilize. No changes were recorded in association with a 25-m drop in pool level in 1965, which was conducted over a four-month period. With a subsequent rapid rise of 25—30 m in pool elevation, further displacements can be noted, but at a much diminished rate, and these too appear to stabilize. A repetition of this process appears to have taken place in 1967, associated with approximately a 16-m pool elevation change, and perhaps also in 1966 with a minor mid-summer pool fluctuation. But the total amount of time-dependent displacements associated with these subsequent events amounts to only a few centimetres, with the maximum total displacement measured since mid-1965 being about 17 cm (control section 5). The slide is now regarded as stabilized with respect to existing reservoir pool conditions.

Laboratory investigations

Rock specimens, 5 cm in diameter, were extracted from borings made on the site of the spillway slide. The samples of soft tuffs were tested in a triaxial chamber, under confining pressures of from 0.4 to 0.8 MN/m^2 in both "natural water content" and "saturated" states. Pore pressures were not measured. The differences observed between the maximum and residual deviator stresses are notable. The angle of internal friction corresponding to residual strength was in no case less than 26° (Table I; Fig. 5).

Discussion

The "limiting equilibrium" stability of the slope was then analyzed, with the position of the failure surface established on the basis of the inclinometer measurements (Fig. 6), material properties estimated from laboratory experiments, and pore fluid pressures estimated according to piezometric measurements. These calculations suggest that the rock mass should not, in fact, have failed! [2] For failure to occur, it seems necessary to assume that the

---

[2] Factor of safety of 1.2 calculated with residual strength values for samples obtained close to failure surface, as detected by inclinometers. No calculations were carried out as a function of pool level.

TABLE I

Triaxial tests on samples of sliding zone — Santa Rosa Dam

Sample No.	σ_3 (MN/m²)	$(\sigma_1 - \sigma_3)$ maximum (MN/m²)	$(\sigma_1 - \sigma_3)$ ultimate (MN/m²)	Max. unit strain (%)	Secant modulus at 50% $(\sigma_1 - \sigma_3)$ maximum (MN/m²)	Initial void ratio	Water content (%) initial	final	Degree of saturation (%)
1	0.6	2.7	1.2	0.9	4.35×10^2	0.28	2.32	10.24	99.5
2	0.6	3.5	1.9	2.2	1.92×10^2	0.33	8.12	8.12	66.7
3	0.8	3.0	1.7	2.9	1.27×10^2	0.35	6.47	15.92	100
4	0.4	5.0	2.0	1.4	4.75×10^2	0.31	5.13	8.36	71.9
5	0.6	20.8	4.4	1.3	1.97×10^3	0.36	5.72	6.17	47.7
6	0.8	3.3	2.5	1.2	3.12×10^2	0.28	5.29	11.63	100
7	0.8	8.7	2.9	2.1	5.65×10^2	0.22	2.01	7.53	91.4

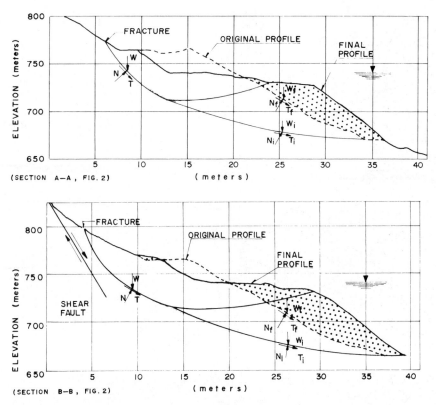

Fig. 6. Santa Rosa Dam. Cross-sections for stability analysis of sliding zone (section locations from Fig. 2).

material along the length of the failure zone exhibits no cohesion whatsoever and can be characterized by an angle of friction of less than 24°. The result of the analysis is not surprising, but it illustrates well the practical difficulty of sampling and testing disturbed and fissured tuffs and inter-layered volcanic units and, above all, the difficulty of sampling the layers of the material crushed by the slide itself. The laboratory tests may have been conducted on materials that were not quite, after all, truly representative of the zone of failure. Perhaps, too, the simple "friction" idealization is insufficient.

CHILAPAN CHANNEL LANDSLIDES

Introduction

The headwater channel from the Catemaco Reservoir to the regulating pool of the hydroelectric plant of Chilapan (State of Veracruz), is a 3.5 km

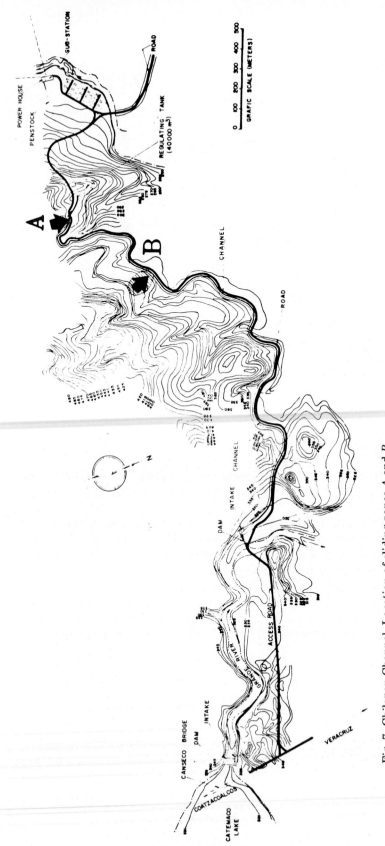

Fig. 7. Chilapan Channel. Location of sliding zones *A* and *B*.

long excavation at the northern slope of a series of highly eroded volcanic cones (Fig. 7) in one of Mexico's tropical jungle areas which has been recently deforested. After excavation of the channel, two landslides occurred; these are denoted by zones *A* and *B* in Fig. 7.

Geologic setting

The rocks affected by the excavation are numerous deeply weathered basaltic flows of Middle to Upper Pleistocene age. They overlie equally weathered units of lahars and volcanic ash. The slopes of the hillsides on the channel are of the order of magnitude of 30° and higher. The slide zones, however, have significantly lower slopes, viz. 10—15° in area *A* and 15—20° in area *B* (Fig. 8). The area is one of high seismicity (see, e.g., Figueroa, 1968).

Description of the landslides

In August 1969, these zones suffered large slides initiated directly by an earthquake, but rendered kinematically possible by the undercut toe due to the channel excavation. In zone *B* the displacement toward the channel was

Fig. 8. Chilapan Channel. Sliding zone *B*.

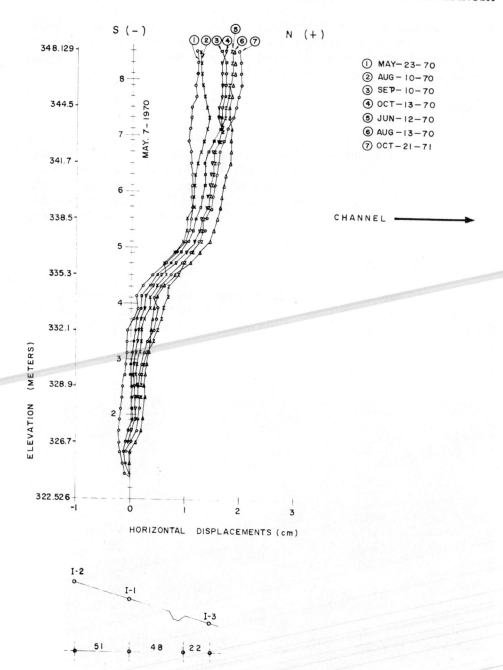

Fig. 9. Chilapan Channel. Inclinometer reading in zone B.

greater than 2 m; a 1 m wide crevasse developed, which delimits the slide zone on its uphill side (Fig. 8).

The sliding mass involved the entire heavily weathered zone of volcanic rock, logged on drillers charts as silt, gravel and boulders. Slide A occurred in weathered rock; slide B occurred in weathered rock and soil. The sliding surface has a depth of the order of magnitude of 15 m as based on the observation of inclinometers placed in zone B (Fig. 9). The surface of slip appears to coincide with a very permeable horizon where a total loss of drilling mud was observed during boring; the mud subsequently appeared in the open test pits excavated downhill. The areas involved in these unstable zones comprise approximately 4800 and 9000 m^2: the slide volumes are thus approximately 70,000 and 135,000 m^3, respectively, for zones A and B.

Discussion

After completion of a cut-and-cover drain drilled at the edge of the excavation channel, in order to reduce percolation and uplift forces, the slope seemed stable. This drain had a discharge of about 10 litres per second. However, four years to the month (August 1973) after the slide event, a second earthquake destroyed and closed the drain; four days later another slide occurred at the same location, causing a new failure of the channel wall. Horizontal and vertical movements of the sliding zone as well as the re-opening of existing crevasses were similar in magnitude to those observed in association with the 1969 slide event.

IXTAPANTONGO POWERHOUSE SLOPE

Introduction

Since 1955, movements of the slope supporting the penstocks for the power plant at Ixtapantongo have been observed. Indeed, the powerhouse structure itself has also been observed to move. Because of the seriousness of this problem, the CFE has since undertaken detailed geotechnical investigations, which include surface and subsurface studies and field instrumentation.

Geologic setting

Surface geology (Fig. 10), combined with exploration holes, yielded the geological sections necessary for the understanding of the penstock and powerhouse foundation areas (Figs. 10—13). The foundation rock is volcanic in nature, formed mainly by fractured Pleistocene andesite and overlying talus deposits, composed of angular andesite fragments in a clay-sand matrix.

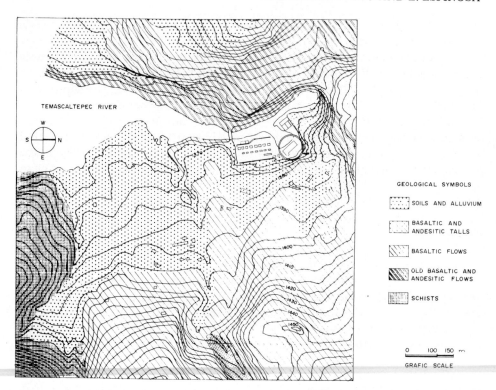

Fig. 10. Surficial geologic map of Ixtapantongo powerhouse and surrounding area. Arrow denotes powerhouse location.

Underlying the andesite mass are metamorphic rocks classified as folded schists, probably of Mesozoic age. No intense weathering is involved.

Instrumentation

The observed downslope movements, though small, continue to create breakage of the surficial terrain and cracking of the penstock supports.

Relative displacement data has been precisely recorded from 1966 to the present time (1974), primarily as a function of relative displacement between the bearing plate of the stiffening rings and the brass plates of the penstock supports (Figs. 14, 15). Furthermore, the profile of an inclinometer (see Fig. 11 for location) detects the depth at which the sliding zone exists (about 7—8 m) and provides a measure of absolute displacements for the period 1966—1971 (Fig. 16). The annual rainfall in the area is about 102 cm with a period of high rainfall from June to September. Observed (open-type) piezometric elevations are generally higher than the sliding plane detected with the inclinometer (Fig. 17). During the boring, partial and total

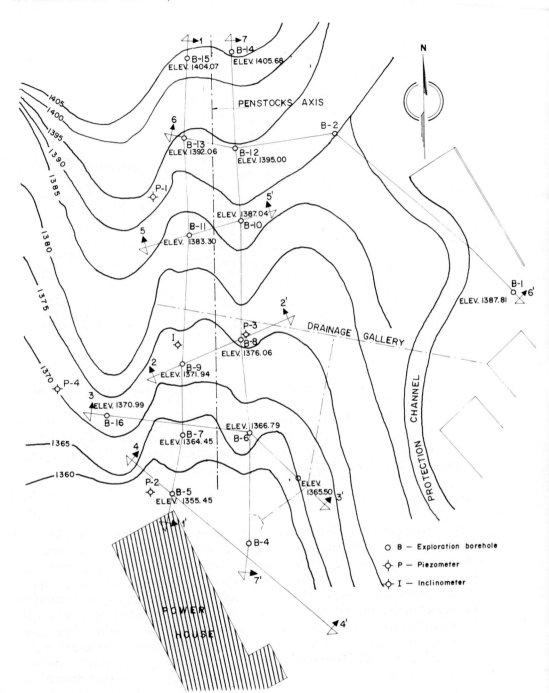

Fig. 11. Ixtapantongo powerhouse. Location of sliding zone and instruments.

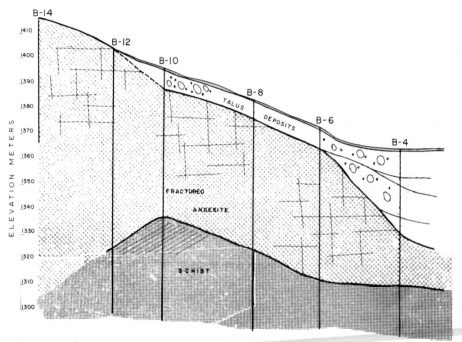

Fig. 12. Ixtapantongo powerhouse. Section through sliding zone (section 7—7', Fig. 11).

losses of drilling mud were observed. No attempt was made to correlate movements with rainfall.

Surface levelling of the powerhouse structure indicated surface uplift at the northwest face of 6—10 mm and settlements of 1.5—3.4 mm at the southwest face. An artesian condition was observed in hole B-4, located between the slope and the foundation of this structure (Fig. 13).

Conclusions

The mass creep at Ixtapantongo involves the andesite talus and the fractured andesite units; the mechanism of sliding may involve an artesian groundwater system within the fractured andesite, with the fine-grained matrix material of the talus sheet providing an effective seal. The surface uplift of the northwest corner of the powerhouse could conceivably be related to this artesian system, but more likely it may reflect upward displacements at the toe of the marginally stable slope. In this regard, the slight settlement observed at the southwest foundation may be a consequence of nearly rigid-block rotation of a powerhouse unit. A drainage gallery was constructed as a remedial measure (Fig. 11). No substantial damage to the powerhouse occurred.

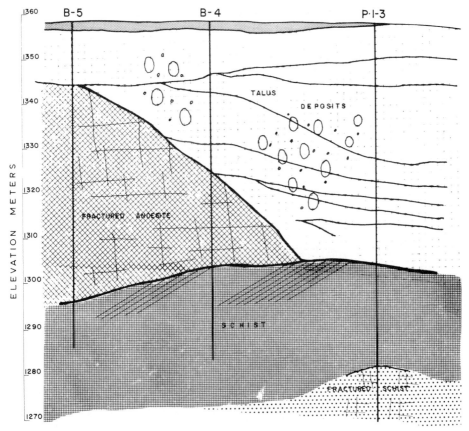

Fig. 13. Ixtapantongo powerhouse. Section through sliding zone (section *4—4'*, Fig. 11).

GENERAL DISCUSSION

If there is a common theme throughout these case histories, it must be the relationship between groundwater and rock slope movements. Although earthquakes seem most directly responsible for the Chilapan slides, for example, it may be noted that a permeable zone seemed co-extensive with the zone of failure, and artesian uplift can be regarded as a distinctly plausible mechanism. The earthquake may have served primarily as a "trigger". Indeed, stability was achieved in 1969 only after completion of a drain installed parallel to the channel excavation. The fact that the second earthquake in 1973 induced slope movements through destruction of the drainage gallery, rather than through direct seismic impulse, underscores the essential role of fluid pressure and slope displacement at this site. Similarly, at Ixtapantongo, careful relative displacement, inclinometer and piezometric measurements, as well as observations during boring, suggest a major role of artesian fluid pressure.

Finally, at Santa Rosa, excellent documentation is given to the concept of a rising reservoir level as a critical case in slope stability. This situation had

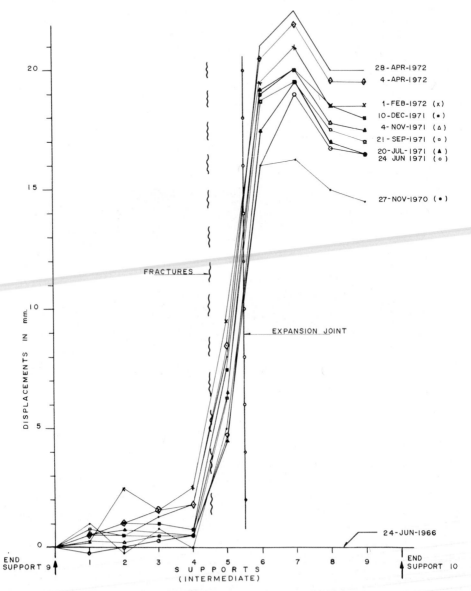

Fig. 14. Ixtapantongo powerhouse. Displacements at supports of penstocks. Displacements observed between the bearing plate of the stiffening rings of penstock No. 3 and the brass support plates.

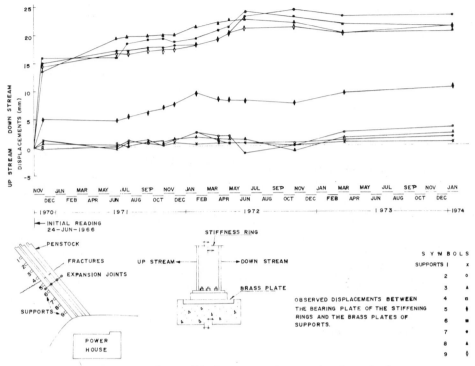

Fig. 15. Ixtapantongo powerhouse. Displacements at supports of penstocks.

been recognized by Jones et al. (1961) in the filling of Roosevelt Reservoir behind Grand Coulee Dam, at Pontesi and Vaiont Reservoirs in Italy (Walters, 1962; Müller, 1968), Gepatsch in Austria (Lauffer et al., 1967), and Bighorn Reservoir in the western United States (Dupree and Taucher, 1974)[3]. Documentation on displacements versus pool elevation are, however, rare. While a comparison of "partial reservoir pool" versus "instant drawdown" calculations suggests the "instant drawdown" idealization as the more critical, in point of fact most actual reservoir drawdowns are conducted at a moderate rate. Hence, the actual rock slope stability is generally significantly greater than that computed for the assumed "instant drawdown" idealization (Kellog, 1948; Lane, 1967).

The gradual stabilization of slope displacements with the maintenance of constant pool level is significant, too; this pattern is analogous to the pattern reported for Gepatsch Reservoir, and contrasts with the complete slope collapse observed (but still not completely understood) at Vaiont. However, as noted by Lane (1967, p. 329; 1974), stability does not necessarily continue

---

[3] See Chapter 6.

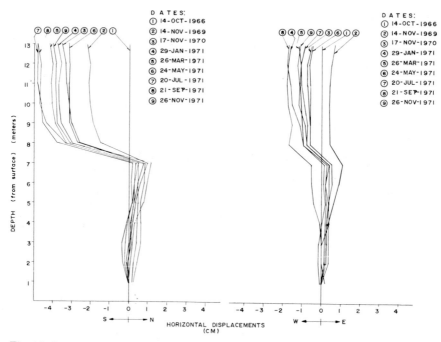

Fig. 16. Ixtapantongo powerhouse. Inclinometer readings.

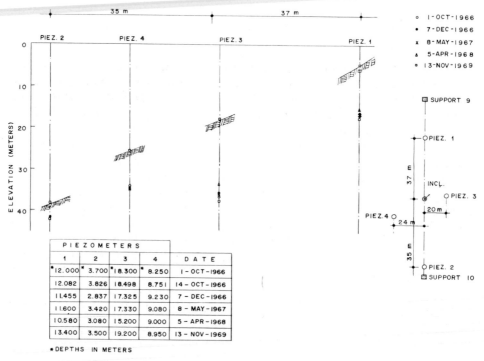

PIEZOMETERS				
1	2	3	4	D A T E
*12.000	*3.700	*18.300	*8.250	1 - OCT - 1966
12.082	3.826	18.498	8.751	14 - OCT - 1966
11.455	2.837	17.325	9.230	7 - DEC - 1966
11.600	3.420	17.330	9.080	8 - MAY - 1967
10.580	3.080	15.200	9.000	5 - APR - 1968
13.400	3.500	19.200	8.950	13 - NOV - 1969

*DEPTHS IN METERS

Fig. 17. Ixtapantongo powerhouse. Piezometer readings.

to decrease as higher reservoir levels are achieved; it is therefore possible for a reservoir slope to become stabilized, once pool elevations rise above a "critical" level associated with a minimum factor of safety, and to be unstable only during periods of reservoir drawdown or rise, when reservoir levels approximate the critical condition.

REFERENCES

Dupree, H.K. and Taucher, G.J., 1974. Bighorn Reservoir landslides, south-central Montana. In: B. Voight and M.A. Voight (Editors), *Rock Mechanics — The American Northwest. 3rd Congr. Exped. Guide, Int. Soc. Rock Mech., Spec. Publ.* Experiment Station, College of Earth and Mineral Sciences, University Park, Pa., pp. 59—63.
Figueroa, J., 1968. La sismicidad an el Estado de Veracruz. *Inst. de Ingeniera, UNAM, México, D.F., Informe,* 167 (April).
Jones, F.O., Embody, D.R. and Peterson, W.L., 1961. Landslides along the Columbia River valley, northeastern Washington, *U.S. Geol. Surv. Prof. Paper,* 367: 98 pp.
Kellog, F.H., 1948. Investigation of drainage rates affecting stability of earth dams. *Trans. Am. Soc. Civ. Eng.,* 113: 1261.
Lane, K.S., 1967. Stability of reservoir slopes. In: *Failure and Breakage of Rock. Proc. 8th Symp. on Rock Mechanics, 1966.* American Institute of Mining, Metallurgy and Petroleum Engineering, New York, N.Y., pp. 321—336.
Lane, K.S., 1974. Stability of reservoir slopes. In: B. Voight and M.A. Voight (Editors), *Rock Mechanics — The American Northwest. 3rd Congr. Exped. Guide, Int. Soc. Rock. Mech., Spec. Publ.* Experiment Station, College of Earth and Mineral Sciences, University Park, Pa., p. 64.
Lauffer, H., Neuhauser, E. and Schober, W., 1967. Der Auftrieb als Ursache von Hangbewegungen beider Füllung des Gepatschspeichers. *Proc. 9th Int. Congr. on Large Dams, Istanbul,* Q. 37, pp. 669—693.
Mooser, F., 1965. *Internal Geological Reports on Santa Rosa Dam.* Dep. of Geology, Comision Federal de Electricidad, México, D.F.
Müller, L., 1968. New considerations on the Vajont slide. *Rock Mech. Eng. Geol.,* 6: 1—91.
Walters, R.C.S., 1962. *Dam Geology.* Butterworth, London.

Chapter 6

BIGHORN RESERVOIR SLIDES, MONTANA, U.S.A.

H.K. DUPREE, GLENN J. TAUCHER and B. VOIGHT

ABSTRACT

Prior to the construction of Yellowtail Dam in 1965, a portion of the Bighorn Canyon was characterized by lower slopes composed of a large pre-historic slide complex in Cambrian shales. Renewed movement occurred in these old slide masses when buoyancy associated with the rising reservoir water affected them. The slides were noticed first in 1966 by aerial surveillance; pool depth was approximately 40 m at that time. Subsequent slide movements have occurred, and these have been correlated with filling periods in the late spring of each year. The slide masses have not interfered with reservoir operation; because of this fact and extremely poor access conditions, detailed field study and ancillary work on the slide areas has not been carried out. Nonetheless, useful geotechnical information is available from nearby highway construction projects, and the mechanical implications concerning these reservoir-activated slide events seem clear.

INTRODUCTION

Yellowtail Dam is located on the Bighorn River about 69 km upstream from Hardin, Montana (Figs. 1, 2). This structure, 151 m high, was built by the U.S. Bureau of Reclamation and completed in 1965.

The reservoir impounded by Yellowtail Dam is called Bighorn Lake; it occupies the deeply incised Bighorn Canyon for a distance of about 70 km upstream from the dam. This canyon crosses the axis of the Bighorn Mountain uplift near its northern end; at this point it attains a depth of about 850 m, approximately 18 km upstream from the dam. Throughout much of the canyon, the canyon width is typically less than 450 m and the sheer canyon walls are formed by one formation, the Madison Limestone of Mississippian age. Only in the reach involving maximum canyon depth, for a distance of approximately 8 km, has the river cut through the Mississippian limestone into the lower sequence of rocks. This lower sequence consists of dolomite, sandstone and shale, with the shale predominating in the lower part of the section.

In this same reach, in part on Crow Indian Reservation land, the canyon attains its maximum rim-to-rim width of approximately 2.4 km. This seemingly excessive width is attributed to geomorphic mechanisms involving re-

Fig. 1. Schematic relief map of the Bighorn Mountains and adjacent areas in northwest Wyoming and adjacent Montana. Reference locations: *1* = Bighorn Canyon and Bighorn Reservoir; *2* = Five-Springs Canyon; *3* = Shell Canyon; *4* = Clark's Fork Valley (Sunlight Basin Highway); *5* = Wind River Canyon in the Owl Creek Mountains.

Fig. 2. South-facing view of Yellowtail Dam and Bighorn Reservoir, crossing the northern Bighorn Mountain uplift (USBR Photo A459-600-219A). A portion of the Pryor Mountain uplift occurs in the right background, rising above the Bighorn Basin (cf. Fig. 1).

moval of support by failure of the lower incompetent shales, which led to subsequent collapse of the carbonate sequence forming the canyon walls.

Thus prior to the construction of the Yellowtail Dam, the lower slopes of the canyon were composed of large, prehistoric slide masses. After completion of the Yellowtail Dam, renewed movement occurred in these old slides when the rising reservoir waters affected them. Slide movements have continued, but at a somewhat lesser rate than initially, since the reservoir was filled in 1967. Numerous slides were reactivated in this process; several of these are quite large, i.e., in the 20—40 × 10^6 m^3 class. The total volume of material moved exceeds 10^8 m^3. However, apart from a brief mention by Lane (1967) and a recent guidebook summary (Dupree and Taucher, 1974), no description nor discussion of these interesting reservoir-activated slides is available in the published literature. To provide more complete information we here consider in detail the geologic and geomorphic history of the Bighorn Canyon area with special emphasis on prehistoric slide phenomena, provide a chronicle of landslide events associated with both fluctuations of reservoir pool level and nearby highway construction projects, and discuss geotechnical topics particularly related to slide events. Division of labor is as follows: Dupree and Taucher are responsible for the discussions of geologic setting and the chronicle of slide events, as well as for all photographs; Voight contributed sections on analogous slides in the Bighorn Basin area, material properties, and the summary discussion.

GEOLOGIC SETTING

Geologic history

The Bighorn Mountains, part of the Rocky Mountain chain, extend from north-central Wyoming into south-central Montana over a distance of about 200 km. Peaks in the central portion of the range attain maximum elevations on the order of 4000 m above mean sea level (m.s.l.), or about 2400 m above the surrounding plains. The mountains comprise a block-uplift complex which developed during the Laramide Orogeny at the close of the Cretaceous period (see, e.g., Darton, 1906; Prucha et al., 1965; Hoppin and Jennings, 1971; Samuelson, 1974; Stearns et al., 1974). The structural trend is north-northwest in the northern section and more nearly north-south in the southern section. At the southern end, the uplift merges with the westward-trending Owl Creek Mountains (Fig. 1). The central part of the uplift seems asymmetrical, with localized thrusting to the east (Hoppin and Jennings, 1971); the northern end of the uplift is a northwestward-plunging "anticlinal" drape fold characterized by a broad top and steep limbs (Darton, 1906; Richards, 1955).

During Paleozoic and Mesozoic times a thick sequence of sediments were deposited over a large area (for a recent summary, see Samuelson, 1974). These sediments, approximately 2700 m in thickness, represent every system from Cambrian to Cretaceous except the Silurian. Uplift and deformation of the Rocky Mountain area, which climaxed in the late Cretaceous and early Tertiary, deformed the accumulated sedimentary rocks over uplifted and rotated blocks of shallowly buried crystalline basement. The ancestral Bighorn "anticline" which formed at this time was partially destroyed by erosion as uplift proceeded. Sediments stripped from the uplifted areas were deposited in surrounding basins; structures adjacent to the uplift were buried, and indeed, portions of the uplifted mass itself were buried in its own debris.

The ancestral Bighorn River developed upon the peneplain formed by Tertiary erosion and deposition, and thus became superimposed upon the buried northern end of the Bighorn Mountains. Rejuvenation of the streams accompanied the regional uplift occurring during later Tertiary time (Mackin, 1937). Increased erosion stripped the soft early Tertiary sediments and entrenched the rivers and streams to their present levels in harder Paleozoic

Fig. 3. South-facing view of entrenched meanders in the southern part of Bighorn Canyon, at the Bighorn Mountain—Bighorn Basin boundary (USBR Photo A459-600-214NA). Note slide at lower left (cf. Fig. 4). Pryor Mountains in right background.

rocks. This entrenchment gives rise to a spectacular canyon where the northward-flowing Bighorn River crosses the northern end of the Bighorn Mountains. The southern portion of this canyon with its incised meanders, and a portion of the southern limb of the Bighorn Mountain monoclinal flexure (drape fold), are shown in Figs. 3 and 4. The greatest entrenchment occurs where the river crosses the broad arch forming the crest of the Bighorn uplift. An oblique view of this arch, looking south from above Yellowtail Dam, is shown in Fig. 3. The crest of the arch is near the distant (south) end of the canyon.

Stratigraphic summary

The reach of greatest river entrenchment begins at the upstream (southerly) flexure of the Bighorn Mountain arch and extends downstream for approximately 8 km. Along this reach the canyon attains its greatest depth (850 m) and width (2400 m). Both upstream and downstream of this section

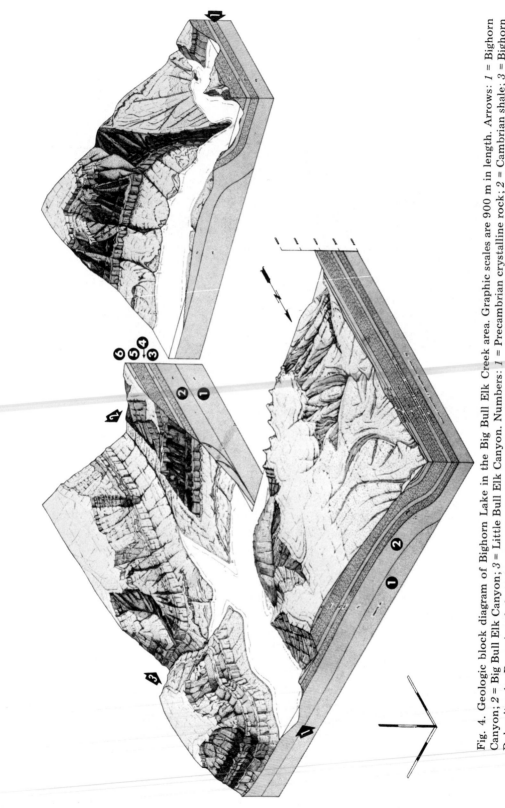

Fig. 4. Geologic block diagram of Bighorn Lake in the Big Bull Elk Creek area. Graphic scales are 900 m in length. Arrows: *1* = Bighorn Canyon; *2* = Big Bull Elk Canyon; *3* = Little Bull Elk Canyon. Numbers: *1* = Precambrian crystalline rock; *2* = Cambrian shale; *3* = Bighorn Dolomite; *4* = Devonian shale and carbonate; *5* = Madison Limestone; *6* = post-Mississippian sequence.

Fig. 8. Peripheral crack, 1968 (cf. Fig. 10) (USBR Photo A459-600-246NA). East side, Big Bull Elk Creek area.

sive enough to permit widespread consistency in geotechnical performance. During Middle Cambrian time the seas moved rapidly eastward, laying down nearshore Flathead (sandstone) facies and offshore glauconitic muds of the Gros Ventre facies. Brief periods occurred in which limestone deposition was favored, and toward the end of the Middle Cambrian the glauconite mud facies was spread from the Powder River Basin to Idaho (Shaw, 1954, p. 36). Throughout northwestern Wyoming, the Cambrian system is thus character-ized by the occurrence of a thick, weak, greenish-grey argillaceous succes-sion, along with a basal coarse clastic unit and interbedded local clastic and carbonate units. Despite local name variations the individual units appear with lithic similarity from the Teton area (Rubey and Hubbert, 1959; Love and Reed, 1971) to the Beartooth Plateau—Sunlight Basin—Rattlesnake Mountain area (area 4, Fig. 1; Lovering, 1929; Dorf and Lochman, 1940; Stearns, 1974), the Pryor uplift (Shaw, 1954), the Owl Creek Mountains (area 5, Fig. 1), and the Bighorn Mountains (Durkee, 1953; Shaw, 1954; Richards, 1955). Individual units become slightly younger in age to the east,

Fig. 9. Slide movement, early 1968 (USBR Photo A459-600-229NA). View faces south on Big Bull Elk Creek (cf. Fig. 4). Little Bull Elk Creek enters on left. Vertical displacement on slide scarp ultimately exceeded 22 m. Compare Fig. 5, left edge of photograph.

a classic example of facies overlap and time transgression; but lithic and geotechnical similarity appears to be very consistent when viewed on a regional scale.

In the Bighorn Mountains, for example, the Cambrian system has presented more construction problems than any other geologic time unit. These construction problems, in the main, involve road construction on the main Bighorn Mountain highway crossings from Sheridan to Cody, Wyoming, on U.S. Route 14A through Five-Springs Canyon (area 2, Fig. 1) and U.S. Route 14 through Shell Canyon (area 3, Fig. 1). Landslides, it may be presumed, affected even the early wagon road routes of the Indian War Period which connected Fort McKenzie (north of Sheridan) to settlers in the Bighorn Basin. The enlarged auto roads constructed prior to World War II were narrow and involved small cuts and fills, but even so, landslides were common. The principal difficulty appears to reside in the extensive regional development of prehistoric landslide terrain (see, e.g., Voight, 1974, fig. 7,

Fig. 10. Continuation of movement on peripheral crack shown in Fig. 8; 1972 (USBR Photo A459-600-273NA). Scarp height about 70 m.

p. 73); these areas were marginally unstable and hence easily triggered by construction activity.

An example of a slide induced by construction has been described by the Wyoming Highway Department (1973, pp. 21—24; cf. Voight, 1974, p. 7). Preliminary studies by the Highway Department suggested that a backslope failure condition could be created by a proposed cut section on Route 14A. Elimination of the cut section by grade raise or a change in alignment did not seem possible because of additional problems associated with the crossing of a then-active slide in an adjoining proposed-fill section.

An apparently satisfactory solution to the problem involved a cut-off of the water flowing at the rubble (carbonate block colluvium)-shale contact. The method selected involved excavation of a portion of the cut section to a point where the potential slide plane could be intercepted by a 2.5-m drainage ditch. In order to maintain a relatively constant grade for a cut-off drain, however, deeper excavation was required in the central portion; this was accomplished by means of a 1 : 1 slope trench.

During installation of the underdrain a backslope failure occurred which was apparently caused by intersecting the slide plane with the underdrain ditch. The slide involved roughly 10^6 m^3, with the rear scarp located about 200 m from the toe. When the failure occurred, 150 m of the proposed 360 m of underdrain had been installed.

This is not an isolated example. In 1965 alone, in some 5.6 km of Route 14 under construction in Shell Canyon, eight major reactivated slides developed in historically unstable terrain; two of these were in the 10^6-m^3 class. The area of ancient slide activity is, however, far more extensive. The roadway at this location lies near the base of the Cambrian shale section (within the "middle shale" unit of the Flathead Formation; Durkee, 1953); thin sandstone and fractured limestone interbeds are common within the shale unit. Regional dip is about 4° into the roadway, and slide movements were associated with spring activity, internal drainage, and other evidence indicating the mechanical effect of water on the slide events. One of the slides was stabilized by surface drainage and modification and highway relocation; at the other location, a 30-cm well was drilled, with a pump set at 60 m. Initial water yield was 1500 m^3/day; this has tapered to about 30 m^3/day. Observation wells and springs have dried up, and the slide mass was sufficiently stabilized by 1974 to allow road reconstruction (Holland and Everitt, 1974, pp. 40—44). The slide mass was instrumented with two inclinometers in 1973, but by that date movements had essentially stabilized and no data on slide displacements versus depth is available. The headscarp is about 18 m high.

Slide planes as indicated in cross-sections by Bauer and Hale (1968) occur in large part parallel to shale-sandstone interbeds and along the colluvium-bedrock contacts. Some rotational movement has been observed, particularly in headscarps, and in a few slides completely curved slip surfaces seem to have been documented (Wyoming Highway Department, 1973, pp. 25—27; Voight, 1974, p. 72). Extensive drill hole investigations by the Wyoming Highway Department and the Bureau of Public Roads were conducted in order to design the drainage facilities. Results include groundwater elevation maps, cross-sections, and test hole logs. Heavy artesian flows were reported from permeable sandstones (Bauer and Hale, 1968, p. 7, cross-sections; cf. Hale and Bauer, 1969, p. 9), and local "permeability barriers" were recognized where facies changes occurred, e.g., from permeable sandstone to far less permeable sandy limestone. Although various types of "triggers" associated with construction suffice to apparently "cause" a landslide event, the greatest single factor here appears to reside in groundwater effects. Remedial measures thus have focused upon elimination or, in the least, reduction of flow of water in permeable interbeds in the Cambrian section.

Somewhat similar conditions have been reported for the Clark's Fork Canyon—Sunlight Basin region (area 4, Fig. 1), where most of the slides originate in Gros Ventre Shale or in colluvial slope debris covering this formation. Most slides are apparently caused by strength reduction due to groundwater effects and to roadway excavation, and many occur adjacent to areas of normal faulting where the shale section has been weakened. Slide areas and suggested remedial measures are discussed by Sherman (1974, pp. 88– 89, 117; Wyoming Highway Department, 1972, pp. 114–119, 138); drill hole investigations have been locally conducted, and more information will be made available in the near future as road construction efforts proceed.

Farther west, landslides are prominent features along the north side of Soda Butte Creek; they are mostly slumps and are especially prevalent throughout the Cambrian section above the Flathead Sandstone. In one large slump near Cooke City, stratigraphic relations are preserved (although the rocks are much fractured), and several mines were developed along a major vein system within the slide mass (Elliott, 1974, pp. 104–105). The toe of this landslide interfingers with glacial till, suggesting that it was a late glacial event in waning stages of the Pleistocene. Many of the Bighorn Mountain slide areas, too, probably represent glacial or periglacial phenomena, although the complete history may be quite complex and numerous periods of movement may be involved (cf. Pierce, 1968).

MATERIAL PROPERTIES OF THE CAMBRIAN SHALES

As has been pointed out, the Cambrian shales have been a persistent source of landslide problems at other locations marginal to the Bighorn Basin; the engineering characteristics of the Cambrian shales have therefore not wholly escaped attention. Additional studies are currently underway, and it seems likely that fairly refined geotechnical information may ultimately become available; the main point to be made here is that such information appears to reflect directly on the problem of the Bighorn Reservoir slides.

Considerable variation exists in grain size distribution of the Cambrian section; the "percent finer" than 200 mesh reported for one section of the Shell Canyon Road, for example, varied from 25 to 95%. Values in excess of 70% are common. Variation from silty clay to clay-rich gravel occur, particularly in weathered and remobilized shale colluvium mixed with carbonate fragments. The amount of reworking seems reflected in the bulk density, which is characterized by a rather large range (Table I).

An X-ray diffraction mineralogical analysis was conducted by Pollok (1974), based primarily on samples of Gros Ventre Shale from the Wind River Canyon (area 5, Fig. 1). The whole rock sample is rich in feldspar, quartz, and clay minerals. The clay fraction is dominated by well-crystallized chlorite and disordered illite; there is some indication that the illite is a mixed layer variety, with minor interstratified smectite. There is probably less than 5 or 10% expandable clay in the -1-μm fraction. Kaolinite, vermiculite, and palygorskite are absent; the interstratified illite-smectite could be

TABLE I

Typical geotechnical data for Cambrian shale sequence (data supplied by W. Sherman)

Liquid limit (%)	Plastic index (%)	Wet density (Mg/m^3)	Moisture (wt.%)	Unit cohesion * (kN/m^2)	Apparent ϕ * (degrees)	AASHO classification
Bighorn Mountain area **						
43	21	1.96	17	22	—	A-7-6 (10)
43	20	2.15	20	64	—	A-7-6 (11)
37	19	2.07	21	52	17	A-6 (12)
31	11	2.26	14	31	27	A-6 (6)
Clark's Fork area ***						
—	—	2.39	8	75	—	—
37	16	1.91	27	50	10	A-6 (4)
39	17	2.12	12	185	—	A-6 (4)
34	13	2.08	14	30	—	A-6 (3)

  * All strength tests are "undrained".
 ** Area *3*, Fig. 1; Route 14A construction sites.
*** Area *4*, Fig. 1; west slope Dead Indian Hill.

identified as glauconite if chemical analyses demonstrated sufficient iron. Thickness of clay platelets was estimated from the width of the basal reflection (chlorite, 16.4 nm; illite, 6.8 nm); the illite seems particularly fine-grained, and exhibits large surface area. Clay aggregate orientation has not been studied.

Plasticity is exhibited by most samples, with the range of values reasonably small. Liquid limits (LL) typically occur in the range 30—43%, with plasticity indices (PI) of 10—20% (see, e.g., Table I). One sample from Shell Canyon gave LL = 53%, PI = 31%, but these values seem unusual. The natural moisture content in the majority of samples was about equal to, or less than, the plastic limit; liquidity indices are therefore, on average, negative.

Strength tests conducted to date have primarily involved unconfined compression or undrained triaxial tests on shelby-tube samples. Scatter is large, as might be expected, due to structural factors within individual samples. Experience has shown that stability analyses based on these strength results are not reliable; this seems in part due to the fact that test specimens have fewer defects and altered zones proportionally than the mass of in-situ material; more important, however, may be the unsuitability of undrained (total stress) analyses to field conditions. At present the friction angles reported for undrained tests are ignored by the Wyoming Highway Department in slide analyses; even so, in the future more attention will have to be given to effective stress forms of analysis, and consequently, to measurement of peak and residual Coulomb parameters and to field measurements or estimates of pore pressure. In the interim, residual friction can only be estimated from plasticity indices; accordingly, the residual friction coefficient is estimated to be about 0.4—0.5 (Voight, 1974, p. 70). This value represents a minimum; roughness of the slip surface in the field would make the apparent field value somewhat more.

Time-dependent (creep) behavior has not yet been investigated, but a study is now underway by B. Lester at Pennsylvania State University in conjunction with studies of a prehistoric "valley anticline" in Cambrian units in the Clark's Fork Canyon area.

DISCUSSION

The stability of previously mobilized Cambrian strata in the Bighorn Canyon was marginal at the time of construction of Yellowtail Dam; little stress difference was apparently required to cause reactivation of slide masses. In the reservoir, the large slides were associated with approximately 40 m of rise in reservoir pool. The association of major slide activity at Bighorn Reservoir with initial and subsequent filling periods is much in line with observations at Grand Coulee, the Panama Canal, Kaunertal, and Vaiont, and with the general interpretations of slope stability for rising reservoirs given by Lane (1967; cf. Breth, 1967).

A detailed assessment of the effects of rainfall and associated cleft-water pressures, superimposed upon the buoyant slide mass submerged under the reservoir pool, cannot be made at the time of writing. Detailed rainfall data for the reservoir area has not yet been made available; in any case, observations of the slide events by aerial surveillance are probably too widely spaced in time to permit close correlation with precipitation data. The first recognition of slide reactivation was made in February, 1966, certainly long before infiltration by snow melt would have been significant. It seems possible, however, that initial displacements could have occurred in late fall, 1965, when infiltration could have been significant, and remained undetected until February.

In accord with highway maintenance experience in similar stratigraphic units 20—40 km due south of Bighorn Canyon, slide stability seems controlled in large part by ground water conditions. Sliding zones probably involve arcuate surfaces behind "oversteepened" slopes, exposed as steep scarps and involving, in part, rotational slumping; these surfaces flatten with depth and often merge with bedding surfaces in interstratified zones of clay shale and sandstone (Fig. 4). The sandstone layers may be presumed to rapidly transmit artesian water pressures, by correlation with borehole observations at Shell Canyon.

Surfaces of detachment are further presumed to occur near the contact of thin sandstone beds within superjacent clay-rich layers. The most difficult factor to examine is the shear resistance across bedding; for reactivation of old slides, however, one may conservatively assume that the residual strength is mobilized over large areas of the slip surface. Very little toe resistance exists in the reservoir area because of previous downcutting of the Bighorn River; most large-scale sliding occurred in association with an apparent bedding dip of only a few degrees. This configuration of failure surfaces remains, of course, conjectural in the absence of subsurface investigation.

Fluid uplift must play an essential mechanical role. Artesian pressure in the sandstone beds and cleft-water pressure in the vicinity of the rear scarp may also have been significant, but it is very doubtful that these phenomena would have been more critical in 1966 than in preceding years. The critical factor seems to have been simply the buoyancy effect.

It can be simply demonstrated that, for the case of a planar block slide on a cohesionless surface, and excluding artesian effects, the buoyancy effect is nil (Jaeger, 1972, p. 357). For a partially curved slip surface configuration, however, buoyancy can be significant, especially if the potential slide mass is marginally stable. By way of example, we consider here a simple two-dimensional composite block model (cf. Jaeger, 1972, p. 358) not unlike the Bighorn Reservoir slides in regard to essential geometric and geotechnical features (Fig. 11).

Two rock mass blocks are considered; W_2 rests on a portion of the slide surface parallel to a bedding plane, assumed horizontal, whereas the inclined portion is inclined at an angle $\beta \cong 45°$. Because of a history of previous displacement the residual Coulomb condition is assumed to have been achieved; all along the slip surface, i.e., cohesion $c_r \cong 0$, friction coefficient $\mu'_r \cong 0.5$, as discussed previously. For simplicity, the uplift force U is assumed to effect only block W_2; the small submerged portion of W_1 is thus neglected.

The factor of safety F_s is given by:

$$F_s = \frac{W_1 \cos \beta \, \mu'_r + (W_2 - U) \, \mu'_r}{W_1 \sin \beta}$$

We assume $W \cong W_1 \cong W_2$; for the initial condition prior to reservoir con-

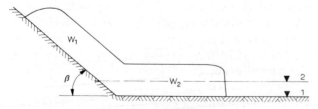

Fig. 11. Sketch of hypothetical slide mass subjected to two reservoir level conditions.

struction (case 1), $U = 0$, hence:

$$F_s \cong \frac{W(0.7)(0.5) + W(0.5)}{W(0.7)}$$

$$F_s \cong 0.85/0.7 \cong 1.2 \tag{case 1}$$

which represents a case of marginal stability in view of the assumptions involved. Next, consider a rise in reservoir pool level (case 2) such that $U = 0.3$ W:

$$F_s = \frac{W(0.7)(0.5) + W(0.7)(0.5)}{W(0.7)}$$

$$F_s = 0.7/0.7 = 1.0 \tag{case 2}$$

Clearly, analysis could be improved by more refined geometric information; for example, if W_1 is considered to be somewhat larger than W_2, the initial safety factor is decreased (i.e., for $W_1 \cong 1.5\ W_2$, $F_s \cong 1.0$ for case 1), and only a seemingly trivial rise in reservoir pool would be sufficient to trigger slope failure. In the absence of borehole information, however, a simple example of this kind appears to be sufficiently informative. The Bighorn Reservoir slides provide an excellent example of slide reactivation associated with a regionally widespread, notoriously weak geologic unit. Almost any disturbance of the existing condition would have sufficed to create anew a condition of instability; the rising reservoir pool behind Yellowtail Dam admirably served this purpose.

REFERENCES

Bauer, E.J. and Hale, J., 1968. Geological report of landslide and subsequent groundwater conditions, stations 770—790, Shell Canyon Road, Bighorn County, Wyo. *Internal Rep., Wyo. Highw. Dep.*, Project ARS-1403 (see also Supplementary Reports dated January 1969 and August 1969).

Breth, H., 1967. The dynamics of a landslide produced by filling a reservoir. *Proc. 9th Int. Congr. on Large Dams, Istanbul*, Q. 32: 37—45.

Darton, N.H., 1906. Geology of the Bald Mountain and Dayton Quadrangles, Montana. *U.S. Geol. Surv. Folio*, No. 141.

Dorf, E. and Lochman, C., 1940. Upper Cambrian formations in southern Montana. *Geol. Soc. Am. Bull.*, 51: 541—546.

Dupree, H.K. and Taucher, G.J., 1974. Bighorn Reservoir landslides, south-central Montana. In: B. Voight and M.A. Voight (Editors), *Rock Mechanics — The American Northwest. 3rd Congr. Exped. Guide, Int. Soc. Rock Mech., Spec. Publ.* Experiment Station, College of Earth and Mineral Sciences, University Park, Pa., pp. 59—63.

Durkee, E.F., 1953. *Cambrian Stratigraphy and Paleontology of the East Flank of the Bighorn Mountains, Johnson and Sheridan Counties, Wyo.* Unpublished Thesis, Univ. of Wyoming, Laramie, Wyo.

Elliott, J.E., 1974. The geology of the Cooke City area, Montana and Wyoming. In: B. Voight and M.A. Voight (Editors), *Rock Mechanics — The American Northwest. 3rd Congr. Exped. Guide, Int. Soc. Rock Mech., Spec. Publ.* Experiment Station, College of Earth and Mineral Sciences, University Park, Pa., pp. 102—107.

Hale, J. and Bauer, E.J., 1969. Engineering geology report on preliminary investigation of the "4000 foot section", stations 795—836, Shell Canyon Road, Bighorn County, Wyo. *Internal Rep. Wyo. Highw. Dep.*, Project ARS-1403.

Holland, T.W. and Everitt, M.C., 1974. The Shell Canyon landslides: a geological engineering case history. In: B. Voight and M.A. Voight (Editors), *Rock Mechanics — The American Northwest. 3rd Congr. Exped. Guide, Int. Soc. Rock Mech., Spec. Publ.* Experiment Station, College of Earth and Mineral Sciences, University Park, Pa., pp. 37—44.

Hoppin, R.A. and Jennings, T.V., 1971. Cenozoic tectonic elements, Bighorn Mountain region, Wyo.-Mont. *Wyo. Geol. Assoc., Guideb., 23rd Annu. Field Conf.*, pp. 39—47.

Jaeger, C., 1972. *Rock Mechanics and Engineering.* University of Cambridge Press, Cambridge, 415 pp.

Lane, K.S., 1967. Stability of reservoir slopes. In: *Failure and Breakage of Rock. Proc. 8th Symp. on Rock Mechanics, 1966.* American Institute of Mining, Metallurgy and Petroleum Engineering, New York, N.Y., pp. 321—336.

Love, J.D. and Reed, J.C., 1971. *Creation of the Teton Landscape.* Grand Teton Natural History Association, Moose, 120 pp.

Lovering, T.S., 1929. The New World or Cooke City Mining District, Park County, Wyo. *U.S. Geol. Surv. Bull.*, 811, 87 pp.

Mackin, J.H., 1937. Erosional history of the Bighorn Basin. *Geol. Soc. Am. Bull.*, 48: 813—894.

Pierce, W.G., 1968. The Carter Mountain landslide area, northwest Wyoming. *U.S. Geol. Surv. Prof. Paper*, 600-D: 235—241.

Pollok, R., 1974. X-ray diffraction analysis of the Gros Ventre Shale. In: B. Voight and M.A. Voight (Editors), *Rock Mechanics — The American Northwest, 3rd Congr. Exped. Guide, Int. Soc. Rock Mech., Spec. Publ.* Experiment Station, College of Earth and Mineral Sciences, University Park, Pa., pp. 45—47.

Prucha, J.J., Graham, J.A. and Nickelsen, R.P., 1965. Basement-controlled deformation in Wyoming Province of Rocky Mountain foreland. *Bull. Am. Assoc. Pet. Geol.*, 49: 966—992.

Richards, P.W., 1955. Geology of the Bighorn Canyon—Hardin area, Montana and Wyoming. *U.S. Geol. Surv. Bull.*, 1026, 93 pp.

Rubey, W.W. and Hubbert, M.K., 1959. Overthrust belt of western Wyoming in light of fluid pressure hypothesis, 2. Role of fluid pressure in mechanics of overthrust faulting. *Geol. Soc. Am. Bull.*, 70: 167—206.

Samuelson, A.C., 1974. Introduction to the geology of the Bighorn Basin and adjacent areas, Wyo. and Mont. In: B. Voight and M.A. Voight (Editors), *Rock Mechanics — The American Northwest. 3rd Congr. Exped. Guide, Int. Soc. Rock Mech., Spec. Publ.* Experiment Station, College of Earth and Mineral Sciences, University Park, Pa., pp. 11—17.

Shaw, A.B., 1954. The Cambrian and Ordovician of the Pryor Mountains, Mont., and northern Bighorn Mountains, Wyo. *Billings Geol. Soc. 5th Annu. Field Conf., Guideb.*, 184 pp.

Sherman, W., 1974. Environmental impact at Sunlight Basin Highway. In: B. Voight and M.A. Voight (Editors), *Rock Mechanics — The American Northwest. 3rd Congr. Exped. Guide, Int. Soc. Rock Mech., Spec. Publ.* Experiment Station, College of Earth and Mineral Sciences, University Park, Pa., pp. 86—91.

Stearns, D.W., Logan, J.M. and Friedman, M., 1974. Structure of Rattlesnake Mountain and related rock mechanics investigations. In: B. Voight and M.A. Voight (Editors),

Rock Mechanics — The American Northwest. 3rd Congr. Exped. Guide, Int. Soc. Rock. Mech., Spec. Publ. Experiment Station, College of Earth and Mineral Sciences, University Park, Pa., pp. 18—25.

Voight, B., 1974. Roadlog: Worland to Wapiti. In: B. Voight and M.A. Voight (Editors), *Rock Mechanics — The American Northwest. 3rd Congr. Exped. Guide, Int. Soc. Rock Mech., Spec. Publ.* Experiment Station, College of Earth and Mineral Sciences, University Park, Pa., pp. 68—79.

Wyoming Highway Department, 1972. *Draft Environmental Impact Statement. Wyo. Highway Project FLH 18-4, Clark's Fork Canyon Road, Park County, Wyo.* Department of Transportation, Federal Highway Administration, 111 pp. (EIS-WY-72-5440-D).

Wyoming Highway Department, 1973. Geology and its relationship to highway construction, Bighorn Mountains, northern Wyoming, *24th Annu. Highw. Geol. Symp. Guideb.*, 32 pp.

Jørstad, F.A., 1956. Fjellskredet ved Tjelle; et 200 års minne. *Naturen*, 80: 323—333.

Kaldhol, H. and Kolderup, N.H., 1937. Skredet i Tafjord 7. April 1934. *Bergens Mus. Årb., 1936, Naturvitensk. Rekke*, 2, No. 11, 15 pp., 4 figs., map.

Kiersch, G.A., 1964. Vaiont reservoir disaster. *Civ. Eng.*, 3: 32—39.

Miller, D.J., 1954. Cataclysmic flood waves in Lituya Bay, Alaska. *Geol. Soc. Am. Bull.*, 65: 1346 (abstract).

Miller, D.J., 1960. Giant waves in Lituya Bay, Alaska. *U.S. Geol. Surv. Prof. Paper*, 354-C: 51—86.

Ogawa, T., 1924. Notes on the volcanic and seismic phenomena in the volcanic district of Shimabara, with a report on the earthquake of December 8th, 1922. *Mem. Coll. Sci., Kyoto Imp. Univ., Ser. B*, 1: 201—254.

Peterson, U., 1965. Regional geology and major ore deposits of central Peru. *Econ. Geol.*, 60: 407—476.

Chapter 8

WEDGE ROCKSLIDES, LIBBY DAM AND LAKE KOOCANUSA, MONTANA

BARRY VOIGHT

ABSTRACT

Wedge rockslides in Precambrian argillite in the left abutment area of Libby Dam were active during prehistoric time and during recent construction. Recent slides include a 6000-m³ wedge, triggered in 1967 by pre-split excavation, and a 33,000-m³ wedge which moved from a 33-m slope following a period of heavy precipitation. The movement history of the latter was recorded by an extensometer from June 1967 to failure in January 1971. The largest of several prehistoric wedges involved about 2×10^6 m³. This slide was emplaced sometime after 10,000 years B.P.; motion analyses suggest a peak emplacement velocity of 22 ± 7 m/s. No satisfactory limiting equilibrium analyses of the prehistoric slides were conducted because of complex geometry. However, suitable back-analyses of several of the smaller wedge slides were conducted, leading to inferred field strengths of approximately $c' = 0$, $\phi' = 35°$ or more. The field strengths are substantially greater than residual strengths as measured in extensive laboratory tests on small samples, but the differences seem explicable in terms of observed roughness patterns of the slide planes.

INTRODUCTION

Libby Dam, a 128-m high multiple-purpose concrete gravity structure in northwestern Montana, impounds the Kootenai River to form Lake Koocanusa (from Kootenai, Canada, and U.S.A.). The dam was constructed in the period 1966—1973 by the Corps of Engineers as part of the Columbia River Basin project (Fig. 1). The lower reach of the reservoir, for about 1 km upstream of the dam, is flanked by moderately steep rock slopes. The left (east) bank slopes particularly are characterized by intersecting discontinuity systems, which form rock wedges plunging as steeply as 33° toward the reservoir. On 31 January 1971, during construction of the dam, the major portion of a rock wedge with volume of about 33,000 m³ slid from an

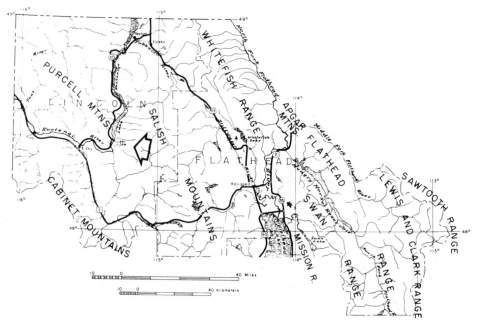

Fig. 1. Map of northwest Montana showing Libby Dam and Lake Koocanusa and surrounding territory.

excavated slope above the left dam abutment (Fig. 2). This failure caused concern in regard to the stability of large areas in the left bank in the immediate vicinity of the dam (Fig. 3). Some potential for sliding had indeed been previously noted with recognition of a large prehistoric slide mass on the Kootenai valley floor; a small wedge slide had also occurred in left abutment excavations in 1967, and numerous small slides had occurred in nearby excavations for the Burlington Northern Railway. Much effort was therefore expended in thoroughly examining the possibilities of large rockslides and slide-induced waves. As a consequence of these studies a large buttress fill was provided at the base of the left bank (Fig. 3), drainage holes were drilled to recognized discontinuities at potential slide zones, and the slopes were monitored and locally reinforced.

The purpose of this paper is to review and summarize available geologic and geotechnical data from unpublished and published sources on the Libby Dam—Lake Koocanusa wedge rockslides, with most emphasis given to the prehistoric events and the slides influenced by construction excavation, rather than to the analysis of remedial support for existing slopes. Despite the great amount of work on this problem, mainly conducted or sponsored by the U.S. Army Corps of Engineers, few readily available reports or summaries have been published (cf. Banks and Strohm, 1974; Davidson and

Fig. 2. View across axis of Libby Dam construction site showing 1971 left abutment slide area. Removal of slide debris was complete by March 1971.

Whalin, 1974; R.W. Galster, p. 225, in Johns, 1974b; Hamel, 1974 [1]) and none completely summarize the issues.

GENERAL SETTING

Topography and surficial geology. The drainage area of the Kootenai lies in the northern Rocky Mountains physiographic province, an uplifted, maturely dissected, glaciated region. At the Libby Dam site, the river flows to the

[1] A summary of this report now appears in the *Proceedings, American Society of Civil Engineers Specialty Conference on Rock Engineering for Foundations and Slopes, August 15—18, 1976, Boulder, Colo.*

Fig. 3. Left bank area in vicinity of nearly completed Libby Dam. Buttress in place upstream from left abutment. Area of potential slides extends from 1971 left abutment slide scar, at right, to 925 and 930 slide scars at left. For identification of slide scars and potential "rib" slides, see Fig. 4.

south in a rock-rimmed valley about 1 km wide. In the vicinity of the dam the left bank is rugged, with broken cliffs, bare rock and talus rising from the original river level at elevation 2130 [2] to a ridgetop at elevation 3100—3800 (Figs. 3, 4; Davidson and Whalin, 1974, p. 10). A rock "rib" approximately normal to the river forms the left abutment; the main dam abuts a low rock knoll on the right abutment.

Overburden on the left side of the valley consists of thin talus deposits; the right side has up to 50 m of overburden that is predominantly glacio-fluvial silty sand or gravel, and minor silt or clay beds (Davidson and Whalin, 1974, p. 10). Most overburden was deposited during or after the last of several periods of glaciation. At one stage in the Pleistocene history, the region was inundated by glacial Lake Kootenai to (maximum) elevation 2500 (Hamel, 1974, p. 7).

Bedrock geology. Bedrock in the vicinity of the dam are Belt (Precambrian) strata of the greenschist metamorphic grade, deposited as sediments 1450—850 million years B.P. in a basin which occupied a large part of western Mon-

---

[2] Elevations are given in feet to correspond with standard topographic map units as indicated in Fig. 4.

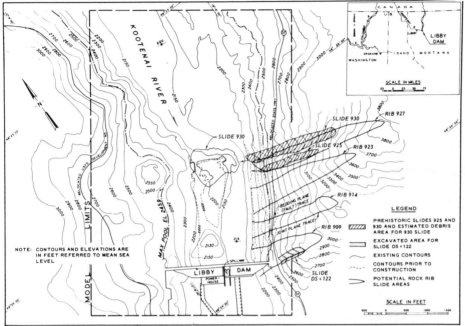

Fig. 4. Topographic map of Libby Dam and vicinity (after Davidson and Whalin, 1974). Major potential "rib" slide areas are indicated. Slides 930 and 925 denote major prehistoric slide scars; debris of slide 930 occupies a portion of the Kootenai River valley floor. Slide DS + 122 refers to the left abutment slide of January 1971. Model limits refer to hydraulic model studies (cf. Fig. 21) described by Davidson and Whalin (1974). Relocated highway routes as indicated.

tana and adjacent territory (Harrison, 1972). The Belt sequence can be subdivided into lower, middle, and upper units (Johns, 1974a; cf. Johns, 1970, 1962); according to Johns, Libby Dam and the lower part of Lake Koocanusa are in the middle unit, the Wallace Formation (Fig. 5). "Wallace rocks are heterogeneous and multicolored, exhibiting subtle and marked changes in mineral composition and texture. Lithologic types include gray and green calcareous argillite and siltite, and gray to gray-black and blue-gray magnesian carbonic rocks. Locally, red argillite and quartzite are prominent in the upper part of the sequence" (Johns, 1974a, pp. 209—210).

A detailed geologic road log for the vicinity of Lake Koocanusa is given by Johns (1974b). In the vicinity of the dam the rocks are dominantly argillite (metamorphosed shale), with subordinate amounts of quartzitic and calcareous argillite and quartzite. Bedding is thin to massive, with strike 150—155° and dip 39—42°SW. Although absent at the damsite, igneous rocks are present nearby in the Belt Series (Boettcher, 1967).

The structural history is moderately complex. That some deformation originated during Precambrian and Paleozoic time can be inferred from such

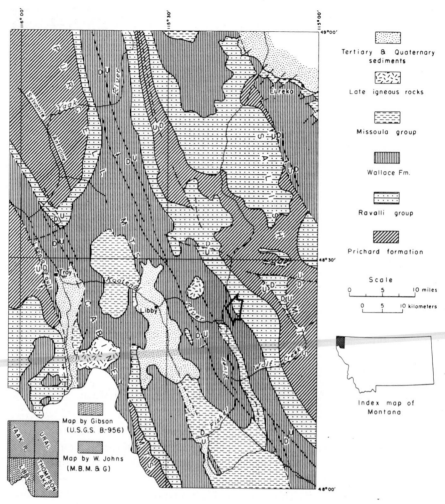

Fig. 5. Geologic map of part of Lincoln County, Montana (after Johns, 1962, 1974b). Libby Dam denoted by arrow.

extensive, partly fault-controlled features as the Rocky Mountain and Purcell Trenches and from the great displacement on several faults (Johns, 1974a). The origin of north- to northwest-trending folds are attributed to the Laramide (Tertiary) orogeny, as are most faults. Normal, reverse and thrust faults parallel traces of north- to northwest-trending fold axial planes; east-striking high-angle faults displace northwest-trending structures and are thus younger.

Following Algermissen (1969) the dam was placed in seismic zone 2 (Hamel, 1974, p. 8), involving a seismic design coefficient of 0.1 (0.1 g

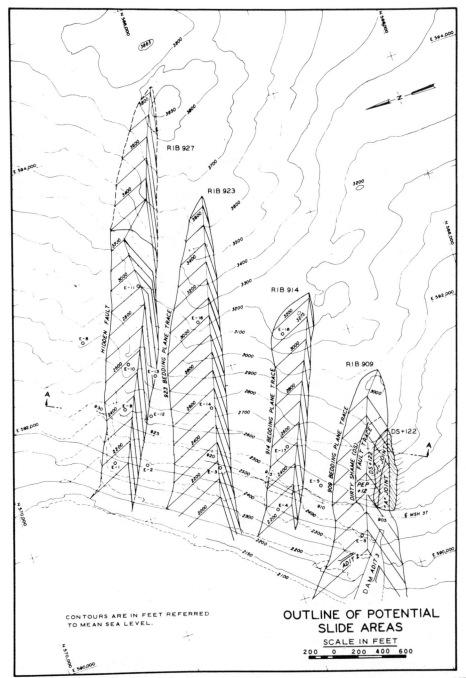

Fig. 6. Topographic and geologic details of left bank area of Libby Dam (after NPS, 1972b; Davidson and Whalin, 1974). Potential rib slide wedges are bounded by bedding plane faults and major joints. Refer to cross-section along line A-A. Core boring locations indicated by E-3, E-4, etc. Scar of prehistoric 930 slide indicated by trough approximately bisected by Hidden fault; 925 slide scar is located a short distance southward, in lower part of rib 927.

equivalent horizontal acceleration). Still, evidence of post-glacial fault movements has been noted nearby at the Flathead Tunnel (Galster, 1974) and
other localities (A.L. Boettcher, personal communication). A seismic station
was installed near the reservoir in 1970 to monitor possible seismic effects of
reservoir filling; some low-magnitude events have been observed but seismicity recorded to 1974 has been insignificant (Johns, 1974b, p. 226; Hamel,
1974, p. 8). The major discontinuities in the vicinity of the left abutment
and left bank are well-defined bedding planes; these have been referred to as
bedding plane faults inasmuch as they display evidence of slip. The bedding
plane faults have been named by reference to designated station numbers
along relocated Montana State Highway (MSH) 37 (Figs. 6 and 7). Several
prominent joint sets are present including an interpreted conjugate shear system of east-west- and northeast-trending joints, a north-south set, and relaxation joints in valley walls (Hamel, 1974, p. 6). Banks and Strohm (1974)
mention the following sets: (1) strike azimuth 10°, dip 70°SE; (2) strike
azimuth 100°, dip 70°NE. The joints are open to considerable depth, possibly due to post-glacial rebound and freeze-thaw mechanisms; their frequency is variable, but they persistently intersect the bedding plane fault system to form wedge-shaped "rock ribs". These ribs, which posed the stability
problem of major concern, have been identified by numbers corresponding
to MSH 37 stations (Figs. 4, 6, 7).

Climate. Mean annual precipitation is about 97 cm, most of which is deposited as snow during the winter. The annual temperature variation is from
below 0° to about 100°F. Freeze-thaw is the dominant rock weathering
mechanism (Hamel, 1974, p. 8).

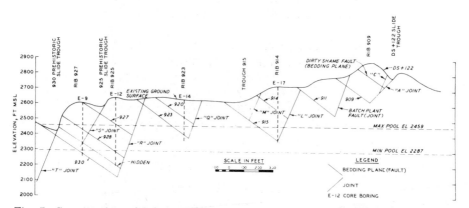

Fig. 7. Cross-section of left bank area of Libby Dam (after NPS, 1972b; Davidson and
Whalin, 1974). Location *A-A* of Fig. 6. Note profiles of potentially unstable rock ribs in
relation to reservoir pool elevations. Left abutment slide of 1971 shown by DS + 122
slide trough. Scars for 930 and 925 prehistoric slides as indicated.

DESCRIPTION OF WEDGE SLIDES

Prehistoric slides

Two large prehistoric "rock rib" slides have been identified. The larger is called the 930 slide (thus corresponding to the relatively planar 930 bedding plane fault on its upstream base; see Figs. 3, 4, 7, 8). The downstream basal portion of the wedge is composed of several east-west-trending fractures. Neither the geometry nor the failure mechanism of the 930 slide is reliably known.

This event was 10,000 years B.P. or younger, for slide debris overlies Kootenai River alluvium on the west valley floor (Figs. 4, 9; cf. Hamel, 1974, fig. 2). Boring data indicated the base of the slide debris to be at elevation 2075. Contours of the debris and a subsequent seismic survey were used to estimate a present volume of 2.4×10^6 m^3 (NPS, 1971a, b; Banks et al., 1972, appendix C); using a 25% swell factor, the in-place slide volume was about 1.9×10^6 m^3 (NPS, 1971a, b) reaching from elevation 2150 to 3370. According to one reconstruction (NPS, 1972b, plate D-2; Hamel, 1974, fig. 3) the lower part of the slide trough turned upstream due to changes in joint attitude (see Table I for summary of geometric data). Accordingly, so-called "kick-out" (upstream-directed release) of the toe of the slide mass was postulated (NPS, 1972b) in order to render the movement kinematically feasible. Without such a mechanism the 930 rock mass would have remained stable.

The smaller 925 slide (Figs. 4, 7, 8, 9) was bounded by the 925 bedding plane fault and by several east-west-trending joints and/or faults. The toe of the 925 slide presumably daylighted in the valley wall about 60 m above

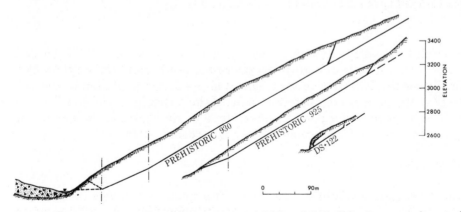

Fig. 8. Reconstructed longitudinal profiles of 930 and 925 slides in comparison to left abutment slide of 1971 (on DS + 122 bedding plane fault). Light dot-dash lines indicate bends in slide axis (after NPS, 1972b).

Fig. 9. Prehistoric 930 and 925 slide scars on left bank, and adjacent 927 and 923 ribs. The 930 slide mass is shown rising above the Kootenai River valley floor (cf. Figs. 4, 8). Emplacement of left bank buttress fill has begun (cf. Fig. 3).

river level; the in-place slide mass had an estimated volume of $2.7 \times 10^5\,\text{m}^3$. The age of the slide is unknown. Its debris has not been found on the valley floor. Either the 925 slide debris is mixed in the debris attributed to the 930 slide (Fig. 9), or it is completely absent; if the latter, the 925 debris could have been removed by a Pleistocene ice lobe or river erosion, in which case suggesting a possible Pleistocene or older age (Hamel, 1974, p. 11).

1967 left abutment slide

The 1967 slide involved displacement of a small rock wedge during highway excavation in the left abutment area of Libby Dam, about 15 m north of the dam axis (Fig. 10). The slide, which occurred because the wedge was undercut during pre-split excavations on a 76° slope, was studied and described by Ward and Galster (1969). Displacement occurred in a few seconds, about 10 minutes after removal of toe support (muck); no water was observed along bounding discontinuities (communication of J.C. Richards, Project Geologist, cited in Hamel, 1974, pp. 15—16). The wedge slid about 3 m and came to rest at the base of the excavation.

The wedge was bounded on its upstream side by the "Dirty Shame" + 122 (DS + 122) bedding plane fault and on its downstream side by two east-west-trending joints. The trough thus defined steepened and widened in a downstream direction (Fig. 10). Geometric data on the slide surfaces and slopes are summarized in Table I; slightly different data have been reported by Ward and Galster (1969) and Hamel (1974, p. 172). The slide wedge had a height of approximately 18 m in the excavated face and an estimated volume of about 6000 m³. A "pocket of soil" behind the wedge suggested that some previous movement of the wedge had occurred (Ward and Galster, 1969; Hamel, 1974).

Seven wire-type multiple position borehole extensometers had been installed in the left abutment area prior to the 1967 excavation. Two of these (L-6, L-7) were installed in the slope, one on each side of the 1967

Fig. 10. Sketch of left abutment highway cut by A.S. Cary, 1969. Scar of 1967 left abutment slide indicated, bounded by DS + 122 bedding plane fault (on left) and two east-west-trending joints. Old Notch 1 located 45 m downstream (to right) of dam axis (ON-1), and Old Notch 2 at right boundary of sketch (ON-2). Extensometer L-7 located to right of 1967 scar. Approximate location of A joint noted.

TABLE I

Data summary for wedge slides (after Hamel, 1974, pp. 172—178; Ward and Galster, 1969; Banks and Strohm, 1974)

Slide	Bedding plane	Joint	Tension crack	Slope face	Slope upland	Face height (m)
Prehistoric 930	N25°W/40°SW	*downslope:* <elev. 2400, N70°W/60°NE *upslope:* >elev. 2400, N82°W/68°NE		N36°E/76°NW	N30°E/40°NW	
1967 left abutment	N24.5°W/42°SW (Hamel) N25°W/42°SW (NPS)	*downslope:* N74.5°E/72°NW (Hamel) N75°E/73°NW (Ward and Galster) *upslope:* N81°W/66°NE (Hamel) N75°W/65°NE (Ward and Galster)				18
1971 left abutment	N26.5°W/41.5°SW (Hamel) N25°W/42°SW (NPS)	N76°E/50°NW (Hamel, NPS)	N09°E/90° located ca. 75 m below crest	N20°E/76°NW	N05°E/29°NW	30—34
Old Notch railroad cut (Station 928)	N25.5°W/40°SW	N73°E/74°NW		N41°E/61°NW	N43.5°E/48°NW	9—11
Old Notch 1 left abutment	N23°W/41°SW	N41.5°E/41°NW		N15°E/64°NW	N15°E/30°NW	15
Old Notch 2	N26.5°W/42°SW	N84°E/73°N		N05°E/76°W	N03°E/29°W	9

slide notch (Fig. 10). These extensometers showed maximum expansion of 10 and 20 mm respectively, during or immediately after excavation.

1971 left abutment slide

In their 1969 report, Ward and Galster discussed a potentially unstable rock wedge adjacent to the 1967 slide trough, bounded by the DS + 122 bedding plane fault and a northwest-trending joint (later designated the B joint). These discontinuities intersected in the slope face, about 3 m above road level, defining a wedge with estimated volume of 50,000 m³. A reinforcement system involving dowels and rock bolts was recommended; the system was not installed (Hamel, 1974, p. 16). A year and a half later a wedge slid out of the cut slope, at approximately 6:04 a.m. (MST) on Sunday, 31 January 1971 (Figs. 11, 12; cf. Fig. 8). This wedge was bounded on its upstream side by the smooth DS + 122 fault (see Table I), on its downstream side by the relatively rough but planar east-northeast A joint, and at its rear by vertical, north-south-trending tension cracks. The slide mass included much of the "DS + 122 and B joint" wedge referred to above; estimates of its volume range from 30,000 to 70,000 m³ (Hamel, 1974, pp. 17, 173—175)[3]. Fortunately no personnel were in the area. Extensive damage was incurred by the contractor's refrigeration plant, electrical substation, and several small sheds and vehicles. Impact of slide debris was recorded at the Libby Seismograph Station (located 8 km north of the dam, on the right bank; Johns, 1974b, p. 226), thus establishing the time of the event. An initial "large" seismic event, which had a duration of 17 seconds, is thought to have resulted from impact on the road by large rock blocks at the front of the slide mass (Hamel, 1974, p. 20). Smaller events, presumably corresponding to impact of smaller blocks, occurred for an additional 24 seconds.

Movement history of the slide wedge was recorded by extensometer L-7 from June 1967 to failure (Fig. 13). The following account is summarized from Hamel (1974, pp. 17—18). The extensometer head moved outward about 21 mm during excavation in June and July 1967. Additional movement of 1.8 mm of a creep-like nature, distributed over the rock mass, occurred from August 1967 to May 1969. Over the period 23—26 May, slip causing 8.6 mm of outward movement occurred on the DS + 122 bedding plane fault. This sudden advance was followed by a quiet period of nine months, followed by a second period of creep over a seven-month interval in which 1.8 mm was measured at the instrument head. About 1.3 mm of this displacement was due to time-dependent slip on DS + 122. On 30 October 1970, 9.4 mm of outward movement occurred due to slip on DS + 122. This was followed by a quiet period of 2.5 months until 14—18

---

[3] The most accurate of these may be Hamel's "best estimate" of actual slide volume, viz. 33,000 m³.

Fig. 11. Left abutment slide of 31 January 1971 (NPS photo 5478, 2 February 1971). Portion of wedge with estimated volume of 33,000 m³ moved on DS + 122 and A joint; duration of slide event about 41 seconds (cf. Figs. 2, 10, 12). Note Old Notch 1 at right boundary of 1971 slide; Old Notch 2 immediately to right of Old Notch 1.

January, when an additional increment of 1 mm was measured at the instrument head. Two weeks of negligible movement preceded failure of the outer portion of the wedge on 31 January. Total measured movement at the instrument head to the date of failure was 43 mm, about 23 mm of which probably resulted from slip on the DS + 122 fault.

An attempt has been made to correlate observed displacements with precipitation and climatic data (Hamel, 1974, pp. 18—19, 126—127):

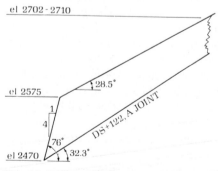

Fig. 12. Longitudinal profile of 1971 left abutment slide (after Hamel, 1974). Compare Figs. 10, 11.

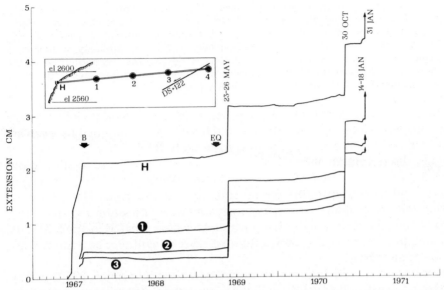

Fig. 13. Movement-time record for extensometer L-7 (after NPS, 1971a; Hamel, 1974). Extension at floating head (*H*) indicated. Anchor locations at *1*, *2*, *3*, and *4*, beyond DS + 122 bedding plane fault, indicated in inset figure. Termination date of blasting indicated by *B*, felt earthquake of 1 April 1969 noted by *EQ*. See text discussion.

"The January 31, 1971, slide was preceded by a period of heavy snow, then rising temperatures, snow melt, and rain . . . there was approximately 1 m [4] of snow on the ground prior to mid-January . . . The period from January 1 through 8 was very cold with temperatures well below freezing (32°F) and little precipitation. Temperatures rose to very slightly below freezing on January 9 to 10. Considerable snow fell from January 8 through 11 . . . Temperatures dropped to well below freezing on January 12 to 15 then rose significantly and generally remained at or above freezing for the second half of January which was unseasonably warm. January 30, the day before the slide, was particularly warm with temperatures in the mid-50's. Snow with water equivalent of about 2.5 cm fell on January 14 through 16. Then, as the temperature rose, the snow changed to rain which continued until January 17. About 1.2 cm of rain fell on January 19 and 20 and another 1 cm of rain and snow fell from January 23 to 26. There was negligible precipitation on January 27 through 29 but 1.2 cm of rain fell on January 30, the day before the slide.

The movement of 1 mm measured by all four sensors of extensometer L-7 from January 14 to 18 occurred at the beginning of the warming period and at the time of the first rain in mid-January. This movement may have been due to thermal expansion of the rock mass or to water pressures in the slope or to a combination of these factors. The absence of movements of similar magnitudes during the Springs of 1968 through 1970 suggests to the writer that the January 1971 move-

---

[4] Units have been converted to metric equivalents by the writer.

ment was due largely to water pressure rather than thermal expansion effects. Though the measured movement of 1 mm on the DS + 122 fault was very small, larger movements may have occurred at that time on other discontinuities, e.g., the A joint. These movements in mid-January 1971 were probably sufficient to open and or widen certain drainage outlets along geologic discontinuities in the slope. It is believed that some drainage and relief of water pressures in the slope then occurred so that failure was temporarily postponed".

Hamel then suggested that ice plugs may have formed on 22—23 January along natural drainage outlets; these could have been maintained by overcast conditions and low mean temperatures from 24 to 29 January. Even without ice plugs, discontinuity permeability may have been too small to allow rapid drainage during the period of heavy infiltration (from snow melt and rain) in January, and throughout this period stability was marginal. The 13 mm of rain that fell on 30 January probably increased water pressures to critical values to trigger the wedge slide. No piezometers were located in the left abutment area prior to the slide, hence fluid pressure magnitudes as a function of time are conjectural.

According to J.C. Richards (communication cited by Hamel, 1974, p. 20), a pocket of dense soil (argillite colluvium and glacial deposits) existed at the base of the tension crack area at the rear of the displaced wedge; a layer of gouge about 1 cm thick (<2.5 cm according to Banks and Strohm, 1974, p. 842) existed on the fault and joint surfaces under the wedge. Removal of remaining debris within the wedge bounded by DS + 122 and A joint was begun in February 1971 and completed in March (Figs. 2, 14, 15; cf. Figs. 10, 11).

Following this slide, extensive re-evaluation of the left bank stability problem was conducted. Four additional extensometers and 57 drain holes were installed in 1971. Several drain holes (including one along C joint and DS + 80 bedding plane fault) issue variable amounts with a time lag of 6—8 hours after rainfall (Hamel, 1974, p. 23). Prestressed rock anchors (tendons) were installed in 1971, with 81 extending from DS + 122 alone; 41 of these were anchored below the Dirty Shame fault (Fig. 16), while the remaining 40 were set below DS + 80.

Old notches and railway excavation slides

Two notches existed in the left abutment area prior to the 1967 excavations (Figs. 10, 11; cf. Hamel, 1974, figs. 7, 9, 10). Old Notch 1, located 45 m downstream of the dam axis, has a height of about 15 m in the slope face; its associated wedge is estimated at 4800 m³ (Hamel, 1974, pp. 13—14). Old Notch 2 has a height of about 9 m in the slope face and a volume of about 800 m³. Each notch is bounded by a relatively planar bedding plane and by a planar joint. The soil in Old Notch 1 may be of glacial origin (Hamel, 1974, p. 14). Another small notch occurs at the downstream edge of

Fig. 14. View toward dam axis down washed V-shaped trough of 1971 left abutment slide. Relatively smooth DS + 122 bedding plane fault on right; relatively rough A joint on left.

the prehistoric 930 slide trough, at station 928. T.E. Wood (oral communication to J.V. Hamel, 1974, p. 12) indicated that the notch was exposed in a cut slope along the former railway line, and thus it probably represents a slide wedge associated with railway construction.

Finally, five small wedge slides in similar rock occurred during slope excavation for the Burlington Northern Railway relocation, about 20 km southeast of Libby Dam. Each is bounded by a bedding plane and a joint; the height of cut slopes are about 10—20 m, and the volumes are in the range 1500—4000 m^3. The slides were due to undercutting by excavation; little water was apparently present (Hamel, 1974, pp. 12—13; Johns, 1974b, fig. 5).

Fig. 15. View up 1971 left abutment slide trough. Smooth DS + 122 on left; undulating, rough A joint on right.

Potential wedge slides

Several major potential slide masses were regarded with concern following the 1971 slide, considered in the context of the 1963 Vaiont disaster (NPS, 1972b, c; cf. summaries by Davidson and Whalin, 1974, pp. 12—14; Hamel, 1974, pp. 24—29). The upstream side of each mass is bounded by one or more bedding plane faults, and each downstream side is bounded (in some cases arbitrarily) by one or more joints and/or faults as given in Figs. 4, 6, 7; these discontinuities define, for each rib, a system of "nested" wedges; the various wedge intersections plunge valleyward at about 28—34°W. Unstained fresh joints in some boreholes indicated relatively recent movement of some wedges (say 10,000 years B.P.) and smeared manganese dentrite coatings on some fractures suggested recurrent movement. These movements could

Fig. 16. Tendon reinforcement on DS + 122 in 1971 left abutment slide trough.

reflect incipient stages of sliding and/or adjustments due to valley development.

Following extensive studies in 1971—1972, a system of 41 "horizontal" (+5°) drain holes were installed, with outlets of several below maximum pool level to prevent outlet freezing. Buttress rock fills were installed in 1973, involving about 1.8×10^6 m³. An instrumentation system, based on a proposal by Shannon and Wilson Inc., (1971), was installed in 1971 and 1972; these included sixteen rod-type borehole extensometers, four inclinometer casings, fourteen vibrating wire electrical piezometer installations, ten pneumatic multiple-tip piezometer installations, and eight installations for rock noise detection (*The Indicator*, 1974, pp. 2—5; Hamel, 1974, pp. 30—38). Initial readings had been obtained from most instruments by 1972 and summaries are included in NPS (1972a, b; 1973). Difficulties with some aspects of the instrumentation systems are reviewed by Hamel (1974, p. 36).

Meaningful rock noise data were not obtained and thus experimentation with the system was discontinued in 1973; most electrical piezometer tips were destroyed by lightning in June 1972 and several pneumatic installations were added.

Filling of Lake Koocanusa began in March 1972. The pool was raised from elevation 2150 (early May) to 2220 (early April), to 2370 (June), and ultimately to maximum elevation of 2405 by late July.

To 1976 no rock movements of concern had been measured. Some downslope creep of near-surface rock blocks was indicated in the spring and summer of 1972 and 1973, possibly in association with seasonal warming and thermal expansion (Hamel, 1974). Slight movement of highly fractured rock in the 914 rib was observed in the first filling of 1972 and the lowering of fall 1972. Lower portions of left bank ribs also showed minor compressional movements as the buttresses were placed in spring 1973. Some minor creep and readjustments have been noted in the left abutment [5]. Piezometer heads in the range 0—6 m above pool level have been measured. Time lag has been noted in deeper portions of the 927 rib, whereas the response of near-surface piezometers has been rapid; e.g., rapid increases of 3—11 m in head followed rainfall of 3.6 cm in June 1972. Excess heads generally dissipated within a few days.

GEOTECHNICAL INVESTIGATIONS

Shear strength and sliding friction

Results of laboratory shear tests and simple small-scale field tilt tests are summarized in Table II. The data are sparse but suggest friction values for rock-on-rock sliding in the range 25—30°, with roughness adding an equivalent friction angle of perhaps 10—15°.

As noted, most shear zones defining unstable or potentially unstable wedges contain gouge material, ranging from unaltered, highly interlocked argillite fragments to sandy or silty clay; test results suggest that index and strength properties of gouge are likely to vary significantly from one discontinuity to another, and even within a given discontinuity (Hamel, 1974, pp. 66, 77). Peak parameters ($c' = 20$—50 kN/m^2, $\phi' = 22$—$25°$) and residual parameters ($c'_r = 10$—20 kN/m^2, $\phi'_r = 20$—$22°$) were suggested for low effective normal stresses with samples of 915 and 930 bedding plane fault gouge.

---

[5] Deformation measurements in left bank slopes during 1972 were interpreted as evidence of downslope creep, and this creep was considered as evidence for possible instability (Hamel, 1974, p. 101). Müller and Logters (1972) suggested that the left bank ribs were only marginally stable due to *Talzuschub* (mountain creep processes; cf. Chapters 1, 13, Volume 1; Chapter 17, this volume).

Subsequent tests by the Missouri River Division (MRD) with low effective normal stresses suggested peak strengths of $c' = 0$, $\phi' = 29-33°$, with the corresponding residual strength, $\phi'_r = 25.5-30°$. Gouge strength thus seems mostly governed by interparticle friction involving argillite fragments rather than the montmorillonite present in the 2-μm fraction (Hamel, 1974, p. 77). (Hamel believed that the relative amount of clay gouge contact along discontinuities was small compared to rock-on-rock contact; Hamel, 1974, p. 77.) The measured residual friction values seem compatible with the presence of layer lattice minerals in the argillite and gouge, notably mica, chlorite, and illite (MRD, 1971; Hamel, 1974, pp. 73—75; cf. Horn and Deere, 1962; Kenney, 1967; Hamel and Gunderson, 1973; Chapters 12, 17, this volume).

Surface roughness

Total shear resistance to sliding includes a roughness-dilatancy component in addition to a material property component (Patton, 1966). Accordingly, field mapping was conducted at several localities to establish typical amplitudes and angles for asperities present along exposed discontinuity surfaces (Banks and De Angulo, 1972; Banks et al., 1972; Banks and Strohm, 1974). Base lines 90 m long were established in the trough of the 1971 left abutment slide (Figs. 14, 15) by stretching a light cord approximately parallel to and about 1 m above the line of bedding plane and joint intersection; offset measurements were then made at intervals of about 0.3 m. According to this study, the DS + 122 bedding plane exhibited a relatively smooth surface. Asperities were rather uniformly spaced at about 1 m, with amplitudes of 0.6—1.9 cm. Many of the surfaces were polished. By comparison the A joint surface was considerably rougher (see Fig. 14), with average amplitude of 0.6 m and asperity angles of 3—14° (mean value 6.5°) (Banks and De Angulo, 1972; Banks and Strohm, 1974, p. 842). Additional observations were reported by Hamel (1974, pp. 68—71), who noted that the exposed A joint was planar on a large scale but rather rough if considered in detail (Fig. 15). Roughness amplitudes range from a few centimetres to over a metre, with the average about that given by Banks and De Angulo. Upslope surfaces of asperities had flatter inclinations (estimated average about 5°, with respect to the mean surface orientation) than their downslope counterparts, and were generally polished; the slide mass apparently rode up over asperities rather than shearing through them. Downslope asperity surfaces were rough. In contrast, the DS + 122 surface was found to be planar on both large and small scales; asperity amplitudes do not exceed about 1.2 cm, and asperity inclinations seem negligible. Surface characteristics of the Dirty Shame and DS + 80 bedding plane faults seemed similar (Hamel, 1974, p. 68).

Discontinuity traces were also examined (Hamel, 1974, pp. 68—71). The DS + 122 fault is represented by a shear zone 8—30 cm thick (average 23 cm), containing numerous subparallel clean fractures at spacings of 1—5 cm.

TABLE II

Strength parameters from small-scale laboratory and field experiments

Material	c' (kN/m^2)	ϕ (deg.)	σ_0' (kN/m^2)	Remarks	References
Argillite Natural surfaces	0	30—80	< 10	Tilt sliding tests with variable surface roughness. Upper bound for interlocked ripple marks. 41° considered typical. Friction value included effect of roughness	Hamel (1974, p. 60), communication by T.E. Ward cited
Argillite Natural surfaces	0	29—30	< 10	Tilt sliding tests with asperity inclinations of 5—6°. Net value given; gross value approx. 35°	Banks and De Angulo (1972), Hamel (1974)
Metasandstone From DS + 122	0	25—26.5	300—1200	Direct shear; water-soaked sample. Higher ϕ' value with increasing σ_0'; one sample only	Banks and De Angulo (1972), Banks and Strohm (1974)
Gouge from 915 fault Undisturbed, saturated	46 0 48	12.4 (R̲) 26 (R̲, NPD) 23 (R̲, Hamel)	100—700 100—700 100—700	Consolidated-undrained (R) triaxial with pore pressure measurements (R̲). Two specimens. Alternative interpretations of R̲ envelope given. SC soil; 28—45% < 200 sieve. LL = 23—27, PL = 13—15, w = 10—27%; typically 70—90% in-situ saturated	NPS (1971a), Hamel (1974, pp. 62—63)

Gouge from 930 fault				Consolidated-drained (S) direct shear. CL soil; 65—68% < 200 sieve. LL = 31, PL = 18.5, w = 11—15%; 80—100% in-situ saturated. Montmorillonite content appreciable	NPS (1971a), Hamel (1974, pp. 63—65)
Undisturbed, 90—94% saturated	29	24.7 (peak)	100—700		
	10	22.0 (residual)	100—700		
Undisturbed (natural water contents), 82—88% saturated	120	23.3 (peak)	300—700		
	62	20.5 (residual)	300—700		
Saturated, using "natural water content" samples	19	19.4 (residual)	300—700		
Undisturbed, saturated	200	17.9 (R̲)	300—700 *	Consolidated-undrained (R) triaxial with pore pressure measurements (R̲). Alternative interpretations of R̲ envelope given	
	0	25 (R̲, NPD)	300—700		
	58	21.5 (R̲, Hamel)			
Gouge from Dirty Shame fault				Repeated direct shear. CH soil; 85% < 200 sieve. LL = 88, PL = 38; SG = 2.71. Fraction —2 μm: 10% montmorillonite, 90% mixture illite-mica. Hamel's sample 3; "waxy clay gouge"	Hamel (1974, pp. 72—75), MRD (1972b)
Remolded —40 sieve, pre-cut sliding surface	0	32 (peak)	600		
	0	30 (residual)	600		
Gouge from A joint				Repeated direct shear. CH soil; 70% < 200 sieve; LL = 52, PL = 17. Fraction —2 μm: 58% montmorillonite, 34% illite-mica, 7% chlorite	Hamel (1974, pp. 72—76), MRD (1972b)
Remolded —4 sieve	0	32.7 (peak)	16—600		
	0	25.5 (residual)	16—600		
Remolded —40 sieve, pre-cut sliding surface	0	29.5 (peak)	16—600	Hamel's sample 4; "soapy clay gouge"	
	0	25.5 (residual)	16—600		

* Values of consolidation cell pressure given for triaxial tests.

A gouge seam was present with a thickness of 0.6—5 cm, consisting of dense angular sand to fine gravel-sized argillite fragments contained locally in a mylonitic-type matrix of silt-sized argillite particles (Hamel, 1974, p. 71). Thin (1 mm) coatings of waxy or soapy clay were locally observed on larger gouge particles and on rock fracture surfaces. By contrast, the DS fault had just one large continuous parting with a gouge zone 1—8 cm thick (average 5 cm).

The A joint was in fact a fracture zone ranging from a few centimetres to a metre in thickness. Thicker portions contain numerous discontinuous sub-parallel fractures and gouge seams or pockets spaced at intervals of 0.6—15 cm (Hamel, 1974, p. 69). Thinner portions contain only a few fractures and no gouge.

Many of these fractures were simply clean smooth breaks in the argillite, whereas others were coated with a film of clay. The gouge varied from dense interlocked sands or gravel-sized argillite fragments to stiff, waxy to soapy clay; the typical thickness of the latter was about 1 cm or less. The gouge presumably resulted from mechanical shearing plus local chemical alteration (Hamel, 1974, p. 69). Gouge seams and pockets were at most 15 cm thick, locally distributed over zones less than a metre long.

Back-analyses of wedge slides

Limiting equilibrium analyses have been performed by the Seattle District, Corps of Engineers [6] (NPS, 1971a, b, 1972a, b, c; cf. Hamel, 1974, pp. 39—44), the U.S. Army Engineer Waterways Experiment Station (Banks et al., 1972; Banks and Strohm, 1974), and Hamel (1974, pp. 51—59 [7]). In the latter work, which is discussed here, the scalar method of Hoek et al. (1972; cf. Hoek and Bray, 1974) was followed; discussion of the computer code is given by Hamel (1974, pp. 211—239). No attempt was made by Hamel to separate the behavior of bedding plane fault and joint surfaces. Most calculations were checked by an independent approach (Hamel, 1974, pp. 183—208) using a modified version of the vector method of Wittke (1964, 1965; cf. Hendron et al., 1971). This second approach was also used here for a

---

[6] The prehistoric 925 and 930 slides were back-analyzed by a very approximate method (NPS, 1972b, plate D-4) which suggested mobilized friction angles of about 23°. However, these slides were underlain by complex (and in part poorly known) *combinations* of geologic discontinuities. Ordinary three-dimensional wedge solution methods are inadequate for their analysis. A multiple-block approximation was devised, but the resulting solution method is of doubtful validity; toe effects and block interlocking mechanisms were inadequately considered, and fluid pressure and seismic loading conditions were probably underestimated; indeed, the actual triggering mechanisms involved in the failure are not known. The calculated friction values are far too low, probably by 10° or more.
[7] Cf. Hamel, 1976, pp. 371—379.

series of equilibrium analyses in which bedding plane and joint surfaces were individually treated.

Assumed geometries are given in Table I. A specific gravity of 2.7 was assumed. The average Coulomb parameters required for static equilibrium were calculated (Fig. 17; Tables III, IV).

First we consider Hamel's results. The influence of water forces on strength was investigated in detail only for the January 1971 event, because water forces undoubtedly influenced its failure; indeed, its average friction angle required for equilibrium assuming *dry* conditions was the lowest of five analyzed cases. The 1971 wedge was investigated both with and without an assumed tension crack, with average piezometric head of 0—9 m (Hamel, 1974, fig. 23, table 5). For the presumed most realistic condition (tension crack assumed), required friction for $c' = 0$ increased from 28 to 37.5° as head increased from 0 to 4.7 m [8]. Each friction value cited corresponds to an average value for *both* slide planes bounding a wedge. Examination of piezometer data suggested to Hamel (1974, p. 57) that the average piezometric head on a large-scale left abutment discontinuity would seldom have exceeded 3—5 m.

So-called "upper and lower bound" zero-cohesion envelopes for effective normal stresses <150 kN/m^2, as suggested from back-analysis by Hamel, are given in Fig. 17. Hamel's upper bound envelope, $\phi = 35°$ (envelope A, Fig. 17) corresponds to Old Notch 1, the 1967 slide and the 1971 slide assuming $h \sim 3$ m; the lower bound envelope, $\phi = 30°$ (envelope B, Fig. 17) corresponds to the Old Notch 2 and 1971 slides, both with zero or insignificant water pressure.

The 1967 left abutment slide provides the only other case for which reasonably complete information on collapse conditions is available. Fluid pressure was probably not involved [9], and the slide mass became detached as soon as it was undercut. The case is equally well explained [10] by a point cor-

---

[8] The tension crack case with average head of 4.7 m corresponds to a tension crack full of static water and a linear decrease in water pressure from the tension crack to the slope face. This distribution is, according to Hoek et al. (1972), the realistic maximum which could ordinarily occur under conditions of very heavy rainfall. The associated friction angle of 37.5°, calculated for equilibrium with $c' = 0$, was assumed by Hamel to represent a reasonable upper bound on friction mobilized in the 1971 slide.

Nonetheless Hamel (1974, p. 56) qualified this by noting that higher water pressures than these could occur if natural drainage outlets were blocked. As previously discussed, climatic conditions of late January 1971 favored the development of ice plugs. Viewed in this light, the assumption of a 9-m head as an upper limit does not seem wholly unreasonable. The associated friction angle of 51.5°, for $c' = 0$, is within the range of results obtained in simple rock block sliding tests by Ward and Galster (1969, table 2).

[9] The 1967 left abutment slide occurred in a dry period with no noticeable water along the slip planes; fluid pressure was probably not involved in this case.

[10] The upstream boundary joint in the 1967 slide was considered as a tension crack for purposes of analysis.

TABLE III

Analytical results for wedge slides (after Hamel, 1974) *

Wedge	Average piezo-metric head (m)	Calculated wedge volume (m³)	Strength parameters for equilibrium	
			ϕ' for $c' = 0$ (deg.)	c' for $\phi' = 0$ (kN/m²)
Old Notch (Station 928)	0	390	33.5	11
Old Notch 1	0	4800	35	39
Old Notch 2	0	830	30	24
1967 left abutment	0	6100	34.5	53
1971 left abutment:				
no tension crack	0	46,000	28	70
	1.5	46,000	31	70
	3	46,000	34.5	70
	9	46,000	59	70
tension crack	0	38,000	28	91
	1.5	38,000	30.5	91
	3	38,000	33.5	91
	4.7	38,000	37.5	91
	9	38,000	51.5	91

* Strength parameters given are average parameters for both sliding surfaces.

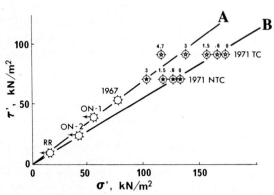

Fig. 17. Shear strength data and envelopes based on back-analyses of small wedge slides (modified after Hamel, 1974). Data points as follows: RR = old Notch at railroad cut, Station 928; ON-1, 2 = Old Notches 1 and 2; 1967 = left abutment slide of 1967; 1971 NTC = left abutment slide of 1971 with no assumed tension crack; 1971 TC = same with assumed tension crack. Numbers above 1971 slide data points refer to average piezometric head assumed in analysis. A, B are upper and lower bound envelopes, respectively, according to Hamel interpretation. Arrows indicate shift of data points if pore pressures were involved in failures.

responding to $\sigma_0' = 80$ kN/m^2 on either envelopes $c' = 0$, $\phi' = 34.5°$, or $c' \sim 10$ kN/m^2, $\phi' = 30°$. However, considered in detail, it seems unlikely that a single envelope can accurately apply to all cases, particularly in view of local differences in roughness, different orders of magnitude in slide volume, and the probability that water forces were perhaps important in wedge slides in addition to the 1971 event. Under these circumstances line A could itself be considered as an approximation of a *lower* bound, because with increase in fluid pressure nearly all data points would shift (probably by different amounts) to the left [9].

Hamel suggested that the strength envelopes calculated for equilibrium of the failed rock wedges represent reasonable estimates of the field residual strength for discontinuities bounding the Libby wedges. Field strength envelopes thus determined were then compared in detail to laboratory-based strength estimates. However, such direct comparisons could be misleading inasmuch as the two planes of sliding were not separated in the equilibrium analyses. Field studies have clearly demonstrated the DS + 122 surface to be much smoother than the A joint, and it may be profitable to try to examine the consequences of this difference in analysis (cf. Banks and Strohm, 1974).

TABLE IV

Effective stress Coulomb parameters for 1971 left abutment slide (slide planes DS + 122 and joint A considered separately)

Average piezometric head (m)	Assumed angle of friction ϕ' (deg.), DS + 122	Angle of friction ϕ' (deg.), joint A	Cohesion c' (kN/m^2), joint A	Equivalent asperity angle (deg.) assuming $\phi'_s = 26°$	
				DS + 122	joint A
0	26	30.5	0 (assumed)	0	4.5
3	26	35 (assumed)	20	0	9
	26	41	0 (assumed)	0	15
9	26	35 (assumed)	110	0	9
	26	66	0 (assumed)	0	40
0	30	26	0 (assumed)	4	0
3	30	36.5	0 (assumed)	4	10.5
9	30	40 (assumed)	90	4	14
	30	64.5	0 (assumed)	4	38.5

* Assumed $c' = 0$ for DS + 122.

Accordingly, results by the modified vector approach are summarized in Table IV. In these analyses, it was assumed that the DS + 122 surface could be characterized by $c' = 0$, $\phi' = 26°$ or $30°$; an average piezometric head was then specified, and either (1) an equilibrium friction angle for joint A was calculated, assuming $c' = 0$, or (2) cohesion for joint A was calculated assuming $\phi' = 35$ or $40°$. These ϕ' values seem reasonable; e.g., asperity angles as great as $14°$ were measured by Banks and De Angulo (the mean value is $6.5°$, but it is not at all clear that mean values are more appropriate than upper bound values for use in analysis); assuming a sliding friction angle (ϕ'_s) of $26°$, $\phi' = \phi'_s$ + asperity angle $\cong 40°$.

The interpreted best estimate, based on the data of Table IV, is about as follows: for DS + 122, $\phi' = 30°$ and head = 3—9 m, lower and upper bounds for the A joint are $c' = 0$, $\phi' = 36.5—64.5°$. Equivalent asperity angles of $10°$ or more are suggested, which seems compatible with field estimates, particularly for the lower part of the range [11].

Theoretical analyses of slide motion

A simple mathematical idealization (Model I) was developed by the Waterways Experiment Station to predict plausible displacement-time relationships for the potential left-bank rib slides (Banks et al., 1972; Banks and Strohm, 1974, p. 846). The prehistoric 930 rock rib slide was also analyzed. The slide masses were assumed to be distributed but continuous (consisting of a series of connected small masses) and to have maintained their shape during the slide process. Equations of motion were developed with the important variables being the number, inclination and frictional resistance of the sliding planes (Banks et al., 1972, appendix C; Banks and Strohm, 1974; cf. Chapter 9, Appendix 2, this volume). Solutions were also obtained (NPS, 1971b) for an idealization involving motion of the centroid of a single block (lumped mass) on two connected planes (Model II). In Model II the mass was assumed concentrated at its centroid, and the entire mass was assumed to accelerate until the centroid passed the junction of the two slide planes; the mass then decelerated until the centroid reached its final position. This approximation is crude inasmuch as in reality deceleration begins as soon as the frontal part of the slide mass passes the junction of the slide planes; Model I was developed specifically in order to account for this refinement.

The geometry of the complex 930 rib changes above elevation 2400 (Table I); e.g., above this elevation the wedge intersection plunges $28°$ at N66°W whereas below it, the intersection plunges $24°$ at N59°W. Because of a wedge factor of 1.36—1.34 (cf. Hoek and Bray, 1974, pp. 185—187), the

---

[11] In these analyses the reader should not forget that the strength is assumed to be uniformly distributed over the sliding surface; this assumption is unlikely to be realized in nature.

corresponding "plane dip surface idealization" dip angles are 21° and 18°, respectively. For simplicity a single plane oriented at the larger value was assumed for analysis (Banks et al., 1972, appendix C; Banks and Strohm, 1974, p. 844). It was assumed further that the base of the slide mass merged with the valley floor (a simplification of actual conditions). The floor was assumed to be 290 m wide, and bounded at its far side by an opposing valley slope of 19°. Previous studies had indicated that the center of gravity had moved about 305 m downslope and 245 m across the valley floor (Fig. 4); the maximum displacement was thus set at 550 m.

The results of motion analyses (Figs. 18, 19) are given in terms of "total stress" friction angles for the detachment slope (ϕ_1), the valley floor (ϕ_2), and the opposing slope (ϕ_3). The solution is non-unique; thus various combinations of ϕ_1, ϕ_2, and ϕ_3 can satisfactorily account for the required 550-m displacement (Fig. 18). Assuming uniform friction, the displacement criterion is satisfied by $\phi_1 = \phi_2 = \phi_3 = 11.5°$ for the distributed mass (Model I) three-plane idealization; this contrasts with $\phi_1 = \phi_2 = \phi_3 = 13°$ as given by the (Model I) two-plane approach (Fig. 18) or 14° as given by a lumped mass (Model II) two-plane approach. The two-plane models discussed here neglect the effects of the opposing valley slope. The criterion is equally satisfied by other appropriate friction combinations, e.g., for the distributed mass (Model I) three-plane solution, $\phi_1 = \phi_3 = 5°$, $\phi_2 = 14.5°$.

An example of acceleration, velocity, and displacement distribution as a function of time is shown in Fig. 20 for the case $\phi_1 = 14°$, in order to illustrate the significant differences between Models I and II (distributed mass and lumped mass). Both cases shown involve two-plane idealizations (opposing slope is neglected) in order to facilitate comparison. The peak velocity for Model I is seen to be about half of that predicted by Model II [13] (cf. Fig. 19).

Motion characteristics for the Model I three-plane case are similar to that shown for the Model I two-plane case; peak velocities are similar, viz. 21 vs. 19 m/s, respectively, for $\phi_1 = 14°$ (Fig. 20).

In the Banks and Strohm (1974) interpretation of these data, the dynamic analysis was preceded by a limiting equilibrium (static) analysis in which the effects of roughness, surface friction and fluid pressure were examined. Once failure commenced, Banks and Strohm assumed that the water force assumed for limiting equilibrium remained exactly constant, the component

---

[12] The angle of Model II corresponds to the Heim and Müller-Bernet approximation ("geometrische Gefälle der Schwerpunkte"; Heim, 1932) as discussed in Volume 1 (cf. Chapter 1).

[13] This comparison is of much interest, inasmuch as most published landslide velocities are based on equations of motion developed on the assumption of motion of a rigid block centroid, i.e., Model II. The conservation of energy approach leads to the same equations. Many of the reported velocities in the literature may be too large, perhaps by a factor of two.

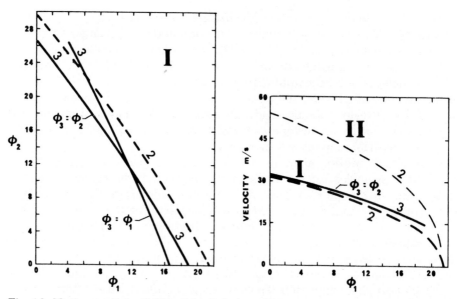

Fig. 18. Motion analysis of 930 slide. Relation of friction values for displacement of 550 m (after Banks et al., 1972). Distributed mass (Model I) three- and two-plane idealizations, denoted by 3 and 2, respectively.

Fig. 19. Variation of peak velocity as function of ϕ_1; 930 slide, 550-m displacement (after Banks et al., 1972). Model I three- and two-plane, and Model II (lumped mass) two-plane solution compared.

of resistance due to roughness totally vanished, and the static value of surface friction force remained in effect for the dynamic phase of slope movement. An "equivalent" friction angle was then calculated so as to produce the same factor of safety (less than unity); downslope acceleration of the mass was then determined with the use of this equivalent friction angle, i.e., $\phi_1 = 18°$. The valley floor and opposing slope friction angles were then estimated by reference to Fig. 18, with the general result that for $\phi_1 = 18°$, $\phi_2 = \phi_3 \leqslant 2°$. These data were then used to estimate peak velocities for the potential left bank rib slides (Banks and Strohm, 1974, pp. 845—847).

    This approach may seem appealing in that measured or calculated physical properties are used in each successive step of analysis; however, particularly for large slides, major uncertainties are involved in all three assumptions stated above. The points of contact of a sliding mass with its subjacent surface are in a state of continuous flux, and it seems unlikely to me that at every given instant the total normal component of body force is impressed upon the sliding surface. The problem is not simply a question of instantaneous surface area of contact, for Amonton friction is presumed to be independent of this factor. Although the matter of static versus dynamic surface

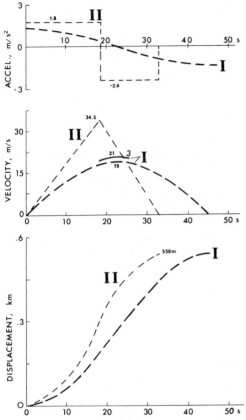

Fig. 20. Comparison of two-plane solution for Models I and II (after Banks et al., 1972). Note great exaggeration of peak velocities by lumped mass approach (Model II). Distributed mass three-plane solution for velocity is only slightly different than two-plane solution. 930 slide.

friction coefficients has hardly been resolved in general terms, the normal force-time history of the moving deformable block may be the more significant factor. If portions of the moving block are virtually in a state of vibra-

TABLE V

Estimated peak velocities for the prehistoric 930 slides

Assumed ϕ values	Peak velocity (m/s)
$\phi_1 = 18°, \phi_2 = \phi_3 = 2°$	15
$\phi_1 = \phi_2 = \phi_3 = 11.5°$	24
$\phi_1 = 5°, \phi_2 = \phi_3 = 20.5°$	29

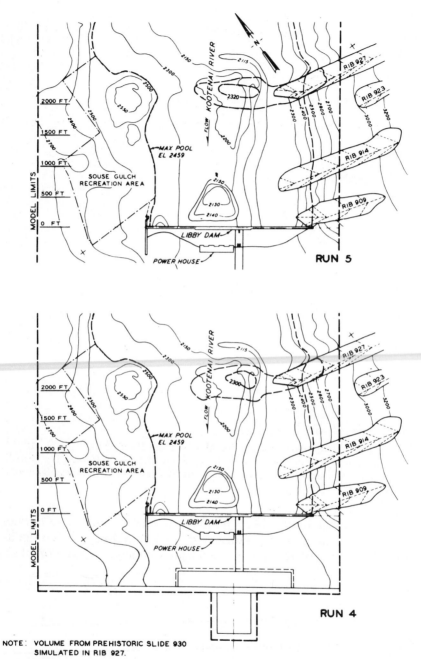

NOTE: VOLUME FROM PREHISTORIC SLIDE 930
      SIMULATED IN RIB 927.

Fig. 21. Model studies of 930 slide mass terminal distribution as a function of peak emplacement velocity (after Davidson and Whalin, 1974). Run 5, velocity 12 m/s; run 4, velocity 22 m/s. Volume from prehistoric slide 930 simulated in rib 927. Compare model limits and slide mass distribution with Fig. 4.

tion with respect to the sliding surface, the average dynamic value of basal resisting force can be much less than the product of static normal force and static surface friction coefficient. Accordingly, the values of ϕ_1 as determined by the Banks and Strohm approach may be too large, and the assumed values for $\phi_2 = \phi_3$ too small. The assumption $\phi_1 = \phi_2 = \phi_3$ seems as least as reasonable as an approximation, recognizing that friction values at the initiation and cessation of motion may both be somewhat higher than the average assumed, and values at high velocity may be lower. Indeed, a rough estimate of $\phi_1 = 5°$ has even been suggested as an appropriate lower limit (Banks et al., 1972, p. 37). The resulting peak velocities as estimated for the 930 slide event by the Model I, three-plane approach are given in Table V. The interpreted best estimate is thus about 22 ± 7 m/s.

Model studies

Because of the potential for left bank sliding and proximity of the Libby Dam, the Waterways Experiment Station conducted a three-dimensional 1 : 120 hydraulic model of the site (see Fig. 4 for model limits) in order to determine the magnitude of wave heights, runup, and overtopping for four potential wedge rockslides. Details have been presented in a comprehensive report by Davidson and Whalin (1974; cf. Chapter 9, this volume). Results showed wave amplitudes and runup as a function of individual slide mass, reservoir pool elevation and slide velocity. The principal conclusion was that full or partial volume slides caused unacceptable conditions at the dam, with critical dependence on slide velocity; buttressing at the toe tended to significantly reduce the velocity, thus resulting in lower wave amplitudes.

As part of the model test program, two tests were conducted using the slide volume from the prehistoric 930 slide in the model 927 rib location (Fig. 21). The experiments were conducted in order to provide a rough indication of the peak emplacement velocity required for the prehistoric mass to come to rest in its presently observed position. No definite conclusions could be drawn from the few tests conducted; in both cases the final resting place for the slide debris was similar to that observed in the Kootenai valley (cf. Figs. 21, 4; Davidson and Whalin, 1974, p. 30).

ACKNOWLEDGEMENTS

This volume would have been incomplete without reference to the Libby problem, and so this article was originally scheduled by individuals actively associated with the work at Libby. When it unfortunately appeared that the submission deadline could not be met, I took on the task to summarize the available information. Although the site is personally known to me from preparation work for an excursion of the 3rd Congress, International Society for Rock Mechanics (1974), and some additional analyses are presented, I did not directly participate in the engineering investigations cited herein. S. Mellon and M. Canich assisted in the analyses. Discussions at the Libby site with T.E. Ward, R.W. Galster, and J.C. Richards are gratefully acknowledged. Important reference material was provided by Ward and Galster, D.C. Banks, J.V. Hamel, W.M. Johns, A. Boettcher and P.M. Douglass.

REFERENCES

Algermissen, S.T., 1969. Seismic risk studies in the United States. *Proc., 4th World Conf. on Earthquake Engineering, Santiago, 1969.*

Banks, D.C. and De Angulo, M., 1972. *Velocity of Potential Landslides, Libby Dam.* U.S. Army Engineer Waterways Experiment Station report to U.S. Army Engineer District, Seattle, Wash.

Banks, D.C. and Strohm, W.E., Jr., 1974. Calculations of rock slide velocities. *3rd Congr. Int. Soc. Rock Mech.,* IB: 839—847.

Banks, D.C., Whalin, R.W., Davidson, D.D. and De Angulo, M., 1972. Velocity of potential landslides and generated waves, Libby Dam and Lake Koocanusa Project, Montana. *U.S. Army Eng. Waterw. Exp. Stn., Tech. Rep.,* No. H-74-15 (preliminary draft).

Boettcher, A.L., 1967. The Rainy Creek alkaline ultramafic igneous complex near Libby, Montana. *J. Geol.,* 75: 526—553.

Davidson, D.D. and Whalin, R.W., 1974. Potential landslide generated water waves, Libby Dam and Lake Koocanusa, Montana. *U.S. Army Eng. Waterw. Exp. Stn., Tech. Rep.,* No. H-74-15, 33 pp. plus appendices (Corps of Engineers, Vicksburg, Miss.).

Galster, R.W., 1974. Engineering geology at the Flathead Tunnel, Libby Project, Montana. In: B. Voight and M.A. Voight (Editors), *Rock Mechanics — The American Northwest. 3rd Congr. Exped. Guide, Int. Soc. Rock Mech., Spec. Publ.* Experiment Station, College of Earth and Mineral Sciences, University Park, Pa., pp. 222—228.

Hamel, J.V., 1974. Rock strength from failure cases: left bank slope stability study, Libby Dam and Lake Koocanusa, Montana; *Mo. River Div., Tech. Rep.,* No. MRD 1-74, 239 pp. (Corps of Engineers, Omaha, Nebr.).

Hamel, J.V. and Gunderson, J.W., 1973. Shear strength of Homestake slimes tailings, *Proc. Am. Soc. Civ. Eng., J. Soil Mech. Found. Div.,* 99 (SM5): 427—432.

Harrison, J.E., 1972. Precambrian Belt Basin of northwestern United States. *Geol. Soc. Am. Bull.,* 83: 1215—1240.

Heim, A., 1932, *Bergsturz und Menschenleben.* Fretz und Wasmuth, Zürich, 218 pp.

Hendron, A.J., Jr., Cording, E.J. and Aiyer, A.K., 1971. Analytical and graphical methods for the analysis of slopes in rock masses. *U.S. Army Eng. Waterw. Exp. Stn., NCG Tech. Rep.,* No. 36, 162 pp.

Hoek, E. and Bray, J.W., 1974. *Rock Slope Engineering.* Institution of Mining and Metallurgy, London, 309 pp.

Hoek, E., Bray, J.W. and Boyd, J.M., 1972. The stability of a rock slope containing a wedge resting on two intersecting discontinuities. *Imp. Coll. Sci. Technol. (London), Rock Mech. Res. Rep.,* No. 17, 63 pp.

Horn, H.M. and Deere, D.U., 1962. Frictional characteristics of minerals. *Geotechnique,* 12: 319—335.

Johns, W.M., 1962. Belt series in Lincoln and Flathead Counties, Montana. *Trans. Am. Inst. Min. Eng.,* 226: 184—192.

Johns, W.M., 1970. Geology and mineral deposits of Lincoln and Flathead Counties, Montana. *Montana, Bur. Mines Geol., Bull.,* 79, 182 pp.

Johns, W.M., 1974a. Belt (Precambrian) sedimentary rocks in northwest Montana. In: B. Voight and M.A. Voight (Editors), *Rock Mechanics — The American Northwest. 3rd Congr. Exped. Guide, Int. Soc. Rock Mech., Spec. Publ.* Experiment Station, College of Earth and Mineral Sciences, University Park, Pa., pp. 208—212.

Johns, W.M., 1974b. Kalispell—Libby Dam—Rocky Mountain trench (road log). In: B. Voight and M.A. Voight (Editors), *Rock Mechanics — The American Northwest. 3rd Congr. Exped. Guide, Int. Soc. Rock Mech., Spec. Publ.* Experiment Station, College of Earth and Mineral Sciences, University Park, Pa., pp. 222—228.

Kenney, T.C., 1967. Influence of mineral composition on the residual strength of natural soils. *Proc., Geotech. Conf., Oslo,* 1: 123—129.

MRD (Missouri River Division), 1971. Petrographic and X-ray diffraction analyses of soil gouge material with gradation and Atterberg limits, Libby Dam slide sample No. 2, A joint. *Mo. River Div. Lab. Rep.,* No. 68/426, 3 pp. + 1 fig. (Corps of Engineers, Omaha, Nebr.).

MRD (Missouri River Division), 1972a. *Geophysical Survey — Ancient Landslide — Libby Dam, Montana.* Missouri River Division Laboratory report to Seattle District, Corps of Engineers, 3 April, 5 pp. (cf. Seattle District comments on report, 25 April 1972).

MRD (Missouri River Division), 1972b. *Repeated Direct Shear and Classification Tests, Libby Dam Slide, Gouge Samples 3 and 4.* Missouri River Division Laboratory, Corps of Engineers, No. 68/426, May 4, 3 pp.

Müller, L. and Logters, G., 1972. *Libby Dam and Lake Koocanusa Project — Left Bank Slope Stability — Reports on Rock Mechanics.* September 4, 22 pp. and December 18, 28 pp.

NPS, 1963. *Design Memorandum 1, Libby Dam Project, Kootenai River, Montana.* Suppl. 2, Appendix B, Geologic Data. Seattle District, Corps of Engineers, January.

NPS, 1970. *Survey of Dangerous Slide Potential (Libby Dam and Lake Koocanusa Project).* Seattle District, Corps of Engineers, April, 11 pp.

NPS, 1971a. *Dam Concrete Review, Left Abutment Slide and Rock Problem.* Special Consultants Meeting, April 19—21, Seattle District, Corps of Engineers.

NPS, 1971b. *Dam Concrete Review, Left Abutment Slide and Rock Problem.* Meeting No. 6, Board of Consultants, July 7—9, Seattle District, Corps of Engineers.

NPS, 1972a. *Dam Concrete Review, Left Abutment Slide and Rock Problem.* Meeting No. 7, Board of Consultants, March 21—23, Seattle District, Corps of Engineers.

NPS, 1972b. Meeting No. 8, Board of Consultants, August 7—11, Seattle District, Corps of Engineers.

NPS, 1972c. Meeting No. 9, Board of Consultants, November 20—21, Seattle District, Corps of Engineers.

NPS, 1973. Meeting No. 10, Board of Consultants, September 12—14, Seattle District, Corps of Engineers.

Patton, F.D., 1966. *Multiple Modes of Shear Failure in Rock and Related Materials.* Ph.D. Thesis, Univ. of Illinois, Urbana, Ill., 282 pp.

Shannon and Wilson, Inc., 1971. *Instrumentation Plan for Left Bank Area, Libby Dam and Lake Koocanusa Project, Libby, Montana.* to Seattle District, Corps of Engineers, July 2, 39 pp.

The Indicator, 1974. How to wire a mountain. 6 (1): 2—5.

Ward, T.E., 1972. *Resumé of Special Geology Meeting, Rock Slide Studies, 3B (1) Area — Libby Dam and Lake Koocanusa Project.* Seattle District, Corps of Engineers, May 3, 3 pp.

Ward, T.E. and Galster, R.W., 1969. *Rock Stability Report — Left Abutment and MSH 37 Cut Area — Libby Dam.* Seattle District, Corps of Engineers, August 28, 4 pp.

Wittke, W., 1964. A numerical method of calculating the stability of slopes in rocks with systems of plane joints. *Rock Mech. Eng. Geol., Suppl.,* I: 103—129.

Wittke, W., 1965. A numerical method of calculating the stability of loaded and unloaded rock slopes. *Rock Mech. Eng. Geol., Suppl.,* II: 52—79.

Chapter 9

OCCURRENCES, PROPERTIES, AND PREDICTIVE MODELS OF LANDSLIDE-GENERATED WATER WAVES

RUDY L. SLINGERLAND and BARRY VOIGHT

ABSTRACT

Large water waves generated by landslides impacting with a body of water are known from Disenchantment and Lituya Bays, Alaska; Vaiont reservoir, Italy; Yanahuin Lake, Peru; Shimabara Bay, Japan; and many fiords in Norway. The combined death toll from these events most likely exceeds 20,000 people. Such waves may be oscillatory, solitary, or bores and nonlinear mathematical theories or linearizing assumptions are thus needed to describe their wave amplitudes, celerities, and periods. In this paper the following approaches are compared: (1) the Noda simulation of a vertically falling and horizontally moving slide by linearized impulsive wave theory and estimation of nonlinear wave properties; (2) the Raney and Butler modification of vertically averaged nonlinear wave equations written for two horizontal dimensions to include three landslide forcing functions, solved numerically over a grid for wave amplitude and celerity; (3) the empirical equations of Kamphuis and Bowering, based on dimensional analysis and two-dimensional experimental data; and (4) an empirical equation developed in this report from three-dimensional experimental data, i.e., $\log(\eta_{max}/d) = a + b \log(KE)$, where a, b = coefficients, η_{max} = predicted wave amplitude, d = water depth, and KE = dimensionless slide kinetic energy. Beyond the slide area changes in waveform depend upon energy losses, water depth and basin geometry and include wave height decrease, refraction, diffraction, reflection, and shoaling. Three-dimensional mathematical and experimental models show wave height decrease to be a simple inverse function of distance if the remaining waveform modifiers are not too severe. Only the Raney and Butler model considers refraction and reflection. Run-up from waves breaking on a shore can be conservatively estimated by the Hall and Watts formula and is a function of initial wave amplitude, water depth, and shore slope. Predicted run-ups are higher than experimental run-ups from three-dimensional models. The 1958 Lituya Bay and 1905 Disenchantment Bay, Alaska events are examined in detail, and wave data are developed from field observations. These data and data based on a Waterways Experiment Station model are compared to wave hindcasts based on various predictive approaches, which yield a large range of predicted wave heights. The most difficult problems are in matching the exact basin geometry and estimating slide dimensions, time history, and mode of emplacement. Nevertheless, the hindcasts show that the mathematical and experimental model approaches do provide useful information upon which to base engineering decisions. In this regard the empirical equation developed in this report is at least as satisfactory as existing methods, and has the advantage of requiring less complicated input data.

INTRODUCTION

Large water waves generated by landslides impacting with a body of water are by now, well documented. The earliest important record of such events in the Western Hemisphere occurred as a consequence of three separate glacier falls from the west side of Disenchantment Bay in Alaska (Figs. 1 and

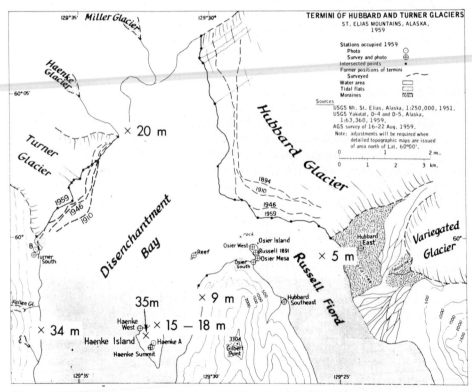

Fig. 1. Disenchantment Bay and vicinity, Alaska (map by W.O. Field); Fallen Glacier is at lower left, Station Reef at map center. Haenke Island is 3.5 km from the shoreline at the foot of Fallen Glacier. For regional location see Fig. 29.

2). In the second of these events in about 1850, waves reportedly killed about one hundred Indians who at the time of the fall, were at a summer seal camp a few kilometres south of Haenke Island (Fig. 1); apparently there was but one survivor (Tarr, 1909, p. 68). A similar event occurred in 1905, although not necessarily involving the same glacier (Fig. 2); fortunately the Indians had left the bay before the glacier fell, "for it is hardly conceivable that their canoes could have lived in the floating ice during the passage of such waves as this glacier avalanche generated . . ." (Tarr, 1909, p. 68; cf. Tarr and Martin, 1914, pp. 166—167).

Probably the most well-studied event in the Western Hemisphere has been the 1958 landslide and resulting waves in Lituya Bay, Alaska. Two of three fishing boats in the bay were sunk and two persons were killed by a 30 m high water wave traveling seaward at about 150—200 km/hr. The shore suffered extensive destruction (see e.g., Figs. 31—34).

Fig. 2. West side of Disenchantment Bay, Alaska. Fallen Glacier (arrow) as photographed in August 1959 from Station Reef (see Fig. 1 for location; photo M-59-P191, courtesy W.O. Field, The American Geographical Society). Large snow accumulations were visible after the glacier slide by 1909 (Tarr and Martin, 1914, p. 167). In 1946 the path of the slide was still visible and the glacier was reforming. By 1959 rocks beneath the glacier still showed where the slide had occurred, although alder thickets were taking hold. The glacier tongue was better formed than in 1946.

Results elsewhere have been even more catastrophic in terms of lives lost. In the 1963 Vaiont reservoir disaster in northern Italy, over 2000 people were drowned by a flood wave which, 1 km downstream from the slide, measured more than 60 m high (Müller, 1964). In Japan, over 15,000 deaths resulted from the 1792 Shimabara Bay catastrophe (Ogawa, 1924, pp. 219—224). In Norway, several hundred fatalities occurred in a series of rockslide-generated wave events dating to, at least, 1731 (Jørstad, 1968; Chapter 3, this volume) and in Peru, several hundred miners drowned in the lakeside 1971 Chungar disaster as described in Chapter 7, this volume. It is thus clearly desirable to predict the occurrences and properties of these abnormal waves. Our purpose here is to summarize the available theoretical and experimental knowledge on landslides and their resulting water waves. Attention is given to important variables, and various predictive models are compared with each other and with results from a Waterways Experiment Station hydraulic model. Finally, some better-known field cases are explored and their wave hindcasts discussed.

GENERAL WAVE DESCRIPTION

Observations of water waves generated by both prototype and model landslides fall into three classes of gravity wave types: oscillatory waves, solitary waves, and bores (Fig. 3). Oscillatory waves are periodic in the direction of travel and have nearly closed elliptical water particle orbitals. Water particles

SMALL AMPLITUDE OSCILLATORY WAVE

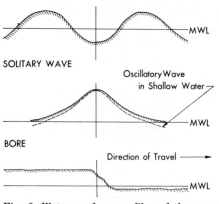

Fig. 3. Water surface profiles of three gravity wave types produced by landslides. Oscillatory waves have closed water particle orbitals whereas solitary waves and bores have a forward translation of mass. Solitary waves travel wholly above the mean water level (MWL).

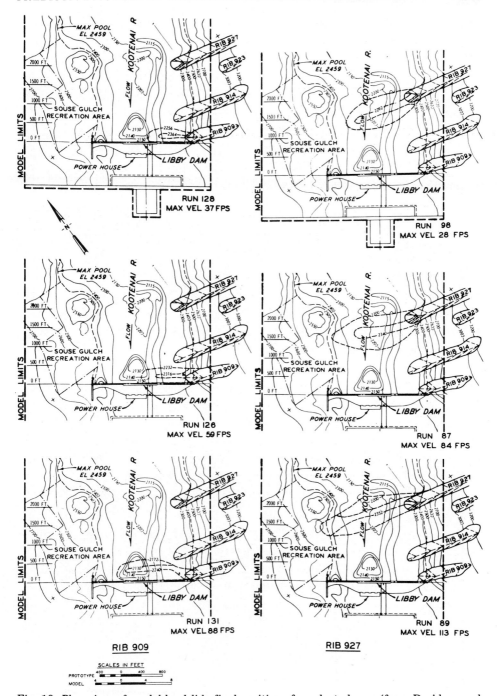

Fig. 19. Plan view of model landslide final positions for selected runs (from Davidson and Whalin, 1974, plates 21, 22).

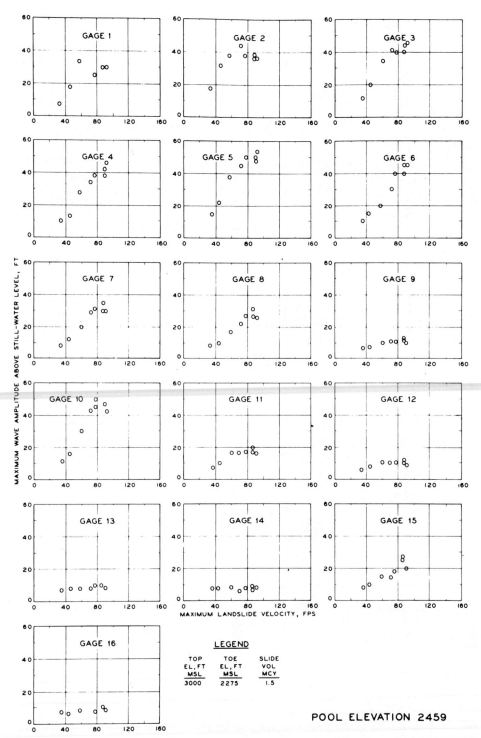

Fig. 20. Summary of maximum wave amplitudes recorded at a gage as a function of the maximum velocity rib 909 slides reached during movement (from Davidson and Whalin, 1974). For gage and rib locations see Fig. 17.

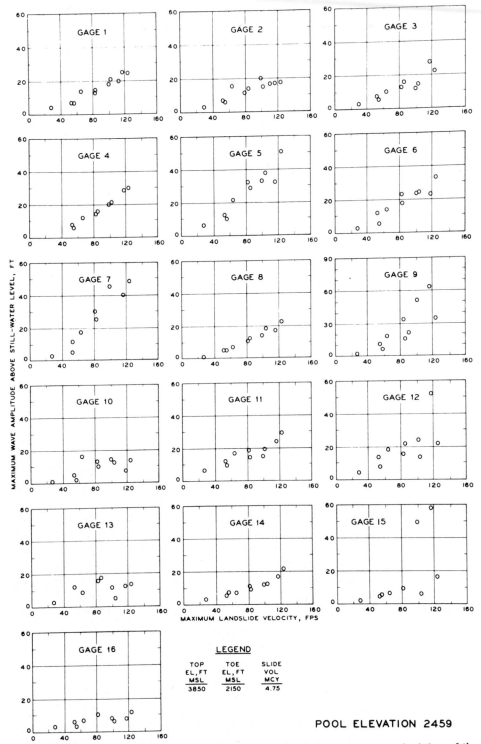

POOL ELEVATION 2459

Fig. 21. Summary of maximum wave amplitudes recorded at a gage as a function of the maximum velocity rib 927 slides reached during movement (from Davidson and Whalin, 1974). For gage and rib locations see Fig. 17.

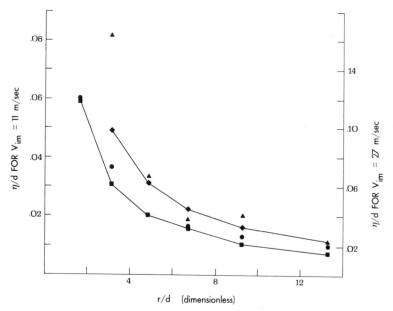

Fig. 22. Dimensionless wave amplitude (η/d) as recorded at selected gages versus dimensionless distance of the gages from the slide (r/d) for rib 909 slides in the WES Lake Koocanusa study. Experimental data, ▲, and function, ◆, are for slide velocity, V_{im} = 27 m/s (right scale) and data denoted by ●, and function, ■, are for V_{im} = 11 m/s (left scale).

uation factor. Kamphuis and Bowering give equation [6] to describe wave height decrease with distance from a slide for their experimental data. If H is measured as 2.7 m at x/d = 13, equation [6] predicts $H \sim 0$ at $x/d \sim 37$. This, however, is an exponential correction instead of the $1/x$ correction probably more appropriate for the WES study, which is more nearly an axially symmetric case. Fig. 22 shows dimensionless wave amplitude versus dimensionless distance as observed from WES runs for the maximum (V_{im} = 27 m/s) and minimum (V_{im} = 11 m/s) slide velocities at site 909, along with d/r functions plotted as before. For V_{min}, $\eta/d = 0.097\ (d/r)$, or $\eta/d|_{r/d \sim 37}$ = 0.00262; if d = 94 m, η = 0.25 m, instead of 0.9 m, the actual value at probe 14. For V_{max}, $\eta/d = 0.307\ (d/r)$ or $\eta/d|_{r/d \sim 37}$ = 0.0083; that is, η = 0.8 m instead of 2.7, the actual value at probe 14 for WES run 129. These amplitudes may be taken as "stable wave heights" since these are solitary waves. Summarizing, predicted values of H_{st} are 29 and 15 m for maximum and minimum rib 909 slide velocities respectively, and observed values corrected to "stable wave heights" are 0.8 and 0.25 m. Thus these two stable wave heights predicted by Kamphuis and Bowering are 36 and 60 times as large, respectively, as observed values from the WES study.

At least part of the difference may be attributed to the different geometries involved in the two cases. Probe 14 in the WES study senses waves traveling at an angle of 90° to the direction of slide emplacement, whereas Kamphuis and Bowering data are for wave heights in front of the slide. Also, as discussed previously, the WES model waves for rib 909 propagate through about 90° arc whereas Kamphuis and Bowering waves are confined to a two-dimensional channel. Finally, the WES slide is porous, whereas the Kamphuis and Bowering slide tray is not. All these differences would produce lower waves in the WES study; nonetheless, the magnitude of the discrepancy is an indication of the difficulty of the prediction problem.

Next, predictions of η_{max} based upon Noda's vertical box-drop theory are compared to maximum wave amplitudes generated by the model slide at rib 909. With water depth at the slide site about 94 m, slide Froude numbers range from 0.4 to 0.9, and $\lambda_m/d \sim 1$. These values fall in regions C and D of Fig. 11, and therefore from Table II, solutions for the maximum wave should be at $x/d = 5$ and $x/d = 0$, respectively. As an example, if slide velocity V equals 11.3 m/s, $F = 0.36$, and using $x/d = 5$ in Fig. 7, $\eta_{max}/\lambda_m = 0.31$, or if $\lambda_m \sim 90$ m, $\eta_{max} \sim 28$ m. Alternatively if slide 909 is modelled as a wall moving horizontally into the reservoir, equation [3] may be used to calculate predicted maximum wave amplitudes.

Predicted values for both the vertical and horizontal models of Noda are plotted in Fig. 23 against the observed maximum amplitudes of probes 2—4 in the WES study. The large jump in predicted values for the vertical box-drop case is a result of shifting from region C to D in Fig. 11 and underlines the fact that these are order of magnitude predictions. Vertical box-drop solutions overestimate maximum wave amplitudes by about a factor of 4, and horizontal solutions overestimate by a factor of 7. Reasons for the discrepancies must certainly include these differences between Noda theory and WES model: (1) the model 909 slide enters the water neither vertically or horizontally, (2) its thickness is less than the water depth, (3) wave energy is distributed in three dimensions, and (4) the slide mass has a porosity. All combine as Fig. 23 illustrates to produce lower wave heights in the WES model than predicted by theory.

Raney and Butler tested their numerical model by comparison with these same WES data. Fig. 24 gives the first-wave amplitudes observed for thirteen probes in the study versus the calculated wave amplitudes of the numerical model. The regression equation has a slope of 1.21 ± 0.46 at the 95% confidence level and $r_p = 0.85$; the fit appears quite favorable. The average difference in the amplitude of the first wave was 25% and an average difference of time of arrival at a probe was only 9%. Raney and Butler (1975, p. 23) conclude, ". . . the numerical model is capable of modeling landslide-generated water waves to a sufficient accuracy to allow overall engineering decisions to be made concerning the possible effects of a potential landslide. The most important parameters to be considered are the volume of the landslide, its velocity, and the final position of the slide in the reservoir."

The rate of decrease of wave height with distance for the WES model data

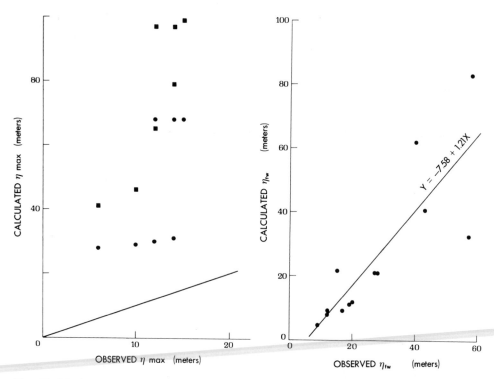

Fig. 23. Maximum wave amplitude observed at gages 2—4 in WES Lake Koocanusa study versus maximum wave amplitudes predicted by Noda theories, for runs using rib 909. Circles are from vertical drop model and squares are for the horizontal moving wall model. The line represents a one to one correspondence.

Fig. 24. Relationship for first-wave amplitudes (η_{fw}) observed at various gages in WES Lake Koocanusa study for one run versus wave amplitudes predicted by Raney and Butler numerical model. Slope of regression equation is 1.21 ± 0.46 and y intercept is -7.6 ± 15.2. Therefore, the possibility the regression equation demonstrates a one to one correspondence cannot be excluded.

(Fig. 22) appears to follow a simple inverse function of distance as predicted by Kranzer and Keller. Irregularities do occur however, due to bathymetric changes and piling up along the dam face. Even without these irregularities there is really no well-defined stable wave height as used in the two-dimensional experimental studies.

All studies predicted that when a wave train is formed, the leading wave is the highest. The WES data do not follow this pattern (Davidson and Whalin, 1974, appendix B, sheet 17). Close to a slide, as many model trials had the first wave highest as not, whereas for probes further away, the first wave was generally *not* the highest. Most likely this is the result of changes in

waveform due to shoaling of waves along the basin margins and wave reinforcement from reflected waves.

As discussed previously, Law and Brebner found a relationship between
wavelength (and thus period) and slide energy parameter whereas Kamphuis
and Bowering did not. The Unoki and Nakano, and Kranzer and Keller
theories also show no dependency. Prins (1958, figs. 3 and 4) showed experi-
mentally that period increased slightly as impulse width (thickness) increased
but was not dependent upon other slide factors. Fig. 15, from Law and Brebner, illustrates that period increases with increasing slide energy even for constant thickness. To explore the relationships between wave period at a point
and slide energy, periods from the WES data of the first two waves and
second and third waves for probe 14, and second and third waves for probe
11 were plotted against slide velocity, all other factors constant. Over the
range of velocities tested, there is no statistically significant correlation
between dimensionless slide kinetic energy and wave period. Wave period
does increase with increasing distance from the slide as predicted by both the
theoretical and experimental studies (Davidson and Whalin, 1974, appendix
B).

Lastly, we compare run-ups predicted by the Hall and Watts empirical
formula to observed maximum run-ups on the reservoir side opposite rib
909. The slope on land nearest probe 6 (see Fig. 17) is about $S = 0.2$. An
example calculation with a wave amplitude $\eta_h = 6$ m at probe 6 in water
depth, $d = 18$ m, is:

$$R_{calc} = [11(0.2)^{0.67}6(6/18)^{1.9(0.2)^{0.35}-1}] = 21 \text{ m}$$

The experimental run-up directly landward of probe 6 is estimated from Fig.
25, run 126, to be about 12 m. In Fig. 26, maximum run-ups on the reservoir side opposite slide 909 are plotted versus wave amplitude at gage 6 (circles). The relationship is a straight line of slope 3 or is slightly convex upwards. Also plotted is the predicted relationship from Hall and Watts which
is virtually a straight line of slope 3.7 falling above the WES data. But this
formula applies only to a solitary wave breaking parallel to a hydraulically
smooth shore of constant slope and not to water surges where all water particles in the water column translate forward causing a water surface bulge at
a boundary. Especially important also is the increase in slope in the WES
model as the run-up increases, which should produce lower observed run-ups
than predicted. However, since this is the maximum run-up, the values are
probably inflated because of convergence in topographic embayments. The
difference in observed versus predicted values could be a result of any one of
these complicating factors. Nevertheless, run-ups predicted from the Hall and
Watts formula compare favorably to those observed in the study.

Summarizing, predicted "stable wave heights" from Kamphuis and Bowering overestimate, by more than an order of magnitude, heights observed in

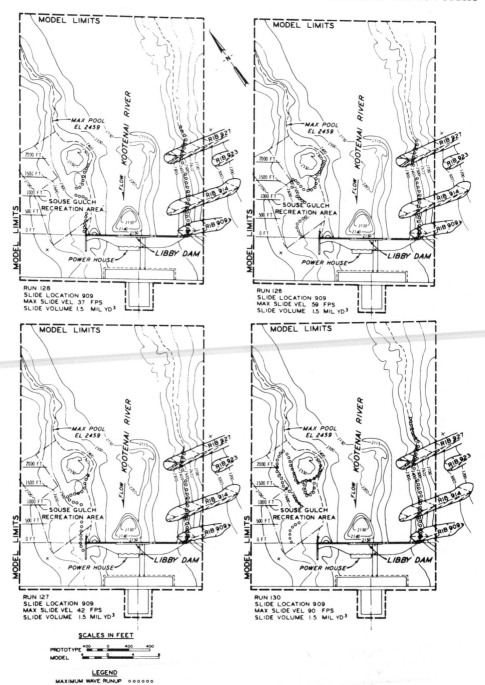

Fig. 25. Plan view of WES Lake Koocanusa model showing run-ups as a function of slide velocity for rib 909 slide (from Davidson and Whalin, 1974, plates C23, C24).

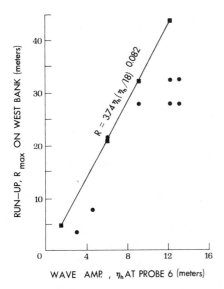

Fig. 26. Plot of maximum run-up on southwest side of model reservoir in WES Lake Koo-canusa study versus wave height at gage 6 (see Fig. 17 for location). Empirical data for rib 909 slides of varying velocities are shown as circles. Relationship of Hall and Watts is also plotted as squares and predicts higher run-ups than observed.

the WES model study; predictions of maximum wave amplitude from both Noda and Kamphuis and Bowering approaches also greatly overestimate observed WES values, but by somewhat less than an order of magnitude.

Raney and Butler numerical model predictions are only an average 25% different from those observed. The rate of decrease of wave height with distance in the WES study follows a simple inverse function of distance from the slide as predicted by Kranzer and Keller theory but contrary to Law and Brebner and Kamphuis and Bowering data. The leading wave is not generally the highest, contrary to predictions by Law and Brebner, Kamphuis and Bowering, and Noda. WES data show no relationship between slide velocity and wave period, supporting Kamphuis and Bowering, Unoki and Nakano, and Kranzer and Keller but contradicting Law and Brebner. Finally, run-ups predicted by Hall and Watts compare favorably with maximum run-ups observed in the WES study.

EMPIRICAL RELATIONSHIPS FROM WATERWAYS EXPERIMENT STATION STUDY

The WES data apply strictly only to one particular basin geometry and a limited range of slide characteristics. The extent to which the landslide material adequately simulates prototype conditions is poorly known and wave

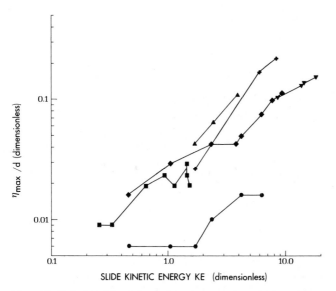

Fig. 27. Relationships of maximum wave amplitude observed at probes in the WES Lake Koocanusa study as a function of dimensionless slide kinetic energy. Symbol key: ● = rib 927 slide, η at probe $14(r/d \sim 7)$; ■ = rib 909 slide, η_{max} at probe $14(r/d \sim 13)$; ♦ = rib 927 slide, 0.002-m³ bags of model landslide material, η_{max} of probes 9—13; ▼ = rib 927 slides, 0.002-m³ bags of model landslide material, water depth of reservoir = 79 m, η_{max} of probes 9—13; + = rib 927 slide, 0.002-m³ bags of model slide material, η_{max} of probes 9—13; ▲ = rib 927 slide, 0.009-m³ bags of model slide material, η_{max} of probes 9—13. Unless otherwise specified, water depth of slide sites is 94 m. Note that different landslide material gives different wave heights for equivalent slide energies.

height values are a composite of reflection and refraction processes. Most disturbing is that duplicate runs do not always give duplicate results. However, the experimental arrangement is sufficiently realistic to make it appealing as a source of predictive equations more general than the experimental and analytical theoretical studies previously discussed. At a minimum these data illustrate the range in wave amplitudes for various slide characteristics in a three-dimensional situation with realistic wave height attenuation functions. Then too, modelling a landslide as discrete bags of shot is probably more realistic than using a tray or box with an imporous planar front. Finally, use of these data allows a choice of two different slide-basin geometries with which to match a particular prototype situation.

In view of these considerations, we have plotted in Fig. 27, dimensionless slide kinetic energy [6] (KE) for various WES slide sites and model slide materials versus dimensionless maximum wave amplitudes at two distances from the slide. Effective slide thicknesses are estimated as: $0.4 < h/d < 0.8$.

---

[6] Dimensionless slide kinetic energy is defined here as $\frac{1}{2}(l \cdot h \cdot w/d^3)(\rho_s/\rho)(V^2/gd)$.

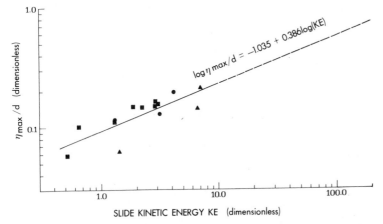

Fig. 28. Relationship of maximum wave amplitudes ($r/d \sim 2$–4) as a function of slide dimensionless kinetic energy. Rib 909 slides, denoted by squares have their KE doubled to simulate propagation of waves through 180°. Rib 923 slides are denoted by circles and rib 923 slides with a reservoir depth of 79 m are denoted by triangles. Normal reservoir depths at the slide sites are 94 m. The regression equation is significant at the 99% level.

The range of wave heights for a given kinetic energy is in part due to various probe r/d values and is also because for a similar slide and r/d, 0.002-m^3 bags of landslide material give lower wave heights than 0.009-m^3 bags. The angle between the sliding direction and radial azimuth from slide front to the probe also appears important.

Fig. 28 is a plot of η_{max}/d (at $r/d \sim 2$–4) vs. dimensionless kinetic energy for slides 909 and 923. For waves which propagate through less than 180°, the kinetic energy should be adjusted accordingly. The 909 slide kinetic energy has thus been doubled to make these data, where waves can disperse only through 90°, comparable to the 923 data where waves propagate through 180°. Slide 927 data were excluded from the plot because maximum wave heights were influenced by a shoaling bottom opposite the slide. A least squares linear regression on the logged data gives $r_p = 0.8$, which is significant at the 99% level. Predictions based on the regression equation:

$$\log(\eta_{max}/d) = a + b \log(KE) \tag{10}$$

are calculated in the next section. (For coefficients a, b, see Appendix 3.)

FIELD STUDIES

Table III gives examples of large waves generated by rock masses sliding into water. Some model studies were made concerning movements of the

TABLE III

Summary of historic slides and associated destructive water waves

Location	Date	Slide material	Dimensions	Volume	Height of slide area above water level	Slope angle
Alaska						
Fallen Glacier, Disenchantment Bay	4 July 1905	glacier ice	1067 m × 808 m × 34 m	29×10^6 m^3	300–900 m	28°
Lituya Bay	9 July 1958	schist	823 m × 970 m × 38 m	30.6×10^6 m^3	200–1000 m	40°
Iceland						
Steinsholt	15 Jan. 1967	avalanche of volcanic rock (basalt, hyaloclastite, ice fragments)	975 m scarp	15×10^6 m^3	150–450 m	5° to vertical
Italy						
Vaiont reservoir	9 Oct. 1963	limestone	2 km width	240×10^6 m^3	—	0—40°
Japan						
Shimabara	21 May 1792	volcanic debris	4 km width	0.5×10^9 m^3	520 m max.	10°

Norway

Rammerfjell, Stranda	8 Jan. 1731	gneiss	270 m × 120 m, thickness not known	324×10^3 m³ if 10 m thick	0—200 m?	—
Tjelle	22 Feb. 1756	granite gneiss	600 m × 250 m × 100 m	15×10^6 m³	400 m	>25°
Ravnefjell, Loen	15 Jan. 1905	gneiss, scree, moraine	100 m × 50 m × 10 m	50×10^3 m³ rock, and 30×10^4 m³ scree, moraine	400—500 m 0—400 m	65° to vertical
Ravnefjell, Loen	20 Sept. 1905	gneiss	—	—	400 m	65° to vertical
Ravnefjell, Loen	13 Sept. 1936	gneiss	400 m × 250 m × 10 m (?)	10^6 m³	400—800 m	65° to vertical
Ravnefjell, Loen	21 Sept. 1936	gneiss	—	—	—	65° to vertical
Ravnefjell, Loen	11 Nov. 1936	gneiss	—	10^6 m³ (?)	—	65° to vertical
Ravnefjell, Loen	22 June 1950	gneiss	—	10^6 m³ (?)	800—900 m	65° to vertical
Tafjord	7 Apr. 1934	gneiss	230 m width	$1—1.5 \times 10^6$ m³ rock, and $1—1.5 \times 10^6$ m³ scree	730 m max.	60°

Peru

Yanahuin Lake, Chungar	18 Mar. 1971	limestone	—	10×10^4 m³	400 m	45°

TABLE III (continued)

Location	Water depth at slide site	Maximum wave height or run-up	Length of affected shoreline or maximum distance where wave noticeable	Damage, fatalities	References
Alaska					
Fallen Glacier, Disenchantment Bay	80 m rough estimate	35 m run-up (see Fig. 1)	>5 km	area uninhabited; reportedly 100 perished in a previous event about 1845	Tarr (1909), Tarr and Martin (1914), Miller (1960), this paper
Lituya Bay	122 m	524 m run-up on opposite shore; >64 m wave in bay	entire bay, i.e., 12 km	2 perished, 2 boats sunk, shoreline devastated; earlier occurrence of giant waves also documented	Miller (1960), this paper
Iceland					
Steinsholt	small glacial lake, area 2×10^5 m²; initial depth unknown (average probably >8—13 m)	lake water thrown out, height >25 m, causing extraordinary flood wave; volume flood wave $\cong 1.5-2.5 \times 10^6$ m³	entire river affected >25 km	enormous rock blocks transported	Kjartansson (1967a,b)
Italy					
Vaiont reservoir	50 m; reservoir ca. 6 × 0.5 km²	270 m on opposite shore; 60-m wave 1 km downstream	entire river valley	>2000 perished; catastrophic destruction	Müller (1964)
Japan					
Shimabara	64 m max.	three giant waves; second wave highest, 10 m runup	damage along 76 km shoreline	>15,000 perished; >6000 houses destroyed	Ogawa (1924, pp. 219—224)

Norway

			damage reported		
Rammerfjell, Stranda	—	100 paces run-up at Stranda, on opposite shoreline	>2 km distant	destroyed settlement of Uren; 17 perished	Strøm (1766), Jørstad (1968)
Tjelle	max. fjord depth >200 m	50 paces on opposite shore; 20 pace run-up at 25 km distance	waves noticeable >40 km	32 perished; deep-water fish thrown 200 paces inland	Schøning (1778), Bugge (1937), Jørstad (1968)
Ravnefjell, Loen	<60 m, 132 m max. lake depth	40 m on opposite shore	entire lake, i.e., >8 km	61 perished; two villages destroyed	Helland (1905), Reusch (1907)
Ravnefjell, Loen	<60 m, 132 m max. lake depth	<15.5 m (?)	—	damage small because of January slide	Reusch (1907)
Ravnefjell, Loen	<60 m, 132 m max. lake depth	74 m opposite shore (see Figs. 37, 38)	>8 km	73 perished	Holmsen (1936), Jørstad (1968)
Ravnefjell, Loen	<60 m, 132 m max. lake depth	2–3 m	>8 km	minor damage	Holmsen (1936)
Ravnefjell, Loen	<60 m, 132 m max. lake depth	>49 m; 17 m at Högrending (2 km distant)	>8 km	considerable damage; no deaths	Holmsen (1936), Jørstad (1968)
Ravnefjell, Loen	<30 m	12–15 m	>8 km	minor; prior debris filled in lake below slide	Jørstad (1968)
Tafjord	max. fjord depth >200 m	62 m run-up adjacent to slide area; 32 m on opposite shore; three waves, last largest	90 km	41 perished; extensive damage for 50 km	Holmsen (1936), Kaldhol-Kolderup (1937), Jørstad (1968)

Peru

Yanahuin Lake, Chungar	lake area 100,000 m² ; average depth 38 m	30 m run-up on opposite shore	entire lake area	400–600 perished; destroyed mine camp	Plafker and Eyzaguirre (Chapter 7, this volume)

Gepatsch reservoir slope in Austria (Lauffer et al., 1967). In addition, at least one site of potential wave hazard is presently being studied in the United States (Baker Lake, Washington; Easterbrook, 1975), and one in Canada involving the Downie prehistoric slide mass (cf. Chapter 10, Volume 1). Slide-induced wave hazards at the Mica Dam in Canada have also been recently examined by model studies (see Appendix 3).

In this section we attempt to quantitatively hindcast waveforms for selected field cases. Unfortunately, few of the cited slides fulfill enough of the simplifying assumptions or are well enough documented to accomplish this. The Vaiont and Steinsholt slides clearly violate assumption (2) of Noda. They are large in relation to water volume of the reservoir. Also, many cases have waves generated whose main direction of travel is highly oblique to the direction of sliding; neither the Noda theory nor the empirical results based on flume experiments account for this, as evidenced by comparison with WES data.

To hindcast wave type, maximum or stable wave height, wavelength, and celerity for a slide, the minimum information needed would be slide velocity (maximum, or at impact), slide width, height, and thickness, slide density, basin bathymetry, and the angle at which the slide entered the water. For estimates of the amount of run-up along a shore, the slope and roughness at that point must be known. The 1958 Gilbert Inlet slide and waves in Lituya Bay are well enough documented to provide reasonable estimates of these values, and that event will be the major hindcast example. Following that we attempt to analyze the Disenchantment Bay glacier avalanche and some better known Norwegian landslides.

Case I: Lituya Bay, Alaska

Geographic and geologic setting

Lituya Bay is a T-shaped inlet that cuts through the coastal lowland and foothills belt flanking the Fairweather Range of the St. Elias Mountains on the south coast of Alaska (Fig. 29). The bay fills and slightly overflows a deep depression only recently occupied by a piedmont glacier of which Lituya, Crillon, and Cascade glaciers are remnants (Figs. 30, 31). Around the head of the bay the walls are glacially oversteepened, and fjord-like, rising to altitudes between 700 and 1100 m in surrounding foothills (Fig. 33). Submarine contours based on U.S. Coast and Geodetic Survey soundings in 1929 and 1940 show a pronounced U-shaped trench with steep walls and a broad flat floor sloping gently downward from the head of the bay to a maximum depth of 220 m (Fig. 30). Minimum depth at the entrance is 10 m at mean lower low water.

Weather records from the nearest stations (Cape Spencer, at Yakutat; Fig. 29) suggest total annual precipitation from 281 to 340 cm and near-annual temperatures about 40° F.

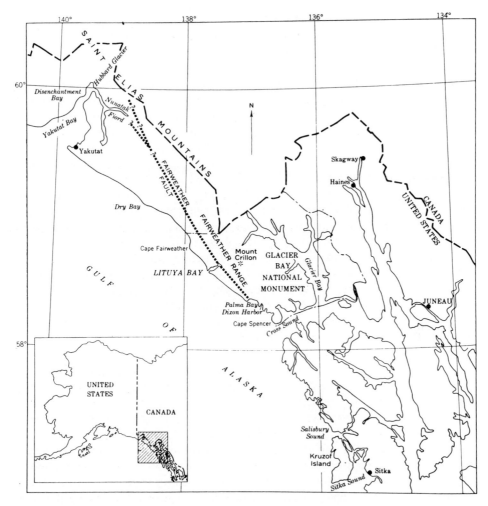

Fig. 29. Map of southeastern Alaska, showing locations of Lituya Bay and Disenchantment Bay (after Miller, 1960).

The Bay transects a geologic province involving sedimentary rocks of Tertiary age. The two arms forming the "T" at the head of Lituya Bay are part of a great trench, the topographic expression of the Fairweather fault (Miller, 1953). This fault in the vicinity of Lituya Bay is vertical or dips steeply to the northeast; along it the crystalline rocks exposed on the northeast side are inferred to have moved up relative to less altered and in part younger rocks to the southwest (Fig. 30). Large-scale systems of inward-dipping, conjugate faults exist in the fissured slopes along the Fairweather Fault, suggestive of downslope extension, i.e., "spreading ridges" as cited by Beck (1968;

cf. Chapter 17, Volume 1). The walls have been buttressed by glaciers until recently; radiocarbon dates on high moraines are less than 1000 years B.P., suggesting retreat of glaciers only in the last millenium (G. Plafker, oral communication, 1975).

Movement along the Fairweather Fault is considered to have been associated with an earthquake that directly preceded the 1958 wave (Tocher and Miller, 1959), with the epicenter located about 12 km east of the fault trace and 21 km southeast of Lituya Bay (Brazee and Jordan, 1958, p. 36; however, see Stauder as cited by Miller, 1960, p. 55, for revised location).

The 1958 Gilbert Inlet rockslide and resulting wave event

Beginning about 10:16 p.m. local time, July 9, 1958, the southwest side and bottom of Gilbert and Crillon Inlets moved northwestward and possibly up relative to the northeast shore at the head of the bay. Total movement as much as 6.4 m horizontally and about 1 m vertically was noted from surface breakage 8—16 km south of Crillon Inlet (Tocher and Miller, 1959). Intense shaking in Lituya Bay continued from 1 to 4 minutes, the range of estimates of two eyewitnesses anchored in the bay. Within 1—2.5 minutes a large mass of rock slid from the northeast wall of Gilbert Inlet (Fig. 31) causing a "deafening crash" reported by one of the eyewitnesses. The rockslide — judged by Miller to be near the borderline between "rockslide" and "rockfall" as defined by Sharpe (1938, pp. 76—78) and Varnes (1958, pp. 20—32, plate 1) — occurred in an area of previously active sliding and gulleying to an altitude of about 914 m on a slope averaging 40°. The rocks are mainly amphibole and biotite schists; bedding and schistosity strike about N50°W and dip steeply northeastward, into the slope. Slide surfaces thus probably predominantly involved joint or fault surfaces transecting bedding. The dimensions of the slide on the slope as mapped by Miller seem fairly accurate, but the thickness of slide mass normal to the slope could be estimated only roughly (Miller, 1960, p. 65). The main mass of the slide presumably involved a prism of rock roughly triangular in cross-section, with width dimensions of 732—915 m, length measured down the slope of 970 m, maximum thickness of 92 m normal to the slope, and a center of gravity at about 609 m altitude (Fig. 32). Miller estimated the volume from these dimensions to be 30.6×10^6 m^3 — about the same size as the Madison Canyon, Montana, slide (see Chapter 4, Volume 1) — and assuming a specific gravity of 2.7, a weight of 82×10^6 metric tons. It is highly probable that the entire mass plunged into Gilbert Inlet as a unit at the time of the earthquake, although the available data require only that the event occur between noon on July 7 and the morning of July 10. Loose rock debris on the fresh scar was still moving at some places on July 10, and small masses of rock were still falling from the steep rock cliffs at the head of the scar.

The impact caused a huge sheet of water to surge up over a high spur on the opposite side of Gilbert Inlet (Figs. 31, 33b, 34); a large gravity wave

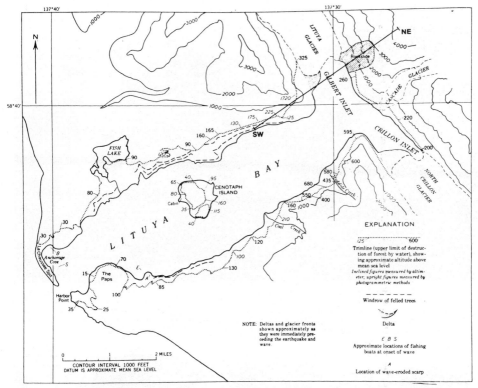

Fig. 31. Lituya Bay, showing setting and effects of 1958 rockslide and giant wave (after Miller, 1960). Cross-section given in Fig. 32.

with a steep front was set into motion, traveling at high velocity. The wave struck first against the south side of the bay near Mudslide Creek (Figs. 31, 33) with maximum run-up over 200 m, and was then reflected towards the

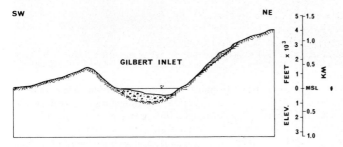

Fig. 32. Cross-section through Gilbert Inlet at head of Lituya Bay, showing rockslide on northeast wall. Surge crossed high spur bounding Gilbert Inlet on southwest.

Fig. 33. (a) Lituya Bay, 1954. Trimlines of 1936 giant waves of unknown origin (*g*) and 1853—1854 (*k*). Lateral moraines (*m*) and end moraine in foreground record recent advance of ice to bay mouth. (b) Lituya Bay, August 1958 (after Miller, 1960). Wave generated on July 9 by rockslide (*r*) destroyed forest to maximum elevation 524 m at *d* and to a maximum distance of 1100 m in from high-tide shoreline at Fish Lake (*F*). Fishing boat anchored at *b* was carried over spit; boat at entrance was sunk, and boat at *e* rode out the wave.

Fig. 34. Rockslide plunged into Gilbert Inlet at lower right corner, shearing off part of Lituya Glacier, and causing water to surge over high spur in photo center (after Miller, 1960). View to west, August 1958. Trimline cuts across old slide scars on spur.

north shore, and again back to the south shore near Coal Creek. Estimates on elapsed time from first sighting of waves to arrival at eyewitnesses positions suggest an average speed of 156—209 km/hr (Miller, 1960, p. 64). Midway between the head of the bay and Cenotaph Island, wave amplitude was about 30 m and the wave crest was 8—15 m wide. After passing Cenotaph Island maximum wave height decreased to perhaps 15—23 m with the back slope of the wave less steep than the front. The wave then traveled over La-Chaussee spit, taking a trolling boat (the "Badger") with it at an estimated height of "two boat lengths" (24 m) above the trees growing on the spit. Following the passage of the giant wave, the bay water returned to about normal water level but continued to surge for about 25 minutes, with steep waves up to 6 m high. The estimated wave speed seems in good agreement with the theoretical speed as calculated from $C = \sqrt{g(d + \eta)}$, where g is acceleration of gravity, d is depth of water, and η is amplitude of wave above sea level.

The highest point on the trimline on the spur was at 524 m altitude (Figs. 31, 34), nearly eight times the maximum height reached by the largest of the celebrated Norwegian slide-generated waves (Table III). The initial report of wave damage at this elevation was thus at first widely doubted (Miller, 1960,

p. 64), but re-examination of the area from the air and on the ground confirmed the initial supposition. Minor sliding of the spur occurred both before
and after the water swath destroyed the forest cover.

Other alternative wave-generating mechanisms were considered, but seem
less acceptable than the rockslide mechanism. An eyewitness account and
configuration of the trimlines indicate approximately radial wave propagation from a point source in Gilbert Inlet. The size of the slide, water depth,
and dimensions of Lituya Bay are compatible with the generation of a wave
similar to a solitary wave (R.L. Wiegel, *in* Miller, 1960, pp. 65—66).

The thoroughness of the destructive effects of the wave are described in
detail by Miller (1960, pp. 60—63). The forest cover was stripped nearly to
the limit of inundation (Fig. 31), to a maximum of 1100 m inland from the
high tide shore line. In most places the trees were washed out and trans-
ported away, leaving bare ground. In some places trees greater than a metre
in diameter were broken off cleanly above the root system. Many of the
felled trees were reduced to bare stems, with limbs and roots removed and
bark stripped by water at high velocity or pressure. The total area over which
the wave was capable of such destruction was about 10 km^2 compared to
total area of inundation of 13 km^2. About 0.3 m of soil on average was
removed between the trimline and shore, amounting to about 3×10^6 m^3.
Two of three fishing boats in the outer part of the bay were sunk, and
two persons were killed.

Wave hindcast for Lituya Bay

Table IV summarizes the available data. A wave amplitude of 30 m at
$r/d \sim 30$—40 was estimated by an eyewitness on a fishing boat at the
entrance to Lituya Bay. This is a precarious position from which to be estimating wave amplitude; possibly a more accurate estimate of wave height
can be back-calculated from observed run-up. Field observation and model
studies by Wiegel (see below) suggest that a large gravity wave moved in a
straight path nearly due south impacting near Mudslide Creek (Fig. 31). Observed prototype run-up at that point is about 183 m, the slope is about 1.1
and the depth offshore is 146 m. The wave amplitude necessary to produce
this amount of run-up is, according to the Hall and Watts formula, $\eta = 64$
m at about $r/d \sim 22$ [7]. This seems a minimum estimate insofar as no roughness due to shore irregularities and vegetation is considered and because the
wave must have struck the shore obliquely at this point.

First, using Noda's theory and his approach to the Lituya problem, if
$V_{im} = 56$ m/s and $\sqrt{gd} = 35$ m/s, the slide Froude number = 1.6. If $\lambda_m = 38$
m, $\lambda_m/d = 0.31$, which places the solution in the nonlinear transition region
B of Fig. 11. Table II suggests for this region that a linear solution be used at

---

[7] See Table V for calculations. The slope component parallel to the direction of wave
advance is 0.7.

x/d = 5 regardless of the actual x/d. Therefore from Fig. 7, η_{max}/λ_m = 0.35 for F = 1.6 and η_{max} is only 13 m. If V_b = 69 m/s is used, η_{max} = 14 m.

These calculations differ from those of Noda (1970, p. 847) in having lower slide velocities and thickness values. The velocity as estimated above seems more appropriate than solutions based on frictionless transport (see Appendix 2); nevertheless, the solution is not particularly sensitive to this parameter. A slide velocity about a third again as large barely increases the wave amplitude a metre. If the slide was 92 m thick (maximum reported value), however, instead of 38 m, λ_m/d = 0.75, which places the solution in solitary wave region D for V_{im}. The linear solution for x/d = 0 (Fig. 8) suggests η_{max}/λ_m = 0.93, hence, η_{max} = 86 m, somewhat greater than the (minimum estimate of) wave amplitude necessary to produce the required run-ups at r/d = 22. For λ_m = 92 m, and V_b, η_{max} = 88 m. Since the model estimates maximum wave amplitude and is a two-dimensional approximation, it should give conservative results far away from surge effects; the amplitude prediction is therefore possibly consistent with prototype estimates. The range of model results thus illustrates the sensitivity of predictions based on the Noda vertical box drop theory to the thickness parameter. Of most importance to the prediction is identification of the appropriate wave characteristics region. For $V_{im}/\sqrt{gd}$ = 2, region B is indicated for $\lambda_m/d <$ 0.43, whereas region D is indicated for $\lambda_m/d >$ 0.43 (Fig. 11). Corresponding solutions [linear solution for x/d = 5 for region B; linear solution for x/d = 0 for region D (Table II)] are discontinuous at region boundaries. Thus for λ_m = 52 m (region B), predicted η_{max} = 0.35(52) = 18 m, whereas for λ_m = 53 m (region D), predicted η_{max} = 0.96(53) = 50 m. For the Lituya case, the observed solitary wave implies that region D is appropriate which in turn suggests the effective slide thickness was greater than 53 m. This thickness value seems reasonable in view of estimates of slide dimensions and the possibility of bulking at the front of the slide with penetration into the water body.

If the actual effective slide thickness, λ_m, was closer to the estimated maximum value of 92 m, in water of depth 122 m the slide could effectively act as a wall moving horizontally into Gilbert Inlet (Fig. 33). This suggests the possibility of using the horizontally moving wall theory of Noda. [8] Equation [3] rewritten is:

$$\eta_{max} = 1.32\, d(V/\sqrt{gd})$$

or for V_{im} and V_b, η_{max} = 261 and 321 m, respectively. These presumably should occur at a distance of x/d = 2 or over 200 m in front of the slide.

From Kamphuis and Bowering, q = $(l/d)(h_k/d)$ = 2.5 and 6.0 for h_k = 38 and 92 m. From equation [5], for h_k = 38 m and V_{im} = 56 m/s, l^r ./d =

---

[8] Some assumptions of Noda theory, especially small displacements relative to depth, are clearly violated. With regard to maximum wave amplitude, Noda solutions indicate minor decay in the range x/d = 2—5.

TABLE IV

Summary of landslide and wave values for the 1958 Lituya Bay and 1905 Disenchantment Bay, Alaska, events

Variable	Lituya Bay		Disenchantment Bay	
	value	description and source	value	description and source
ρ_s	2.7	assumed	1.0	assumed
w	823 m	width—average of 732 and 915 m values given in Miller (1960, p. 65)	366 m 808 m	initial width at water entry (Miller, 1960, p 66; Tarr, 1909)
h, λ_m or h_k	38 m 92 m	calculated average thickness maximum thickness of slide roughly estimated by Miller (1960)	34 m 75 m	at water entry [+] initial width
l	970 m	length measured along slope from Figs. 31, 33	1067 m	length of glacier estimated from Miller (1960)
Volume	30.6×10^6 m^3	calculated from estimated dimensions by Miller (1960, p. 65)	29×10^6 m^3	calculated from above dimensions
i	40°	slope angle measured from Figs. 30, 33	20° 28°	last 400 m to shore average value from Fig. 36
$\tan \phi_s$	0.25	average value based on run-out data of other slides (Appendix 2)	0.25	estimated average value
s	356 m 545 m	distance of sliding along slope: to point where slide front hits water to point where slide front hits bottom *	743 m 1418 m	distance of sliding of front of glacier to water level distance of centroid to water level

	Value	Value 2	Description
d	122 m		water depth at slide front from bathymetric charts
	140 m		average depth of bay
	146 m		offshore depth in vicinity of Mudslide Creek
V_{im}	56 m/s	60 m/s	velocity of impact of front of slide with water ** / impact of slide front ‡‡
V_b	69 m/s		velocity of impact of slide front with bay bottom ignoring velocity decrease due to water drag
V_c	83 m/s		when centroid meets mean water level, ignoring drag
η_{max}	100—224 m	26—77 m	estimated (see Table V) / estimated (see Table V)
H_{st}	64 m	4—6 m	estimated from run-up near Mudslide Creek / estimated (see Table V)
	30 m	5 m	estimated from eyewitness for wave midway between head of Bay and Cenotaph Island / eyewitness (Tarr and Martin, 1914)
	15—23 m		after passing Cenotaph Island
C	156—209 km/hr		wave celerity calculated from reports

* If water depth = 122 m, and distance from slide front to water is 356 m then s = 356 + 189 = 545 m.

** To calculate slide velocity: From Appendix 2, equation [A-25], $V_{im} = [2 \cdot 9.8 \cdot 356(0.643 - 0.25 \cdot 0.766)]^{1/2} = 56$ m/s. Calculation of slide velocities is subjective. Estimates of velocities at impact for this slide have been 109 m/s (Noda, 1970) using free-fall equations for a slide centroid at 609 m elevation, and 110 m/s (Law and Brebner, 1968), assumptions unstated, but probably also assuming no frictional loss.

‡ In a personal communication dated March 1976, W.O. Field stated "My guess is that the glacier would average less than 100 m in thickness, possibly closer to 50 m". Taking 75 m as a best estimate, and assuming thinning proportional to spreading over a half mile front, slide thickness = 75 · 366/808 = 34 m.

‡‡ Calculated as from Lituya Bay, using equation [A-25], with assumed tan ϕ_s = 0.25, $V_{im} = [2 \cdot 9.8 \cdot 743 (\sin 28° - 0.25 \cos 28°)]^{1/2} = 60$ m/s.

$(1.6)^{0.7}(0.31 + 0.2 \log 2.5) = 0.54$ or $H_{st} = 0.54(122) = 66$ m. Similarly for $V_b = 69$ m/s, $H_{st} = 77$ m. If $h_k = 92$ m, for V_{im}, $H_{st} = 79$ m and for V_b, $H_{st} = 92$ m. These values based upon $d = 122$ m, and others based upon $d = 140$ m [9] are compatible with estimates based on run-up; however, the approximation is two dimensional, and $\theta > 30°$, both of which should make upper-bound predictions. We note that q is beyond the range of experimental data $(0.05 \leqslant q \leqslant 1.0)$.

For an estimate of maximum wave heights, arbitrarily taken at $x/d = 4$ unless otherwise stated, for $h_k = 92$ m, $d = 122$ m and V_b, equation [6] gives:

$$H_{max}/d = 0.75 + 0.35\, e^{-0.08(4)} = 1.00$$

Thus $H_{max} = 122$ m. Similarly for $h_k = 38$ m, $d = 122$ m and V_b, $H_{max} = 108$ m.

We may also use equation [10] to predict maximum wave amplitudes; h/d is in the same range as the model data (0.3—0.8). Dimensionless kinetic energy of the Gilbert Inlet slide for V_{im} is:

$$\frac{1}{2}\left(\frac{30.6 \times 10^6}{122^3}\right) 2.7\left(\frac{56^2}{9.8 \cdot 122}\right) = 60$$

This situation is in some ways geometrically analogous to the WES rib 909 slide where waves could propagate only through $90°$; when applied to equation [10] the kinetic energy should possibly be doubled, giving $\eta_{max} = 1.69(122) = 206$ m. For V_b, $KE = 91 \times 2 = 182$ and from equation [10] $\eta_{max} = 2.27(122) = 277$ m. These amplitudes seem large, but certainly credible in view of surge and run-up observations. Non-doubled KE suggests amplitudes of 126 and 169 m.

The hindcasts have been summarized in Table V. Since the best slide velocity estimate is probably between V_{im} and V_b, the two wave heights given should bracket the true value. The predicted maximum wave heights range from 13 m for a thin slide modeled by Noda vertical box-drop theory to 321 m estimated by Noda's horizontally moving wall theory, a twenty-five fold increase. The larger estimates are enormous waves, but then Figs. 31 and 34 show that the trimline on the spur opposite the slide was at 524 m, an impressive elevation for water wave surge to strip forest cover.

The estimated minimum stable wave height at $r/d \sim 14$—30 is 64 m. Footnote [‡] of Table V shows for this case that if wave height attenuation follows an inverse function of distance from the slide, the wave height at $r/d \sim 4$ should have been 224 m [10]; this solution is a kind of upper bound, inasmuch

---

[9] See Table IV for explanation of various depths.

[10] According to Ippen (1966), a solitary wave will begin to break when $(H/d)_{max} = 0.78$, or, if for this case $d = 122$ m, the maximum nonbreaking solitary wave would be 95 m in height. Therefore we mean to imply here only that the back-calculated wave height would have *ideally* been 224 m. However, a larger wave could form that need not be solitary nor stable.

as wave propagation was not radial but was restrained by the irregular geometry of the bay. A lower bound is given by the function describing two-dimensional attenuation, i.e. equation [6], which gives for a maximum wave height at $x/d \sim 4$, $H/140 = (64/140) + 0.35\ e^{-0.08(4)}$ or $H_{max} = 100$ m.

Theory may be adequate to predict the enormous surge wave on the spur at Gilbert Inlet (Fig. 34; see footnote, Table V). Wiegel (*in* Miller, 1960, pp. 65—66) constructed a 1 : 1000 scale model and conducted model experiments. His results suggested that the prototype slide must have fallen virtually as a unit, and very rapidly; if these conditions were met experiments showed that a sheet of water washed up the opposite slope to an elevation about three times water depth. At the same time a large gravity wave "several hundred feet high"[11] moved in a southerly direction, causing a peak rise in the vicinity of Mudslide Creek much as observed. The wave then swung around into the main portion of Lituya Bay, due to refraction and diffraction. Movements of the main wave and tail were additionally complicated due to reflections, but scale modelling apparently produced a good approximation to the Lituya event.

Judging from these data the maximum wave height should have been at least about 100 m; it might have been twice that high. The correct order of magnitude is therefore predicted by Kamphuis and Bowering and *KE* empirical function hindcasts, with the former giving values near the lower-bound estimate. Some Noda solutions are similar; however these predictions are very sensitive to assumed values of the thickness, as discussed previously.

Cenotaph Island, in the center of Lituya Bay, provides a prototype case most similar to the model for the comparison of observed versus predicted wave run-ups. There, an estimated 30-m solitary wave traveling down the bay in water about 140 m deep, shoals on a fairly uniform slope ($S = 0.1$; depth and slope calculated from U.S. Coast and Geodetic Survey Chart 8508, 1972; cf. Fig. 30). Using the Hall and Watts formula:

$$R/30.5 = 11(0.1)^{0.67}(30.5/140)^{[1.9(0.1)^{0.35}-1]}$$

or

$R = 90$ m

From Fig. 31 the trimline on Cenotaph Island was about 29—49 m above mean sea level. Considering that the trimline is the upper limit of forest destruction, this value should be less than the predicted run-up because of roughness and energy dissipation from trees. Therefore, the predicted value seems satisfactory.

Wiegel (*in* Miller, 1960, p. 67) estimated the energy of a solitary wave 30

---

[11] Note that this figure is in excellent agreement with wave heights back-calculated from run-ups at Mudslide Creek and the spur at Gilbert Inlet.

TABLE V

Summary of predicted and observed wave heights and run-ups for Lituya Bay and Disenchantment Bay events

	Lituya Bay				Disenchantment Bay			
	η_{max} (m)		H_{st} (m)		η_{max} (m)		H_{st} (m)	
Wave height:								
Velocity used:	V_{im}	V_b	V_{im}	V_b	V_{im}	V_c	V_{im}	V_c
Method								
Noda vertical box drop								
$\lambda_m = 38$ m	13	14	—	—	32	34	—	—
$\lambda_m = 92$ m	86	88	—	—	—	—	—	—
Noda horizontal moving wall								
$\lambda_m = 92$ m	261	321	—		—		—	
Kamphuis and Bowering								
	$h_k = 38$ m				$h_k = 34$ m			
$d = 122$	97	108	66	77	83	99	63	79
$d = 140$	104	114	68	79	—	—	—	—
	$h_k = 92$ m							
$d = 122$	110	122	79	92	—			
$d = 140$	119	132	83	96				

($\lambda_m = 34$ m for Disenchantment Bay)

(equation [10]; Appendix 3)

estimate from KE	126	169	—	—	143	226	—	—
estimate from $2KE$	266	277	—	—	—	—	—

Actual values

Lower bound of wave amplitudes back-calculated from run-up:
100 ** , 64 * (Mudslide Creek), —, —, 26 ** , 4—6
188 *** , 157 (Gilbert Inlet), —

Maximum wave amplitude based on inverse formula ‡ : 224, —, 77, —

Observed "stable" wave height (H_{st}) and η_{max} back-calculated from H_{st} : 66 ** , 30, 25 ** , 5

* Value of 64 m appears appropriate along the section of the shore near Mudslide Creek 2—4.4 km or $14 < r/d < 30$ from the slide. The calculation using equation [9] is:

$$183/\eta_h = 3.05(0.7)^{-0.13}\,(\eta_h/146)^{1.15(0.7)^{0.02}-1}$$

or $\eta_h = 64$ m. The slope is taken parallel to assumed direction of wave advance. If run-up is assumed perpendicular to shore, wave amplitude is 68 m; apparently the longer distance over which the run-up must travel in the first case more than compensates for the steeper slope in the second.

** Value extrapolated to $r/d \sim 4$ by equation [6]; e.g., using $H_{st} = 64$ m for Lituya Bay and $d = 140$ m, $\eta_{max} = 100$ m.
*** For the 524-m run-up on a 0.64 slope at Gilbert Inlet, $\eta = 157$ m at $x/d = 6$—10; $\eta_{max} \approx 188$ m at $x/d = 4$ by equation [6].
‡ Value extrapolated to $r/d \sim 4$ by simple inverse formula; e.g., if $\eta = 64$ m in lower Lituya Bay:

$$\eta = k/r \text{ or } k = \eta r_{r/d \sim 14} = 1.25 \times 10^5 \text{ m}^2. \text{ At } r/d \sim 4, \eta = 1.25 \times 10^5/4 \cdot 140 = 224 \text{ m.}$$

m high in water 120 m deep with a channel width of 2400 m to be about 8.2 $\times 10^{12}$ J. [12] Wiegel estimated the potential energy of the slide at 4.6×10^{14} J (which for assumed free fall is the same as the kinetic energy [13]), and suggested that the total wave energy of the first solitary wave was about 2% of the kinetic energy of the slide upon impact. However, this estimate is based upon a questionable free fall velocity. If V_b = 64 m/s is used, slide kinetic energy is 1.69×10^{14} J and the wave energy is about 5% of this value.

Case II: Fallen Glacier, Disenchantment Bay, Alaska

While conducting studies of Alaska tidewater glaciers in 1905, Ralph S. Tarr documented an unusual glacier fall and its resulting waves (Tarr, 1909, pp. 67—68):

"On the western wall of Disenchantment Bay, between Black and Turner glaciers, three small glaciers were perched in short, steep hanging valleys [Fig. 35; see Figs. 1 and 29 for location]. Their slope was so steep and they had such an appearance of instability that they attracted particular attention. The southernmost of these was estimated to have a length of approximately a mile, its chief supply coming from a steep mountain crest from which the snow slides into a cirquelike amphitheater about halfway down the slope. The glacier was photographed from the crest of Haenke Island by Russell, 1890; by Brabazon, of the Canadian Boundary Commission, in 1895; and by Gilbert in 1899. Attracted by the steep inclination of the three perched glaciers, the Survey party photographed this mountain side from the bay on July 3, 1905, which happened to be the last day in the life of the southernmost of the three.

This glacier, which I will call Fallen Glacier, lay for the most part in a cirquelike amphitheatre with steeply rising mountain walls at its head [Figs. 35,2,36]. The amphitheatre has a narrow mouth, out of which the crevassed terminus of the glacier protruded, the lower end terminating at an elevation of about 1000 feet above the fiord, from which it was separated by an ice-steepened rock slope. Aside from its apparently instable position, there was so little to attract special attention to this glacier that no detailed observations were made on it.

A moderate rain fell during the night of July 3 and continued during July 4. On the latter day, when working in Russell Fiord, about 15 miles [14] from Fallen Glacier, I was surprised by the appearance of a series of waves far too pronounced and

---

[12] From Ippen (1966), for a solitary wave:

total energy = $[(8/3\sqrt{3})\gamma H^{3/2} d^{3/2}$ ft-lbs/ft of crest width] · crest width

$= (8/3\sqrt{3}) \cdot 62.4 \cdot 100^{3/2} \cdot 400^{3/2} \cdot (8000) = 6 \times 10^{12}$ ft-lbs $= 8.2 \times 10^{12}$ J

[13] Kinetic energy $= \frac{1}{2}mV^2 = \frac{1}{2} \cdot 2.7 \times 10^3 \cdot (30.6 \times 10^6) \cdot 106^2 = 4.6 \times 10^{14}$ J.

[14] The value of 5 miles appears possibly more consistent with these observations and photographs taken on July 5 (Tarr and Martin, 1914, plate XLVII).

Fig. 35. (a) Hanging glaciers on west side of Disenchantment Bay, from Haenke Island (1891 photo by I.C. Russell, No. 523, U.S. Geological Survey; see Fig. 1 for location). Fallen Glacier immediately above fisherman. (b) Comparison of above, photographed in August 1959 from station Haenke A (see Fig. 1) near summit of Haenke Island (F-59-R100, courtesy W.O. Field, the American Geographical Society). The glacier has reformed.

lasting far too long a time to be ascribed to iceberg origin. The water rose and fell from 15 to 20 feet, and the disturbance lasted for fully half an hour. At the time I could think only of earthquake origin for the waves, but the next day one of the Indian guides returning from Yakutat, reported the falling of a glacier in Disenchantment Bay. Later in the season, returning to the west side of Disenchantment

Fig. 36. Portion of Yakutat (D-5) 15′ Quadrangle, 1959, U.S. Geological Survey, showing Fallen Glacier (arrow) and Disenchantment Bay. Note sparse bathymetric data, based on 1906 surveys.

Bay, it was found that the glacier which had fallen was the southernmost of the three small glaciers described above.

The valley was almost completely emptied of ice, there remaining only a mere remnant of the steeply perched neve area and some minor ice fragments near the edge of the cirque. The entire glacier had evidently shot out of its valley, tumbled a thousand feet down the steep slope, and entered the fiord, generating a series of pronounced waves. The walls and bottom of the cirque were bare of ice and distinct evidence of the avalanche was present on the sides of the narrow throat of the amphitheatre out of which the glacier shot. Emerging from this throat the avalanche had spread out fan-shaped, sweeping all soil away, and near the fiord killing the alders over an area half a mile in width. Since the fiord is evidently deep at this point, only a small remnant of the avalanche was visible at the time of visit, most of the ice having floated away and the debris sunk to the bottom. The coast was pushed out slightly, with a new shore line of angular rock debris, beneath which ice evidently remained, since in places the surface was freshly faulted by slumping.

The water wave generated by this avalanche was of great height near its source. A half mile south of Fallen Glacier the wave rose 110 feet, breaking off alder bushes at that height [Fig. 1 gives these locations]. Three miles north of it, near Turner Glacier, vegetation was killed by the wave to a height of 65 feet. About an equal distance, on Haenke Island, the wave swept to a height of 50 or 60 feet on the north end, and 115 feet on the northwest end of the island, washing out good-sized alders at that level; but the latter unusual elevation was due to an especially favorable topography which developed high breakers."

Table IV summarizes data on the glacier fall. No documentation of observed maximum wave amplitude is available and Tarr's estimate of stable wave amplitude is for waves which have filtered past Station Reef (Figs. 1, 2) through Russell Fiord. Therefore we back-calculate wave amplitudes from observed run-ups close to the glacier.

The observed run-up northwest of Gilbert Point (Fig. 1) was 9 m. The slope as measured from the Yakutat, Alaska 15' quadrangle map (Fig. 36) is 0.57 and the depth offshore from that point is 81 m. Thus from equation [9], η = 4 m plus the height necessary to correct for roughness due to alders. Similarly, a run-up of 35 m [15] on the northwestern tip of Haenke Island on a slope of 0.3 requires a wave amplitude of 12 m in water of 81 m, and on the northern tip, a run-up of 16.5 m on a slope of 0.21, requires an amplitude of 6 m. Thus, including Tarr's observation, wave amplitudes at distances of r/d = 46, 51, 73, and 132, are respectively 12, 6, 4, and 5 m.

Equation [6] gives an estimate of η_{max} at $r/d \sim 4$, for η = 6 m, as:

$$\eta_{max} = [\tfrac{6}{80} + 0.35 \ e^{-0.08(4)}] \cdot 80 = 26 \text{ m}$$

If 5 m is assumed to be the stable wave amplitude, calculation gives: η_{max} = 25 m. However, this equation is based on two-dimensional models [16]; if the inverse distance formula of footnote $^{+}$, Table V, is used, $\eta_{max} \approx$ 77 m for r/d = 51. Thus the true maximum wave amplitude from the glacier fall was greater than 25 m and probably closer to 77 m.

Noda's solution for the vertical box-drop model follows:

$$\lambda_m / d = 34/80 = 0.43$$

and for V_{im} = 60 m/s:

$$V_{im}/\sqrt{gd} = 60/\sqrt{9.8 \cdot 80} = 2.1$$

---

[15] Value possibly influenced by wave concentration; see Tarr quotation.
[16] Application of this equation obviously has its limitations. A 24-m wave is predicted with no stable wave height at all.

Therefore, the solution falls in region D of Fig. 11 and:

$\eta_{max}/\lambda_m = 0.95$, or $\eta_{max} = 32$ m

For $V_c = 83$ m/s, $V/\sqrt{gd} = 3.0$, and $\eta_{max} = 34$ m.

Irrespective of velocity, $\eta_{max} \leqslant \lambda_m$, thus illustrating a peculiarity of the Noda solution. By the Kamphuis and Bowering method H_{st} values of 63 and 79 m are predicted, which are an order of magnitude greater than the estimates of wave amplitude based on observed run-up. Back calculations for η_{max} using the two-dimensional wave height attenuation function yield 83 and 99 m.

The empirical regression (equation [10]) from this report predicts for a dimensionless kinetic energy [17] $= 130-249$, $\eta_{max} = 143-226$ m.

The lower-bound maximum wave height ($H = \eta$ for solitary wave) as estimated by the two-dimensional wave height attenuation function, and the height predicted by the Noda function are similar. However, the prediction of greater maximum wave amplitude by equation [10] for V_{im} seems on the whole more acceptable, if conservative. Predictions based on the Kamphuis and Bowering methods lead to apparently reasonable estimates of maximum wave amplitudes, but overestimate wave amplitudes at large r/d by about an order of magnitude.

Case III: Norwegian events

To further illustrate the problems of application, we compare predicted wave heights for the catastrophic winter 1756 Tjelle event in Langfjord, Norway, in which 32 perished. The slide had dimensions roughly of length = 250 m, width = 600 m, and thickness = 100 m for a volume of 15×10^6 m^3. Following continuous rain for eight days and nights it slid about 696 m down a slope of approximately 25° into water perhaps 100 m deep causing severe turbulence over the entire fjord, including run-up "50 paces", or about 40 m high (Jørstad, 1968, p. 21). Calculating an impact velocity as before (see Appendix 2) with $\tan \phi_s = 0.25$ gives $V_{im} = 52$ m/s, slide Froude number = 1.7 and $\lambda_m/d = 1.0$. Noda theory predicts maximum wave height, η_{max} = 92 m. From Kamphuis and Bowering, $q = (h_k/d)(l/d) = 2.5$, and $H_{st} = 57$ m. Since the "observed wave height" is probably a run-up of a far-travelled wave, the Noda solution may be of the correct order whereas the Kamphuis-Bowering "stable wave" estimate is too high. This overestimation can be explained, at least partially, since the Kamphuis-Bowering model is two-dimensional whereas the Tjelle slide radiated wave energy through 180°. This then suggests using equation [10], the empirical relationship from this report. For

---

[17] $KE = \frac{1}{2}(29 \times 10^6/80^3) \cdot (1/1) \cdot (60^2/9.8 \cdot 80) = 130.$

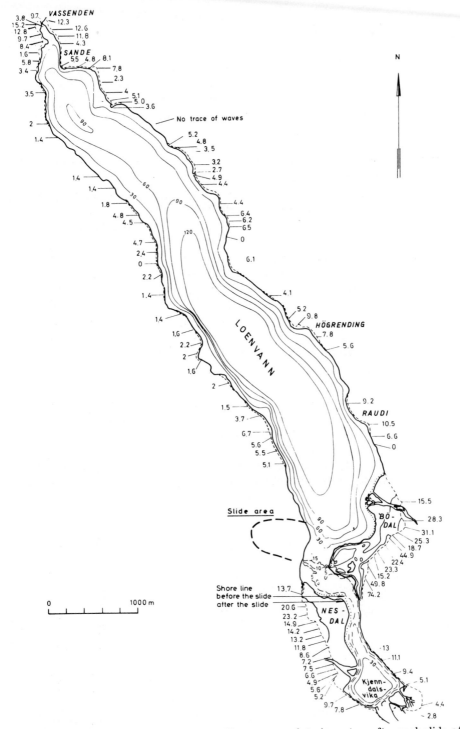

Fig. 37. Loenvann, southwest Norway. Wave run-up data in metres after rock slide of 13 September 1936 (after Jørstad, 1968, based on map by Th. J. Selmer and G. Saetre). The slide involved an exfoliation sheet perhaps 10 m thick and 400 m high, comprising a volume of about 10^6 m^3, released from 400—800 m above sealevel and moving on about a 65° slope; 73 persons lost their lives. Maximum surge height was 74 m, directly opposite the slide area.

a dimensionless slide kinetic energy equal to 56 and wave propagation through 180°, maximum predicted wave height equals 98 m. However, the depth is only approximate, and equation [10] is rather sensitive to this parameter since a 30% decrease of depth almost doubles η. Note also that this application seems clearcut compared to other Norwegian case histories, like the 1905 Loen and 1934 Tafjord slides which moved both scree and glacial debris below them (see Chapter 3, this volume) and for which velocity and effective thickness values would be much more difficult to estimate, or for the 1936 Loen slides (see Fig. 37) involving collapsed exfoliation sheets which fell in an irregular bay of variable depth.

However, the Ravnefjell slide of September 13, 1936 into Loenvann, Norway, does provide a test of the wave celerity equation. According to Jørstad (1968), from eyewitness accounts of when waves passed towns along the lake, wave celerity was between 15 and 30 m/s. From $C = \sqrt{g(d + \eta)}$, with mean water depth equal to 69 m, C = 26 m/s which is a close agreement (Jørstad, 1968, p. 26).

Wave run-up has been mapped along the shores of Loenvann for the September 13, 1936 landslide (Jørstad, 1968) and is given in Figs. 37 and 38. The previous two-dimensional studies have shown an exponential or geometric decrease to a stable wave height with distance. Since run-up is proportional to wave height offshore, we should expect roughly the same relationship between run-up and distance, even though these waves must be refracted 90° before breaking perpendicular to shore. Inspection of Figs. 37 and 38 suggests that if one excludes the surging effect directly across from the slide, the decrease to a stable run-up is obscure and of more importance in run-up variation is refraction, diffraction, reflection, a shadowing effect behind promontories, and convergence of wave energy at the end of the lake.

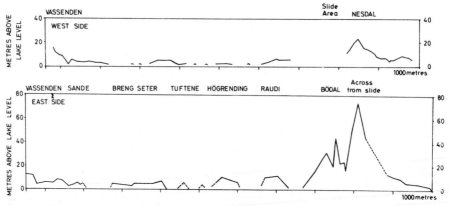

Fig. 38. Wave run-up data from 13 September 1936 rockslide in Loenvann (after Jørstad, 1968). Ordinate is observed run-up height, abscissa is trace of shoreline. Cf. Appendix 3.

Eie et al. (1971) claim that maximum run-up in Fig. 38 decreases as $1/x$, but did not consider data for $x > 6$ km. [18]

CONCLUDING REMARKS

We have discussed and compared with experimental and field data, two theoretical and three empirical models of water waves generated by landslides. These models generally agree that to predict an initial waveform, the important variables include some measure of slide energy, thickness (variable with slide displacement) and depth of water at the slide site. The angle at which the slide enters the water and the angle of the front of the slide are of lesser importance.

Comparison of these models with experimental data from the Waterways Experimental Station Lake Koocanusa study demonstrates that Kamphuis and Bowering overestimate "stable" wave height by more than an order of magnitude, and the Noda horizontal wall and vertical box drop and Kamphuis and Bowering solutions severely overestimate maximum wave heights. In addition, Noda's estimates for wave heights of nonlinear waves are very sensitive to slide thickness such that an increase in slide thickness of less than one metre can increase a wave height prediction threefold. Field hindcasts suggest that vertical Noda solutions are not always conservative. The Kamphuis and Bowering method for field hindcasts provided reasonable estimates of maximum wave amplitude, but tended to greatly overestimate wave amplitudes at large r/d where wave propagation was relatively unchanneled, as at Disenchantment Bay. For a first estimate of potential maximum wave height, the empirical equation [10] of this report seems about as satisfactory as the Kamphuis and Bowering model, and has the advantage of requiring less complicated data. Both seem more satisfactory than the vertical drop method of Noda.

Regardless of the method of estimating maximum wave amplitudes, an estimate of wave amplitude at large distance from the slide site can be calculated from either the simple inverse function (see footnote [+], Table V) or the two-dimensional channel approach of equation [6], depending on basin geometry.

For detailed wave analysis, scale models or numerical methods must be used. The advantages and potential of the Raney and Butler numerical model seem substantial. For the WES data this model predicts wave heights of less than 25% difference from those observed. Unlike the analytical models, it adequately deals with wave nonlinearity and it considers complex basin geometries so refraction-induced wave convergence and shadowing can be simulated. In the future, even run-up equations could be coupled to the model.

---

[18] See also Lied et al. (1976) and Jørstad (1968), for discussions of snow avalanches in Norwegian water bodies.

Fig. 39. Scar of rockfall at Stegane, Årdal (Norway), 21 June 1948 (photo courtesy F. Jørstad). Estimated volume 30,000 m^3 (ca. 100 m height × 50 m × 6 m); maximum wave height estimated at 3—5? m in Sognefjord, with waves noticeable at 6 km distance (see Jørstad, 1968, table 3). Base of rock fall near water level, so that initial impact velocity was very small. Mode of collapse (e.g., toppling versus simple free-fall) is, however, uncertain. Use of case history to test wave models is not straightforward, despite relatively complete observational data.

However, our present inadequacy in estimating potential slide dimensions, velocities, and modes of emplacement gives wide limits of confidence to all models discussed. The 1948 Stegane, Norway, rockfall (Fig. 39) illustrates these problems. Although its dimensions might be fairly well known, its velocity (V_{im} = 0?) and mode of emplacement (vertical drop versus buckling or toppling about its bottom edge) are problematical, and an unacceptably wide range of waveforms could be hindcast. If the Raney and Butler model is used, the history of slide emplacement must be especially well known, which includes characterizing geometric variations due to bulking and deformation during transport, as well as velocity slowdown due to slide-fluid interaction. However, the difficulties involved in predictions of high accuracy do not necessarily diminish the practical value of the methods.

Scale or numerical models, for example, can readily provide information on physical locations especially susceptible to (or free of) wave attack even if fine details of slide emplacement are not well known. Parameter studies, e.g., velocity versus amplitude, can aid engineering judgment where only broad ranges of parameter variation can be specified.

In other cases detailed results are not needed nor warranted for engineering judgment. The following "hypothetical" case history is for the reader's consideration. A mine is located near the shore of a small lake; directly opposite the camp, set 400 m above lake level at the top of a 45° talus cone, is a steep rock cliff. A noticeable increase in frequency of small rock falls from the cliff face and widening joints suggest growing instability of a large part of this cliff, with the potentially unstable portion involving perhaps 10^5 m^3. A rapid assessment of the slide-induced wave potential follows:

(1) $V_{im} \simeq [2 \cdot 9.8 \cdot 566 \cdot (0.707 - 0.25 \cdot 0.707)]^{1/2} \simeq 77$ m/s

(2) Assuming water depth in the slide impact area to be about 30 m:

$$KE = \tfrac{1}{2}(10^5/30^3)\,(2.7/1)\,(77^2/9.8 \cdot 30) \simeq 100$$

(3) From equation [10], $\eta_{max}/d > 1$,

∴ $\eta_{max} > 30$ m (to one significant figure)

Noting the possibility of run-up significantly in excess of deep water wave amplitude, the conclusion is drawn that facilities and mine shafts within 50 m or so of the shoreline are endangered. The recommended solution is to evacuate humans, animals, and to bring other moveable valuables to positions of safety, until the slide mass can be released under as much control as the situation permits (see, e.g., Bjerrum and Jørstad, 1968, pp. 7—8). Emphasis in this case history is placed on sound judgment following well-founded but rapid analysis, implemented immediately in order to prevent loss of human life. Failure to deal with such an alarm quickly can lead to drastic consequences, as described in Chapter 7 of this volume for Chungar, Peru.

ACKNOWLEDGEMENTS

We very much appreciate assistance and information provided by William O. Field, American Geographical Society, Finn Jørstad and the late Laurits Bjerrum, Norges Geotekniske Institutt, Don Banks, Waterways Experiment Station, George Plafker, U.S. Geological Survey, J. Douglas Breen, Realand Associates, S. Thorarinsson, University of Iceland, S.P. Jakobsson, Museum of Natural History, Reykjavik, and W.P. Harland, CASECO Ltd.

APPENDIX 1. DERIVATION OF THE LONG-WAVE LANDSLIDE NUMERICAL MODEL

This derivation follows those of Leendertse (1967) and Raney and Butler (1975). From conservation of mass the continuity equation in a rectilinear Cartesian coordinate system (Fig. A-1) is, if fluid density is assumed constant:

$$\frac{\partial u}{\partial x} + \frac{\partial v}{\partial y} + \frac{\partial w}{\partial z} = 0 \qquad\qquad [A-1]$$

To eliminate the z direction equation [A-1] is integrated over the z-axis from $-d$ to η, giving:

$$\int_{-d}^{\eta} \left[\frac{\partial u}{\partial x} + \frac{\partial v}{\partial y} + \frac{\partial w}{\partial z} \right] dz = 0 \qquad\qquad [A-2]$$

Term by term:

$$\int_{-d}^{\eta} \frac{\partial u}{\partial x} dz = (d + \eta) \frac{\partial \overline{u}}{\partial x} \qquad\qquad [A-3]$$

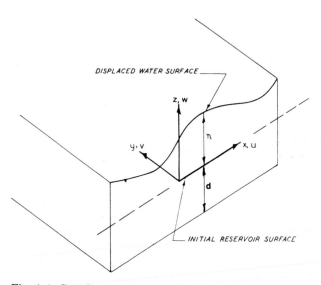

Fig. A-1. Coordinate system for Raney-Butler model.

$$\int_{-d}^{\eta} \frac{\partial v}{\partial y} \, dz = (d + \eta) \frac{\partial \bar{v}}{\partial y} \qquad\qquad\qquad [A-4]$$

and

$$\int_{-d}^{\eta} \frac{\partial w}{\partial z} \, dz = w \Big|_{-d}^{\eta} \qquad\qquad\qquad\qquad [A-5]$$

where $\bar{u}$ and $\bar{v}$ are the horizontal velocity components averaged over the whole water column. $w|_{z=\eta}$ is equivalent to the total derivative of $\eta(x, y, t)$ with respect to time, or:

$$w \Big|_{z=\eta} = \frac{d\eta}{dt} = \frac{\partial \eta}{\partial t} + \frac{\partial \eta}{\partial x} \cdot \frac{dx}{dt} + \frac{\partial \eta}{\partial y} \cdot \frac{dy}{dt}$$

Since $u_s = dx/dt$ and $v_s = dy/dt$, where the subscript s denotes the horizontal velocity component at the free surface:

$$w \Big|_{z=\eta} = \frac{\partial \eta}{\partial t} + u_s \frac{\partial \eta}{\partial x} + v_s \frac{\partial \eta}{\partial y} \qquad\qquad [A-6]$$

Similarly:

$$w \Big|_{z=d} = u_f \frac{\partial (d)}{\partial x} + v_f \frac{\partial (d)}{\partial y} \qquad\qquad [A-7]$$

where $\partial(d)/\partial x$ and $\partial(d)/\partial y$ are the bottom slopes in the x and y direction respectively, and the subscript f denotes the horizontal velocity components at the floor of the water body.

If the horizontal velocity distribution is uniform, i.e., $u = \bar{u} = u_s = u_f$, and $v = \bar{v} = v_s = v_f$, recombining equations [A-3], [A-4], [A-6] and [A-7] gives:

$$\frac{\partial \eta}{\partial t} + \frac{\partial}{\partial x} [(d + \eta)u] + \frac{\partial}{\partial y} [(d + \eta)v] = 0 \qquad\qquad [A-8]$$

which is the vertically averaged equation of continuity written for long waves.

Since the landslide will vary the water depth over part of the study area, the water depth is time dependent, and $\partial \eta/\partial t$ becomes $\partial(\eta + d)/\partial t$ or adopting the notation of Raney and Butler, the continuity equation becomes:

$$\frac{\partial \eta}{\partial t} - \frac{\partial \zeta}{\partial t} + \frac{\partial}{\partial x} [(d + \eta)u] + \frac{\partial}{\partial y} [(d + \eta)v] = 0 \qquad\qquad [A-9]$$

which is their equation 3 (Raney and Butler, 1975, p. 8).

For the appropriate equations of motion we start with the Navier-Stokes equations written in the rectilinear Cartesian coordinate system of Fig. A-1. If w and its rates of change with respect to the horizontal coordinates are assumed small, the Navier-Stokes equation for the z direction reduces to:

$$0 = -\frac{1}{\rho}\frac{dP}{dz} - gz$$

which is the hydrostatic equation

Integrating over z to η, where the pressure at the surface is P_s or atmospheric:

$$\int_P^{P_s} dP = -\int_z^\eta \rho g\, dz$$

or

$$P = \rho g[\eta(x,\, y,\, t) - z] + P_s$$

This gives:

$$\frac{\partial P}{\partial x} = \rho g\frac{\partial \eta}{\partial x} + \frac{\partial P_s}{\partial x}$$

$$\frac{\partial P}{\partial y} = \rho g\frac{\partial \eta}{\partial y} + \frac{\partial P_s}{\partial y}$$

and if P_s is constant over the area in question:

$$\frac{\partial P_s}{\partial x} = \frac{\partial P_s}{\partial y} = 0$$

and:

$$\frac{\partial P}{\partial x} = \rho g\frac{\partial \eta}{\partial x} \qquad\qquad\qquad\qquad [A\text{-}10]$$

$$\frac{\partial P}{\partial y} = \rho g\frac{\partial \eta}{\partial y} \qquad\qquad\qquad\qquad [A\text{-}11]$$

Now considering the Navier-Stokes equation for the x direction:

$$\frac{\partial u}{\partial t} + u\frac{\partial u}{\partial x} + v\frac{\partial u}{\partial y} + w\frac{\partial u}{\partial z} = -\frac{1}{\rho}\frac{\partial P}{\partial x} + \frac{1}{\rho}\left(\frac{\partial \tau_{xx}}{\partial x} + \frac{\partial \tau_{yx}}{\partial y} + \frac{\partial \tau_{zx}}{\partial z}\right) + F_x \qquad \text{[A-12]}$$

where F_x represents the force which the landslide exerts per unit mass of water. If the horizontal stresses, τ_{xx} and τ_{xy} are small with respect to τ_{xz}, they may be dropped. Substituting equation [A-10] for $\partial P/\partial x$ gives:

$$\frac{\partial u}{\partial t} + u\frac{\partial u}{\partial x} + v\frac{\partial u}{\partial y} + w\frac{\partial u}{\partial z} = -g\frac{\partial \eta}{\partial x} + \frac{1}{\rho}\frac{\partial \tau_{zx}}{\partial z} + F_x \qquad \text{[A-13]}$$

Proceeding as before with the continuity equation, equation [A-13] and the similar y-direction equation are integrated over $z = -d$ to η. After a significant amount of rearranging and use of the Leibnitz rule one obtains for the x direction:

$$\frac{\partial u}{\partial t} + u\frac{\partial u}{\partial x} + v\frac{\partial u}{\partial y} = -g\frac{\partial \eta}{\partial x} + \frac{1}{\rho(d+\eta)}(\tau_{sx} - \tau_{fx}) + F_x \qquad \text{[A-14]}$$

where τ_{sx} and τ_{fx} are the shear stresses at the top and floor of the fluid column, respectively. If no wind stress is considered, $\tau_{sx} = 0$. For the bottom, Raney and Butler assume the Chezy relationship between shear stress and fluid velocity:

$$\tau_{fx} = \rho g\, u\frac{(u^2 + v^2)^{1/2}}{C_R^2} \qquad \text{[A-15]}$$

where $C_R = (1.49/N)\,(d+\eta)^{1/6}$ and $N = $ Manning friction factor.
Substituting equation [A-15] in [A-14] gives:

$$\frac{\partial u}{\partial t} + u\frac{\partial u}{\partial x} + v\frac{\partial u}{\partial y} = -g\frac{\partial \eta}{\partial x} - \frac{g}{(d+\eta)}\,u\frac{(u^2 + v^2)^{1/2}}{C_R^2} + F_x \qquad \text{[A-16]}$$

Similarly for the y direction:

$$\frac{\partial v}{\partial t} + u\frac{\partial v}{\partial x} + v\frac{\partial v}{\partial y} = -g\frac{\partial \eta}{\partial y} - \frac{g}{(d+\eta)}\,v\frac{(u^2 + v^2)^{1/2}}{C_R^2} + F_y \qquad \text{[A-17]}$$

Equations [A-16] and [A-17] are equations 1 and 2 of Raney and Butler (1975, p. 8) with their equations 4 and 5 included.

The force of the landslide on the water, F_x and F_y are considered by Raney and Butler to consist of three components, one due to displacement of the water, and viscous drag and pressure drag forces. The water displacement component has already been accounted for in the $\partial\xi/\partial t$ term of the continuity equation. Considering the others, the viscous drag equation is:

$$F_D = C_D A \rho \frac{V_r^2}{2} \qquad \text{[A-18]}$$

where F_D = viscous drag force, C_D = coefficient of drag, A = surface area of slide, ρ = fluid density, and V_r = relative velocity between slide and water.

The drag force per unit mass of water is:

$$\frac{F_D}{\text{mass}} = \frac{C_D A \rho V_r^2}{2\rho \nabla_w} \qquad \text{[A-19]}$$

where ∇_w is the volume of water acted on by F_D. Raney and Butler take this balance over one grid cell of the finite difference scheme. Thus:

$$\frac{F_D}{\text{mass}} = \frac{C_D A V_r^2}{2A_c d} \qquad \text{[A-20]}$$

where $\nabla_w = A_c d$ and A_c = grid cell area, d = water depth.

They assume further that $A/A_c \approx 1$ and thus:

$$\frac{F_D}{\text{mass}} = \alpha(V_{slide} - V_{water})^2$$

where $\alpha = C_D/2d = 0.004/2(200) = 1 \times 10^{-5}$ ft^{-1} if C_D for turbulent flow over a flat plate $= 4.0 \times 10^{-3}$ and the average depth in their model is 60 m (200 ft).

Similarly the pressure drag due to the front of the slide is:

$$\frac{F_P}{\text{mass}} = \frac{C_P \rho V_r^2 A_z}{2\rho A_c d}$$

where A_z is the vertical cross-section of the leading edge of the slide.

If $A_z/A_c \sim 1$, then:

$$\frac{F_P}{\text{mass}} = \beta(V_{slide} - V_{water})^2$$

where $\beta \approx C_P/2d = 1/400 = 2.5 \times 10^{-3}$ ft^{-1} if $C_P \sim 1$.

Thus the final equations to be solved in the Raney and Butler numerical model are:

(1) the x-direction equation of motion

$$\frac{\partial u}{\partial t} + u\frac{\partial u}{\partial x} + v\frac{\partial u}{\partial y} + g\frac{\partial \eta}{\partial x} = -\frac{g}{(d+\eta)}\frac{u(u^2+v^2)^{1/2}}{C_R^2} + (\alpha+\beta)(V_x - u)^2 \qquad \text{[A-21]}$$

(2) the y-direction equation of motion

$$\frac{\partial v}{\partial t} + u \frac{\partial v}{\partial x} + v \frac{\partial v}{\partial y} + g \frac{\partial \eta}{\partial y} = - \frac{g}{(d + \eta)} \frac{v(u^2 + v^2)^{1/2}}{C_R^2} + (\alpha + \beta)(V_y - v)^2 \qquad [\text{A-22}]$$

(3) the continuity equation

$$\frac{\partial \eta}{\partial t} - \frac{\partial \zeta}{\partial t} + \frac{\partial}{\partial x}[(d + \eta)u] + \frac{\partial}{\partial y}[(d + \eta)v] = 0 \qquad [\text{A-23}]$$

APPENDIX 2. VELOCITY ESTIMATES

A problem of major interest is calculating the velocity of a slide mass either at impact or during entry, inasmuch as wave predictive procedures require such input. This must be determined on an individual basis because each slide locality has its own set of geometric factors and material properties. The literature, of course, contains abundant estimates of slide velocities. However, the reader must be cautioned that many of these estimates are of questionable accuracy. Some, based on distance and estimated travel time, purport to be average velocity estimates. Others, based on energy considerations or block kinematics, may be erroneously high because of incorrect assumptions of work expended, incorrect friction estimates, and/or inaccurate idealization of the slide mass.

For example, the basic equation governing sliding of a rigid block on an inclined plane is as follows,

$$V = V_0 + gt(\sin i - \tan \phi_s \cos i) \qquad [\text{A-24}]$$

or

$$V = V_0 + [2gs(\sin i - \tan \phi_s \cos i)]^{1/2} \qquad [\text{A-25}]$$

where V is the velocity at elapsed time t and downslope distance s, g is gravitational acceleration, i is slope angle of the slide plane, and ϕ_s is the angle of dynamic sliding friction.

As applied to slides the equation dates from Heim (1932) and his colleague Müller-Bernet (see, e.g., Chapter 1, Volume 1). In the solution for a single plane, acceleration is constant and velocity increases linearly with time. Initial velocity V_0 is assumed small and may thus be usually ignored. A curved or complex slide surface can be idealized as a series of planar segments, however, in which case V_0 can be taken as an initial velocity term for each separately analyzed segment. Commonly it is assumed that the entire

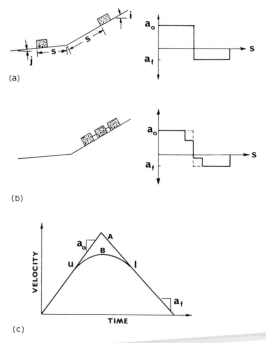

(a)

(b)

(c)

Fig. A-2. Comparison of motion analyses of (a) rigid block (lumped mass) and (b) connected segment block models (after Banks et al., 1972). Slope angles i and j; distance of transport, s. Initial acceleration a_0, final acceleration, a_f. In (a) acceleration is assumed constant for each slope segment. In (b) acceleration varies as individual block segments pass change in slope; value is constant when *all* blocks are on upper or lower slopes. In velocity-time plot (c), models are identical except for region between point u, where mass starts to slide on lower plane, and point l, where entire mass rests on lower plane. Peak velocity predicted by (a) overestimates peak velocity of actual slide; solution (b) approaches correct solution as size of segments decreases.

mass accelerates uniformly until the centroid passes the junction of the planar segments. The result of this assumption is *to overestimate the maximum velocity* achieved by the slide mass, e.g., by *as much as 100%*. In an improved solution the slide mass can be treated as a series of connected segments rather than as a simple mass concentrated at its centroid (Banks et al., 1972).

A solution for the equation of motion based on the kinematics of a segmented mass of total length l sliding on two connected planes follows. Geometric symbols are given in Fig. A-2. Slide displacement is s, elapsed time is t, friction coefficient on plane with slope i is $\tan \theta_{s(i)}$.

$$s = \frac{k_1}{k_2} (1 - \cos \sqrt{k_2}\, t) \qquad\qquad [\text{A-26}]$$

$$\frac{ds}{dt} = \frac{k_1}{\sqrt{k_2}} \sin \sqrt{k_2}\, t \qquad\qquad\qquad\qquad [\text{A-27}]$$

$$\frac{d^2 s}{dt^2} = k_1 \cos \sqrt{k_2}\, t \qquad\qquad\qquad\qquad [\text{A-28}]$$

where:

$$k_1 = g(\sin i - \tan \phi_{s(i)} \cos i)$$

$$k_2 = [(\sin i - \tan \phi_{s(i)} \cos i) - (\sin j - \tan \phi_{s(j)} \cos j)]\frac{g}{l}$$

Given an appropriate set of equations of motion and a reasonably defined geometry, the problem still remains to specify the appropriate material properties — most notably the dynamic coefficients of friction. *There is as yet no suitable way of readily obtaining this information in the field.* Static or quasi-static friction coefficients are *not* appropriate, and perturbations concerning surface roughness are complexly involved.

In the absence of reliable measurement procedures, we recommend analysis based on average friction values reported in the literature for rock slides. The average effective friction angle is approximately given by the line connecting the mass centers of the slide mass before and after the sliding event; most of the friction-coefficient data determined in this manner fall into the range 0.25 ± 0.15 (cf. Scheidegger, 1973; Banks and Strohm, 1974). These values are not wholly independent of the model assumed for motion analysis (e.g., rigid block versus segmented block on multiple planes); however, the variation of friction coefficient due to model idealization is usually very small.

For determination of displacement-time relationships for landslides penetrating water bodies, the equations of motion must be modified to include appropriate drag forces as indicated in Appendix 1.

APPENDIX 3. DIMENSIONLESS KINETIC ENERGY MODEL

When this chapter was in proof we were fortunate to obtain a copy of the report, "Hydraulic Model Studies, Wave Action Generated by Slides into Mica Reservoir (British Columbia)," by Western Canada Hydraulic Laboratories, Ltd. This study included slides with an order of magnitude larger dimensionless kinetic energy than the WES study, and therefore enabled us to refine our estimate of the coefficients for equation [10], the regression equation of dimensionless wave amplitude versus dimensionless kinetic energy.

Fig. A-3 is a plot of the WES data as used in Fig. 28 (but here all η/d values have been extrapolated by the inverse formula to $r/d = 4$) plus six values from the Mica Reservoir study (all extrapolated to $r/d = 4$). These are all first-wave amplitudes for sites approximately in front of a slide.

The regression equation:

$$\log(\eta_{max}/d) = a + b \log(KE) \qquad\qquad [10]$$

has for Fig. A-3, coefiicients $a = -1.25$ and $b = +0.71$, with 97% of the variation in $\log(\eta_{max}/d)$ "explained" by variation in $\log(KE)$ *. The relationship thus seems to be linear over two orders of magnitude of $\log(KE)$. For these data, $1 < KE < 100$, and $0.3 < h/d < 0.8$.

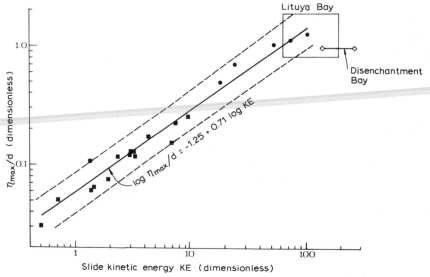

Fig. A-3. Relationship of maximum wave amplitudes ($r/d \sim 4$) as a function of slide dimensionless kinetic energy. WES Lake Koocanusa data denoted by squares, Mica Reservoir data by circles. The regression equation is significant at the 97% level: 95% confidence limits are indicated by dashed lines. For comparison, best estimates of actual wave amplitudes and associated kinetic energies are indicated for Lituya Bay (box) and Disenchantment Bay (line connecting open squares) events (Table V).

---

* Note that spurious correlation between these variables is possible due to the appearance of depth in the denominator of both. For example, if the variances of dimensional wave amplitude, dimensional slide kinetic energy, and water depth were all equal, and there was no correlation between wave amplitude and kinetic energy, the correlation coefficient due to depth alone would be 0.69 (Kenneth Potter, personal communication). However, the variance of water depth in these data is much less than that of the other variables and this effect is minimized here.

On the other hand, three problems must be noted with this empirical model. First the regression equation [10] is quite sensitive to water depths, a condition troublesome for irregular bathymetries. Second, the model has not accounted for variations in wave amplitude at radially equidistant points from a slide. Amplitudes directly in front of a slide are empirically known to be highest, indicating the importance of slide configuration and orientation. Third, the present coefficients of equation [10] are not conservative for broad slides at low dimensionless kinetic energy. Perhaps later data will suggest separate design curves for slides acting as line sources versus point sources.

Analysis of other empirical data from the Mica Reservoir model study also shows that:

(1) decrease of wave amplitude is inversely proportional to distance,

(2) the first wave near a slide is always the highest, but far away, the highest waves often occur later in the train,

(3) at radially equidistant points from a slide, wave amplitudes strongly decrease with increasing angular deviation from the direction of slide entry (this appears to be especially important for wider slides),

(4) celerity of the first wave follows $C = \sqrt{g(d + \eta)}$. These results are consistent with our previous conclusions.

REFERENCES

Banks, D.C. and Strohm, E., 1974. Calculations of rockslide velocities. In: *Advances in Geomechanics*. U.S. National Academy of Sciences, Washington, D.C., pp. 839—847.

Banks, D.C., Whalin, R.W., Davidson, D.D. and De Angulo, M., 1972. Velocity of potential landslides and generated waves, Libby Dam and Lake Koocanusa Project, Montana. *U.S. Army Waterw. Exp. Stn., Tech. Rep.*, No. H-74-15 (preliminary draft).

Bjerrum, L. and Jørstad, F., 1968. Stability of rock slopes in Norway. *Norw. Geotech. Inst. Publ.*, 79: 1—11.

Brazee, R.J. and Jordan, J.N., 1958. Preliminary notes on the southeastern Alaska earthquake. *Earthquake Notes*, 29: 36—40.

Bugge, A., 1937. Fjellskred fra topografisk og geologisk synspunkt. *Nor. Geogr. Tidsskr.*, 6: 342—360.

Davidson, D.D. and Whalin, R.W., 1974. Potential landslide-generated water waves, Libby Dam and Lake Koocanusa, Montana. *U.S. Army Eng. Waterw. Exp. Stn., Tech. Rep.*, No. H-74-15 (Corps of Engineers, Vicksburg, Miss.).

Dronkers, J.J., 1964. *Tidal Computations*. North-Holland, Amsterdam, 518 pp.

Easterbrook, D.J., 1975. Mount Baker eruptions. *Geology*, 3: 679—682.

Eie, J., Solberg, G., Tvinnereim, K. and Tørum, A., 1971. Waves generated by landslides. 1st Int. Conf. Port and Ocean Engineering under Arctic Conditions, Tech. Univ. Norway, Trondheim, 1: 489—513.

Goldsmith, V., Morris, W.D., Byrne, R.J. and Whitlock, C.H., 1974. Wave climate model of the mid-Atlantic continental shelf and shoreline, 1. Model description, shelf geomorphology, and preliminary data analysis. *VIMS SRAMSOE*, 38.

Grantham, K.N., 1953. Wave run-up on sloping structures. *Trans. Am. Geophys. Union*, 34: 720—724.

Hall, J.V. and Watts, G.M., 1953. Laboratory investigation of the vertical rise of solitary waves on impermeable slopes. *Beach Erosion Board, Corps Eng., Tech. Memo.*, No. 33 (U.S. Department of the Army, Washington, D.C.).

Heim, A., 1932. *Bergsturz und Menschenleben*. Fretz und Wasmuth, Zürich, 218 pp.

Helland, A., 1905. Raset paa Ravnefjeld i Loen. *Naturen*, 29: 161—172.

Hoek, E. and Bray, J.W., 1974. *Rock Slope Engineering*. Institution of Mining and Metallurgy, London, 309 pp.

Holmsen, G., 1936. De siste bergskred i Tafjord og Loen, Norge. *Sven. Geogr. Årsb.*, pp. 171—190.

Ippen, A.T. (Editor), 1966. *Estuary and Coastline Hydrodynamics*. McGraw-Hill, New York, N.Y., 744 pp.

Johnson, J.W. and Bermel, K.J., 1949. Impulsive waves in shallow water as generated by falling weights. *Trans. Am. Geophys. Union*, 30: 223—230.

Jørstad, F., 1968. Waves generated by landslides in Norwegian fjords and lakes. *Norw. Geotech. Inst. Publ.*, 79: 13—32.

Kaldhol, H. and Kolderup, N.-H., 1937. Skredet i Tafjord 7, April 1934. *Bergens Mus. Årb. 1936*, Rekke 11, 15 pp.

Kamphuis, J.W. and Bowering, R.J., 1972, Impulse waves generated by landslides. *Proc., 12th Coastal Engineering Conf., Am. Soc. Civ. Eng.*, 1: 575—588.

Kjartansson, G., 1967a. Steinsholtslaupid 15 Januar 1967. *Náttúrufraedingurinn*, 37: 120—169 (in Icelandic).

Kjartansson, G., 1967b. The Steinsholt Hlaup, central-south Iceland, on January 15, 1967. *Jökull*, 17: 249—262.

Kranzer, H.C. and Keller, J.B., 1959. Water waves produced by explosions. *J. Appl. Phys.*, 30: 398—407.

Lamb, Sir Horace, 1945. *Hydrodynamics*. Dover, New York, N.Y., 6th ed., 738 pp.

Lauffer, H., Neuhauser, E. and Schober, W., 1967. Uplifts responsible for slope movements during the filling of Gepatsch Reservoir. Commission Internationale des Grands Barrages, Istanboul, Q. 32, R. 41, p. 669—693.

Law, L. and Brebner, A., 1968. On water waves generated by landslides. *3rd Australas. Conf. on Hydraulics and Fluid Mechanics, Sydney*, Paper 2561, pp. 155—159.

Leendertse, J.J., 1967. Aspects of a computational model for long-period water-wave propagation. *Rand Corp., Memo.*, No. RM-5294-PL, Santa Monica. Calif.

LeMehaute, B., Koh, R.C.Y. and Hwang, L., 1968. A synthesis on wave run-up. *Proc. Am. Soc. Civ. Eng., J. Waterw. Harbors Div.*, 94: 77—92.

Lied, W., Palmstrom, A., Schieldrop, B., and Torblaa, I., 1976. Dam Tunsbergdalsvatn: a dam subjected to waves generated by avalanches and to extreme floods from a glacier lake. Commission Internationale des Grands Barrages, Mexico, Q. 47, R. 9, pp. 861—875.

McCulloch, D.S., 1966. Slide-induced waves, seiching, and ground fracturing caused by the earthquake of March 27, 1967 at Kenai Lake, Alaska. *U.S. Geol. Surv., Prof. Paper*, 543-A, 41 pp.

Miller, D.J., 1953. Preliminary geologic map of Tertiary rocks in the southeastern part of the Lituya district, Alaska. *U.S. Geol. Surv., Open-file Rep.*

Miller, D.J., 1960. Giant waves in Lituya Bay, Alaska. *U.S. Geol. Surv., Prof. Paper*, 354-C: 51—83.

Miller, D.J. and White, R.V., 1966. A single-impulse system for generating solitary, undulating surge, and gravity shock waves in the laboratory. *Fluid Dyn. Sediment. Transp. Lab. Rep.*, No. 15, Univ. of Chicago, Chicago, Ill.

Müller, L., 1964. The rockslide in the Vaiont Valley. *Rock Mech. Eng. Geol.*, 2: 148—212.

Noda, E., 1969. Theory of water waves generated by a time-dependent boundary displacement, *Tech. Rep.*, No. HEL-16-5 Univ. of California, Berkeley, Calif.

Noda, E., 1970. Water waves generated by landslides. *Proc., Am. Soc. Civ. Eng., J. Waterw. Harbors Div.*, 96 (WW4): 835—855.

Ogawa, T., 1924. Notes on the volcanic and seismic phenomena in the volcanic district of Shimabara, with a report on the earthquake of December 8, 1922. *Mem. Coll. Sci., Kyoto Imp. Univ., Ser. B*, 1 (2).

Prins, J.E., 1958. Characteristics of waves generated by a local disturbance. *Trans. Am. Geophys. Union*, 39: 5: 865—874.

Raney, D.C. and Butler, H.L., 1975. A numerical model for predicting the effects of landslide generated water waves. *U.S. Army Eng. Waterw. Exp. Stn., Res. Rep.*, No. H-75-1.

Reusch, H., 1907. Skredet i Loen 15 de januar 1905. *Nor. Geol. Unders.*, 45, 20 pp.

Scheidegger, A.E., 1973. On the prediction and reach of catastrophic landslides. *Rock Mech.*, 5: 231—236.

Schøning, G., 1778. *Reise som gjennem en Deel af Norge i de Aar 1773, 1774, 1775 paa hans Majestets Kongens Bekostning er giort og beskreven, 2.* Gyldendal, Kiøbenhaven, 148 pp.

Sharpe, C.F.S., 1938. *Landslides and Related Phenomena.* Columbia Univ. Press, New York, N.Y., 137 pp.

Strøm, H., 1766. *Physisk og Oeconomisk Beskrivelse over Fogderiet Sondmor, beliggende i Bergens Stift i Norge, 2.* Sorøe. 509 pp.

Tarr, R.S., 1909. The Yakutat Bay region, Alaska; physiography and glacial geology. *U.S. Geol. Surv., Prof. Paper*, 64, 183 pp.

Tarr, R.S. and Martin, L., 1914. *Alaskan Glacier Studies.* National Geographic Society, Washington, D.C., 498 pp.

Tocher, D. and Miller, D.J., 1959. Field observations on effects of Alaskan earthquake of 10 July, 1958. *Science*, 129: 394—395.

Unoki, S. and Nakano, M., 1953, On the Cauchy-Poisson waves caused by the eruption of a submarine volcano (2nd paper). *Oceanogr. Mag.* 5 (1): 1—13.

Varnes, D.J., 1958, Landslide types and processes. *Highw. Res. Board Spec. Rep.* 29, *Natl. Acad. Sci.-Natl. Res. Counc. Publ.*, 544: 20—47.

Wiegel, R.L., 1964. *Oceanographical Engineering*, Prentice-Hall, Englewood Cliffs, N.J., 532 pp.

Wiegel, R.L., Noda, E., Kuba, E.M., Gee, D.M. and Tornberg, G.F., 1970. Water waves generated by landslides in reservoirs. *Proc., Am. Soc. Civ. Eng., J. Waterw. Harbor Div.*, 96 (WW2): 307—333.

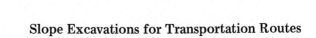

Slope Excavations for Transportation Routes

Chapter 10

PARACTI ROCKSLIDE, BOLIVIA

GEORGE F. SOWERS and B. ROGER CARTER

ABSTRACT

A rockslide of 60,000 m^3 occurred suddenly during construction of a new highway in west-central Bolivia, grinding seven men to bits and destroying construction equipment valued at several hundred thousand dollars. The movement took place along bedding planes that dip from 42° to 45°. Old slide scars form the mountain face, dipping generally at about 32°, but locally steeper in the area of failure. Excavation of a bench 18 m deep into the mountain face exposed several shaley seams within the dipping sandstone. The sliding occurred at night, without warning, during dry weather, months after construction had commenced and days after the face height had reached 15 m. Failure appeared to have been triggered by expansion and contraction of the rock, causing accumulating slippage downhill on an undulating shale bed, ultimately leading to sudden, catastrophic movement.

INTRODUCTION

The limited food productivity of the Andes region with its high, dry intermountain valleys and the lack of direct access to international water transport has caused Bolivia to build a series of highways leading eastward from the highlands to the headwater rivers of the Amazon lowlands. One such highway leads northeast from Cochabamba, at an altitude of about 3000 m, to Villa Tunari (a village) at elevation 300 m on the Chapare River which leads to the upper Amazon. The direct distance is about 75 km; the actual road length is nearly twice that. An existing road built decades ago is even longer and so steep and narrow that it strangles the economic development of the potentially rich Amazon region for the benefit of Bolivia. The location of the new road is shown in Fig. 1a.

Between Cochabamba and the Amazon region lie rugged mountains with steep slopes and narrow valleys. The peaks rise to about 5000 m; the pass in the new highway is at 4500 m. Because of the rough terrain, highway construction is difficult. Fill is impossible on many mountainsides because the slopes are too steep; cutting is hazardous because many of the slopes are barely stable; the valleys are too narrow and are susceptible to flooding during the infrequent but intense rains. The old road generally followed the ridge crests like a snake lying atop a crooked tree limb. Designing a road for modern grades and without hairpin turns required a bolder design: viaducts in valleys, deep fills between ridges, tunnels through thin ridges and innumerable deep cuts into the mountainsides. Most cuts were stable; however, a large number of landslides was inevitable in such a design. While most were minor, several were hazardous and expensive. Although the slide described in this paper was small, it killed seven men and precipitated fear among the

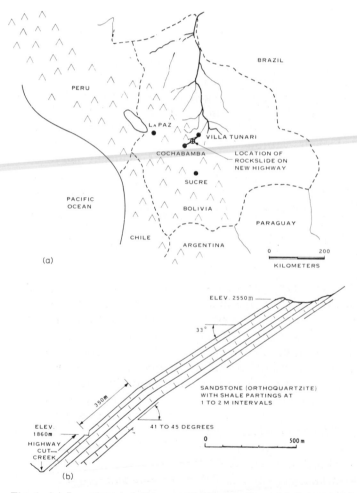

Fig. 1. (a) Location and (b) general cross-section of failure area showing bedding looking north.

workers, with political overtones that threatened progress on the highway. Therefore, the investigation described in this paper was more detailed than the size of the failure appears to justify. The information gained provides an insight into the mechanisms of bedding plane failures.

GEOLOGY OF THE REGION

Detailed geologic data are lacking in the area because the absence of mineral resources has discouraged exploration. Moreover, the route was so inaccessible that regularly spaced borings were not made for highway design. Air photographs were the basis for topographic mapping, but were not used for

Fig. 2. General view of mountainside showing road cut, waste downhill and former landslide scars. New failure is delineated by dark arcuate shadow, concave downhill, in center of photograph.

Fig. 3. Undulations in sandstone-shale contact, surface of block 1.

air photo interpretation of geologic structure nor soil deposits. Limited geologic reconnaissance was made by foot along a preliminary slash line to aid in
the final design.

Formations in slide area

The area between Cochabamba and the Amazon Basin is underlain by
sedimentary formations, predominantly sandstones and shales, with localized
limestones and small extrusive bodies. These have been subject to high pressure and are locally metamorphosed. The sandstones have been converted
into orthoquartzites and some shale to phyllite. The original bedding has
been tilted and faulted; the predominate strike of the folds and faults is
parallel to the high Andes, approximately north-south.

Structure

In the immediate area of the failure, the formations dip to the west, as shown in Fig. 1b. The dips range from 30° to more than 45°; in isolated areas the dips are nearly vertical. The west mountain face is the dip slope (Fig. 2). The dipping beds have been forced into a series of gentle corrugations, anticlines and synclines, whose axes are perpendicular to the strike of the regional dip. In the immediate vicinity of the slide, km 85+400, the dips decrease somewhat up the mountainside and the corrugations converge to form a broad, fluted dome. The general conditions can be seen in Fig. 2. Toward the left (north end) of the photograph, the corrugations are steeper and the dips locally are close to 90°, with the local strike nearly at right angles to the regional north-south strike of the formations in the mountainside.

The mountainside of Figs. 1b and 2 is predominately of well-indurated sandstone and orthoquartzite. Interbedded at intervals of 1 to 2 m are shale or phyllite seams, 5—15 cm thick. The strata are relatively uniform in thickness. However, the beds undulate in small waves whose crests parallel the general dip, with wavelengths (crest to crest of the undulations) of about 30

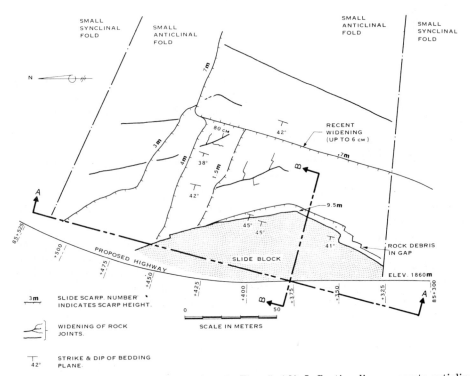

Fig. 4. Plan of slide area (cross-sections in Figs. 5, 10). Inflection lines separate anticlinal from synclinal areas.

cm. These undulations can be seen in Fig. 3. The dip immediately above the road in the failure area ranges from 42° to 45°; further up the mountain it is flatter, about 32°. It becomes somewhat less, about 30°, to the south and west toward Cochabamba.

Weathering and jointing

The sandstone-quartzite is generally sound and very hard. Locally it is weathered and stained by iron oxide. There are numerous widely spaced joints, some of which can be seen in the lines of higher shrubs (Figs. 2, 9). Three sets predominate, perpendicular to the bedding. One set is more or less perpendicular to the strike of the folds; the second trends north-south, roughly 45° from the fold strike; the third parallels the fold strike.

The shales and phyllites are locally slickensided parallel to the dip, with smaller slickensides perpendicular to the axes of the corrugations. This suggests interbed shear along the weaker contacts during the episodes of folding that produced the regional and local structures.

DESCRIPTION OF CONSTRUCTION

Road design

The plan of the area which failed is shown in Fig. 4, between km 85+350 and 85+500. The cross-section through the approximate center of movement, between 85+375 and 85+400 is shown in Fig. 5. Fig. 5a illustrates the design cross-section and the stages in construction; the remainder of the figure shows the sequence of failure.

The mountainside in that area was bare sandstone. The surface, a bedding plane, dipped down west-northwest at about 45° (Figs. 4, 5 and 6). The slope was too steep for filling without a retaining structure. Therefore, the entire 9 m width of right-of-way was formed by drilling, blasting and bull-dozing the broken rock down the slope to the west.

The design was based on the successful performance of the old road nearby which had been similarly constructed in that area. However, the dip of the formations that defined the mountain slope on the old road was less than 30°. The dip of the sandstone on the new alignment was also generally less than 30°, but locally much steeper because of the transverse folding and warping.

Bench construction

The excavation was made in lifts of about 3 m. The drill holes were 9 cm in diameter, spaced 2—3 m apart. The loading per hole was about 1 kg of

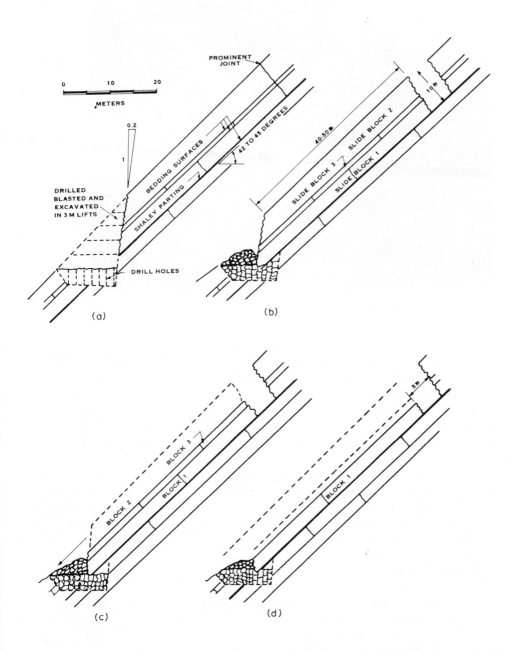

Fig. 5. Schematic cross-sections of the failure area and road profile, section *B-B* (see Fig. 4). (a) Before sliding. Drilling for final excavation lift. (b) Initial movement, blocks 1, 2 and 3 simultaneously. (c) Guillotine action of block 2, seconds after initial movement. (d) Sliding of block 3, two days after initial movement.

Fig. 6. View of failure, looking north, showing block 1 remaining intact.

dynamite for a primer and 6 kg of ammonium nitrate (plus fuel oil = ANFO). From 50 to 100 holes were detonated at one time using five delays or 10—20 holes per delay. The powder factor was about 0.4 kg/m³ of rock or somewhat less than 1 lb per cubic yard. This is not excessive for the hardness of rock excavated. The amount of explosive per delay or detonation should not have produced undue vibration in the formations. That is confirmed by the lack of any serious rock dislocations during or immediately following blasting.

A small slide occurred in colluvium that occupied a small trough-like syncline whose axis dipped downhill at about 35°, between 85+200 and 85+ 350, just south of the future rockslide zone. The scar of bare rock in the syncline left by this movement can be seen at the right in Fig. 2. There was no injury or equipment damage and work continued without incident.

Construction timing

Construction progressed along the mountain face for a distance of about 2 km for six months. The bench was initially cut deep enough to provide a pilot road 5—6 m wide that permitted commencing a tunnel through the end of the ridge. Because the tunnel construction was more critical to the project schedule, the remaining benching was slow. The drilling was in progress for the last 3 m lift at the time of the rockslide, six months after the benching had commenced.

The slope above the bench was checked daily and after blasting for opening of joints or the formation of cracks. Some progressive joint width increase was noticed as much as 100 m upslope, but no changes were found within six hours following blasting.

THE FAILURE

Occurrence

The rockslide occurred between 85+325 and 85+460 on July 9, 1969, at about 10 : 30 p.m. local time. A night crew of drillers was at work at the south end of the slide zone, preparing the last lift of excavation. Without warning, seven men, as well as air compressors, drills and bulldozers, were swept down the mountainside and ground into small pieces by the angular blocks of quartzite weighing up to 5 tonnes. At 4:30 p.m. a small shot (a few kilograms) had been made to break up some irregular rock projections that interfered with drilling; otherwise, there had been no blasting for five days. The weather was dry and there had been no significant rain for several months. The shape of the failure zone is shown in plan in Fig. 4 and in cross-section in Fig. 5. Fig. 7 shows its general appearance looking northwest.

Fig. 7. Measuring dip and dip undulations on the upper surface of block 1 within the failure zone.

Stages of failure

Failure took place in several stages (Fig. 5). Initially, a block of rock, near-ly triangular in shape, 160 m long, 40 m wide (in plan) and 10—12 m thick, broke loose along existing joints and slipped down onto the road excavation (Fig. 5b). Sliding took place along a shale seam exposed by the road excava-tion. When the block struck the unexcavated bench, it broke along a second shale seam. The upper portion of the mass, denoted slide block 2, continued downward while blocks 1 and 3 remained in place. Most of the men and equipment were swept down by block 2, which cut across the bench like a giant guillotine blade; finally fragmenting, it continued down the slope to the creek, 200 m below. The breaking rock slab formed angular faces that crushed and ground the men and equipment as it slid downward.

The movement was so rapid that a tremor was felt 4 km opposite the slide; it was not felt at the construction camp 2 km away along the mountain slope. No earthquakes were reported in the region.

Two days later, another portion of the displaced rock slab, block 3, slid down (Fig. 5c and d). The area for months afterwards remained as shown in Figs. 5d and 6. It did not change materially for four years afterwards except for the construction of the permanent road on a rock wall above the former bench. The surface of block 1 is shown in Fig. 3. A slot 10 m deep uphill,

Fig. 8. Slot between the mountainside and the remaining intact portion of the sliding rock, block 1, looking south.

5 m deep on the face of block 1 and 5 m wide remained between the intact mountainside and block 1 as shown in Figs. 7 and 8.

DETAILS OF THE SLIDE ZONE

The detailed structure of the slide zone can be seen in Figs. 3 through 9. The authors made direct measurements of the rock bedding and joints above the slide zone, and on the main shear surface below block 1 exposed in the slot (Fig. 8). These were correlated with the upper surface of block 1 as shown in Fig. 7 and were shown to be similar. The contractor made a ground survey of the major features; these were correlated with oblique aerial photographs by the authors.

Evidence of previous failures

As can be seen in Figs. 2 and 9, the slide event in question is only one of a continuing series of episodes. The mountainside on both sides of the slide area is a succession of slide scars. All did not occur at the same time. Their relative age is seen in the different levels of vegetation growth; some areas are nearly bare while others are almost completely covered. The barest areas, indicating more recent sliding, are in the vicinity of the new slide. There are

Fig. 9. Aerial view of mountainside above the rockslide showing the rock joints delineated by lines of shrubs.

scarps 1—4 m high defining the lateral limits of the old slides, just north of the present slide (Fig. 9). These scarps parallel the strike of the anticline.

The major joint patterns can be seen in lines of more vigorous vegetation growth (Figs. 2, 9). In between major fractures are smaller joints that are obscured by rock fragments and vegetation on a smaller scale. The major joints are 30—50 m apart, the smaller joints are spaced 0.3—3 m. Three sets are seen. Two predominate sets strike obliquely across the mountain face (Figs. 2, 4 and 9); a trace of the third, less obvious set, runs downhill parallel to the steepest dip.

Scarps

The left or north side of the slide scarps (Figs. 4, 7 and 8) is parallel to one of the predominate joints. The right side, although appearing curved, is made up of a series of zig-zags where the scarp follows two sets of joints. The face of the scarp (Fig. 8) is stained and slightly weathered, with some small roots extending to nearly its full depth. This confirms that the fracture existed long before construction began, as a part of the joint system.

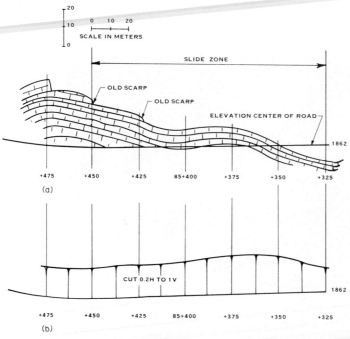

Fig. 10. Section and profile of bench face in cut, looking east (*A-A*, Fig. 4). (a) Cross-section of anticline perpendicular to strike of fold looking up-dip. (b) Profile of center-line of highway and top of rock cut.

Slide surfaces

The surface of sliding of the initial failure can be seen in the slot below the scarp (Fig. 8). It is undulating and shaley. Samples of the shear surface were tested in direct shear. The angle of friction ranged from 27° to 34° for contact areas much smaller than the crest to crest width of the undulations. Large samples, similar in size to the undulations, had friction angles of from 29° to 42°. When wet, the shear surfaces had friction angles of 24—25°. The top of slide block 1 is still intact (Figs. 3, 6). It is similar, but somewhat less shaley and with slightly higher friction angles.

The surface of initial sliding was examined in detail. There was some dampness despite months of dry weather. There was, however, no flow nor water seepage. It appeared that moisture had entered the joints uphill and had followed the shaley seam, which acted as an aquiclude. Beautiful quartz crystals were noted lining several flat vugs 10—20 cm in diameter and 1 cm thick in the shale. The vugs were dry inside. Loud cracking and popping sounds, like small-arms fire, could be heard as the rock in the scarp expanded in the sunshine. Similar sounds could be heard late in the day as the rock cooled. The noises were frightening; if the authors had shown any signs of their fear, the workmen assigned to them would have disappeared despite the attractiveness of the jewel-like quartz they collected while the authors sampled and measured the shear surface.

Structural details

The road cut is almost uniformly 18 m deep between km 85+500 and 85+ 325 because it curves to follow the ground surface contours (Fig. 10). To the north and south, however, small synclines reduce the cutting materially.

From the strike and dip measurements as well as from their low-level helicopter aerial photography, the authors developed a cross-section of the formations perpendicular to the axis of the fold, looking up-dip. The slide area is a gentle anticline whose axis is perpendicular to the regional strike. The apex of the slide block is at the anticline high. Movement began at the south end, with westward sliding and some rotation about the north end at km 85 + 475, as suggested by striations on the failure surface.

MECHANISMS OF MOVEMENT

Primary cause

From the observations of the slide, it was obvious that failure was caused by a weaker shale seam interbedded in the sandstone that was exposed by the notch or bench excavation (Fig. 5). The shale exhibited old slickensides;

the angles of friction were significantly less than the dip of the formation. The pertinent questions were: (1) why did the slope remain intact for so long a time after its support was destroyed by excavations, and (2) what finally triggered the failure? The answer to both questions seems to lie in the undulating bedding planes and slickensided surfaces.

Shelving and temporary stability

A simplified section of the slide zone is shown in Fig. 11. The undulations form flat shelves whose dips are smaller than the angle of internal friction. These flat areas or shelves render the slope stable although the average dip exceeds the angle of friction. Patton (1966) described similar irregularities as "teeth". Goldstein et al. (1966) described the part played by such irregularities in failure of fractured rock.

Initiating forces

Excavation destroyed the downhill support. The joints upslope defined planes of tensile weakness. During the day, the block defined by the joints

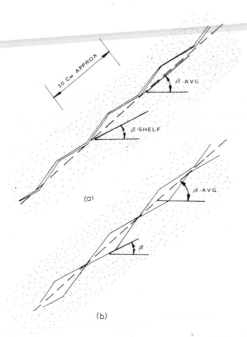

Fig. 11. Schematic sections of slide surface suggesting mechanism of rock block stability. (a) Undisturbed slope: stable if shelf angle, β_{shelf}, is flatter than angle of friction, ϕ, although average slope, β_{avg}, exceeds angle of friction ϕ. (b) Slope block displaced: load no longer supported by shelves; if β_{avg} exceeds ϕ, slope fails by block gliding.

above and the undulating surface expanded in the hot summer sun. The sum of forces and movement was downslope. At night it shrank; however, because the vector sum of forces upslope is less than the forces downslope during expansion, the net effect was a small movement downslope. Such slow movement was observed in the gradual opening at the joints above the failure zone — movement that was independent of the construction operations. Eventually, the movement accumulated downslope until the shelf contact was virtually eliminated as shown in Fig. 11b. When that point was reached, the angle of point contact of the undulations equalled the average dip which, in turn, exceeded the angle of friction. Sudden failure ensued. This mechanism is suggested by the fact that failure occurred during a period of rapid rock cooling — 10 : 30 p.m. on a clear night following a hot, sunny day. The proposed mechanism explains the long period between loss of support by excavation and failure. A similar mechanism for creep was suggested by Tamburi (1974).

Other mechanisms

A number of other triggering mechanisms were investigated. Although the area is seismically active, the nearest seismograph recorded no significant tremors that might have initiated the sliding. Although damp areas were found and moisture on the interface cannot be eliminated as a factor, there had been no rainfall that might have provided an increase in moisture or interface water pressure that could have initiated movement.

The vibration of blasting could not have triggered the slide because the movement occurred five days after the last significant shot and six hours after an insignificant trimming shot. Blasting could have aggravated the "shelf movement"; however, crude measurements before and within six hours after each shot found no measurable changes. Equipment vibration might have aggravated any shelf movement. However, at the time of failure, the only machines operating were air compressors and air track drills. The air compressors at the north end of the zone did not produce noticeable rock vibration. The air track drills produced some vibration; it was not noticeable beyond about 3 m.

Cause of failure

The failure seems to have been inherent in the structure of the mountainside. This view is confirmed by the numerous old scars and scarps in the immediate area. The only uncertainty involves the question as to why the failure did not occur months earlier, when the bench was first cut, or during the earliest rainy season.

A second mystery is why the slide did not encompass more of the mountainside. The joint pattern and the anticline in the failure area provides an answer. Just beyond each end of the failure area are synclines where the rock cuts are small. The next higher major points (Figs. 4, 9) project beyond these limits. Thus, the zones of rock defined by the joints formed a broad inverted V-arch across the failure area, restraining movement above to some degree. However, if the old scars are any clue, more of the mountainside will slide in the future, as occurred in 1973.

PROPOSED CORRECTIVE MEASURES

Rock removal. The most obvious measure would be to remove the rock above the shale seam upslope to the change in formation dip, a distance of 350 m as shown in Fig. 1. This would improve safety but not eliminate future failures because the topography (Fig. 2) shows slide scars to the mountain top, about 1.3 km.

Buttress support. A second method would be to partially fill the scarp slot with concrete buttresses. Thus, slide block 1 and the buttresses would pro-vide support for the dipping rock above. The major disadvantage is the secu-rity of the toe of slide block 1 resting on the drilled and fractured rock bench.

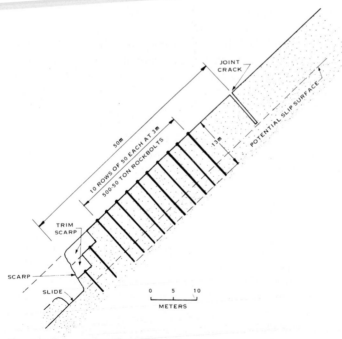

Fig. 12. Remedial rock bolting.

Rock bolting. A third method, with several variants, would be to provide rock bolt support for the rock "arch" above the slide zone (Fig. 12). Tensioning the bolts would enhance the rock friction and minimize the continuing movement.

Drainage. Drain holes drilled to the joints would probably enhance the stability by providing exits for any water that might accumulate from runoff on the slope. Sealing the open joints above the slide zone similarly would be helpful during the rainy season. However, the effectiveness of water control cannot be evaluated because water was not shown to be a major factor in the slide.

Subsequent events

Unfortunately, economics and time dictated that no corrective measures be taken. The road was built on a masonry wall across the toe of slide block 1. In 1973, 40,000 m³ of rock fell onto the road in the same area. It was removed. There have been no problems during the ensuing two years except superficial dropouts from the scarp. It is the authors' opinion, however, that the future will bring additional movements.

ACKNOWLEDGEMENTS

This investigation was made by the authors for the joint contractors: Constructores Jones y Associados of Cochabamba (and Charlotte, N.C.) and Bartos y Cia, La Pez, under the direction of Mr. L.B. Jenkins of Jones and Sr. Jorge Bartos of Bartos y Cia. The project design was by Tippets, Abbett, McCarthy, Stratton Engineers. An independent report on the slide was prepared by H.W. Burke, Consulting Geologist, who suggested the concrete buttress remedial method. The entire project was under the Ministry of National Economics of Bolivia.

REFERENCES

Goldstein, M., Goosev, B., Pyrogovsky, N., Talinev, R. and Turovskaya, A., 1966. Investigation of mechanical properties of cracked rock. *Proc., 1st Int. Conf. on Rock Mechanics, Lisbon, 1966*, 1: 519—523.
Patton, F.D., 1966. Multiple modes of shear failure in rock. *Proc., 1st Int. Conf. on Rock Mechanics, Lisbon, 1966*, 1: 509—513.
Tamburi, A.J., 1974. Creep of single rocks on bedrock. *Geol. Soc. Am. Bull.*, 85: 351—356.

Chapter 11

LANDSLIDES OF BRAZIL

A.J. DA COSTA NUNES, ANA MARGARIDA MARIA COSTA COUTO e FONSECA and
ROY E. HUNT

ABSTRACT

Landslides and erosion are particularly troublesome geologic hazards in
Brazil, as the local natural conditions of geology, topography and climate are
all favorable for their occurrence. The majority of landslides, generally in
the form of slides, avalanches, mudslides, rockfalls and block glides, occur
in coastal mountain ranges subjected to frequent heavy rains and cloud-
bursts. The coastal mountains are formed primarily of granite, gneiss, and
migmatite, covered with thick deposits of residual and colluvial soils. In the
northeast and southern areas of the country, landslides occur in formations
of marl and clayshale with expansive properties.

Methods of slope stabilization utilized are numerous, and one of the most
successful is the use of reinforced concrete curtain walls with earth or rock
anchors. These have been used frequently to retain high, steep slopes in residual and colluvial soils and weathered rock, as well as slopes in expansive
materials. Vertical cuts to 26 m in height in residual and colluvial material
have been successfully retained with anchored curtain walls. In this article
the problem of landsliding in Brazil is approached from the phenomenological aspect rather than the analytical side; numerous typical landslides are
described, and appropriate solutions are outlined.

INTRODUCTION

The problem of landslides and stabilization of slopes is common to all
countries with mountainous or hilly terrain, but it is particularly troublesome in Brazil. Many of the major cities in Brazil are located either within or
on the lower slopes of a long chain of mountains adjacent to the litoral zone
(Fig. 1). Heavy rains and cloud bursts are common in the mountains. Urban
growth, causing development to spread from the lowlands onto the slopes,
and modern highway and railway design and construction requiring huge
side-hill cuts and fills, have caused many problems of slope stabilization.

Local geological and climatological factors tend to severely aggravate the problem. The semi-tropical to tropical climate of Brazil has resulted in thick deposits of residual soils and deeply weathered rock, and slope movements

Fig. 1. Geologic map of Brazil, showing locations of landslide-prone areas described in text. (1) Paraná Basin, subdivided as follows: A = Estrada Nova, Irati, Morro Pelado, and Rio do Rato Formations; B = Botucatu Formation; C = Guabirotuba Formation; D = other rocks (see also, Bahia region). (2) Recôncavo Basin, including Ilhas, Candeias, S. Sebastião, and Marizal Formations. (3) Areas underlain by crystalline rock and associated residual materials.

have resulted in numerous areas with colluvial soil deposits. The subject is now well known, having been discussed by Terzaghi (1953), who suggested to Brazilian engineers that they assume the task of studying residual soils and weathered rock masses. Tropical rains of great intensity have caused a remarkable number of landslides, many very large in volume, including a special type of landslide — the "hydraulic excavation" (Vargas, 1967b; Da Costa Nunes, 1969c; Jones, 1973).

In this paper the problems of landslides in Brazil and their correction are discussed from the phenomenological point of view. Indeed, mathematical analysis is often not appropriate because of the heterogeneous nature of residual soils which commonly retain the discontinuities and other textural features of the original rock (Da Costa Nunes, 1971; Patton and Hendron, 1974). In addition, a critical plane along which rupture is likely to occur is at the contact between the residual soil and the subjacent rock surface; this is often a poorly defined zone, inasmuch as the soil changes gradually to fractured rock. Because of the characteristically higher permeability within this zone, high pore pressures develop during periods of heavy rain (Kanji, 1970, 1972).

Although landslides and erosion in the residual soils comprise the most common stabilization problems in Brazil, there are also serious stabilization problems resulting from rockfalls and expansive soils. These, too, are discussed herein.

CLASSIFICATION

The criteria for classification of landslides, so well developed by Terzaghi and Varnes, among others, will be followed in this paper with emphasis on the classification presented by Vargas (1967b) for residual soils. The problems of landslides in Brazil will be discussed herein as follows:

(1) Sliding of exfoliation slabs, weathered rock, and residual deposits derived from Brazilian crystalline rocks.

(2) Sliding of colluvial soils or colluvial-residual complexes derived from Brazilian crystalline rocks.

(3) Sliding of marl formations of the Recôncavo Basin of northeast Brazil, which are associated with the Barreiras Series.

(4) Sliding of expansive formations, primarily in the southern part of the country.

(5) Intensive erosion of a regional character.

LANDSLIDES IN CRYSTALLINE ROCK AND ASSOCIATED RESIDUAL AND COLLUVIAL FORMATIONS

Brazilian crystalline rocks are extremely variable, but granite, gneiss and migmatite are the most common types. In some areas, in particular Minas

TABLE I

Typical properties of Brazilian rocks (from Ruiz, 1966)

Rocks	γ_a	γ_r	Ab	P_a	S	σ_c	E_d
Basalt with sandstone	2.25	2.82	6.1	13.7	68.1	29	285.45
Brown compact basalt	2.92	3.00	0.7	2.1	78.8	104	699.24
Black compact basalt	2.82	2.97	0.5	1.4	28.0	228	572.33
Botucatu sandstone	2.32	2.44	1.2	2.8	56.0	75	436.70
Diabase	2.96	3.04	0.4	1.2	40.0	148	918.44
Gneiss	2.75	2.69	0.1	0.2	14.3	134	703.77
Granite	2.56	2.62	0.5	1.3	40.0	116	583.60
Granitic gneiss	2.68	2.74	0.6	1.6	72.5	73	435.10
Granulite	2.58	2.63	0.7	1.7	89.1	84	221.60
Light gray compact basalt	2.52	2.71	1.7	4.2	63.0	114	566.70
Limestone	2.74	2.79	0.1	0.4	21.5	100	499.60
Pyroclastic basalt	2.44	2.97	4.8	11.6	62.0	65	393.24
Porphyritic granite	2.68	2.74	0.4	1.0	27.0	97	470.10
Quartzite	2.63	2.67	0.3	0.9	75.0	210	213.40
Tourmaline granite	2.62	2.65	0.3	0.8	66.6	109	601.40

γ_a = apparent specific gravity in Mg/m^3; γ_r = real specific gravity in Mg/m^3; Ab = water absorption in percent; P_a = apparent porosity in percent; S = void filling in percent; σ_c = compressive strength in MN/m^2; E_d = dynamic Young's modulus in $10^2\ MN/m^2$.

Gerais, schist and phyllite occur. The rock structure varies from the lightly to heavily jointed, and from the sound to highly altered (Serafim and Da Costa Nunes, 1966). Slickensided surfaces are common in the highly altered zones. Some typical engineering properties of Brazilian rocks are given in Tables I and II.

Residual and colluvial formations derived from decomposition of these

TABLE II

Water absorption (alteration index) versus Young's modulus for a typical Brazilian gneiss (from Serafim and Da Costa Nunes, 1966)

Water absorption (%)	Young's modulus ($\times\ 10^3\ MN/m^2$)
0.140	100
0.145	90
0.160	80
0.170	70
0.185	60
0.210	50
0.250	40
0.310	30
0.410	20
0.490	10

TABLE III

Characteristics of residual and colluvial Brazilian soils

A. From Vargas (1967b)

Site location	Soil type	LL	PI	w	Void ratio	Degree of saturation (%)	c	ϕ
Santos	residual	30—45	NP—18	19—27	0.73—1.07	61—70	20—40	±35
Tapera *	residual	30—80	NP—40	20—35	±1.00	50—60	50—80	32—38
Rio de Janeiro	residual	20—55	5—25	8—26	0.8 —2.3	50—80	0—40	33—43

B. From Da Costa Nunes and Ferreira (1971)

Site location	Soil	γ_{nat}	γ_{sat}	c_{nat}	c_{sat}	ϕ_{nat}	ϕ_{sat}
Rio de Janeiro	residual	1.3—1.7	1.5—2.1	20—100	0—50	35—60	15—45
	colluvial	1.2—1.7	1.4—2.2	10—100	0—50	30—60	10—45
Paraná	residual	1.2—1.7	1.6—1.8	20— 40	0—30	25—50	20—40

LL = liquid limit in percent; PI = plasticity index in percent; w = natural water content in percent; c = cohesion in kN/m^2; ϕ = angle of internal friction in degrees; γ_{nat} = natural density in Mg/m^3; γ_{sat} = saturated density in Mg/m^3; c_{nat} = cohesion in kN/m^2, natural water content; c_{sat} = cohesion in kN/m^2, saturated state; ϕ_{nat} = angle of internal friction in degrees, natural water content; ϕ_{sat} = angle of internal friction in degrees, saturated state.

* Northwest of Santos.

rocks are heterogeneous mixtures of soils often containing boulders and large blocks of rock. At times the mass of soil and rock slide as a unit, and at other times individual large blocks break free and fall. Residual soil thickness varies with the type of underlying rock. For granites, for example, residual cover can vary from non-existent to many metres in thickness, whereas for gneisses with high feldspar content, residual soils can reach 30 to 100 m in thickness. It is often difficult to differentiate between slides occurring in residual deposits from those occurring in colluvial deposits, as these deposits are similar in composition and are locally gradational. In addition, in some locations they are associated with urban landfills composed of similar material. Geotechnical properties of these formations have been described in numerous works published in Brazil (Vargas 1967a, 1971b; Da Costa Nunes and Ferreira, 1971; Sandroni, 1973); typical values are summarized in Table III. Dynamic aspects of sliding of residual soil deposits have been approached by Fernandes (1974) in studies made for Brazil's first nuclear power plant at Angra dos Reis. More important landslides that have occurred recently in these types of formations are described below.

Slides involving exfoliation slabs, fragmented rocks and blocks of "intact" rock in residual soils

Examples of these phenomena include falling blocks on the Estrada Grajau-Jacarepaguá in Rio de Janeiro (Fig. 2); landslides of blocks that destroyed the Viaduct Annes Dias in Rio de Janeiro; landslides of blocks on the Rua Benjamin Batista (Fig. 3); and similar slides on the Morro da Providência (Fig. 4), Morro Santa Marta (Fig. 5) and the Morro do Encontro (Fig. 6).

The downslope movement of boulders and large blocks of rock occur frequently when construction causes erosion and loosening of the soil around the rock mass. Fig. 2 illustrates a typical solution involving adequate slope drainage and an anchored retaining wall. Large boulders encased in the wall are shown in the photo.

Another example of the sliding of blocks of rock is shown in Fig. 3, which illustrates the slope of the Morro do Corcovado that slid onto Rua Benjamin Batista in Rio de Janeiro, in December 1974. For many years since the 1930's this slope has been the location of many slides; the most recent is apparent in Fig. 3. The General instability of this slope since the 1950's is caused primarily by exploration for sandy residual soil used for pavement base course; this exploration requires removal of a layer from the base of the residual soil cover, and hence loss of support for superjacent rock. The existence of a talweg on the slope at this site caused the loss of support for the layer of altered rock. The slide demonstrated the surface instability, after alteration, of a great complex of crystalline rock. Sheeting joints cut across the planes of foliation of the original gneiss, approximately parallel to the slope. Blocks of rock break loose from the exfoliating slabs, along the joints,

Fig. 2. Estrada Grajau-Jacarepaguá, Rio de Janeiro. Anchored wall to retain loose colluvial soil with boulders; note boulders encased in wall.

resulting in a disintegrated mass downslope. The scale of the phenomena can
be estimated by comparison with the apartment building also shown on the
photo.

Often these phenomena are concentrated in areas of the local mountains
where exploration for rock with explosives has extensively loosened the
slopes, thus leaving the residual soil over the rock vulnerable to sliding. Such
a slide is illustrated in Fig. 4. The slide occurred in a rock mass on the Morro
da Providência in Rio de Janeiro in December 1968. Overlying the rock, at
the top of a cut made for quarrying operations, is a thin cap of residual soil,
on which houses have been constructed. The progressive use of explosives
without geologic study of the locale resulted in a slide involving a substantial
mass of rock material coincident with foliation planes of the gneiss bedrock.

Fig. 3. Slope of the Morro do Corcovado, Rio de Janeiro. Location of slide onto Rua Benjamin Batista, December 1974. Photo 21 February 1975. Exfoliating granite slabs and talweg are clearly shown. Apartment houses provide scale, lower left.

The slide occurred almost instantaneously, without observable warning. The soil mantle remained on the rock mass as it slid, together with a few small houses.

The slope of the Morro Santa Marta, a mountain along Rua Alzira Cortes in Rio de Janeiro, is the location of a very old rock quarry; its unstable slopes have caused serious problems to buildings located at its base (Fig. 5). In 1966 and 1967, during heavy summer rains, the sliding of a mantle of soil overburden and altered rock situated on top of the mountain caused serious damage to these buildings. The removal of blocks of rocks and overhangs extending from the face, either by explosions or from natural climatic forces, would have caused grave danger to the buildings immediately below. A terrace and drainage system installed on top of the mountain eliminated the instability of the mantle of soil and altered rock. The construction of impact walls to low heights was designed to provide protection against the fall of blocks of rocks; concrete supports for overhanging or loose blocks, and anchorages in other areas were installed to provide local stabilization.

Typical older landslides of this type are those of the mountains near Santos (Vargas and Pichler, 1957), while more recent ones are described by Cer-

Fig. 4. Morro da Providência, Rio de Janeiro. Block glide of mass of rock and residual soil along foliation planes of gneiss bedrock occurred in right center of photo in December 1968. Photo 8 January 1969.

queira (1967). The phenomenon is, in general, aggravated by erosion of the residual soil matrix, leaving interior blocks without support. Around the residential groupings of hillside "favelas" (slums), slide events occur with great frequency because of the modification of drainage conditions on the slopes as caused by disorderly construction (Da Costa Nunes, 1967). A particularly important example occurred at the Morro do Encontro in the city of Rio de Janeiro (Fig. 6). The fall of blocks from this hill, invaded by a "favela", menaced a populated suburb and expensive residences in Vila Isabel along the Rua Visconde de Santa Isabel. Stabilization was accomplished by the consulting engineering firm, Tecnosolo S/A, for the former Departamento de

Fig. 5. Morro Santa Marta, Rio de Janeiro. During 1966 and 1967, slides of soil overburden and weathered rock caused serious damage to buildings at the base of the mountain. Note drainage system and terraces constructed on top of the mountain. Photo 22 April 1969.

Estradas de Rodagem da Guanabara (DER-GB), by using several techniques including drainage of the slopes, cementing of the blocks, and anchorages.

Slides of predominantly residual soil mantle locally stressed with artificial fills, with slip generally occurring along the rock surface

A major accident of this type in Brazil, measured either in terms of volume or in tragic consequences, occurred in the Bairro Jardim Laranjeiras in February 1967; 110 people died (Fig. 7). The landslide occurred in a mass of residual soil of altered granite at the contact with gneiss, following very

Fig. 6. Morro do Encontro, Rio de Janeiro. Unstable boulders on steep slope retained by anchored walls and grillages; slope gunited for drainage and erosion control.

heavy rainfall. The slide was aggravated by the removal of substantial material from the foot of the potential slide area for use in clandestine fills, and by indiscriminate exploration for fine gravel. The saturated mass of soil broke loose and formed an avalanche which destroyed in rapid succession three buildings, two of which were apartment houses, thus causing in a few seconds the most tragic accident of this type in the country. The region was subsequently stabilized by extensive work accomplished by the DER-GB and the Instituto de Geotécnica da Guanabara, today united in the new State of Rio de Janeiro. Stabilization methods included the removal of unstable formations, principally blocks of rock and talus, concreting of space between blocks, and construction of anchorages and curtain walls for containment. It should be noted that Brazil was one of the first countries to pioneer the use of permanent anchorages in soil, initiating the technique around 1957 (Da Costa Nunes et al., 1960; Da Costa Nunes and Velloso, 1963).

Some important aspects of the slide on the slope above Bairro Jardim Laranjeiras are visible in Fig. 7. Note, e.g., the scar on the upper part of the slope, from which the mass of earth broke loose to become an avalanche. The scar is conchoidal in shape, following a plane extending downward along the rock surface. Geotechnical studies made after the accident revealed that the scar area represented a geologic anomaly of highly altered zones of gneiss, on both sides of which was granite. Cuts made on the top of the slope created basins in which water accumulated; this water subsequently pene-

Fig. 7. Bairro Jardim-Laranjeiras, Rio de Janeiro. This avalanche occurred on 18 February 1967 and resulted in the destruction of two apartment houses and the death of 110 people. The slide scar is clearly shown. Photo 20 March 1967.

trated into the joint system of the gneiss. Pore water pressures at the contact between the residual soil cover and the rock increased, thus "loosening" the soil mass and allowing it to slide down slope in the form of an avalanche.

A recent photo of the slide in Bairro Jardim Laranjeiras shows the anchored curtain wall and slope drainage used to stabilize the area (Fig. 8). The slide scar in altered rock is shown in the center of the photo, while the right side of the photo shows exfoliating slabs of hard granite remaining after the residual colluvial soil cover was stripped.

Another example of soil mantle sliding on a rock surface is illustrated in Fig. 9, which illustrates the slide in ancient "talus" that occurred in April

Fig. 8. Recent photo of Bairro Jardim-Laranjeiras showing anchored curtain wall and slope drainage. Slide scar in photo center, exfoliating granite slabs on right. Residual soils have been cleaned from slope as a safety measure.

1966 on the slope of Morro dos Urubus, Bairro de Terra Nova, Rio de Janeiro. Slide scars at the top of the slope indicated subsidence on the order of 3 m. The entire mass of material, from the base of the slope upward, moved within a few hours resulting in an upheaval (pressure ridge) at the toe, of about 3 m. The phenomenon occurred about three months after heavy rainfall and was aided by rivulets of water at various locations on the slope comprising the zone of the slide.

Formations of weathered rock and associated complexes of colluvial and residual soils with slide activity caused by unusually intense rain fall

A phenomenon characteristic of this type of event, and of most importance, is the violent erosion that always occurs (Vargas, 1971a; Da Costa Nunes, 1969a; Jones, 1973). As described by geologist Fred Jones of the U.S. Geological Survey (1973), in the most extensive work to date on the subject, very intense rains provoke erosion of the slope causing the process of avalanche or "hydraulic excavation" that destroys virtually all obstacles downslope. Vegetation is carried away, and the moving mass, including huge boulders, comes to rest in the lower part of the valley; a characteristic triangular-shaped scar is left on the hill side (Fig. 10).

The phenomenon occurred in many hundreds of locations almost simul-

Fig. 9. Morro dos Urubus, Bairro de Terra Nova, Rio de Janeiro. Slide in ancient "talus" at the lower part of the slope occurred in April 1966, three months after heavy rains. Photo taken in 1966 shows exposed granite of ancient slide scar.

taneously in the State of Rio de Janeiro in 1966, and once again in 1967. The total loss of life from both landslides and floods has been estimated to be as high as 1000 in 1966, and 1700 in 1967. Property and industrial damages have been described as "inestimable" (Jones, 1973). The phenomenon is very characteristically manifested in the Serra das Araras described by Da Costa Nunes (1967) and Jones (1973), and in Caraguatatuba, studied principally by Vargas (1967b). The avalanches resulted in immense human and material loss in the Serra das Araras mountain region, where traffic is heavy over the most important roadway in Brazil, the Rio—São Paulo highway, and where, in addition, there are important hydroelectric installations. In the opinion of Fred Jones, "Probably no other generating complex ever suffered such intense devastation from a single rainstorm as the Lages Rio Light installation."

Jones' summary provides insight into the intensity and enormity of the event:

"On the night of January 22 and 23, 1967, a landslide disaster of unbelievable magnitude struck the Serra das Araras region of Brazil. Beginning at about 11 : 00 p.m.,

Fig. 10. Serra das Araras, State of Rio de Janeiro. Unusually heavy rains during 1966 and 1967 resulted in hundreds of avalanches in the mountains about 60 km from the city of Rio de Janeiro. Property damage and loss of life from avalanches and floods were tremendous. Areas where avalanches stripped vegetation from slopes are apparent.

an electrical storm and cloudburst of 3½ hours duration laid waste by landslides and fierce erosion a greater land mass than any ever recorded in geological literature. The area laid waste was 25 kilometers in length and 7 to 8 kilometers in maximum width. A large part of the area of heavy destruction was on the steep slopes of the Serra das Araras escarpment. Thunderbolts from the lightning and the collapse of the hills shook the region like an earthquake. Landslides numbering in the tens of thousands turned the green vegetation-covered hills into wastelands and the valleys into seas of mud. Within the area of most intensive sliding are the Rio Light S/A generating complex, the principal power supply for Rio de Janeiro, and a section of the Presidente Dutra Highway, the main arterial between Rio de Janeiro and São Paulo. Landslides in the steep canyons above the Rio Light S/A generating complex turned to mudflows in the valley bottom and buried the main power units."

Landslides onto the roadways numbered many hundreds, occurring along the oldest lane of the Rodovia Presidente Dutra (Rio—São Paulo Highway), a roadway that had not suffered any landslides in the previous 39 years of its existence (see, e.g., Jones, 1937, pp. 29—34).

The techniques of recuperation of the Serra das Araras resulted in a valuable collection of experience. For the first time anchored curtain walls for highways were used on a large scale. In the most part these were executed by starting construction from the top of the wall and continuing to the bottom by the Brazilian method (Da Costa Nunes, 1969b; Deere and Patton, 1971). The method was used soon after on the highway across the mountains from Curitiba to Paranagua, State of Paraná (Lubina, 1968). Today, anchored curtain walls are being constructed to retain vertical cuts in colluvial and residual soils to heights of 26 m.

This type of mass movement of land, by violent erosion, presents an enormous geological hazard; even with the precautions that have been thus far adopted and with our understanding of the problem, such events are unpredictable and inevitable and will lead to further material losses and deaths. Appropriate solutions lie perhaps in a civil defense warning system and extensive and costly protection for the works of man (Da Costa Nunes, 1971) menaced by the potentially unstable slopes.

The rainfall associated with these disastrous events is incredible, as summarized by Jones (1973, pp. 5, 6):

> During the 1966 season, two exceptionally heavy rains fell; they were the storms of January 10, 11, and 12, and March 26 and 27. On January 10, 1966, a cold front moved into the city of Rio de Janeiro area and remained stationary there for 3 days. At the meteorological observatory in the center of Rio, the oldest station in the city, which has about 80 years of recorded observations, the depth of rainfall in the 3 days reached a total of 484 mm (millimeters) (19.05 in). The normal rainfall for January is 171 mm (6.73 in), and the previous maximum rainfall recorded for any one month was 473 mm (18.62 in) in January 1962. During the 3-day January 1966 storm, the Alta da Boa Vista station recorded 675 mm (26.57 in) of precipitation.
>
> The storm of March 26 and 27 began about 3 p.m. Precipitation was most intense during the early part of the storm when 240 mm (8.45 in) of rain fell in 6 hours. The intensity reached 100 mm (3.94 in) per hour. During the 18 hours of total duration there was 320 mm (12.6 in) of precipitation. Two characteristics set this storm apart from the January storm — there was a greater amount of rainfall in a shorter period of time and a greater amount for a single day.''

Data from the 1967 storm are as follows (Jones, 1973, p. 22):

> "In the area of the generating complex, there were three rain gages. During the storm, the rainfall recorded at these gages was as follows: Fazenda da Rosa, 275 mm (10.83 in); Ipe Acampamento, 225 mm (8.85 in); and the Lajes Creek dam, 218 mm (8.58 in). Between 30 and 50 minutes after the beginning of the storm the Lajes Creek dam station recorded intensities varying from 100 to 114 mm per hour."

Thus the landslides in the Serra das Araras well illustrate the effect of uncontrolled erosion on slopes. Another example, much smaller in magnitude, but nevertheless typical, is shown in Fig. 11. On this slope, along Avenida Niemeyer in Rio de Janeiro, there was no provision for internal drainage of the mass, which enabled it to become saturated. The initial movement of the slope caused the superficial drainage system to break, thus allowing even more water to enter the mass. The result was the landslide shown which occurred during the heavy summer rains of 1972.

Fig. 11. Slope along Avenida Niemeyer, Rio de Janeiro. Landslide caused by inadequate slope drainage occurred during heavy summer rains of 1972.

Another example of the effect of uncontrolled erosion is a landslide on the Ilha do Governador (Estrada do Rio Jequiá), Rio de Janeiro, caused primarily by the opening of two cuts that functioned as drainage basins on top of the slope (Fig. 12). The Estrada do Rio Jequiá had been constructed by making a small cut in the toe of a steep slope. The construction of buildings along the street required increasing the cut into the slope, thus aggravating its instability. Runoff from intense rainfall during January 1966 entered the slope through the open drainage basins at the top of the slope, directly causing the landslide below.

SLIDES IN THE MARL FORMATIONS OF THE RECONCAVO BASIN

In and around the important northeastern city of Salvador (Fig. 1) with substantial urban, tourist and industrial development, there occur three types of geologic formations susceptible to landsliding:

(1) Complex crystalline rocks, predominantly gneisses, with problems of slope stability similar to the central south, in particular the State of Rio de Janeiro.

Fig. 12. Estrada do Rio Jequiá, Ilha do Governador, Rio de Janeiro. Slump occurred during heavy summer rains in January 1966. Runoff entered the slide zone in open drainage ditches at the top of slope. Excavations at the toe of slope had decreased slope stability. Photo 6 June 1968.

(2) Barreiras Series of Tertiary Age, formed of soils (sandy clayey silts), with frequent limonite concretions. This series is quite subject to erosion, in particular in the more silty layers (Fig. 13). Erosion by "piping" is very common, but in general, significant problems of slope stability do not occur from this activity. At the contact with the marly phases, however, the problems of slope instability increase because the Barreiras has, in general, higher permeability which leads water to the expansive components of the marl.

(3) Marly phases of the Cretaceous formations which include chiefly the Ilhas, Candeias, São Sebastião and Marizal Series. These formations have similar lithologic characteristics that generally include strata of sandstone, siltstone and shale, with various amounts of limestone (Machado, 1958; Sobral and Menezes, 1958; Da Costa Nunes, 1973).

The shales of the Cretaceous formations contain expansive clay minerals, with montmorillonite predominating. The residual soils of these formations are called "massapes" (an indian name). These soils cause very difficult problems to civil engineering works, in particular to highway engineering (Fig. 14). The more striking aspects, to those not familiar with the problem of

Fig. 13. Piping erosion in the Barreiras Series, Salvador, Bahia.

Fig. 14. Progressive rotational slide in the "massape", or Cretaceous clay shales, Salvador, Bahia. Slide occurred during August 1975 in a cut made for a new highway. Slope drainage, grading and an anchored wall were prescribed to stabilize the slope.

TABLE IV

Geotechnical indices of swelling soils of the Reconcavo Basin, Salvador, Bahia

Property	
Clay content (%)	34—72
Silt content (%)	19—31
Liquid limit (%)	40—61
Plasticity index (%)	12—43
Clay mineral	montmorillonite
Silica sesquioxide ratio	3 : 1
AASHO Classification	A-7-5

stabilizing these non-saturated soils, is the rapid deterioration of masses that are apparently very resistant at the moment a cut is made. Slopes containing shales and sandstones, that require explosives for their efficient excavation, subsequently collapse and flow as mud after a few hours of intense rain (which occurs with high frequency in this region). The subject is summarized in the contribution presented at the Third Congress on Expansive Soils at Haifa 1973 (Da Costa Nunes, 1973). Some geotechnical characteristics of these soils are presented in Table IV. The plasticity of the formations are indicated in Fig. 15 in accordance with Casagrande's A-line, to enable comparison with other expansive soil formations of the world.

The work of stabilizing these deposits must include, necessarily, special attention to drainage; water is the primary cause of disintegration, especially in the marly materials. The process of stabilization by anchorages also has been used frequently in these formations, with the pioneer work in the region being that of the Reservoir do Bonfim in 1959. Recently the performance of anchored curtain walls was evaluated (Da Costa Nunes, 1974), and it was observed that pre-tensioning holds the swelling pressure of the formations, thus assuring their stability.

SLIDES IN EXPANSIVE FORMATIONS IN SOUTHERN BRAZIL

The development of the states of Paraná, Santa Catarina, and Rio Grande do Sul, relatively more recent as compared to the south central region, has contributed to the fact that landslide problems have only recently been recognized. The broad nature of this presentation does not permit discussing fine details of the geologic formations and the problems involved in the southern region. The more characteristic phenomena associated with the southern regions located in the large basin of Paraná (Fig. 1) are, however, as follows:

(1) The fall of blocks in Santa Catarina and Paraná caused by differential

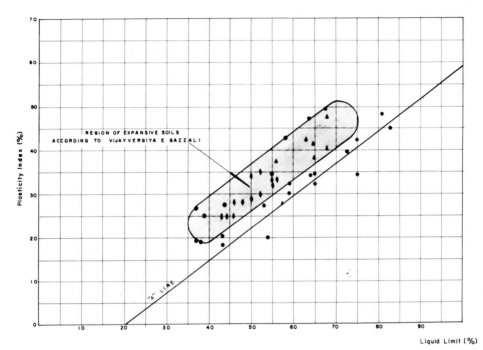

Fig. 15. Plasticity index vs. liquid limit (plasticity) diagram. Data for expansive materials as follows: dots, Mari Series, Salvador; open triangles, Guabirotuba Formation, Paraná; diamonds, Bom Retiro Series, Rio Grande do Sul.

erosion and loss of support in alternate strata of sandstones, siltstones and shales, from the formations Botucatu, Rio do Rato (Morro Pelado Member), Estrada Nova, and Guabirotuba (Fernandes et al., 1974; Tecnosolo, 1975). Expansibility of thin strata of montmorillonitic clay shales in masses of these generally hard formations, especially in the Guabirotuba Formation causes slides in formations seemingly very resistant.

(2) In Rio Grande do Sul, similar problems have occurred in the Bom Retiro do Sul and Irati Formations (Coulon, 1975; Filho, 1975), which in particular were associated with the stability of navigation dams.

Characteristics of residual soils from these formations are summarized in Table V (see also Fig. 15).

The problem of the stabilization of loose blocks of sandstone undermined by the activity of the underlying expansive soils (to prevent falling due to loss of support) still has not had adequate experimentation. In recent projects (Tecnosolo S/A, 1975), solutions were proposed involving protection of stretches of slopes in shales by means of gunite with a supporting mesh, and the construction of impact walls.

In the case of expansibility of a clay shale mass, little experience has been

TABLE V

Characteristics of Brazilian expansive soils from the southern states (Soil Mechanics Laboratory, Tecnosolo S/A)

Property	Guabirotuba Formation	Bom Retiro Series, Rio Grande do Sul
Clay content (%)	42—68	26—34
Silt content (%)	16—22	34—51
Liquid limit (%)	58—68	44—50
Plasticity index (%)	27—49	25—35
Clay mineral	montmorillonite and kaolinite	—

thus far developed. The most promising solution proposed to date involves covering the slope with non-expansive material, after grading the slopes (Fig. 16). The solution with anchored walls is effective and provides considerable support over the slope containing the expansion pressure. It is, however, an expensive solution for large areas.

INTENSE EROSION OF A REGIONAL CHARACTER

Due to the rapidity of Brazilian development and its great territorial extension, there has not been until recently the tradition of soil conservation that exists in certain countries. Many European countries, for example,

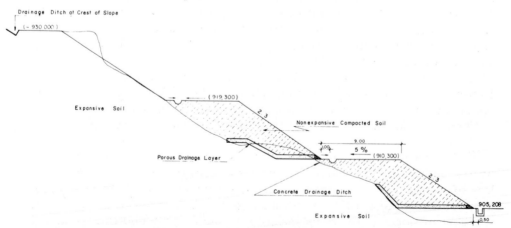

Fig. 16. Sketch of stabilization design of expansive soil slope in southern Brazil (Curitiba, Paraná).

with limited areas of land suitable for development, have many centuries of experience of protecting their lands, especially on hilly terrain. In Brazil, as a result of inadequate drainage and protection against erosion, there have been severe problems in areas where the forest has been cut on a large scale to enable agricultural development.

Fine-grained granular soils with low cohesion (Fig. 13) are particularly susceptible to erosion. This problem is very serious in the north of Paraná. The stabilization of slopes in the northeast part of the state of Paraná is predominantly affected by the erosion of silty residual soils derived from basalt. In other regions, important subjects include erosion of the Barreiras Series and the non-cohesive components of the marly phases of the Cretaceous formations of the Recôncavo Basin.

In our point of view, engineers and geologists in Brazil have not given the erosion problem the attention it deserves. Literature on the subject is almost limited to publications from outside of Brazil (American Society of Civil Engineers, 1966; Nascimento et al., 1973). Brazilian publications are meager (ABCP 1961; OEA 1973; Paraná, 1971). Studies related to construction of highways in Africa, for example, provide more information than has yet been provided by Brazilians, despite the importance of the problem in our latitudes (see also Highway Research Board, 1973).

Primary protection against erosion is obviously by drainage control. Soil cement has been utilized to fill ravines and gullies in order to increase the resistance to erosion of critical areas (Da Costa Nunes, 1966).

DISCUSSION

As this brief overview has shown, Brazil, one of the largest countries of the world, has many regions where landslides present significant geological hazards. The major factors present in Brazil that contribute to landslides are geology, climate, topography and urbanization. The climate has resulted in the chemical and physical decomposition of rock, e.g., from simple "softening" to exfoliation of hard granites, formation of residual deposits up to 100 m in thickness in gneiss, and the disintegration of clay shales. Heavy rains falling on moderate to steep slopes cause severe erosion; pore pressure increases beyond stable values, resulting in landslides. Urban development on a large scale has required construction to encroach upon slopes, further aggravating potentially unstable conditions.

Examination of the slopes in almost any hilly or mountainous area reveals numerous indications of movements, varying from hillside creep to large ancient slide scars. It is almost impossible to predict where landslides might occur, primarily because the torrential rains are often concentrated in a particular area and may not reoccur at the same location for a period of many years. It is safe to assume, however, that where geologic conditions are favorable to landslides, almost any construction activity that cuts into a slope, or removes protective vegetation, or worsens natural drainage conditions, will result in sliding at some future date unless protective measures are provided.

The protective measures to be utilized, of course, depend upon the size of the unstable mass and whether it involves rock or soil formations or both. In unstable rock formations, solutions commonly used include: (1) underpinning loose blocks and ledges; (2) wire mesh to retain loose blocks or to control their fall; (3) anchorages of loose masses or blocks; (4) impact walls for protection from falling loose blocks; (5) gunite to retard weathering; and (6) anchored grillages to retain soft rock and retard weathering.

On steep high slopes involving a large mass of earth, anchored concrete curtain walls have proved to be the most satisfactory method of retention. Failures of various types of walls in residual or colluvial soils are common because the driving force of a large saturated mass moving in the form of a large slope failure is tremendous. Anchored walls have been found to provide support to slopes cut into even expansive soil or rock formations.

Drainage, benching, and flattening of slopes are measures commonly used. Because of the frequency of heavy rains or extended rainy seasons, however, there is no assurance that a cut slope will remain stable; almost certainly pore water conditions within the slope will change. In colluvial deposits and other unstable formations, slopes cut steeper than the natural slopes can be expected to be unstable because even the natural slopes are unstable, as pointed out previously.

With constantly increasing urbanization, landslides are continuously becoming one of the major construction hazards in Brazil; engineers are becoming more and more aware of the necessity for thorough geological studies and explorations prior to preparation of designs. They are also aware that the methods of exploration and design constantly require improvement and innovation to properly cope with the slope stability problem.

REFERENCES

ABCP, 1961. O problema das voçorocas. *Assoc. Bras. Cimento Portland, Eng.*, XX(227).
Alves, de Lima, G.A., 1967. Recuperoção de emergência de estruturas de arrimo ferroviárias. *Simpósio Sobre Proteção Contra Calamidades Públicas.* Clube'de Engenharia, Rio de Janeiro.
Antonio, W. de S., 1972. Sistematização da drenagem de encostas em rodovias. *Reunião das Organizações Rodoviárias, 2, Brasília.*
American Society of Civil Engineers, 1966. Abstracted bibliography on erosion on cohesive materials. *Proc. Am. Soc. Civ. Eng.*, 92(HY2), November.
Associação dos Antigos Alunos da Escola Politecnica, 1966. *Cursos de Extensão Universitária em Estabilização de Taludes e Construção em Encostas.* Escola de Engenharia, Rio de Janeiro.
Assumpção, L.C.X., 1973. *Aspectos Sobre a Incidência da Erosão e seu Controle pelo DER/PR.* DER, Divisão de Pesquisas Rodoviárias, Curitiba.

Barata, F.E., 1964. Estabilidade dos taludes dos cortes. *Construção*, 8(92): 29—36.

Barata, F.E., 1969. Landslides in the tropical region of Rio de Janeiro. *7th Int. Conf. Soil Mech. Found. Eng., Mexico*, 2: 507—516.

Bock, E.I., 1967. Cortinas atirantadas de grande altura na construção de edificio em encosta. *Simposio Sobre Proteção contra Calamidades Públicas*. Clube de Engenharia, Rio de Janeiro.

Camacho, M.E. and Rezende, S.H., 1968. Estabilização da ombreira direita da barragem do funil. *Semin. Nac. de Grandes Barragens 5°, Rio de Janeiro*.

Cerqueira, C. de A.G., 1967. Exemplo de proteção contra o deslizamento de placas de esfoliação junto a construções. *Simposio Sobre Proteção contra Calamidades Públicas*. Clube de Engenharia, Rio de Janeiro.

Conselho Nacional de Pesquisas, 1967. *Os Movimentos de Encostas no Estado da Guanabara a Regiões Circunvizinhas*. Relatório da Comissão de Especialistas, Rio de Janeiro.

Coulon, F.K., 1975. Mapa geotécnico das folhas de Morretes e Montenegro. Resumo da Tese de Mestrado em Geociências. *Tóp. Geomec.*, 19 pp.

Da Costa Nunes, A.J., 1958b. *Curso de Mecânica dos Solos e Fundações*. Globo, Porto Alegre.

Da Costa Nunes, A.J., 1959. Estabilização de encostas em rodovias. *Simp. I.P.R., 2°, Rio de Janeiro*.

Da Costa Nunes, A.J., 1965a. Uso da potensão em barragens e estruturas auxiliares. Relatório. *Semin. Nac. de Grandes Barragens, 4°, Saneamento*, 20(28): 41—50.

Da Costa Nunes, A.J., 1965b. Discussion. *6th Int. Conf. Soil Mech. Found. Eng., Montreal*, 3: 526.

Da Costa Nunes, A.J., 1966a. *Relatório Sobre Aspectos Geológicos Geotécnicos e Hidrológicos de Comissão Interuniversitaria Sobre Calamidades no Estado da Guanabara e Estados Vizinhos*. M.E.C., Rio de Janeiro.

Da Costa Nunes, A.J., 1966b. Aspectos concernentes à mecânica dos solos e das rochas. *Semana de Debates Sobre o Problema das Favelas*. Clube de Engenharia, Rio de Janeiro.

Da Costa Nunes, A.J., 1966c. Estabilidade de taludes em rocha. *Congr. Bras. Mec. Solos, 3°, Belo Horizonte*.

Da Costa Nunes, A.J., 1966d. Discussion — theme 3. *1st Congr. Int. Soc. Rock Mech., Lisbon, 1966*, 3: 301.

Da Costa Nunes, A.J., 1966e. Discussion — theme 6. *1st Congr. Int. Soc. Rock Mech., Lisbon, 1966*, 3: 448.

Da Costa Nunes, A.J., 1966f. Slope stabilization improvements in the techniques of prestressed anchorages in rocks and soils. *1st Congr. Int. Soc. Rock Mech., Lisbon, 1966*, 2: 141—146.

Da Costa Nunes, A.J., 1967. Análise dos deslizamentos de terra havidos no país nos ultimos anos. *Simpósio Sobre Proteção Contra Calamidades Publicas*. Clube de Engenharia, Rio de Janeiro.

Da Costa Nunes, A.J., 1969a. Deslizamentos generalizados nas encostas de solos residuais e coluviais devidos a chuvas intensas. *Jornadas Eng. Arquit. do Ultramar, 2°*.

Da Costa Nunes, A.J., 1969b. *Estabilização de Encostas em Rodovias*. I.P.R., Rio de Janeiro.

Da Costa Nunes, A.J., 1969c. Landslides in soils of decomposed rock due to intense rainstorms. *7th Int. Conf. Soil Mech. Found. Eng., Mexico*, 2: 547—554.

Da Costa Nunes, A.J., 1970a. Diskussionen Ber. 10. *Ländertreffen Int. Bür. Gebirgsmech.*

Da Costa Nunes, A.J., 1970b. Mecânica das rochas e geologia applicada. Estabilidade de taludes naturais e artificiais. *Congr. Bras. Mec. Solos Eng. Fund., 4°, Rio de Janeiro*.

Da Costa Nunes, A.J., 1971. Fatores geomorfologicos e climaticos na estabilidade de taludes de estrades. *Rev. Lat. Am. Geotec.*, 3.

Da Costa Nunes, A.J., 1973. Discussion. *Proc. 3rd Int. Conf. on Expansive Soils, Haifa*, 2.

444   A.J. DA COSTA NUNES, A.M.M. COSTA COUTO e FONSECA AND R.E. HUNT

Da Costa Nunes, A.J., 1974. Estabilidade de taludes — rocha e solo. Relatório — tema 3 (state-of-the-art). *Congr. Bras. Mec. Solos, 5°, São Paulo,* 111.

Da Costa Nunes, A.J. and Velloso, D. de A., 1962. A ancoragem em rocha como técnica de construção em encostas. *Estrutura* 46: 11—16.

Da Costa Nunes, A.J. and Velloso, D. de A., 1963. Estabilidade de taludes em capas residuais de origem granito-gnaissica. *Congr. Panam. Mec. Solos Eng. Fund., 2°.*

Da Costa Nunes, A.J. and Velloso, D. de A., 1966. Stabilisation de talus de sols residuels d'origine granito-gneissique. *Tech. Travaux,* 43(5-6): 383—94.

Da Costa Nunes, A.J. and Menezes, M.S., 1967. Problemas geotécnicos na construção de base naval em solos sensíveis. *Jornadas Luso-Bras. Eng. Civ., 2°,* Comunicação 3.

Da Costa Nunes, A.J. and Ferreira, M.S., 1971. Panorama dos problemas de encostas em estradas. *Jornadas Luso-Bras. Eng. Civ., 3°,* Tema Português IV-7.

Da Costa Nunes, A.J. and Fonseca, A.M. C.C.e, 1972. A experiência brasileira e mundial de estruturas de arrimo rodoviarias ancoradas e sua normalização. *Reunião das Organizações Rodoviárias, 2°, Brasilia.*

Da Costa Nunes, A.J., Fonseca, A.M. C.C.e, 1974. Normalização de estruturas de arrimo ancoradas. *Geotecnia,* 9: 5—34.

Da Costa Nunes, A.J., Cerqueira, C.A.G. and Novaes, J.L.M., 1972. *Aproveitamento de Terreno Acidentados com Cortinas Ancoradas,* 1. EXPO-ENCO, São Paulo.

Da Costa Nunes, A.J., Velloso, D. de A. and Sarto, F., 1960. A consolidação de taludes rochosos por meio de chumbadores. *Rev. Club Eng. Juiz de Fora.*

Da Cruz, P.T., 1967. *Estabilidade de Taludes.* Escola Politécnica, São Paulo.

Da Rocha Filho, I.P., 1966. *Projeto e Execução de Estruturas de Arrimo nas Encostas.* Curso de Estabilização de Taludes.

Da Silveira, I., 1954. Cálculo da capacidade de carga de terrenos inclinados; métodos de trabalhos virtuais. *Congr. Bras. Mec. Solos, Porto Alegre.*

Dantas, H.S., 1967a. Estudo comparativo de custo de estruturas de arrimo. *Simpósio Sobre Pesquisas Rodoviárias, 3°.*

Dantas, H.S., 1967b. Obras de estabilização da encosta de Laranjeiras. *Rev. Eng. Est. Guanabara,* 34 (1-4): 22—23.

Dantas, H.S., 1967c. The behaviour of low-cost roads in Brazil. *3rd World Road Congr., Tokyo.*

Deere, D.U. and Patton, F.D., 1971. Slope stability in residual soils. *Proc. 4th Panam. Conf. Soil Mech. Found. Eng., San Juan, Porto Rico,* 1: 87—170.

Faria, P. and Montetro, P.F., 1967. Estabilização e proteção de taludes na garagem de bondes da CTC na rua Vitória em Sta. Tereza. *Simpósio Sobre Proteção contra Calamidades Públicas.* Clube de Engenharia, Rio de Janeiro.

Fernandes, C.E.M., 1972. A estabilização de taludes no aproveitamento hidroelectrico do Funil. *Geotecnia,* 4 pp.

Fernandes, C.E.M., Teixeira, H.A.S., Cadman, J.D. and Barroso, J.A., 1974. *Estudos Geotécnicos Relativos à Estabilidade dos Taludes Marginais à Rodovia BR-277-373-PR-Trecho: Ponta Grossa—Foz do Iguaçu—Serra da Esperança—Paraná.* Relatório Final. Inst. de Geociências, Rio de Janeiro.

Fernandes, C.E.M., 1974. Estabilidade de capas sobre rocha nas encostas sob solicitação dinâmica. *Semin. Nac. de Grandes Barragens, 9°, 1974.*

Filardi, F., 1966. Obras de recuperação do bairro de Sta. Tereza, apos as chuvas de janeiro. *Rev. Eng. Est. Guanabara,* 33(1-4): 40—43.

Fonseca, A.M. C.C. e, 1969a. Apresentação esquemática dos tipos de soluções adotados nas encostas do Estado da Guanabara pelo Instituto de Geotécnica. *Semana Paulista de Geologia Aplicada, 1°, São Paulo.*

Fonseca, A.M. C.C. e, 1969b. Relato sobre causas e problemas nas encostas da Guanabara. *Semana Paulista de Geologia Aplicada, 1°, São Paulo.*

Fonseca, A.M. C.C. e, 1974. Obras de construção das encostas do Estado da Guanabara, analise dos problemas, desempenho, e eficiência das soluções. *Congr. Bras. Mec. Solos Eng. Fund., 2°, São Paulo*, Tema 3 — Estabilidade de taludes de rocha e solo.

Freire, E.S.M., 1965. Movimentos coletivos de solos e rochas e sua moderna sistematica. *Construção*, 8(95).

Guanabara, Leis, Decretos, 1965. Decreto no 417 de 14.7.75. Dispõe sobre o licenciamento de construções em terrenos acidentados e nas bases de encostas dos morros e dá outras providências. D.O. Port. 19.7.86.

Guanabara, Leis, Decretos, 1974. Secretaria de Obras Publicas. Portaria "N" (13) de 30.10.64.

Highway Research Board, 1973. Soil Erosion: Causes and Mechanisms; Prevention and Control. *Highw. Res. Board Spec. Rep.*, No. 135.

Hudson, M.N., 1961. An introduction to the mechanics of soil erosion under conditions of sub-tropical rainfall. *Proc. Trans. Sci. Assoc. Rhodesia*, XLIX, Part 1.

Instituto de Geotécnica, 1967. Documentário fotográfico das obras de limpeza e contenção de encostas na Guanabara. *Rev. Eng. Est. Guanabara*, 34(1-4): 44—88.

Jones, F.O., 1973. Landslides of Rio de Janeiro and the Serra das Araras escarpment, Brazil. *U.S. Geol. Surv. Prof. Paper*, 697, 42 pp.

Kanji, M.A., 1970. *Shear Strength of Soil-Rock Interfaces*. M.S. Thesis, Dep. of Geology, Univ. of Illinois, Urbana, Ill.

Kanji, M.A., 1972. *Resistência ao Cizalhamento de Contatos Solorocha*. Ph.D. Thesis, Int. de Geociências, Univ. of São Paulo, São Paulo.

Lacerda, W.A., 1967. *Métodos Modernos de Observação e Controle dos Movimentos dos Taludes de Terra*. Curso de Extensão Univ.

LNEC, 1959. As lateritas do Ultramar Portugues. *LNEC Mem.*, 141 (Lisbon).

Lubina, A.F., 1968. Brazilian link in Pan-American Highway. *Civ. Eng.*, October, pp. 56—59.

Machado, J.A., 1958. *O Problema da Pavimentação nos Solos do Recôncavo Baiano*. Rodovia, May.

Motta Filho, A.R. de O., 1970. *As Pedras Crescem*. SI da ACAR, Rio de Janeiro.

Mousinho de Mets, M.R. and Silva, J.X., 1968. Considerações morfológicas a propósito dos movimentos de massas ocorridos no Rio de Janeiro. *Rev. Bras. Geogr.*, 30(1), January/March.

Nascimento, Ú., 1952. Estudo da regularização e proteção das barrocas de Luanda. *LNEC Publ.*, 30 (Lisbon).

Nascimento, Ú., 1967. Simpósio sobre estabilidade e consolidação de taludes. *Jornadas Luso-Bras. Eng. Civ., 2°, Rio de Janeiro, São Paulo*.

Nascimento, Ú. et al., 1973. Consolidação de taludes. *LNEC Curso*, 142 (Lisbon).

OEA, 1973. *Bacio do Rio da Prata; Estudo para Planificação e Desenvolvimento*. Rep. Fed. do Brasil, Noroeste do Est. do Paraná. Relatório do estudo realizado pelo Escrit. de Desenv. Regional durante o período 1970—1972.

Paraná (Estado), DER, Divisão de Pesquisas Rodoviárias, 1971. *Rodovia: BR 376*. Trecho: Maringá-Nova Esperança. Sub-Trehco: Contorno de Mandaguaçú. Projeto de aterro, obras complementares e controle da erosão. DER, Curitiba.

Patton, F.P. and Hendron, A.J., Jr., 1974. General report on "mass movements". *2nd Int. Congr. Int. Assoc. Eng. Geol., São Paulo*.

Pichler, E., 1957. Aspectos geológicos dos escorregamentos de Santos. *Bol. Soc. Bras. Geol.*, 6(2).

Rodrigues, L.F. V.C., 1967. Acidentes ocorridos em janeiro de 1967 na Serra das Araras. *Simpósio Sobre Proteção contra Calamidades Públicas*. Clube de Engenharia, Rio de Janeiro.

Ruiz, M.D., 1966. Some technological characteristics of twenty-six brazilian rock types. *1st Congr. Int. Soc. Roch Mech., Lisbon, 1966*, 1: 115—120.

Sandroni, S.S., 1973. *Resistência ao Cisalhamento dos Soles Residuais das Encostas da Guanabara*. P.U.C., Rio de Janeiro.

Santos Júnior, A., 1967. Caracteristicas hidrológicas da Serra do Mar. *Jornadas Luso-Bras. Eng. Civ., 2ªs, Rio de Janeiro, São Paulo.*

Serafim, J.L. and Da Costa Nunes, A.J., 1966. Studies of dam foundations under a residual cover. *1st Congr. Int. Soc. Rock Mech., Lisbon, 1966*, 2: 639—644.

Silva Filho, B.C., 1974. Alguns dados sobre o intemperismo e a mineralogia das argilas dos basaltos e dos seus solos residuais (1ª pte). *Tóp. Geomec.*, 16 pp.

Silva Filho, B.C., 1975. Alguns dados sobre o intemperismo e a mineralogia das argilas dos basaltos e dos seus solos residuais (2ª pte). *Tóp. Geomec.*, 17 pp.

Sobral, H.S., 1956. *Contribuição ao Estudo dos Massapês como Solos para Construção*. Tese Esc. de Belas Artes, Univ. Bahia.

Sobral, H.S. and Menezes, M.S., 1958. *Influência da Umidade no Comportamento dos Massapes*. I.P.T., São Paulo.

Stemberg, H.O., 1949. Floods and landslides in the Paraiba Valley, December 1948. Influence of the destructive exploitation of the land. *3rd Congr. Int. Geogr., Lisbon.*

SUDESUL, 1974. *Estudo para o Desenvolvimento Regional do Noroeste do Estado do Paraná.* Documento informativo. Ministerio do Interior(SUDESUL/DNOS/Governo do Paraná/OEA. SUDESUL, Porto Alegre.

SURSAN, Grupo de Técnicos do Estado da Guanabara, 1966. *Os Aguaceiros e as Encostas da Guanabara.* Relator: Icarahy da Silveira.

Tecnosolo, A/S, 1975. *Projeto de Estabilização de Taludes.* BR 101/116/470, DNER, Sta. Catarina.

Terzaghi, K., 1950. Mechanism of landslides. In: S. Page (Editor), *Application of Geology to Engineering Practice (Berkey Volume).* Geological Society of America, Washington, D.C., pp. 83—123.

Terzaghi, K., 1953. Opening address. *3rd Int. Conf. Soil Mech. Found. Eng., Zurich*, 3.

UFRJ, 1967. *Seminário Interuniversitário para o Exame das Conseqüencias das Chuvas e Enchentes de Janeiro de 1966, na Régiǎo da Guanabara e Áreas Vizinhas.* Rio de Janeiro.

Vargas, M., 1953. Correlation between angle of internal friction and angle of shearing resistence in consolidated quick triaxial compression tests on residual clays. *3rd Int. Conf. Soil Mech. Found. Eng., Zurich.*

Vargas, M., 1963. General discussion. *2nd Panam. Congr. Soil Mech. Found Eng.*

Vargas, M., 1966. Estabilização de taludes em encostas de gneisses decompostos. *Congr. Bras. Mec. Solos, 3°, Belo Horizonte.*

Vargas, M., 1967a. Design and construction of large cutting in residual soil. *Proc. 3rd Panam. Conf. Soil. Mec. Found. Eng., Caracas*, 2: 243—254.

Vargas, M., 1967b. Estabilização de taludes; deslizamentos apontam soluções. *O. Dirigente Construtor.* 4(2).

Vargas, M., 1971a. Discussion on slope stability of residual soils. *Proc. 4th Panam. Conf. Soil Mech. Found. Eng., San Juan, Porto Rico.*

Vargas, M., 1971b. Geotecnica dos solos residuais. *Rev. Lat. Am. Geotec.*, 1(1).

Vargas, M. and Pichler, E., 1957. Residual soil and rock slides in Santos, Brasil. *4th Int. Conf. Soil Mech. Found. Eng.*

Velloso, D. de A., 1966. Empuxos de terra sobre suportes temporários e permanentes e estabilidade de taludes. *Congr. Bras. Mec. Solos, 3°, Belo Horizonte.*

Vijavergiya, V.N. and Ghazzaly, O.I., 1973. Prediction of swelling potential for natural clays. *Proc. 3rd Int. Conf. on Expansive Soils, Haifa, 1973*, 1.

Chapter 12

SLOPE STABILITY IN THE APPALACHIAN PLATEAU, PENNSYLVANIA AND WEST VIRGINIA, U.S.A.

RICHARD E. GRAY, HARRY F. FERGUSON and JAMES V. HAMEL

ABSTRACT

The Appalachian Plateau region of Pennsylvania and West Virginia is a maturely dissected plateau with deep valleys, moderate to steep slopes, and local relief on the order of 120—150 m. Rocks of the region are generally flat-lying interbedded Paleozoic age claystone, shale, and sandstone with a few limestone and coal seams. Most of the region was not glaciated during the Pleistocene Epoch; during that time the major rivers entrenched their valleys, then filled them with glacial outwash or lake silt and clay deposits. Valley wall stress relief and joint development accompanied downcutting of the rivers and soil and rock masses presumably slumped from valley walls under periglacial conditions. Weathering of rocks in valley walls plus down-slope creep and sliding of the weathering products, particularly those derived from claystone and shale, have continued to the present. As a result of these processes, colluvial masses of various thicknesses and lateral extents exist in states of marginal equilibrium on many, if not most, slopes in the region.

Deep-seated rockslides are relatively rare. Where such slides occur, they typically involve excavated slopes in which large wedges of rock, separated from the valley walls by near-vertical stress relief joints, slide along weak claystone or shale beds. Water pressures in the slopes are usually significant contributing factors in such slides.

Rockfalls are common where rock strata are exposed in slopes. Differential weathering and erosion remove low-strength claystone and shale leaving unsupported ledges of stronger shale, sandstone, and limestone which ultimately fall. Typical rockfall volumes are small, e.g., on the order of 100 m^3 or less.

Most of the present slope stability problems in the region involve slump-type slides, or slow earthflows of colluvial soil and rock masses. The precarious equilibrium of these masses is frequently upset by man's activities, e.g., removal of toe support, loading the slope, or changing surface and subsurface drainage. Abnormally high precipitation also initiates movement of colluvial masses.

INTRODUCTION

The Appalachian Plateau (Fig. 1) with its steep hillsides, thick soil cover and precipitation of 890—1140 mm per year, with the greatest amounts occurring in the late winter and early spring, has long been recognized as an area of major landslide severity. This paper presents the writers' experience in western Pennsylvania and northern West Virginia where urbanization and large construction projects aggravate the marginally stable slope conditions, resulting in numerous landslides. General aspects of slope formation are discussed in which the effect of past periglacial climate on existing conditions is emphasized. The type, relative number, and size of landslides are summarized and geotechnical data on rock and associated colluvial soil are presented. Six case histories, involving two rockfalls, one rock wedge slide and three large colluvial slopes are presented to illustrate the types of slope problems encountered in the Appalachian Plateau. Locations of the case history sites are given in Fig. 2.

PHYSIOGRAPHY AND GEOLOGY

The Appalachian Plateau physiographic province (Fig. 1) extends from north-central New York State to northern Alabama and attains a maximum

Fig. 1. Appalachian Plateau physiographic province.

the toes of slopes in colluvial masses. Colluvial soils tend to be 1.5—9 m thick on slopes and generally increase in thickness (to a maximum of about 30 m) near the toes of slopes. Colluvial soils are generally stiff to hard and individual samples have relatively high shear strengths. However, creep or sliding processes (or both) during slope development has generally reduced the shear strength along movement surfaces to residual or near-residual levels. These movement surfaces may occur at several levels within the colluvial mass but there is always a movement surface at the soil-rock interface (Deere and Patton, 1971). As the slope materials seek equilibrium between stress and strength, the soil mantle moves downslope and the mean slope angle decreases until a relatively flat slope angle, compatible with a state of marginal equilibrium, has been achieved. This natural slope-flattening process accounts for the relatively thick soil cover of mature colluvial slopes, particularly at the base of the slopes. Deere and Patton (1971) have suggested that there are no stable natural slopes in the Appalachian Plateau where the inclination exceeds 12—14°. Movements have been reported on slopes as flat as 10° by Terzaghi and Peck (1948, p. 357), whereas Gray and Donovan (1971) demonstrated that several mature colluvial slopes, with evidence of pre-existing failure surfaces, had slope angles ranging from 7° to 10°.

The writers do not have any documented data on rates of creep of colluvial slopes in the area. Field observations suggest, however, that colluvial slopes may creep at rates of a few centimetres per year. With the exception of creep, large colluvial masses appear stable.

LANDSLIDING

The Appalachian Plateau province seems among the most severe for landsliding within the United States (Ladd, 1927, 1928; Sharpe and Dosch, 1942; Ackenheil, 1954; Eckel, 1958; Baker and Chieruzzi, 1959). Because a soil mantle blankets much of the rock surface, most landslides are in soil, the most common being slump-type slides or slow earthflows which range in size up to several million cubic metres. Rockfalls, the next most common type of slide, are typically much smaller with maximum volumes on the order of a hundred cubic metres. Other types of slide movements are relatively rare.

Numerous small slump or slow earthflow slides occur during seasonal wet periods or due to local stream erosion, with catastrophic hydrological events being of major significance. For example, the great amount of precipitation associated with Hurricane Agnes in June 1972 caused a significant number of such slides. However, most slides are a direct result of man's disturbance of natural conditions. Frequent causes of sliding are (1) removal of toe support, (2) surcharging slopes by the placing of fill embankments, or (3) a change in surface and subsurface waterflow. The largest slides usually result from disturbance of ancient landslide masses in soils and/or rock. These ancient landslides appear to have occurred in the main under periglacial conditions. Limited radiocarbon dating (Philbrick, 1961; D'Appolonia et al., 1967) suggest a Pleistocene age for some of these deposits. Peltier (1950) and Denny (1956) found fossil periglacial features close to the front of the maximum advance of the Wisconsin glaciation in Pennsylvania and strongly support the influence of Pleistocene periglacial processes on slopes. Rapp (1967), in a study of a portion of central Pennsylvania just east of the Appalachian Plateau, concurred with the above authors. Carson and Kirkby (1972) review the effect of periglacial processes on slope profiles in areas currently experiencing humid temperate climates and conclude the effect was not as great as indicated by some studies.

Where interbedded strong and weak rock strata are exposed, differential weathering and erosion result in the weaker rock being removed, leaving the more resistant rock as overhanging ledges. The result of this process is small but often dangerous rockfalls. Cuts containing hard sandstone or limestone

Fig. 5. Ames Limestone undercut by weathering in Pittsburgh Redbeds claystone — case history 1.

beds underlain by relatively low-strength shale or claystone are common throughout the area. Weathering causes relatively rapid decomposition and spalling of the softer rock, leaving unsupported ledges of limestone and sandstone. Rates of undercutting of about 60—180 mm per year have been reported by Philbrick (1959), based on observations over a period of several years for a highway cut with claystone underlying massive Morgantown Sandstone. Average rates of undercutting of 13—38 mm per year (over a period of two years) were measured by Bonk (1964) at two carefully prepared test sites in excavated slopes in Pittsburgh Redbeds claystone. Average rates of undercutting of 30—150 mm per year were reported by Bonk (1964) for Pittsburgh Redbeds exposed for periods of two to ten years in highway cuts. In relatively short periods of time, weathering can progress to the point where a resistant rock can no longer sustain the tensile stress developed by its cantilevered weight and the ledge falls (Fig. 5).

Deep-seated rock slides are relatively rare. Those with which the writers are familiar have typically involved excavated slopes in which large wedges of rock, separated from the valley walls by near-vertical stress relief joints, slide along or through weak claystone or shale beds. Water pressures in the slopes are usually significant contributing factors in such slides (case history 3).

GEOTECHNICAL DATA

The engineering properties of intact rock materials of the northern Appalachian Plateau vary widely with lithology and degree of weathering. Almost no data on the permeability and deformability of these rocks have been published and relatively few strength data are available in the literature. Ranges of unit weight and strength values for the common rock types are summarized in Table I. Most of the rocks of the Appalachian Plateau are of very low to medium strength and low to medium modulus ratio according to the classification system of Deere (1968).

As with other rocks, the engineering behavior of the sedimentary rocks of the Appalachian Plateau is controlled largely by geologic discontinuities; e.g., faults, joints, bedding contacts, weak beds or beds affected by fluid pressure conditions, and old sliding surfaces, rather than by properties of intact rock materials. Even the engineering behavior of colluvium derived from these rocks seems to be governed mainly by old failure surfaces. Due to a history of previous shearing, these old failure surfaces generally have much lower shear strengths and much higher permeabilities and deformabilities than other portions of the colluvium or the rocks from which the colluvium was derived.

Index properties and strength data on colluvium derived from claystones of the Appalachian Plateau have been given by D'Appolonia et al. (1967), Hooper (1969), Hamel (1969), and Hamel and Flint (1969, 1972). This col-

TABLE I

Ranges of properties, cyclic sedimentary rocks of northern Appalachian Plateau (intact, unweathered specimens)

Rock type	Dry unit weight (Mg/m³ [pcf])	Unconfined compressive strength (MN/m² [psi])	Angle of internal friction (degrees)	Cohesion intercept (MN/m² [psi])
Limestone	2.65—2.72 [165—170]	110 [16,700]	—	—
Sandstone	2.30—2.72 [144—170]	44—130 [6280—18,900]	37—67	1—4 [200—640]
Sandy shale	2.46—2.77 [154—173]	25—110 [3600—16,400]	29—69	1—18 [180—2600]
Silty shale	2.31—2.85 [145—178]	3—56 [400—8100]	10—45	0.3—14 [50—2000]
Clay shale	2.15—2.65 [134—165]	0.2—21 [30—3000]	6—33	0—10 [0—1500]
Siltstone	2.39—2.66 [149—166]	14—84 [2000—12,200]	19—48	3—21 [500—3000]
Claystone (indurated clay)	2.08—2.72 [130—170]	0.3—41 [50—6000]	10—60	0—4 [0—600]

After Pittsburgh District, U.S. Army Corps of Engineers (1938), McKelvey (1940), Philbrick (1960), Kenty and Meloy (1965), Mellinger (1966, 1969), Underwood (1967), Gray (1969), and Hamel (1972).

luvium ranges from massive blocks of relatively intact claystone to silty or sandy clay soil with rock fragments. Dry unit weights typically range from about 1.9 to 2.2 Mg/m³.

Table II contains a summary of index properties for claystone-derived colluvium from the failure surfaces of ancient and recent landslides at eleven locations along a 1.6-km section of Interstate Route 79 (formerly Interstate Route 279) near Pittsburgh, Pennsylvania (Hamel, 1969; Hamel and Flint, 1969, 1972; Flint and Hamel, 1971). Slides at this site are described in case history 4. The failure surface materials are well-graded mixtures of sand-, silt-, and clay-size particles with 40—80% finer than the No. 200 sieve (0.074 mm). About half the particles in each sample are in the fine sand to silt size range so that the materials generally fall at or near the border between SM-SC and ML-CL soils in the Unified Soil Classification System. X-ray diffraction analyses indicated that these failure surface materials contain quartz, kaolinite, illite, and expandable lattice clay minerals. The expandable lattice minerals (possibly vermiculite and one or more minerals of the smectite group) appear to occur preferentially along the failure surfaces of the slides.

TABLE II

Index properties of claystone failure surface materials from eleven locations along Interstate Route 79 (case history 4)

Property	Range of values	Average value
Natural water content (%)	17—31	24
Liquid limit (%)	27—41	35
Plastic limit (%)	19—29	24
Plasticity index (%)	8—13	11
Clay fraction, −2 μm (%)	14—29	21
Specific gravity of solids	2.74—2.80	2.77

Note: X-ray diffraction analyses indicated that these materials contain quartz, kaolinite, illite, and expandable-lattice clay minerals (possibly vermiculite and one or more minerals of the smectite group).

The colluvium generally exhibits strain-softening behavior (Skempton, 1964) and its residual (large displacement) shear strength is generally less than half its peak (small displacement) strength at a given effective normal stress. For effective normal stresses of less than about 350 kN/m², the peak strength of claystone colluvium is commonly characterized by cohesion intercepts of 7—35 kN/m² and friction angles of 20—25° while the residual strength is usually characterized by negligible cohesion intercepts and friction angles of 8—20°. Measured residual friction angles for most claystone-derived colluvium are on the order of 11—16°. The writers' experience in calculation of strength data from colluvial slide masses (Hamel, 1969; Hamel and Flint, 1969, 1972; Gray and Donovan, 1971) indicates that shear strengths characterized by residual level friction angles of 13—16°, with zero cohesion intercept, are commonly mobilized in place.

CASE HISTORIES

Case history 1

An example of differential weathering resulting in a rock fall is a cut on the Penn-Lincoln Parkway (Interstate Route 376) in Pittsburgh, Pennsylvania (Fig. 2). Here the hard Ames Limestone (Fig. 5), 0.5 m thick and approximately 6 m above road level, is underlain by the weak Pittsburgh Redbeds claystone (Conemaugh Group, Fig. 4). Slaking of the claystone and subsequent falls of the undercut limestone frequently prevented use of the hillside lane of the road from the time of its completion in 1955 until 1963 when a barrier was constructed to prevent the falling limestone blocks from reaching the road.

A comparison of the original cut slope with a 1963 survey showed the face of the slope at a point immediately below the overhanging Ames Limestone to have retreated more than 2 m. This corresponds to an average rate of slope retreat of 0.25 m per year.

Case history 2

A 6.4 km long rockfall area was located along Pennsylvania Route 930 on the west side of the Ohio River opposite Ambridge, Pennsylvania (Fig. 2). The valley wall at this location is about 120 m high with a mean inclination of 30—35°. A 15 m high nearly vertical sidehill cut was made for the highway in 1922 (Fig. 6; Ackenheil, 1954). Approximately 5 m of soft clay shale exposed in the cut at road level is overlain by about 2 m of soft claystone. The claystone, in turn, is overlain by about 6 m of hard sandy shale with numerous vertical joints parallel to the valley wall. The soft shale and claystone rapidly weathered away, undercutting blocks of sandy shale. According to Ackenheil (1954), at least nine major rockfalls occurred during the period from 1932 to 1954. The most serious rockfall occurred on December 22, 1942, when a 115-m³ block of sandy shale crushed a bus; twenty-two persons were killed and four were injured. The slope was later redesigned to minimize rockfalls and it was reconstructed in 1956.

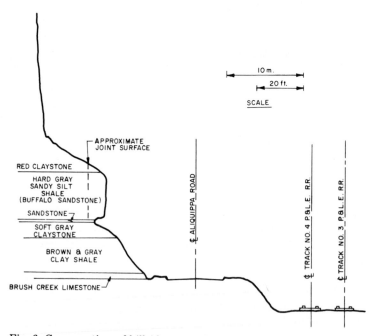

Fig. 6. Cross-section of hillside — case history 2.

Case history 3

The slide at Brilliant Cut (Philbrick, 1953, 1960; Ackenheil, 1954; Hamel, 1969, 1972) is, to the writers' knowledge, the only large rockslide which has occurred in the northern Appalachian Plateau in many years. This slide occurred in 1941 in a railroad cut adjacent to the Allegheny River in Pittsburgh, Pennsylvania (Fig. 2). The cut, which was made in the nose of a hill in 1930-31, had a height of 50 m and an average inclination of 45° (Fig. 7). Sometime during the 1930's, a vertical joint trending approximately parallel to the valley wall and probably due to valley wall stress relief began to open at the crest of the cut slope. This joint had opened to a width of about 0.5 m by 1940.

Early on the morning of March 20, 1941, a rock mass of more than 80,000 m³ slumped from the hill (Fig. 8). The head of the slide mass dropped about 5 m and the toe displaced three sets of railroad tracks, causing a train to be derailed. The failure was apparently triggered by water pressures in the slope; natural drainage outlets in the slope face were blocked with ice. A period of snow, rain, and snow melt immediately preceded the slide. Surface water infiltrating the upland behind the slope and the upper part of the slope normally drained vertically through joints and then laterally to the slope face through relatively permeable coal and limestone beds (Fig.

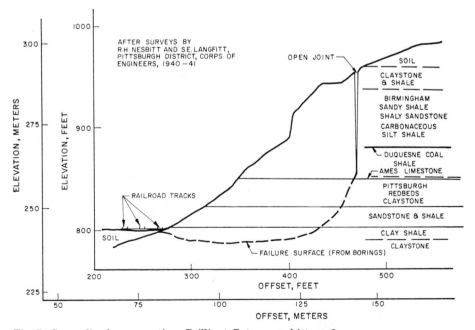

Fig. 7. Generalized cross-section, Brilliant Cut — case history 3.

Fig. 8. Brilliant Cut slide — case history 3. Note icicles at groundwater table below carbonaceous shale.

7). On the day of the slide, however, the slope face had icicles at the levels of these natural drainage outlets (Fig. 8).

Approximately 70% of the failure surface passed through layers of soft clay shale and claystone (indurated clay) at the base of the cut (Fig. 7). Hamel (1969, 1972) used the Morgenstern-Price (1965, 1967) method of stability analysis to calculate the set of Mohr-Coulomb shear strength parameters required in the basal formations for limiting equilibrium of the slide mass. With $c' = 0$ and a 30-m vertical joint full of water at the rear of the failure mass, $\phi' = 32°$ was required in the basal shale and claystone for equilibrium. Additional calculations have shown that $\phi' = 27°$ was required for equilibrium with $c' = 0$ and the joint only half full of water. These calculated friction angles of 27—32° for average effective normal stresses of 350—420 kN/m^2 on the failure surface correspond to peak or near-peak strengths measured in field and laboratory direct shear tests on clay shales and claystones similar to those at Brilliant Cut. Progressive failure of the slope related to opening of the vertical joint at the rear of the failure mass during the 1930's may have reduced the strength of the basal clay shale and claystone somewhat. It appears unlikely, however, that pre-slide movement was sufficient to reduce the strength to a residual level.

In the early 1940's, the Pittsburgh District, U.S. Army Corps of Engineers, was designing a 90 m high rock cut for the spillway of the Youghiogheny

Dam located about 100 km southeast of Pittsburgh on the Youghiogheny River (Philbrick, 1953, 1960). The Youghiogheny Spillway Cut was in the same stratigraphic interval as Brilliant Cut, then almost ten years old. When the Brilliant slide occurred as the clear result of inadequate drainage, it was recognized that drainage of the Youghiogheny Spillway Cut was necessary to insure long-term stability. A system of horizontal drain holes 75 mm in diameter by 76 m long was therefore installed in the slope. The Youghiogheny Spillway Cut has remained stable, with only minor surface ravelling, up to the present time.

Case history 4

A section of Interstate Route 79 (formerly I-279) about 16 km west of Pittsburgh, Pennsylvania, passes through a zone of colluvium in the wall of a tributary valley of the Ohio River (Fig. 2). Numerous landslides were initiated along ancient slide surfaces when the toes of deep colluvial masses were excavated during highway construction in 1968-69 (Hamel, 1969; Hamel and Flint, 1969, 1972; Flint and Hamel, 1971). A typical slope cross-section is shown in Fig. 9. The colluvium, which was produced by large-scale ancient mass movements in the Pittsburgh Redbeds claystone and overlying Morgantown Sandstone, was among the thickest which the writers have encountered anywhere in the Appalachian Plateau. Colluvium thicknesses ranged from less than 2 m to more than 30 m and averaged about 15 m. The upper part

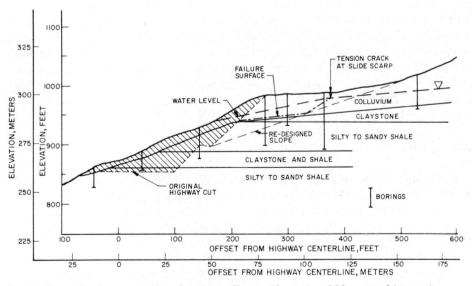

Fig. 9. Generalized cross-section, Interstate Route 79, station 908 — case history 4.

of the colluvium was a heterogeneous mixture of angular, gravel- to boulder-size sandstone fragments with variable amounts of sand, silt, and clay. Some localized zones of this material consisted almost exclusively of highly interlocked sandstone boulders. The lower part of the colluvium consisted mainly of claystone and clay from the Pittsburgh Redbeds. A perched water table existed on the impervious lower portion of the colluvium (Fig. 9).

The shear zones and failure surfaces in the claystone-derived colluvium were studied in detail (Hamel, 1969; Hamel and Flint, 1969, 1972). Most shear zones could be subdivided into three parts. The actual surface of sliding was generally a 6—50 mm thick seam of damp to wet, soft to medium stiff silty clay with locally variable amounts of sand and sand-size shale and claystone fragments. (Index properties for some of these failure surface materials are given in Table II.) The failure surface clay seam was usually located near mid-height of a shear zone, the upper and lower portions of which consisted of mixtures of silty clay and angular sand- to gravel-size claystone and shale fragments. Thicknesses of the upper and lower portions of the shear zones varied considerably with location but were typically on the order of 50—75 mm.

Shear strength values were calculated for limiting equilibrium of two of the slide masses (Hamel, 1969; Hamel and Flint, 1969, 1972) using the Morgenstern-Price (1965, 1967) method of stability analysis. For the most probable failure surfaces, which had average effective normal stresses on the order of 140—210 kN/m^2, shear strength characterized by $\phi' = 14$—$15°$ with $c' = 0$ were required for equilibrium. These calculated friction angles of 14—15° are in excellent agreement with the typical residual friction angles of 13—16° measured in multiple reversal laboratory direct shear tests on undisturbed and remolded specimens of clay obtained from the failure surfaces of the slides. Mobilization of residual level strengths in these slides is, of course, consistent with their geologic histories. Shearing displacements in the ancient slides reduced failure zone strengths to residual values. Failure zones apparently did not regain strength after the period of ancient landsliding had ceased and residual level strengths existed when recent slope excavation began.

Case history 5

Slope instability which resulted from excavation of the toe of a slope at Weirton, West Virginia (Fig. 2) exemplifies stability problems associated with colluvial slopes throughout the region. In 1956, the lower portion of the Weirton slope (Fig. 10) was excavated to develop a coal storage facility. This excavating initiated movement upslope in an area occupied by public roads, private property, and buildings. After initial sliding was observed above the excavation, remedial measures were undertaken, including sealing of tension cracks to prevent surface runoff from entering the slide mass and

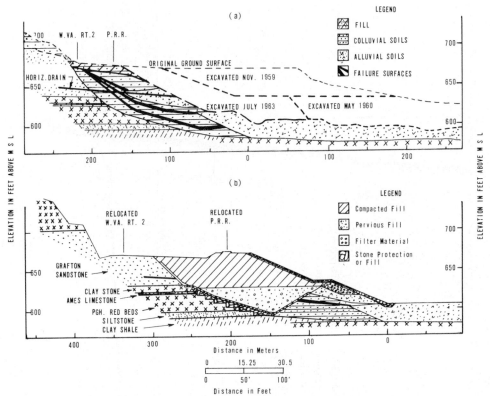

Fig. 13. Generalized cross-section, Pike Island slide — case history 6; (a) excavation history, (b) reconstructed slope.

Horizontal drains (Fig. 13) were later drilled through the slide mass and approximately 12 m into the base of the sandstone to intercept water in the vertical fractures. They were installed on 15-m centers and were equipped with flap valves to prevent entrance of water during high-river stages. These drainage measures also slowed the slide movement but did not stop it. Several types of instrumentation were installed to monitor movement of the slide. These included surface alignment pins, piezometers and thin-walled plastic pipe failure surface detectors. All but the surface alignment pins had short lives due to the continuing motion of the slide mass. A plot of the slide motion from several surface alignment pins during portions of 1962 and 1963 are shown in Fig. 14.

Slide movement continued from 1960 to 1963 at variable rates depending on construction activities. Movement from June to November 1962 was 76 mm with the rate increasing to about 5 mm per day with flooding of the cofferdam at the end of construction in early November 1962 (Fig. 14). With

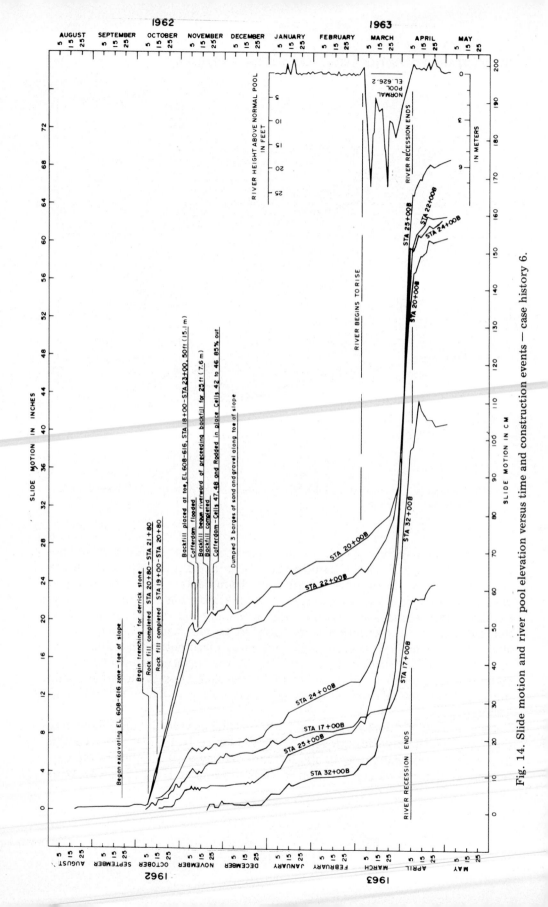

Fig. 14. Slide motion and river pool elevation versus time and construction events — case history 6.

this increase in motion, the head crack of the slide extended downstream (Figs. 11 and 12). A sand and gravel berm was then placed in the river to buttress the toe of the slide. This berm also slowed, but did not stop, slide movement.

In March 1963, flood waters raised the river level 7 m above normal pool and caused the slide movement to accelerate to rates on the order of 125—150 mm per day (Fig. 14). Until that time, the slide mass had been essentially intact but, with increased movement in the spring of 1963, the mass broke into a series of segments, some of which rotated slightly toward the river. It was then decided that stabilization and preservation of the slope could only be accomplished by relocation of the highway to intact rock and by providing support of the railroad on new fill keyed into rock with a riverward buttress and good under-drainage (Fig. 13). This fill was constructed and the highway and railroad were relocated in 1964. The reconstructed slope has remained stable up to the present time.

REFERENCES

Ackenheil, A.C., 1954. *A Soil Mechanics and Engineering Geology Analysis of Landslides in the Area of Pittsburgh, Pennsylvania.* Ph.D. Thesis, Univ. of Pittsburgh, Pittsburgh, Pa.

Baker, F.F. and Chieruzzi, R., 1959. Regional concept of landslide occurrence. *Highw. Res. Board, Bull.*, 216: 1—16.

Bjerrum, L., 1967. Progressive failure in slopes of over-consolidated plastic clay and clay shales. *Proc. Am. Soc. Civ. Eng., J. Soil Mech. Found. Div.*, 93(SM5): 1—49.

Bonk, J.G., 1964. *The Weathering of Pittsburgh Redbeds.* M.S. Thesis, Univ. of Pittsburgh, Pittsburgh, Pa.

Carson, M.A. and Kirkby, M.J., 1972. *Hillslope Form and Process.* Cambridge Univ. Press, London, 475 pp.

Dahl, H.D. and Parsons, R.C., 1971. Ground control studies in the Humphrey No. 7 mine, Christopher Coal Division, Consolidation Coal Company. *Am. Inst. Min. Metall. Eng., Preprint,* SM4: 447—473.

D'Appolonia, E., Alperstein, R. and D'Appolonia, D.J., 1967. Behavior of a colluvial slope. *Proc. Am. Soc. Civ. Eng., J. Soil Mech. Found. Div.*, 93(SM4): 447—473.

Deere, D.U., 1968. Geological considerations. In: K.G. Stagg and O.C. Zienkiewicz (Editors), *Rock Mechanics in Engineering Practice.* Wiley, New York, N.Y., pp. 1—20.

Deere, D.U. and Patton, F.D., 1971. Slope stability in residual soils. *Proc. 4th Panam. Conf. on Soil Mechanics and Foundation Engineering.* American Society of Civil Engineers, New York, N.Y., 1: 87—170.

Denny, C.S., 1956. Surficial geology and geomorphology of Potter County, Pennsylvania. *U.S. Geol. Surv. Prof. Paper*, 288: 72.

Eckel, E.B. (Editor), 1958. Landslides and engineering practice. *Highw. Res. Board, Spec. Rep.*, No. 29, 232 pp.

Ferguson, H.F., 1967. Valley stress release in the Allegheny Plateau. *Bull. Assoc. Eng. Geol.*, 4: 63—71.

Ferguson, H.F., 1974. Geologic observations and geotechnical effects of valley stress relief in the Allegheny Plateau. Paper presented at *Am. Soc. Civ. Eng. Water Resourc. Eng. Meet., Los Angeles, Calif., January 1974*, p. 31.

Flint, N.K. and Hamel, J.V., 1971. Engineering geology at two sites on Interstate 279 and Interstate 79 northwest of Pittsburgh, Pennsylvania. In: R.D. Thompson (Editor), *Environmental Geology in the Pittsburgh Area. Geol. Soc. Am., Annu. Meet., November 1971*, Guideb. for Field Trip No. 6, pp. 36—45.

Gray, R.E., 1969. Shear and bond strength of shales. Paper presented at *Conf. on Engineering in Appalachian Shales, West Virginia Univ., Morgantown, W.Va., June 1969*.

Gray, R.E. and Donovan, T.D., 1971. Discussion of slope stability in residual soils, by D.U. Deere and F.D. Patton. *Proc., 4th Panam. Conf. on Soil Mechanics and Foundation Engineering*. American Society of Civil Engineers, New York, N.Y., 127—130.

Hamel, J.V., 1969. *Stability of slopes in Soft, Altered Rocks*. Ph.D. Thesis, Univ. of Pittsburgh, Pittsburgh, Pa. (No. 70-23, 232 — Univ. Microfilms, Ann Arbor, Mich.)

Hamel, J.V., 1972. The slide at Brilliant Cut. In: E.J. Cording (Editor), *Stability of Rock Slopes. Proc., 13th Symp. on Rock Mechanics, Univ. of Illinois, Urbana, Ill., 1971*. American Society of Civil Engineers, New York, N.Y., pp. 487—510.

Hamel, J.V. and Flint, N.K., 1969. *A Slope Stability Study on Interstate Routes 279 and 79 near Pittsburgh, Pennsylvania*. Report by Departments of Civil Engineering and Earth and Planetary Sciences, Univ. of Pittsburgh to Pennsylvania Department of Highways and U.S. Department of Transportation, Bureau of Public Roads.

Hamel, J.V. and Flint, N.K., 1972. Failure of colluvial slope. *Proc. Am. Soc. Civ. Eng., J. Soil Mech. Found. Div.*, 98(SM2): 167—180.

Hooper, J.R., 1969. Slope movements of residual clays in Southeastern Ohio. *Proc. Symp. on Landslides, Ohio Univ., Athens, Ohio, February 1969*. pp. 33—57.

Kenty, J.D. and Meloy, C.R., 1965. Shear evaluation of weak rock foundations. *Ohio River Div., U.S. Army Corps Eng., Tech. Rep.*, No. 3-45 (draft copy).

Ladd, G.E., 1927-1928. Landslides and their relation to highways. A report of observations made in West Virginia and Ohio to determine the cause of slides and devise means of control. *Public Roads*, Part 1, 8(2): 21—35; Part 2, 9(8): 153—163.

McKelvey, K.M., 1940. *Final Report on Physical Properties of Rock, Youghiogheny Dam, Lower Site*. Pittsburgh District, U.S. Army Corps of Engineers, Pittsburgh, Pa., 11 pp.

Mellinger, F.M., 1966. Laboratory and in situ tests of shale. Paper presented at *Am. Soc. Civ. Eng. Natl. Meet. Water Resour. Eng., Denver, Colo., May 1966*.

Mellinger, F.M., 1969. Engineering properties of Appalachian Shales. Paper presented at *Conf. on Engineering in Appalachian Shales, West Virginia Univ., Morgantown, W.Va., June 1969*.

Morgenstern, N.R. and Price, V.E., 1965. The analysis of the stability of general slip surfaces. *Geotechnique*, 15: 79—93.

Morgenstern, N.R. and Price, V.E., 1967. A numerical method for solving the equations of stability of general slip surfaces. *Comput. J.*, 9: 388—393.

Nickelsen, R.P. and Hough, V.N.D., 1967. Jointing in the Appalachian Plateau of Pennsylvania. *Geol. Soc. Am. Bull.*, 78: 609—630.

Peltier, L.C., 1950. The geographic cycle in periglacial regions as it is related to climatic geomorphology. *Ann. Assoc. Amer. Geogr.*, 40: 214—36.

Philbrick, S.S., 1953. Design of deep rock cuts in the Conemaugh Formation. *Proc. 4th Symp. on Geology as Applied to Highway Engineering*. State Road Commission of West Virginia, pp. 79—88.

Philbrick, S.S., 1959. Engineering geology of the Pittsburgh area. *Geol. Soc. Am., Pittsburgh Meet.*, Guideb. for Field Trips, pp. 191—203.

Philbrick, S.S., 1960. Cyclic sediments and engineering geology. *Proc., 21st Int. Geol. Congr.*, Part 20, pp. 49—63.

Philbrick, S.S., 1961. Old landslides in the Upper Ohio Valley. Paper presented at *Geol. Soc. Am. Annu. Meet.*

Pittsburgh District, U.S. Army Corps of Engineers, 1938. *Foundation Report, Mahoning Dam*. Pittsburgh, Pa., 17 pp.

Rapp, A., 1967. Pleistocene activity and Holocene stability of hillslopes, with examples from Scandinavia and Pennsylvania. In: *L'Evolution des Versants. Int. Symp. Geomorphol., Liege, June 1966.* Univ. of Liege, Liege, pp. 230—242.

Rodgers, J., 1970. *The Tectonics of the Appalachians.* Wiley-Interscience, New York, N.Y., 271 pp.

Sharpe, C.F.S. and Dosch, E.F., 1942. Relation of soil-creep to earthflow in the Appalachian plateaus. *J. Geomorphol.,* 5: 312—324.

Skempton, A.W., 1964. Long-term stability of clay slopes. *Geotechnique,* 14: 77—101.

Terzaghi, K. and Peck, R.B., 1948. *Soil Mechanics in Engineering Practice.* Wiley, New York, N.Y., 566 pp.

Thornbury, W.D., 1965. *Regional Geomorphology of the United States.* Wiley, New York, N.Y., 609 pp.

Underwood, L.B., 1967. Classification and identification of shales. *Proc. Am. Soc. Civ. Eng., J. Soil Mech. Found. Div.,* 93(SM6): 97—116.

U.S. Army Engineer District, Pittsburgh, 1963. Relocation and treatment of West Virginia Route 2 subsidence adjacent to lower approach to locks, Pike Island locks and dam, Ohio River, West Virginia and Ohio. *U.S. Army Eng. Dist., Pittsburgh, Design Memo.,* No. 19.

Voight, B., 1974. A mechanism for "locking-in" orogenic stress. *Am. J. Sci.,* 274: 662—665.

Chapter 13

LOVELAND BASIN SLIDE, COLORADO, U.S.A.

FITZHUGH T. LEE and WALTER MYSTKOWSKI

ABSTRACT

The Loveland Basin landslide occurred in early 1963 in a high mountain valley near the Continental Divide, about 88 km west of Denver, Colorado; it took place during construction of the east plaza area for the Eisenhower Memorial Tunnel. The landslide developed in Precambrian bedrock consisting of granite with inclusions of metasedimentary rocks that are chiefly biotite gneiss and schist. Thin deposits of surficial material, consisting of moraine and soil mixed with talus, overlie the bedrock.

The entire area is within the Loveland Pass fault zone. Individual faults within this zone are characterized by masses of breccia and gouge, ranging in width from less than 0.3 to 300 m; these masses are separated by masses of less intensely sheared rock. A pre-existing fault zone partly determined the shape of the landslide.

The landslide was originally outlined by distinct scarps that were subdued by 1974. The upper two-thirds of the landslide was downdropped in relation to the surrounding rock, but the lower third of the landslide was thrust upward and outward over the surrounding topography. Within the landslide mass, movement was expressed by tension fractures, which were most numerous near the margins of the mass.

Early investigations used information from seismic refraction and resistivity surveys, geologic mapping, and measurements of surface control points to arrive at the volume and weight of the landslide mass. Later investigations employed borehole inclinometers, borehole extensometers, and surface measurements.

Buttress loads placed on the toe of the landslide in 1963 and 1968 did not stop movement, and most drainage measures also were ineffective. A third, large buttress was completed in late 1971, and surface and subsurface information indicates that the landslide has been relatively stable since then. Several stability analyses suggest that the present buttress affords a safety factor of about 2. The creep rate should be established by continued instrument monitoring. Analogies with other slope deformations and borehole instrumentation suggest that zones of deformation became progressively deeper.

The importance of a continuous program of stability investigation, guided by informed and experienced personnel, is emphasized.

INTRODUCTION

The Loveland Basin landslide that occurred in the spring of 1963 was caused by an excavation made for the approach road to the east portal of the Eisenhower Memorial Tunnel. The site of the tunnel is about 89 km west of

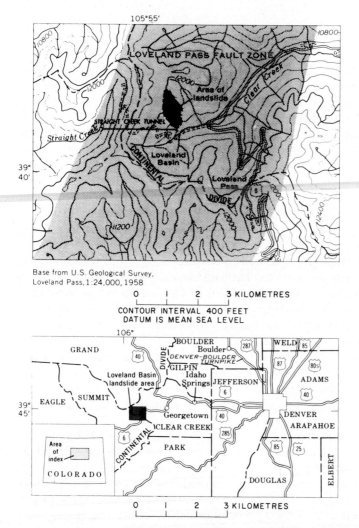

Fig. 1. Index map of the Loveland Basin landslide area, Colorado (from Robinson et al., 1972). Stipple indicates area of Loveland Pass fault zone.

Denver (Fig. 1), on the north side of Loveland Basin and northwest of the Loveland Basin ski area.

A contract for the construction of the 3.2 km of Interstate 70 that serves as the east approach road for the Eisenhower Memorial Tunnel was awarded in August 1962. In June 1963, much of the construction under this contract had been completed, and the cut for the east portal of the Eisenhower Memorial Tunnel (originally called the Straight Creek Tunnel) was nearly finished; the slide became active about this time.

Action was taken immediately to stabilize the landslide, because the Colorado Division of Highways (then the Colorado Department of Highways) planned to let a contract early in the fall of 1963 for the construction of the pilot bore for the tunnel (Robinson et al., 1974). The landslide was above and north of the proposed east portal of the pilot bore. Before remedial procedures could be planned, the volume and mass of the landslide had to be determined and, if possible, its rate of movement had to be evaluated. Joint conferences of the Colorado Division of Highways, the Federal Highway Administration (then the Bureau of Public Roads), and the U.S. Geological Survey resulted in decisions to map the landslide geologically and to try to determine the thickness of the slide by geophysical methods. Drilling the slide was thought to be impractical because of the relative inaccessibility and the nature of the geology of the slide area. R. Woodward Moore, assisted by other members of the Federal Highway Administration and by members of the Colorado Division of Highways, made resistivity measurements in the slide area. R.D. Carroll and J.H. Scott, assisted by other members of the U.S. Geological Survey, made seismic measurements on the slide. C.S. Robinson and F.T. Lee mapped the geology of the slide area and surveyed the resistivity and seismic lines by planetable methods. Control for the planetable mapping was established by a transit survey around the slide by the Colorado Division of Highways. At that time the Division of Highways also had a topographic map of the landslide area prepared from aerial photographs taken in 1961.

Approximately three weeks' time in July 1963 was available in which to obtain the geologic and geophysical data. During this time, the Colorado Division of Highways had two holes drilled in the plaza area at the bottom of the cut (Fig. 2). The Division inserted plastic casing in these holes and probed the holes daily for movement. A first-order grid on 15.2-m (50-ft) centers was established in the plaza area and records kept of the movement of these points.

The geological and geophysical data were compiled and interpreted, and a report giving the volume and mass of the slide was prepared and submitted by Robinson et al. (1964) to the Federal Highway Administration and the Colorado Division of Highways in September 1963. These data were then used by R.A. Bohman, Federal Highway Administration, to analyze the stability of the landslide and to calculate the load necessary to stabilize the slide. Late in September 1963, compacted fill material was placed at the base

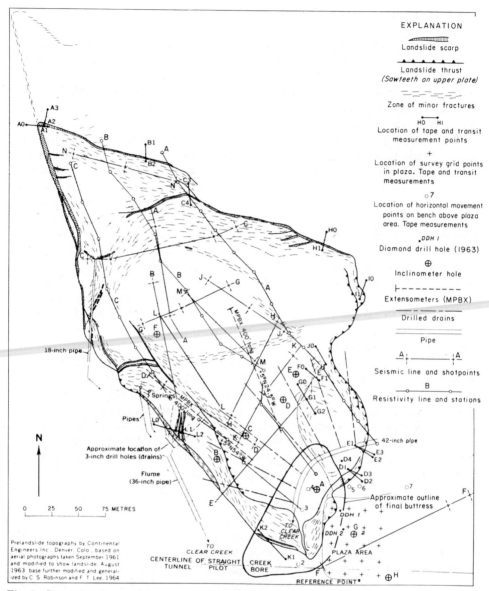

Fig. 2. Generalized geologic map of the Loveland Basin landslide, showing the location of measurement points, instrument holes, geophysical traverses, and surface and subsurface drainage pipes (modified from Robinson et al., 1972).

of the slide to serve as a buttress, and placement was complete by early October 1963. Low-angle drainage holes were drilled into the slip zone at various locations near the western and southern perimeter of the slide, and these were lined with perforated casing. The construction of French drains and the installation of corrugated-metal pipe culverts provided additional drainage. The contract for the construction of the pilot bore was awarded late in October. For several years, surface inspections of the slide were made and some surface stations were monitored, mainly by J.D. Post of the Colorado Division of Highways.

A design contract for the Eisenhower Memorial Tunnel that included measures to monitor and control the landslide was awarded to TAMS (Tippets-Abbett-McCarthy-Stratton, Engineers and Architects) in 1965. During 1966, eight inclinometer holes were cored in the landslide to depths of from 26 to 74 m (Fig. 2), and grooved aluminum casing was inserted and backfilled. Several control measures were carried out in 1966. Seven perforated drain pipes were installed 91 m into the slide, draining into a 107-mm pipe that also diverted surface water from the western side of the landslide to the headwaters of Clear Creek (Fig. 2). Open cracks on the slide surface were filled to prevent surface water from entering the slide mass. However, deeper zones of movement were found than those previously detected in 1963 and additional buttress material was placed on the toe of the slide. Slide movement continued, principally in the summer months, and accelerated disturbingly in 1969-70. In early 1971, two six-point multiple-position borehole extensometers (MPBX's) were installed from the buttress berm upward into the slide at 5° for 122 m. The MPBX's were helpful as a check on the inclinometer determinations of slip-zone location and rate of movement of the slide. A third buttress load was completed in November 1971. Surface and borehole measurements and visual inspection of the slide mass have detected no substantial movement after this final buttress was emplaced. The total buttress is 249,600 m³, many times the 36,100 m³ emplaced in 1963.

Because this landslide was a continuing hazard to important and costly tunnel construction, it has been monitored and treated discontinuously for more than eleven years. From the available data, it has been possible to form a reasonably accurate understanding of the behavior of this heterogeneous mass over a significant period of time.

GEOLOGIC SETTING

The bedrock of the area consists of metasedimentary rocks of the Idaho Springs Formation and Silver Plume Granite, both of Precambrian age (Fig. 3); the bedrock has been extensively, but not uniformly, faulted and altered. During Quaternary time, the stream valleys at the altitude of Loveland Basin (Fig. 1) were glaciated and their sides were steepened. When the glaciers retreated, they deposited morainal material in the bottoms of the valleys and plastered it along oversteepened valley sides. Since the time of the glaciers, the morainal deposits locally have moved downslope; the oversteepened and unstable walls have retreated, aided by pressure-release jointing parallel to the slopes. Fig. 4 shows the terrain surrounding the landslide area.

Many slopes in the high mountain valleys are not stable; construction procedures frequently result in the failure of these jointed or sheared bedrock slopes, which are loaded with surficial material.

Geologic mapping in the vicinity of the landslide (Robinson and Lee,

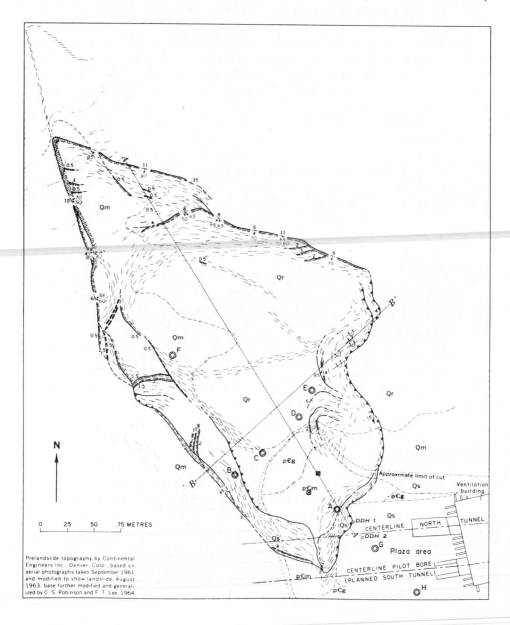

1962) showed that the bedrock is 75% granite and 25% metasedimentary rocks. Inclusions of metasedimentary rocks in the granite range from less than 0.3 m to about 300 m in maximum dimension.

Metasedimentary rocks

The metasedimentary rocks in the landslide area (Fig. 3) consist of several types of fine-grained biotite gneiss that are typically interbanded with granitic material. The most common varieties of the gneiss are biotite-quartz-microcline gneiss and biotite-quartz-plagioclase gneiss. The rocks have a distinct foliation as a result of the concentration and orientation of the minerals in layers that range in thickness from about 0.2 cm to several centimetres. Layers of granitic material — principally quartz and microcline — of irregular thickness occur in the metasedimentary rocks. These layers commonly constitute 25% of the rock.

The metasedimentary rocks are commonly altered — more extensively than is the granitic material. Most of the minerals, with the exception of quartz, in these rocks are altered to clay over a distance of from less than 2 cm to several centimetres on each side of the joints and faults. Where large bodies of metasedimentary rock are within a shear zone, the rock is commonly altered to masses of chloritic clay in which the foliation, though con-

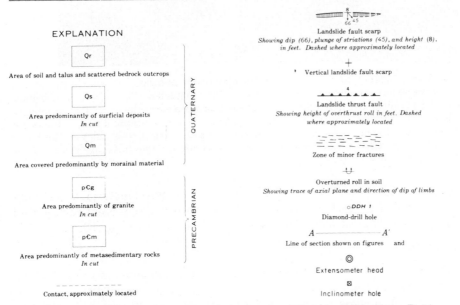

Fig. 3. Geologic and topographic map of the Loveland Basin landslide (from Robinson et al., 1972).

Fig. 4. View, looking west, of the Continental Divide showing the east portal of the Eisenhower Memorial Tunnel (August 1973). The completed (final) buttress is behind the portal building. The Seven Sisters avalanche tracks are along the north-facing slope on the left of the picture above U.S. Highway 6. The eventual westbound lane of Interstate Highway 70 proceeds to the tunnel along the course of Clear Creek valley in the lower right. The Loveland Basin ski area buildings lie between the two highways in the center of the photograph.

torted, can still be recognized. The interlayered granitic material is generally less altered and forms layers of sheared but relatively competent rock within the clay mass. Typical shear-zone conditions are shown in Fig. 13.

Granitic rocks

Petrographic analyses of samples of the granite collected in the area show that the granite typical of the area consists of about equal percentages of microcline, plagioclase feldspar (sodic oligoclase), and quartz, and from 1 to 15% biotite. Muscovite is a common accessory mineral. Other accessory minerals are apatite, zircon, magnetite, and pyrite. The size of mineral grains ranges from less than 1 to 10 mm; the average grain size is about 3 mm.

Most microcline grains in the granite are subparallel, giving the rock an indistinct foliation. Most of the biotite is not oriented; however, in biotite-rich varieties of the granite the orientation is parallel to the microcline grains. The foliation of the granite is believed to result from flow at the time of the rock's emplacement. Microshearing of the grains locally imparts an indistinct cataclastic foliation. In the granite outcrops along a stream west of the landslide, feldspar grains are offset by many microfractures, and biotite grains are alined parallel with the microfractures. These features give the rock a cataclastic foliation that is generally not parallel to the flow foliation.

The mineral grains of the granite are commonly altered adjacent to faults and joints. The ferromagnesian minerals and the feldspars are commonly partially altered to clay minerals at distances ranging from less than 0.5 cm to several centimetres from the faults and joints. In this area there is no evidence of sulfide mineral deposition; therefore, the authors believe that the alteration is primarily the result of the action of groundwater rather than of hydrothermal solution.

Surficial deposits

The landslide surface is mantled by a thin layer of morainal material, soil, and talus (Fig. 3). Morainal materials occur along the west side, across the southeast end, and to the east of the landslide. Geophysical investigations indicated that the morainal material along the west side of the landslide was deposited, in part, in a preglacial valley. The morainal material ranges in thickness from less than 0.3 m to about 5 m, and consists of an unsorted mixture of clay, sand, gravel, and boulders. In general, it is well compacted.

The higher parts of the landslide to the east are mantled by less than 0.3 m to about 1 m of soil mixed with talus. The soil is sandy and silty, and in the forested areas is masked by a few centimetres of decayed organic material and evergreen needles. Mixed with this soil, in varying amounts, are talus blocks ranging in largest dimension from less than 2 cm to about 1 m. Talus forms a small cone below the steep slope on the northeast margin of the landslide between altitudes of 3508 and 3530 m (11,510 and 11,580 ft).

Structure

The landslide is within the northeast-trending Loveland Pass fault zone (Fig. 1). The fault zone, which is about 5 km wide, consists of numerous faults and shear zones that are separated by relatively unsheared rock. Individual faults or shear zones range in width from less than 0.3 to 300 m. The Loveland Pass fault zone is considered to be an area of Precambrian shearing and faulting that was resheared during the Laramide orogeny in a manner similar to that described by Tweto and Sims (1963).

The extent of the Loveland Basin landslide is principally controlled by pre-existing faults that form a wide shear zone, a part of the Loveland Pass fault zone. The shear zone in the area of the landslide was previously mapped by Robinson and Lee (1962) as about 183 m in width and about 1220 m in length. It trends about N15°E and is terminated on the west by the fault that forms the western boundary of the landslide. This fault can be traced for more than half a kilometre and extends northwest of the northern end of the landslide.

The northeastern boundary of the landslide was determined by two pre-existing en echelon faults. The northern fault was not recognized before the development of the landslide. Fig. 5 shows this scarp as it appeared in August 1974. The scarp of the southern fault was exposed in the vicinity of

Fig. 5. View, looking east, of old fault scarp along northern margin of landslide tnat had more than 3 m of new displacement in 1963. This photograph, taken in August 1974, shows the downthrown block on the right. The abrupt change in slope ("trench") at base of scarp slope allowed increased amounts of water to enter the landslide.

Fig. 6 Landslide fault scarp at the northwest border of the landslide, August 1963. Note striations.

the curve in the fault near the northeast boundary of the landslide (Fig. 3). Before 1963, this scarp was about 1.8 m high, and the fault surface showed silicified slickensides with striations bearing N25°W and plunging 25° southeast indicating that the south side was downdropped relative to the north side.

The landslide is outlined by landslide fault scarps and thrusts. The scarps occur where the landslide mass has been displaced downward. The thrusts occur where the landslide mass moved up and over the pre-existing land surface.

The landslide fault scarps are on the western and northeastern sides of the landslide. At the time of geologic mapping, the heights of the scarps, which ranged from less than 0.3 to 3.4 m, were measured; they are shown in Fig. 3 by numbers adjacent to the scarps. The dip of the scarps and the bearing and plunge of striations on the scarp surface were also measured and are shown in Fig. 3. A field inspection of the landslide in August 1974 showed that weathering had removed the striations and subdued the scarps (Figs. 6 and 7).

Fig. 7. Same view as in Fig. 6 taken in August 1974.

The landslide thrust faults occur principally along the southeast margin of the slide and near the southwest margin. The heights of the thrusts range from less than 0.3 m to about 2 m (Fig. 3), and rolls in the soil with amplitudes of about 0.3 m exist in the landslide mass adjacent to the thrusts. During the initial period of landslide movement the thrusts were connected to the scarps by open fractures that had little or no vertical displacement. These thrusts were the most spectacular feature of the landslide during the early days of the investigation. At their margins, flowers and small trees could be seen being overturned and covered by the mass of moving soil and rock of the landslide. A photograph of the thrust sheet near the southeast corner of the slide (Fig. 8) shows a small tree being pushed over by the landslide mass.

During the early period of landslide movement (1963), the ground surface

Fig. 8. Landslide thrust fault at southeast border of slide.

in the central part of the landslide showed little evidence that the mass was moving. However, near the west margin and upslope (northwest) from the main thrust, there developed some relatively small scarps and thrusts and some zones of small fractures that locally cut across the landslide mass. Along the southwest margin of the landslide is a zone of scarps, open fractures, and a thrust where the landslide mass was considerably disturbed; probably the greatest amount of movement took place here (Fig. 3). Upslope from the southeastern marginal thrust is another zone of scarps and open fractures that is probably the result of tension caused by the development of the thrust. Associated with the other marginal scarps and thrusts, and, locally, in the middle of the landslide mass, are zones of minor fractures. The fractures in these zones range in length from a few centimetres to several metres and in width from less than 2.5 cm to about 7.5 cm. Few of these minor fractures show vertical displacement.

Surface deformations as evidenced by new fractures, particularly in the upper central part of the landslide, were noted by Colorado Division of Highways personnel in the spring and summer of 1970.

The magnitude, orientation, and location of the scarps, thrusts, and fractures indicated that the landslide did not move as a single homogeneous mass, but rather as several units, each somewhat dependent upon the others. The principal movement was parallel to the long axis of the landslide, but the southwest side of the slide moved faster, or farther, than the east side.

GEOPHYSICAL INVESTIGATIONS

Seismic refraction and electrical resistivity surveys were made on the surface of the landslide in July 1963 (Fig. 2). The electrical resistivity investigation was supervised by R.W. Moore of the Federal Highway Administration, and the seismic refraction work was directed by J.H. Scott and R.D. Carroll of the U.S. Geological Survey. The purpose of the measurements was to gain a better understanding of the depth and configuration of the slip surface. At the time of these early investigations, the information was needed in a few weeks as a basis for relocation of the pilot bore portal. Drilling was desirable, mainly for instrumentation, but it was impractical at that time. Core drilling done later supported the early conclusion that a slip surface could not be recognized with certainty from drill samples because of the heterogeneous mixture of sheared and altered metasedimentary and granitic bedrock and of blocks of unaltered and unsheared metasedimentary rocks and granite.

Although geophysical methods lack the precision of drilling in the detec-

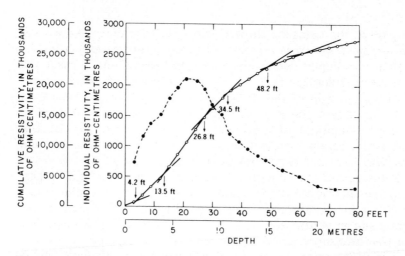

Fig. 9. Resistivity-depth curve (dashed line) and cumulative resistivity curve (solid line) for talus and disturbed sheared bedrock overlying sheared bedrock in the Loveland Basin landslide used to empirically determine depths to subsurface layers (from Robinson et al., 1972).

tion of subsurface conditions, this may be an advantage where no discrete failure surface exists and where movement occurs throughout one or more significant zones. The obvious key is the character of the zone and whether or not a detectable contrast exists.

Resistivity measurements

Resistivity measurement is a logical technique for use in studies of landslides in which excess water is likely to be associated with the zone of sliding; it has been used successfully in the past (Moore, 1957; Trantina, 1963; Takada, 1964; Takeuchi, 1971). When porous talus material and sheared bedrock slide out over less pervious material such as undisturbed sheared bedrock, one would expect evidence in the plotted resistivity curve of the effect of the excess water at the base of the disturbed material. The change in slope shown in the cumulative resistivity curve of Fig. 9 at a depth of about 10.7 m is an example of the type of effect to be expected when the less pervious, deeper material is less resistant to current flow than the overlying slide material (Moore, 1957). The 10.7-m depth correlated well with the depths at which slope changes were detected in eight other depth tests made along a section through the center of the moving mass. Such slope changes were taken as evidence of a probable slip zone.

The three lines of tests from which the three sections were prepared were spaced at intervals of 60—90 m across the slide area from east to west. The cross-section shown in Fig. 10 is typical of the three sections plotted from the data obtained in the several tests, showing the relation of the existing ground surface to the inferred position of the slip surface.

Fig. 10 shows all the significant subsurface resistivity changes along line *A* (Fig. 2). The longer the dashed lines, the more laterally persistent were the resistivity layers. The most significant and the most persistent resistivity layers are shown by the heavier dashed lines. These heavier lines also indicate the location of the probable slip surface. The authors believe that the subsurface conditions at the depths indicated by these lines are the weak zones that are the most continuous and therefore the most susceptible to movement. Near the toe of the slide, two such zones or surfaces were defined from the resistivity measurements. Probably two separate slip surfaces were present, one above the other, in this part of the slide. Depths to the slip surface(s) ranged from 0 to about 30 m in the central-western part of the slide.

Seismic refraction measurements

Because of the high altitude, steep slopes, and inaccessibility to motor transport, portability was a major consideration in the selection of seismic equipment; consequently, a portable seismic instrument (Portaseis ER-75) was used that recorded seismic signals on Polaroid film. The entire recording

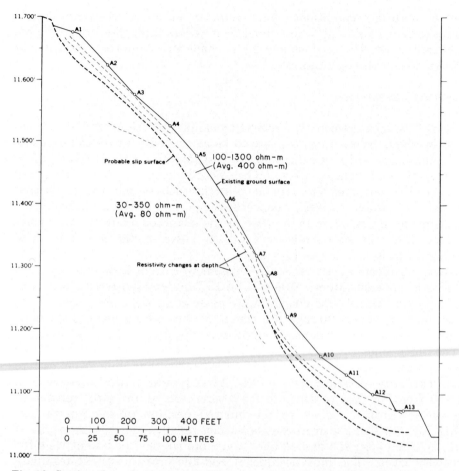

Fig. 10. Section through east part of the slide (resistivity line A, Fig. 2; from Robinson et al., 1972). A1—A13, resistivity-depth test locations. Vertical exaggeration, 2×.

apparatus was contained within a 11-kg package. Standard seismic cables were used, and twelve geophones were spaced at 7.6-m intervals. Measurements were made along thirteen seismic lines in the landslide area (Fig. 2). Shot points were located at distances of 7.6 and 15.2 m, or 7.6 and 30.5 m from the ends of the lines. On several lines a shot point was located at the midpoint of the line to obtain more information on variations in the overburden velocity.

Holes for the emplacement of the seismic charges were driven with a punch bar, and charges, averaging five sticks of dynamite (40% gel) per hole, were placed at depths of 0.5—0.9 m. Energy input to the ground ranged from good to poor, depending on the nature of the near-surface material. The aerated and disturbed condition of the overburden material above the slip zone yielded poor first arrivals on many lines. Several lines crossed fracture zones visibly open at the surface to depths of a few centimetres to several metres. The poor energy-transmission characteristics of the overburden would have made sledge-hammer refraction methods unsuccessful.

Interpretation of the results obtained on several lines was hampered both by severe attenuation of high-frequency energy and by weak energy in the first arrivals. Interpretation was further complicated by occasional irregular

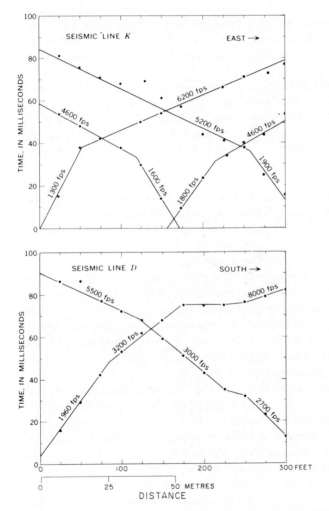

Fig. 11. Time-distance graphs obtained on seismic lines K and D (Fig. 2; from Robinson et al., 1972).

refracting surfaces, as well as by abrupt velocity variations within the near-surface layer. These two features were sometimes inseparable, and caused considerable scatter of points on several travel-time plots. Two examples, typical of the extremes of difference in the travel-time plots, are shown in Fig. 11.

Travel-time curves for line K (Fig. 11) are such that a unique interpretation may be made of the thickness of the low-velocity layer in this area. The travel-time curves for line D, on the other hand, exhibit the effects of complex subsurface conditions. On several other lines, especially those parallel with the long axis of the landslide, travel-time curves indicate complex seismic ray paths. These complex paths possibly may be explained by the relatively competent zones of high velocity that existed within the low-velocity layer. These competent zones may have acted as isolated buttresses within the sliding mass, inhibiting the development and movement of the slide.

The results of the seismic refraction survey indicated that, in general, two velocity layers existed within the landslide — an upper layer of soil and disturbed incompetent bedrock in which the velocity ranged from 305 to 823 m/s, overlying a second layer which probably represented less disturbed but incompetent bedrock, and which had a velocity ranging from 610 to 1981 m/s. An exception to this generalization occurred on line F in the plaza area (Fig. 2). Velocities in the overlying material along this line ranged from 1676 to 2042 m/s to a depth of about 9 m. The material composing the underlying high-velocity layer was characterized by velocities in the range of 4267 to 4572 m/s; these suggested the presence of firm granitic bedrock. Drill-core data subsequently furnished by the Colorado Division of Highways substantiated this interpretation.

Thickness of the low-velocity layer of the slide ranged from 2.4 to 20 m (Fig. 12). This layer was thin in the eastern and southern parts of the slide and thick in the western and central parts. Interpretation of the seismic data in the vicinity of the old preglacial stream channel on the western margin of the slide was hampered by velocity variations in the subsurface bedrock and in the surficial material. The data suggested, however, that the low-velocity layer was thickest near the western edge of the slide.

The thickness of the low-velocity layer (Fig. 12) was considered to represent the probable minimum amount of material involved in landslide movement. Data from other investigative methods used in the early study of the landslide suggested that a greater volume of material might be involved than was indicated by the seismic data. However, at many locations the seismic data were in general agreement with estimates made by other means.

The results indicated that the seismic refraction method is helpful in delineating landslides of a complex nature similar to that of the Loveland Basin landslide. At some locations, however, the reliability of the interpretation of seismic data was doubtful without independent criteria for

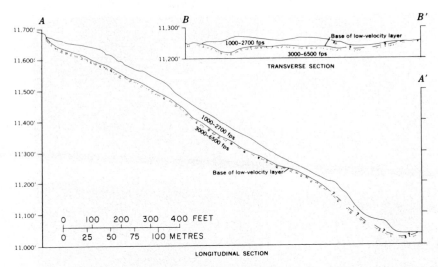

Fig. 12. Longitudinal and transverse sections through landslide, approximately along seismic lines *A*, *D*, *M*, and *E* (Fig. 2; from Robinson et al., 1972).

checking the thickness of the sliding mass. By using independent criteria for estimating the geometry of the sliding mass — in this instance, seismic and electrical resistivity soundings coupled with detailed geologic mapping — a greater measure of confidence was achieved in making calculations.

HYDROLOGIC CONDITIONS AND DRAINAGE MEASURES

The heterogeneous nature of the sheared and altered bedrock is a major control on the permeability of the landslide mass and has a direct bearing on any drainage schemes. A large part of the landslide is composed of fault gouge. Samples of this material having identical porosities were taken from the Straight Creek Tunnel pilot bore; these samples had hydraulic conductivities of 1.5×10^{-2} m^3/day per square metre for granitic gouge and 1.1×10^{-5} m^3/day per square metre for metasedimentary gouge. Such a large difference indicates nonuniform groundwater movement. Fig. 13 shows the east portal cut at an early stage. Water was able to move downward through the upper zone of fractured and sheared granitic material, but seeped or flowed out into the cut at the contact of the granitic material and the relatively impermeable metasedimentary material.

Conditions similar to those described above — typical shear-zone conditions — are present throughout the landslide mass. For example, significant intervals of "sound" rock were encountered in only one of the six inclinometer boreholes (hole F). The difficulty in draining the landslide mass is ex-

Fig. 13. Unfinished cut near east portal of Eisenhower Memorial Tunnel prior to development of landslide. Depth of cut is about 46 m. Blocky rock in upper part of cut is granite (pЄg); remainder is largely sheared and altered metasedimentary rocks (pЄm) and morainal material (Qm). The rock exposed here represents typical shear-zone conditions.

pressed in the results from 77 drain holes drilled into or beneath the slide from the north tunnel wall in 1969. Only 20 of the holes produced measurable water. Interestingly, holes drilled in April were dry, half of the holes drilled in May encountered water, and nearly all of the holes drilled in June encountered water. Rates were estimated to be as high as 80—110 m³/day from holes drilled in June, although most of these holes yielded only a trickle. Most waterflow decreased rapidly with time.

Pumping of wells

Pumps were placed in inclinometer holes B, C, E, and F in April, 1969, and operated during the summers of 1969 and 1970 in an effort to drain and thereby stabilize the slide. When pumping ceased, water levels rapidly re-

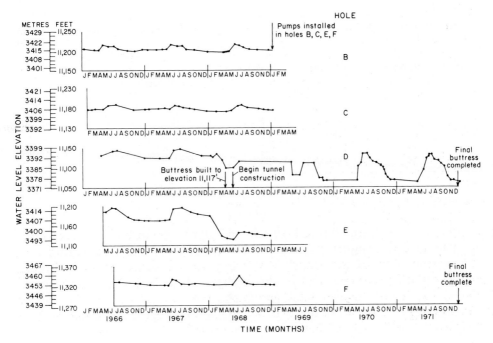

Fig. 14. Water levels in inclinometer holes.

turned to a high level. Pumping one hole had little effect on water levels in other holes, although the water level in hole D (Fig. 2) declined below previous levels during the winters of 1969-70 and 1970-71. Summer levels remained high. At a meeting on May 4, 1970, of several individuals representing the agencies involved (Colorado Division of Highways, Colorado Geological Survey, Colorado Bureau of Mines, and the Federal Highway Administration), there was general agreement that it was more important to monitor landslide movement than to continue to pump the small amount of water being obtained from the inclinometer holes. Accordingly, all of the wells were again used to monitor displacement by August 1970. Water levels in the inclinometer holes are shown in Fig. 14.

Drains

In mid-September 1963, five nearly horizontal 7.6-cm holes were drilled into the west side of the landslide from a location about 7.6 m northwest of a flume that is west of the landslide (Fig. 2). The flume had been installed during the construction of the cut to divert a small stream around the area of the cut. A tabulation of the lengths, bearings, and inclinations of the drill holes and of the initial flows from each is found in Table I.

TABLE I

Lengths, bearings, inclinations, and initial flows from drill holes in western part of the Loveland Basin landslide

Hole	Length (m)	Bearing	Inclination	Initial flow (m³/day)
1	30	N25°W	+10°	5.5
2	51	N10°E	+5°(?)	0
3	15	N40°E	+5°	*
4	18	N30°E	+12°	2.8
5	34	N16°E	+21°	11

* Plugged.

The holes were drilled into bedrock, and pipes were inserted into the holes for part of their lengths. In addition, the water from two small springs near the west margin of the landslide was impounded and conducted through pipes to the large flume west of the landslide. In July 1964, water was running from two of the drill holes, and one of the springs that had been tapped was flowing at an estimated rate of 44 m³/day. In August 1974, one of these holes was yielding about 5.4 m³/day, one was barely dripping, and another was dry.

In mid-October 1963, a 46-cm diameter pipe was laid on the ground near the west margin of the slide (Fig. 2), to divert surface runoff from snow that collects during winter months in a depression on and above the present head of the landslide. The snow accumulation amounts to as much as 6.4 m, and the snow commonly persists into July of each year. It was anticipated that the runoff from this area of snow accumulation would be collected by the pipe and then be diverted off the landslide to the large flume west of the landslide. Actually, very little water was diverted, because the numerous fractures around the inlet caused the runoff to go underground.

In 1967, seven near-horizontal perforated drainpipes were installed in drilled holes in the lower part of the slide (Fig. 2). These holes emptied into a 1.07-m diameter pipe, which eventually diverted the water collected to Clear Creek. Soon after completion, these holes produced only traces or trickles of water.

Runoff and filling of fractures

The initial geological investigation (Robinson et al., 1964) revealed numerous open fractures on the landslide surface as well as boundary scarps (Fig. 3) that allowed easy access of water. It was several years after the landslide occurred before efforts were made to fill these openings. Ridges due to slumps or scarps also impeded surface runoff. In our opinion, such measures were of great importance to landslide stabilization. Whereas drainage measures were largely ineffective, *reducing the water going into the slide* is unquestionably feasible and effective. No landslide mass should be permitted to gain more water after failure than it gained before failure. Runoff should be encouraged wherever possible. The following example illustrates this problem. The surface area of the landslide is approximately 65,400 m². The snowpack during the winter of 1969—70 was estimated to be from 1.5 to 2 times the 15-year average. The average annual precipitation, mostly snow, yields about 64 cm of moisture. Also, thundershowers are common during the summer months. A total precipitation of 120 cm is probably a reasonable estimate for the year October 1969 through September 1970. If we assume that only 40% of the precipitation went into the landslide, then about 50 cm of water was added to the landslide, or approximately 95 m³/day in this high-precipitation year. Such a significant addition of water to the landslide mass very likely triggered the accelerated movement during the summer of 1970. Considering the difficulty with which this mass can be drained, procedures to increase runoff are helpful.

LANDSLIDE DEFORMATION MEASUREMENTS

Surface surveys

The Colorado Division of Highways, upon recognition of the landslide and after conferences with the Federal Highway Administration and the U.S. Geological Survey, established surveyed reference points on and around the Loveland Basin landslide to determine the amount and direction of movement of the landslide. The Division's early work was under the direction of J.D. Post, who made periodic records of the movement of the landslide through 1966.

Early in July 1963, a transit and tape survey was run around the periphery of the slide. After this survey was run, 15 pairs of transit and tape-measurement points, at or within the slide margin and at 32 other points, were established across the slide mass, to be observed periodically by transit. The locations of some of the pairs of transit and tape-measurement points across the breaks in the slide are shown in Fig. 2. Fig. 15 shows the relative change in distance between the pairs of points between July 1963 and January 1965.

The measurement points were established after major movement of the landslide had occurred; consequently, the total vertical change in distance between a pair of measurement points can be approximated only by adding the vertical change as measured since July 1, 1963, to the height of the scarps as shown in Fig. 3. A systematic schedule of observations of the measurement points could not be made throughout the year because of deep winter snows. As a result of the movement and breakup of the landslide

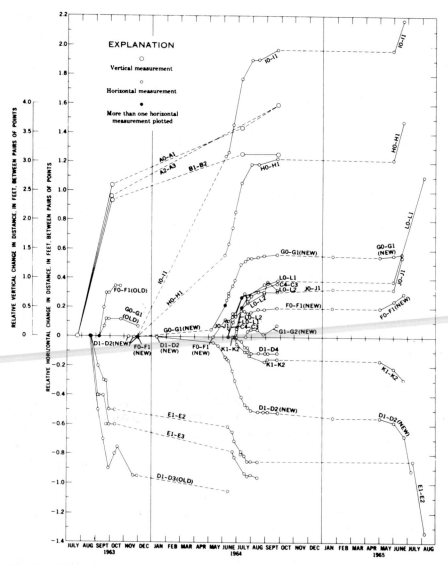

Fig. 15. Relative vertical and horizontal changes in distance between pairs of points out-side and within the landslide (from Robinson et al., 1972). Locations of surveyed lines are shown in Fig. 3. Dashed lines indicate inferred change.

mass, some points were destroyed and had to be re-established (new points, Fig. 15). Fig. 15 shows that both the horizontal and vertical movements of the slide mass were greatest between July and November of 1963. Movement continued, principally during the summer months, in 1964 and 1965. A but-

TABLE II

Horizontal movements of points on bench above plaza area, July 11 to September 3, 1963

Point	Total horizontal movement (m)
2	0.01
3	0.91
4	0.81
5	0.02
6	0.01
7	0.01

tress load was placed on the toe of the slide in September and October 1963. Much of the slide mass was frozen during the winter and spring months (1963—1965), and little movement took place. Noticeable movement occurred on the east and west sides of the slide in June and July 1964, and then decreased as the water table lowered. The central part of the slide did not move as much as the margins — which were deformed more during the original movement.

On July 11, 1963, six points were established on the first cut bench above the plaza area (Fig. 2). The horizontal distances from these points to a reference point south of the landslide area (Fig. 2) were established with a 152-m steel tape at a tension of 27 kg. Measurements between these points and the reference point were made between July 11 and September 3, 1963. The total horizontal movements of these points during this period are shown in Table II. These measurements show that the horizontal movement of the cut bench at the front of the landslide during this period was chiefly restricted to the northwest side (points 3 and 4) and that the maximum movement averaged about 0.57 m per month.

At the start of the landslide investigation the location of the toe of the landslide in the plaza area was in considerable doubt. This uncertainty was partly due to the fact that the contractor was continuing to lower the grade in the plaza area. A survey grid, shown in Fig. 2, was established in the northwest corner of the plaza to determine the limit of movement of the landslide so that the portal of the Straight Creek pilot bore could be relocated in stable ground. The points of the grid were X's on steel caps threaded on 4.76-cm steel pipes grouted into holes drilled into the bedrock of the plaza floor. The grid system was completed on August 12, 1963, and daily horizontal and vertical measurements (referenced to a point south of the plaza, Fig. 2) were made from August 12 to August 23, 1963. During this short period, maximum upward movements of 0.07 m occurred at two points, one in the northwest and one in the southwest part of the plaza.

Borehole measurements

Initial exploration

From August 5 to 10, 1963, two NX-size (7.6-cm) core holes were drilled and logged in the plaza area (Fig. 2). These holes were drilled to determine whether a slip surface of the landslide existed below the plaza level or whether a slip surface would develop below the plaza with time, and also to determine rock conditions. Hole 1 was drilled in competent granite to a total depth of 27.4 m. It was cased, after reaming, for the first 3.05 m using uncemented 10.2-cm casing. Hole 2 was drilled to a depth of 25.2 m, also in competent granite, and was left uncased. No slip plane was recognized in either drill hole. Plastic pipe, 2.54 cm in diameter, was placed in each hole, and the space between the pipe and the wall of the hole was packed with cinders. In order to test for bending of the plastic tubing, a 76-cm-long, 0.95-cm diameter rod was lowered on a nylon line to the bottom of the holes.

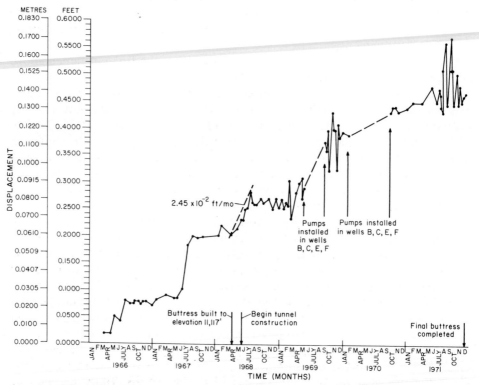

Fig. 16. Time-displacement curve for inclinometer hole B. Vector displacement of top of hole (average) S38°E.

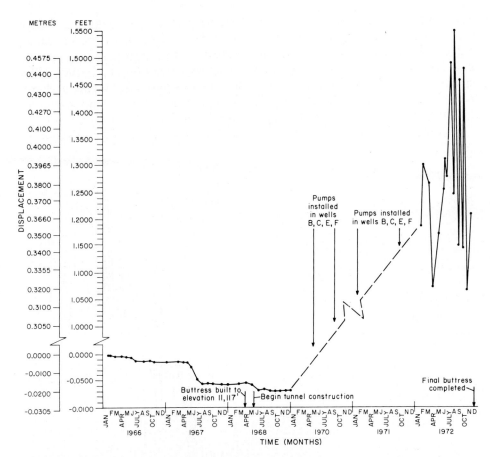

Fig. 17. Time-displacement curve for inclinometer hole C. Vector displacement of top of hole (average) S38°E.

These tests were made daily until mid-August 1963, and then at various times until the start of construction of the rockfill buttress in the first week of September. During this time interval there was no indication that the plastic tubing had been bent. Unfortunately, these casings were not extended up through the buttress to monitor possible later movement.

Inclinometer measurements

Concern continued for the stability of the landslide and, more importantly, for the safety of the east portal and plaza area after completion of the pilot bore in 1965. The contract for the design of the Eisenhower Memorial Tunnel, which was awarded to TAMS in 1965, included landslide instru-

mentation and the design of landslide-control measures. Additional data
were required concerning slip-surface geometry, thickness of the landslide,
and direction and rates of movement. Most of this information could be
provided by slope-inclinometer measurements. Accordingly, during 1966
eight vertical holes were drilled and cased for this purpose, ranging in depth
from 26.4 to 73.8 m (Fig. 2). When initial readings were made in these holes
the landslide had been intermittently active for nearly 3 years.

Readings prior to construction of the tunnel were made twice monthly

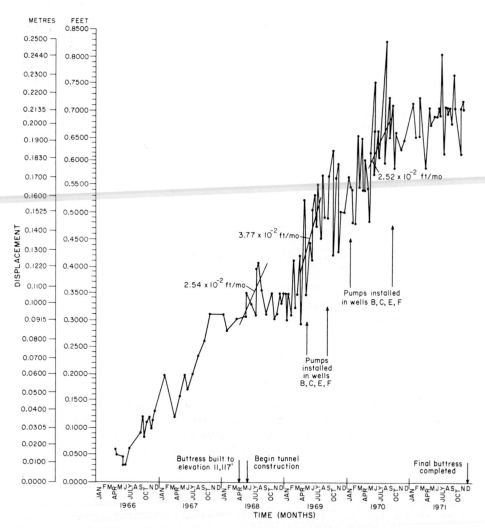

Fig. 18. Time-displacement curve for inclinometer hole D. Vector displacement on top of
hole (average) S38°E.

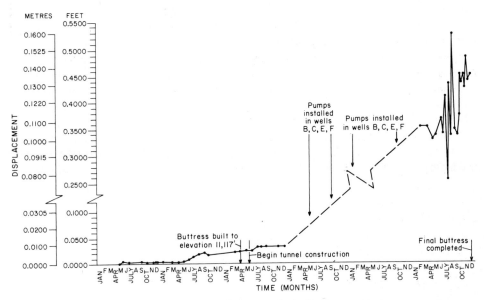

Fig. 19. Time-displacement curve for inclinometer hole E. Vector displacement of top of hole (average) S38°E.

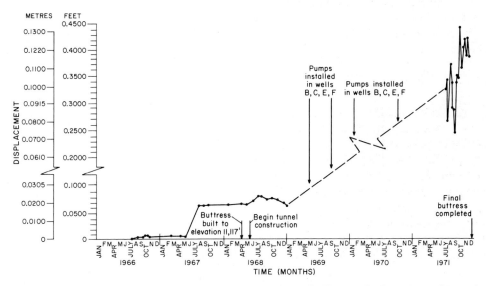

Fig. 20. Time-displacement curve for inclinometer hole F. Vector displacement of top of hole (average) S38°E.

during the thaw season and once monthly during the freeze season. When
construction began, in May 1968, the slide was monitored on a daily basis
during the thaw season and once a week during the freeze season. The inclinometer proved to be a reliable instrument for determining slip surfaces and
rates of movement. However, because of scatter of data, a large number of
time-consuming readings (one hole required up to four hours) were needed
to obtain trends of movement; this limited results to an after-the-fact study.
The location of slip surfaces, on the other hand, could be determined from a
few readings taken during a period of activity. Computed displacements have
an accuracy of about 0.15 cm. Instrument accuracy is about 0.061 cm for
a 61-cm measurement interval. All displacements were calculated relative to

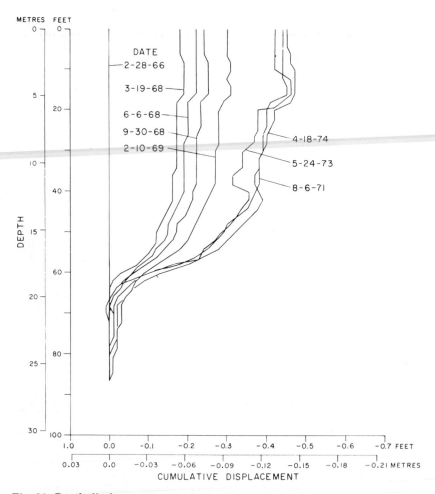

Fig. 21. Depth-displacement curve for inclinometer hole B.

the initial position of the bottom of the casing. In the plots of displacement, negative displacement is to the south or west.

Time-displacement plots of data from several of the inclinometer holes (Figs. 16—20) revealed accelerated movement in some holes during part of 1966 and an alarming increase in movement in holes B and D in 1967 (Figs. 16 and 18). Also, the deepest movement on the depth-displacement curves appeared to be greater than in 1963 (Figs. 21—24). Construction of an additional buttress was undertaken in 1967 and completed to an altitude of 3388 m (11,117 ft) in March 1968. However, movement in most holes continued until 1971, when a final buttress was completed to an altitude of 3398 m (11,150 ft).

Throughout the eight-year period of measurement, it was evident that more than one and in most places at least two slip surfaces, or zones, were present. The depth-displacement curves bear this out. For example, in inclinometer hole D (Fig. 23) there were possibly six zones of movement: at 0—5 m (surficial creep), 12—15 m (poorly developed), and well-developed zones at 30—38 m, 43—45 m, 48—50 m, and 54—57 m. In general, deeper zones of movement were initiated and became more pronounced with time until the final buttress was in place. Movements were both deepest (which agreed with the early study) and of greatest magnitude in the lower central part of the landslide.

Study of the inclinometer data indicated that slide activity was closely related to thaw and freeze seasons. During the spring and summer months, movement of the slide mass accelerated as the water level rose; it stabilized during the fall and winter months as the water level declined. Exceptions to this normal activity were cause for great concern, and immediate stabilization efforts were necessary. Such an exception occurred when slide activity continued throughout the winter of 1969—70. There was no doubt then that the second buttress (1968) was insufficient and should be increased to the design specified by TAMS in 1966. A maximum displacement rate of 1.54 cm/month in hole D occurred in early 1970. As was noted earlier, new surface fractures had developed in the central upper part of the landslide. Of further concern was the penetration of the base of the slide by the tunnel excavation on February 21, 1969. Although care was taken in blasting through this zone, steel supports deformed eastward near the crown line, possibly owing to slide movement.

In holes B and C (Figs. 21 and 22) the direction of displacement changed, probably in late 1971 or early 1972, although it may not have been synchronous in both holes. This change occurred during the period of final buttress construction and perhaps was an indication of the stabilizing effect of the buttress. Because the displacement changed over a considerable interval in each hole the change was not caused by the movement of isolated joint blocks, but by the deformation of a larger mass.

A new inclinometer hole was drilled in 1972, at the approximate location

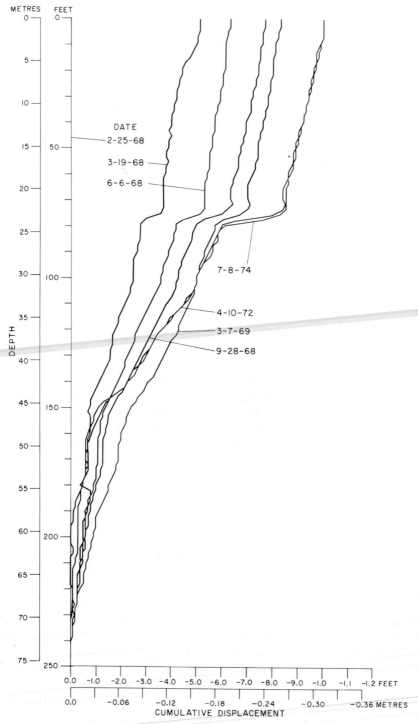

Fig. 22. Depth-displacement curve for inclinometer hole C.

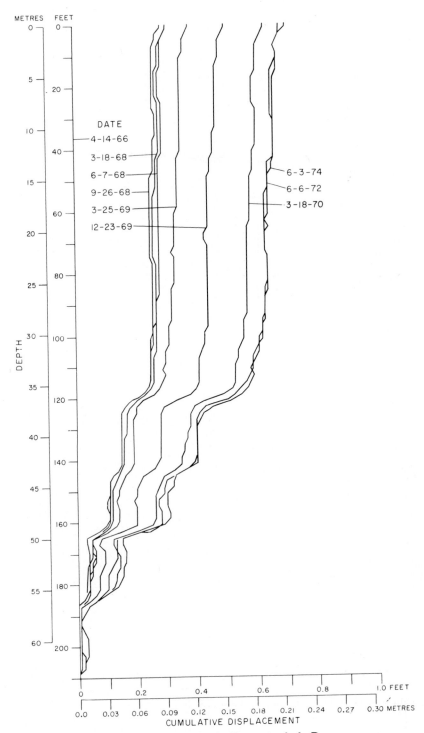

Fig. 23. Depth-displacement curve for inclinometer hole D.

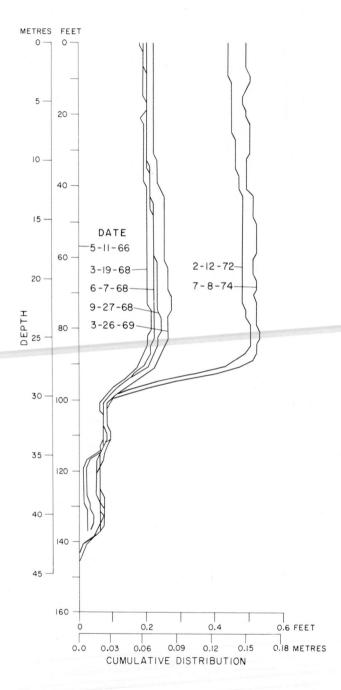

Fig. 24. Depth-displacement curve for inclinometer hole E.

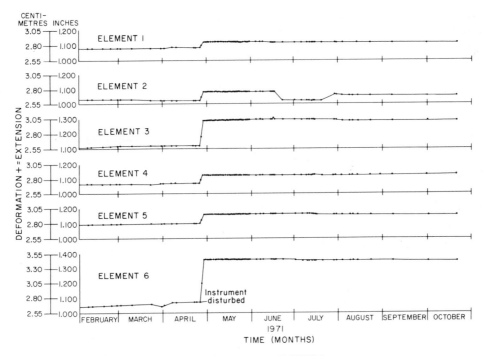

Fig. 25. Time-deformation curves for extensometer MPBX-1.

of hole A, to a depth of 74 m. No movement has been detected in this hole.

Extensometer measurements

In order to determine rapidly the activity of the slide mass, two six-point, multiple-position borehole (wire) extensometers were installed into the slide from the buttress in early 1971 (Fig. 2). The holes were drilled 5° upward, and were 122 m long. Permanent dial gages were installed to reduce data-reduction and reading error. A time-deformation plot for 1971 from one of the instruments is shown in Fig. 25. There was negligible detectable landslide movement during the critical May-August period. However, the length of the extensometer sensor wires reduced the sensitivity of the instrument, and magnitudes of deformations were only approximate.

STABILITY ANALYSES

Prior to the preparation of a report by Robinson et al. (1964), a stability analysis of the Loveland Basin landslide by R.A. Bohman, Federal Highway

Administration, was made from only preliminary electrical resistivity data, measured dimensions, and assumed physical properties of the sliding material. This analysis indicated that a fill containing approximately 76,000 m³ of material properly placed in the plaza area at the toe of the landslide would

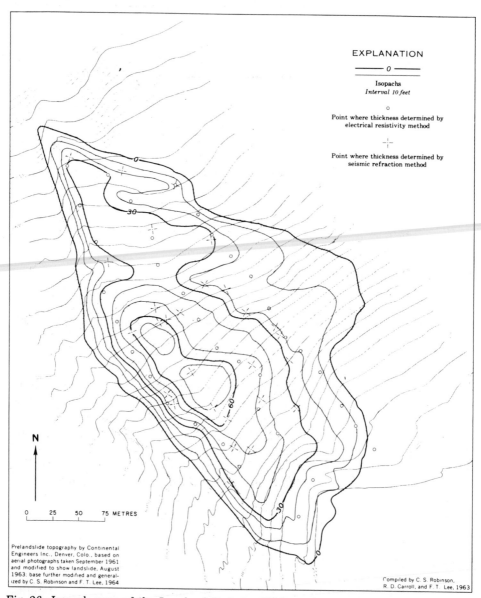

Fig. 26. Isopach map of the Loveland Basin landslide to the base of the slip zone as of 1963 (from Robinson et al., 1972).

provide a factor of safety against further movement along the full length of the landslide of approximately 1.4. The fill was expected to act primarily as a buttress to receive pressure of the sliding material. The stability analysis of the buttress, after its construction, was based on the delineation of the slide and slip zone by Robinson et al. (1964), as illustrated by the isopach map in Fig. 26. Such maps were used to calculate the volume and weight of moving material and to locate drill holes for drainage and instrumentation.

As a result of analysis of all available data and after numerous meetings, field inspections, and office projections, the decision was made to shift the position of the Straight Creek Tunnel pilot bore to the south, to lower the pilot bore and approach grade, and to shift the portal to the east of the location originally planned, in order to place the portal of the pilot bore as far from the active landslide as possible and to develop all material possible for constructing the recommended buttress. The original plaza area, final centerline location, and other pertinent features are shown in Fig. 2.

The buttress was constructed in September and October, 1963, and contained a volume of 47,220 m^3 of compacted material consisting primarily of shot and ripped rock obtained from the highway section immediately east of the pilot bore portal, where the highway grade was lowered approximately 4.9 m. This amount of material was all that was readily available at the time, and was far less than that recommended for the buttress. A second analysis based on the actual buttress was considered essential. This analysis by Bohman followed the method of Baker and Yoder (1958) and did not consider the effect of water, although it was recognized that the influence of water was important and needed more attention. An average safety factor against failure at an altitude of 3377 m (11,080 ft) was 1.00, and additional movement was expected (Robinson et al., 1972, p. 39). Details of Bohman's analysis are in Robinson et al. (1972).

One method of minimizing the possibility of failure above the 1963 buttress was the construction of another buttress on top of the existing one. The desirability of doing this was recognized and discussed during earlier stages of investigation and analysis. Lack of material and cost were the primary reasons given for not building an additional second-story buttress during the period July-October, 1963. Another reason was the general belief that shear failure along a plane at an approximate altitude of 3377 m would not result in an instantaneous movement of the entire landslide mass that might override the buttress and spill into the new plaza or portal area. Rather, it was hoped that such a shear failure would result in a slow movement of landslide material out onto the surface of the buttress, thereby forming a natural second-story buttress that would effectively resist further movement before it reached the edge of the buttress, where it could spill over into the plaza area. Any movement of material onto the buttress would result in a decreased driving force and increased resistance to further movement.

By September 1966, inclinometer measurements and surface surveys indicated that slide movement had not been stopped by the 1963 buttress, and
that movement was deeper than that determined in 1963. TAMS (1966) reviewed earlier reports concerning the landslide and analyzed the inclinometer
data available through September 1966. They concluded that the landslide
probably had a lower slip zone or slip surface as deep as 76 m below the surface. In their analysis, based on a sliding block method, TAMS assumed that
the water level was at the surface and that the landslide had a saturated unit
weight of 2.4 Mg/m³. It was further assumed that the slide mass was a statistically homogeneous, granular material, and that its shear strength was controlled by frictional characteristics. An effective angle of internal friction (ϕ)
of 34° was used in the analysis, which was based on a potential sliding surface at a maximum depth of 76 m. TAMS found that to maintain a safety
factor of 1.8 under these conditions would require an additional buttress of
279,000 m³, built to an altitude of 3399 m (11,150 ft).

An additional buttress was completed to an altitude of 3388 m in April
1968 (Fig. 27). This additional material was considerably less than was recommended by TAMS. Those involved believed that this added load would
provide a more secure buttress against failure until the buttress could be
completed to TAMS' requirements. The plan was to construct an additional
buttress with rock excavated for the north tunnel. Construction of this tunnel began in May 1968.

Because of the continued slide movement mentioned previously, further
study of the landslide was undertaken by the tunnel contractor. Stability
analyses, using an infinite slope analysis and a sliding wedge analysis, were

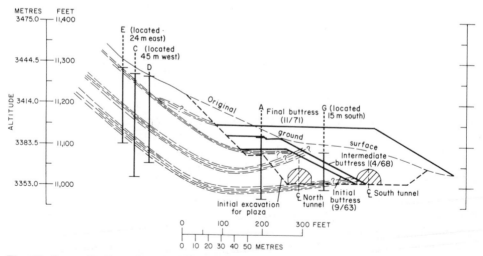

Fig. 27. Generalized section through the toe of the landslide showing the approximate
location of original ground surface, buttresses, and major slip zones.

performed by D.U. Deere and F.D. Patton (written communication, 1969). Both analyses were based on fall-winter water conditions, during which time water levels are generally 6—9 m below summer levels. The infinite slope analysis yielded a safety factor of 1.06, assuming a ϕ of 34° and provided that the water level in the toe was lowered about 6 m. The sliding wedge analysis was probably the more conservative and realistic of the two analyses, using a ϕ of 31°. This value was back-calculated by using a safety factor of 1.0 for the ground configuration existing in 1963 and the maximum observed groundwater levels. This value of ϕ was then used to compute the factor of safety for various conditions of drainage and for various buttresses. With no change in water level or buttress, a factor of safety of 1.06 was obtained. With a 306,000-m³ total buttress, the factor of safety increased to 1.56. If the toe of the slide were drained, the factor of safety of the slide would be increased by 0.1. Deere and Patton recognized that additional slide movement during the next period of high groundwater levels was a likely possibility. They recommended additional drainage and buttress construction.

The buttress of 326,000 m³ recommended by TAMS in 1966 was not completed until November 1971. This was the third increment of buttress construction, and brought the buttress altitude to 3400 m (Fig. 27). The Colorado Division of Highways made an analysis of the stability of the final buttress by the limit-equilibrium method (Morgenstern and Price, 1965), using $\phi = 34°$, cohesion = 0, and a unit weight of 2.5 Mg/m³ for the landslide, and $\phi = 30°$, cohesion = 68.9 kN/m², and a unit weight of 2.4 Mg/m³ for the buttress. This analysis indicated a safety factor of 1.75 against deep movement (about 43—50 m, inclinometer hole D) and a safety factor of 2.81 against shallow movement (about 30—38 m, inclinometer hole D). There has been no detectable evidence of movement of the slide since the fall of 1971.

It is not known how well the assumed properties in these analyses represent actual conditions along zones of movement. Although few physical properties were determined, it would have been extremely difficult, if not impossible, to obtain meaningful in-situ or laboratory test results at the necessary locations in such a heterogeneous mass. The "back-calculated" analysis made by Deere and Patton has merit, but its reliability is dependent upon the extrapolation of past conditions to future behavior.

The 1963 buttress did not stabilize the landslide, nor did the 1968 buttress. Thus, there was a nine-year period of intermittent deformation. This slow deformation, approaching creep, conforms to the concept of slow, progressive deformation of nonhomogeneous, slope-forming material (Terzaghi, 1950, pp.85—87). Terzaghi pointed out that if a load acts on such material for many years, the material undergoes very important and very intricate deformations that reflect all the details of the internal structure of the mass, something a sudden failure could never achieve. Qualitatively expressed, drag effects (forces) produced by the gravity action of joint blocks on each other progressively disturb deeper zones in the irregularly jointed and sheared rock mass, continuing until shearing stresses in the unstable material are reduced by buttress loading, drainage, etc. It is doubtful that in any slope the shearing stresses can be reduced below the shearing stresses at the fundamental strength, the stress at which creep is presumed to begin.

The phenomenon of landslide failures becoming progressively deeper has been studied in Europe (Hofmann, 1973). Hofmann used physical models that simulated jointed rock systems and that considered joint fabric, friction angle, inclination of strata, and rigid and yielding blocks. The mode of failure was similar in each case: early displacements led to the development of a zone of shallow failure ("Hakenwerfen") at the front of the slope, followed by the formation of deep zones of weakness ("Talzuschubë"); curved failure zones were produced.

Ter-Stepanian and Goldstein (1969) have identified "2- and 3-storied" landslides in crushed argillites and sandstones near the Caucasian coast of the Black Sea. They also found that weaker materials (clays) were susceptible to "multi-storied" sliding.

CONCLUSIONS

Surface and subsurface information indicates that there has been no substantial movement in the Loveland Basin landslide since the summer of 1971, which coincides with the completion of the large buttress berm. The several stability analyses indicate that the landslide has an average safety factor of approximately 2. This is probably a higher safety factor than that of the original slope. However, surface surveys and monitoring of underground instruments are necessary for several years to establish a creep rate and to detect new fracturing or settlement.

The isopach map (Fig. 26) developed in 1963 from several kinds of information showed a maximum depth to the base of the slip zone of approximately 23 m. Deeper displacements of up to 67 m were apparent in several of the inclinometer holes by late 1966. Two possibilities for this discrepancy are recognized: (1) the deeper slip surfaces or zones existed in 1963 but were not detected, or (2) movement became deeper as new surfaces or incipient surfaces of sliding were activated. Persistent thick slip zones, such as between approximately 25 and 31 m in hole E (Fig. 24), should have been detectable with the techniques used in 1963, because the more shallow but less persistent zones were detected. That is, the slip zone as defined in 1963 (Fig. 26) was apparently less active (although detectable) than were the deeper slip zones after 1966 (Figs. 21—24). It is entirely possible, however, that the activity of the early shallow zone of deformation had decreased by the time borehole instrumentation was installed.

The Loveland Basin landslide occurred in a slope whose fabric had been long established by several episodes of tectonic deformation. Zones of weakness, in which ancient movement had occurred, existed before the present landslide developed. This ancient movement is expressed by the pre-landslide faults, mentioned earlier, showing old slickensides in the direction of landslide movement. We suggest that such slopes have a "memory" of past deformations that strongly influences future deformations. A thorough knowledge of the structural development and land-forming processes in these high mountain valleys in glaciated Precambrian terrain is prerequisite to understanding slope behavior.

ACKNOWLEDGEMENTS

Many individuals participated in the investigation of the Loveland Basin landslide during the 10 years that it was of nearly constant concern. The present writers are in great debt to those individuals, mainly with the Colorado Division of Highways, who, through many months of severe weather, endeavored to make and analyze the necessary measurements. These persons primarily include J.D. Post, H. Ueblacker, J.E. Gay, D. Pitts, and L. Farnsworth. The expertise of several consultants contributed to our understanding of this complex landslide. These include D.U. Deere, F.D. Patton, C.S. Robinson, and R.L. Schiffman. J.W. Rold, Colorado Geological Survey, recognized the severity of the problem and was instrumental in expediting the efforts of several agencies. R.A. Bohman, Federal Highway Administration, was helpful in several discussions of the behavior of the landslide.

REFERENCES

Baker, R.F. and Yoder, E.J., 1958. Stability analyses and design of control measures. In: E.B. Eckel (Editor), Landslides and Engineering Practice. Highw. Res. Board, Spec. Rep., No. 29; Natl. Acad. Sci.-Natl. Res. Counc. Publ., 544: 189—216.
Hofmann, H., 1973. Modellversuche zur Hangtektonik. Geol. Rundsch., 62: 16—29.
Moore, R.W., 1957. Application of electrical resistivity measurements to subsurface investigations. Proc. Southeast. Assoc. State Highw. Officials, pp. 192—210; Public Roads, 29(7): 163—169.
Morgenstern, N.R. and Price, V.E., 1965. The analysis of the stability of general slip surfaces. Geotechnique, 15: 79—93.
Robinson, C.S. and Lee, F.T., 1962. Geology of the Straight Creek Tunnel site, Clear Creek and Summit Counties, Colorado, and its predicted effect on tunnel construction. U.S. Geol. Surv. Open-File Rep., 41 pp.
Robinson, C.S., Carroll, R.D. and Lee, F.T., 1964. Preliminary report on the geologic and geophysical investigations of the Loveland Basin landslide, Clear Creek County, Colorado. U.S. Geol. Surv. Open-File Rep., 5 pp.
Robinson, C.S., Lee, F.T. et al., 1972. Geological, geophysical, and engineering investi-

gations of the Loveland Basin landslide, Clear Creek County, Colorado, 1963-65. *U.S. Geol. Surv. Prof. Paper*, 673, 43 pp.

Robinson, C.S., Lee, F.T., Scott, J.H., Carroll, R.D., Hurr, R.T., Richards, D.B., Mattei, F.A., Hartmann, B.E. and Abel, J.F., Jr., 1974. Engineering geologic, geophysical, hydrologic and rock-mechanics investigations of the Straight Creek Tunnel site and pilot bore, Colorado. *U.S. Geol. Surv. Prof. Paper*, 815, 134 pp.

Takada, Y., 1964. On the landslide mechanism of the Tertiary type landslide in the thaw time. *Disaster Prev. Res. Inst., Kyoto Univ., Bull.*, 14: 11—21.

Takeuchi, A., 1971. Fractured zone type landslide and electrical resistivity survey, 1. On the usefulness of the electrical resistivity survey at fracture zone type landslide areas. *Disaster Prev. Res. Inst., Kyoto Univ., Bull.*, 21: 75—98.

TAMS (Tippetts-Abbett-McCarthy-Stratton, Engineers and Architects), 1966. East portal slide — Measures for control. *TAMS Design Rep.*, No. 13, prepared for Colorado Department of Highways and Bureau of Public Roads, 14 pp., appendix, 21 pls.

Ter-Stepanian, G.I. and Goldstein, M.N., 1969. Multi-storied landslides and strength of soft clays. *Proc. 7th Int. Conf. Soil Mech. Found. Eng., Mexico 1969*, 2: 693—700.

Terzaghi, K., 1950. Mechanism of landslides. In: S. Paige (Editor), *Application of Geology to Engineering Practice (Berkey Volume)*. Geological Society of America, Washington, D.C., pp. 83—123.

Trantina, J.A., 1963. Investigation of landslides by seismic and electrical resistivity methods. *Symp. on Field Testing of Soils. Am. Soc. Test. Mater., Spec. Tech. Publ.*, 322: 120—134.

Tweto, O. and Sims, P.K., 1963. Precambrian ancestry of the Colorado mineral belt. *Geol. Soc. Am. Bull.*, 74: 991—1014.

Chapter 14

LANDSLIDES IN ARGILLACEOUS ROCK, PRAIRIE PROVINCES, CANADA

S. THOMSON and N.R. MORGENSTERN

ABSTRACT

In the Prairie Provinces of western Canada many of the river valleys are post-Pleistocene in age and are incised into Upper Cretaceous argillaceous bedrock. This report discusses the landslides which are prevalent along the valley walls.

The bedrock of the area comprises essentially flat-lying, poorly indurated sandstones, siltstones, claystones and clay shales with interspersed beds of coal and pure bentonite. The presence of bentonite as admixtures in some of the beds was shown to be the most important single geologic factor affecting the shearing resistance.

Rapid erosion of postglacial channels into the sedimentary bedrock removed considerable load from the strata underlying the valley floor. Load removal was accompanied by a rebound that gave rise to a gentle anticlinal structure beneath the valley bottom and a gentle upwarping of the strata comprising the valley walls. This upward flexing gave rise to interbed slip which provided enough deformation to reduce the angle of shearing resistance from peak to some lesser value. This weakened zone exerts an obvious influence on valley wall stability.

In an airphoto analysis of landslides along some major rivers in southern Alberta it was noted that marine and brackish to freshwater formations were more prone to sliding than the clastic and deltaic bedrock. Other factors such as surface drainage, the presence of preglacial valleys, and river erosion also influence valley wall stability.

Specific case studies of landslides are presented to illustrate analyses and the dominant role played by geology in landslide activity. Minor geologic details, such as thin bentonite seams, sheared zones resulting from old landslides, and softened zones due to interbed slip are difficult to detect in a drilling program or from geomorphic evidence, yet their influence on slope stability is profound. Subsurface pore water pressures demand careful field investigation.

INTRODUCTION

The stability of slopes, natural or man-made, has posed problems of general interest for a considerable period. When transportation systems encounter slopes, problems of special interest to the geotechnical engineer are generated. In the Prairie Provinces of western Canada, valley crossings are of particular concern. Many of the valleys are post-Pleistocene in age and are incised into Upper Cretaceous argillaceous bedrock. Landslides are ubiquitous in this region; their characteristics are discussed here.

The areas in which most of the work reported here has been carried out have been designated as the Alberta Plain and the Saskatchewan Plain (Douglas, 1970). Together these form a large part of the Interior Plain of western Canada. The Alberta Plain stretches from the Athabasca River in central Alberta southeastward toward the International Boundary. It is largely composed of late Mesozoic strata, although some strata of Tertiary age occur in the west and southwest. The undulating to rolling surface is interrupted by a few widely separated groups of low, rounded hills. Much of the plain is about 760 m in elevation; the river valleys are sharply incised 60—120 m and the hills rise to 1000 m or higher.

The eastern edge of the Alberta Plain, referred to as the Missouri Coteau, constitutes a step down to the Saskatchewan Plain. The coteau, well marked at the International Boundary by a line of low hills partly formed by Tertiary sediments, gradually becomes indefinite northwestward. East of the coteau, the surface of the Saskatchewan Plain, formed on flat-lying Mesozoic sediments and lower and smoother than the Alberta Plain, ranges from 450 to 800 m above sea level and the relief is about 90 m in the hillier parts. For the most part, the larger rivers are entrenched in the bedrock.

GEOLOGIC HISTORY

Deposition of sediments over the plains area of western Canada, and parts of North and South Dakota and Montana, was influenced by widespread transgressions and regressions of epeiric seas. Stratigraphically, this portion of the basin is characterized by shale, siltstone, and sandstone formations which were laid down in or near the shallow epeiric seas. Vertical variation from mainly marine shale at the base of the Upper Cretaceous to continental sandstone at the top is common. In addition, superimposed on the vertical stratigraphic change is a lateral facies change from predominantly marine shale in the southeast to continental clastics in the northwest. Neither lateral nor vertical changes are uniform but reflect alternating transgressive and regressive phases caused by the continuously variable interplay of subsidence rate and sediment supply. Essentially continuous deposition continued throughout the Upper Cretaceous epoch. Ash beds, such as the Knee

Hills Tuff, attest to volcanic activity late in the period. Bentonite occurs widely in the Upper Cretaceous rocks of western Canada. Important deposits are in the Battle Formation of southern Saskatchewan and in the Bearpaw and Edmonton Formations of southern and central Alberta (Williams and Burke, 1964).

Sedimentation continued without interruption through the uppermost Cretaceous into Paleocene time resulting in Tertiary rocks that are almost wholly continental. The bulk of these sediments are quartzose clastics derived from the highlands of the Cordillera. The deposits formed a series of coalescent, interfingering alluvial fans and fluvial sediments. Irregularities in deposition assisted by minor differential epeirogenic warping of the surface of aggradation brought about decreased stream velocities and ponding in places, in which shales and, rarely, limestone were deposited.

In central and southern Alberta, uplift introduced a period of erosion which ended when the source region to the west underwent renewed uplift, spreading coarse clastics over the eroded areas. In North Dakota, southwestern Manitoba and possibly southeastern Saskatchewan, downwarping produced the last far-reaching marine invasion of the continental interior in earliest late Paleocene.

The Eocene was the time of major uplift and resulted in the formation of the Rocky Mountains. Erosion on a grand scale accompanied and followed the uplift of the mountains. The dissection of the Interior Plains initiated by this uplift resulted in removal of Eocene and Oligocene deposits from all but a small area of southeastern Alberta and southwestern Saskatchewan.

Miocene and Pliocene time saw continued erosion of the Interior Plain. Late Tertiary and early Quaternary erosion was so extensive that only small scattered patches of Miocene and/or Pliocene fluvial deposits are left on isolated uplands. Of all the great streams that once carried sediments from British Columbia eastward to the Cretaceous sea and, later, across the Interior Plains only one, and possibly two, remained at the end of Eocene time. The Liard River (and probably the Peace River also) is antecedent to the growth of the Rocky Mountains: both rivers drain parts of the eroded core of the Cordilleran geanticline (Taylor et al., 1964).

The Quaternary affected essentially the entire area of western Canada. Glacial ice and its associated processes were largely responsible for sediment deposition and landscape modification. Neither glacial erosion nor deposition altered the pre-existing landforms to any great extent, and variations in drift thickness can generally be attributed to preglacial topography.

A study of the preglacial drainage map shows an increased derangement westward toward the mountain front. As westerly parts of the valleys became free of ice, downstream areas were still blocked and therefore streams were diverted. Upon ice retreat, the diverted streams had become established to the point that they no longer recovered their former courses. Preglacial valleys often became infilled and in many instances there is little or no surface expression of them. After the ice fronts became inactive, final deglaciation of western Canada was accomplished by ice stagnation and melting in place. Whether the four major ice advances that occurred during the Pleistocene of North America were also present in western Canada is problematical. The existence of pre-Wisconsin deposits in the Alberta Plain has not been established and the deposits now present in the surface are all likely Wisconsin in age (Barton et al., 1964).

In summarizing the geology of the Upper Cretaceous, Tertiary and Quaternary, the following are the major points of geotechnical interest. The predominantly flat-lying sedimentary rocks are comprised of sandstones, siltstones, claystones or mudstones and shales with interspersed beds of coal and pure bentonite. The vertical variation and horizontal facies change of these sedimentary rocks is significant and were brought about by transgressing and regressing seas. Marine, brackish and freshwater deposits interfinger with continental clastics. This wide range of sediments has presumably not been subjected to tectonic forces but only to the weight of overlying sediments. Subsequent to their deposition, they were eroded and it has been estimated that about 600 m of sediments has been removed from the area (Rutherford, 1928). In general, the sedimentary rocks of the Interior Plain are poorly indurated and are sometimes referred to as "soft rocks". Those containing clay minerals are subject to rebound and swelling when load is removed and water becomes available to them. A few of the strata, particularly the sandstones, may be considered competent due to cementation by silica or carbonates. The presence of bentonite both as seams and as admixtures to various beds is of consequence in the engineering characteristics of a depositional sequence. The preglacial river valleys, infilled by Pleistocene deposits, influence the stability along parts of the postglacial or present river valleys. The rapid entrenchment of these young valleys is an important factor in valley wall stability.

GEOTECHNICAL CHARACTERISTICS

Locker (1969, 1973) undertook a study in mid-central Alberta to determine the geotechnical properties of the claystones and siltstones of the Upper Cretaceous Belly River and Edmonton Formations and the Paleocene Paskapoo Formation, and the relationship between these properties and the geological characteristics of the materials. It had been observed, for example, that the propensity for landslides in the bedrock decreases in a westerly direction from the plains area to the margin of the foothills of the Rocky Mountains. However, the gross lithology and inferred depositional environments of the sediments is similar over this distance and hence it may be postulated that post-depositional (including diagenetic and denudational) processes have affected the mass engineering properties of these materials.

The gentle westerly dip of the strata results in wide outcrop bands of the sedimentary rocks; these were sampled in an east to west direction, nearly perpendicular to the strike of the beds. Samples were obtained from outcrops and from cores obtained from drill holes.

The average percentages of sand-, silt-, and clay-sizes in the fine-grained rocks of central Alberta were 14% sand, 55% silt, and 31% clay. There is, however, considerable deviation from this average composition across the area studied. Clay aggregate orientation was found to be generally poorly developed in the fine-grained rocks, possibly due to the silt content and also to the fact that much of the clay is of diagenetic origin, having been deposited originally as volcanic ash. Subsequent alteration of the ash to clay minerals would not necessarily lead to the formation of oriented aggregates. The clay fraction of the fine-grained rocks was dominated by montmorillonite and illite. Montmorillonite accounted for an average of 23% of the total rock composition, with a range of 5—92%, in the eastern part of the area, whereas in the western portion the average montmorillonite content was 9% with a range of 4—38%. Kaolinite and chlorite were minor constituents. Well-developed fissility was found to be uncommon in the fine-grained soft rocks of central Alberta, most of which are siltstones or claystones.

All of the fine-grained rocks exhibited plasticity except those samples of the Lower Cretaceous rocks included in this particular part of the study. The range of plasticity values is high. The natural moisture content in the majority of the samples was less than the plastic limit and the liquidity indices are, therefore, all negative. Some typical plasticity characteristics are shown in Table I.

Direct shear tests were carried out on undisturbed core hole samples of the various fine-grained soft rocks. Typical data are given in Fig. 1 and Table

TABLE I

Plasticity characteristics of Upper Cretaceous rocks

Formation	Soil description	Depth (m)	Liquid limit (%)	Plasticity index (%)	Natural moisture content (%)
Belly River	clayey siltstone	11	61	40	15.5
Belly River	clayey siltstone	12	60	31	21
Paskapoo	clayey siltstone	16.5	40	16	11
Edmonton	sandy siltstone	21	47	17	13
Saunders	clayey siltstone	13.5	31	11	9
Belly River	clayey siltstone	12	60	31	21
Belly River	clayey siltstone	16	70	42	12
Paskapoo	sandy siltstone	16.5	39	18	14

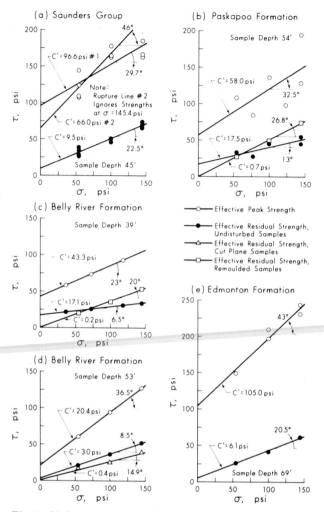

Fig. 1. Mohr rupture envelopes from direct shear testing of soft rocks from central Alberta (after Locker, 1969).

II. It should be noted that there is considerable scatter in the strength data; this scatter seems to be due to a combination of factors. The samples are neither truly homogeneous nor isotropic in that there is a variation in grain size on both a microscopic and macroscopic scale. Variations in shearing resistance within a particular rock type arise from the location of the failure plane with respect to weak zones, which coincide with micro-fissures, laminations or micro-seams of bentonite, or to structural reinforcement supplied by natural cements or by zones of recrystallization. Size, type and number

TABLE II

Typical shear strength parameters of fine-grained, soft, sedimentary rocks of Alberta (from Locker, 1969)

Material	Peak		Residual	
	ϕ' (degrees)	c' (kN/m^2)	ϕ'_r (degrees)	c'_r (kN/m^2) *
Siltstones	40	420	25	70
Clayey siltstones	35	140	20	385
Bentonite	14	42	8	0

* Based on intercept from tests at high normal pressures.

of irregularities on the failure surface vary from sample to sample of the same material and this contributes to strength variations.

In a general manner, the clay content of the soft rocks studied by Locker was shown to be the most important single geologic factor affecting the peak angle of shearing resistance. The dominant clay mineral is montmorillonite, and the addition of a relatively small amount of this mineral caused a marked decrease in the peak friction angle of sandy materials. The distribution of the clay sizes within a sample is of consequence; instances were noted where concentrated seams of montmorillonite, irrespective of thickness, controlled the peak strength of a fine-grained rock. In this latter case, the peak angle had a low value even though the clay content of the whole sample may have been quite small.

The magnitude of the drop in shear resistance from peak to residual varied over the study area. Samples from the plains exhibited a relatively moderate strength decrease for materials strained beyond peak strength values; strength decrease continued as a function of displacement until the residual value was obtained. Samples from near the foothills area exhibited more brittle behavior in that the strength value decreases to the residual value almost immediately after the peak strength has been exceeded.

Sinclair and Brooker (1967) carried out a series of triaxial and direct shear tests on undisturbed samples of fine-grained bentonitic claystones from strata of the Upper Cretaceous Edmonton Formation; these samples were collected in connection with a slope stability study. The average clay content of these rocks was about 30%; the dominant mineral was montmorillonite. Cell pressures were applied in the range of 0.1—3.5 MN/m^2. Up to about 1 kN/m^2 confining pressure, the Mohr envelope for a claystone was very nearly a straight line exhibiting a peak effective angle of shearing resistance of 20°. At higher confining pressures the slope of the line decreased to a peak angle of 8.5°. The residual angle of shearing resistance of the montmorillonite-rich samples was 8.5°, as deduced from direct shear tests.

REGIONAL LANDSLIDE STUDIES

The present drainage of the Interior Plain was formed during the retreat phases of continental glaciation some 25,000—12,000 years ago (Gravenor and Bayrock, 1961; Barton et al., 1964). The large volumes of meltwater rapidly eroded deep, steep-sided channels which were frequently ice marginal or glacial lake spillways. Once entrenched in these postglacial channels, the rivers could not return to their preglacial valleys when the Pleistocene ice disappeared. Many of these postglacial channels and some preglacial channels are presently occupied by small streams, referred to as misfit or underfit streams. Some sections of the modern rivers have re-excavated their preglacial channels. The geologically very rapid erosion of the deep, steep-walled postglacial channels was accompanied by extensive landsliding. The present-day rivers are widening their valleys by lateral erosion and meander development. During the last 6000 years, vertical erosion or valley deepening appears to have been slight (Westgate et al., 1969). This lateral erosion was responsible for the initiation of new landslides or re-initiation of old slide masses, both largely by the mechanism of toe erosion.

The rapid erosion of postglacial channels through the thin mantle of Pleistocene deposits and into the sedimentary bedrock of the Interior Plain removed a considerable load from strata underlying the valley floor. Load removal was accompanied by a rebound that gave rise to an upwarping of the strata in the valley walls and a gentle anticlinal structure beneath the valley floor (Matheson, 1972; Matheson and Thomson, 1973). The up-warping of the strata comprising the valley walls manifests itself as a raised valley rim with gullies and tributary streams that parallel the main river for some distance before joining it. Table III and Fig. 2 illustrate these phenomena.

Of importance in landslide studies is the interbed slip that accompanies the upward flexing of the strata in the valley wall. This interbed movement creates gouge zones of softened material and provides enough deformation to reduce the angle of shearing resistance from peak to some lesser value which may approach the residual angle. The upward tilt of the beds at their outcrop on the valley wall is a matter of a metre or so; hence the beds remain essentially flat-lying but have associated with them a thin weakened zone. This thin zone is often difficult to find during a subsurface exploration program, but if present can exert an obvious influence on valley wall stability. Since the flexure that the particular bed undergoes decreases inward from the face of the valley wall, so must the deformation caused by interbed slip. The width of this zone of influence is not known, but it is suggested that the width of the slide mass perpendicular to the valley may be controlled by this weakening effect. The thin, essentially horizontal weak zone may also be a likely factor in the explanation of the fact that the majority of the landslides are block movements and not rotational slumps.

TABLE III

Compilation of rim survey results (from Matheson, 1972)

River	Bedrock	Overburden		Valley characteristics			Rim height (m)
		type	thickness (m)	width (m)	depth (m)	slopes	
1. Pembina	Paskapoo and Edmonton Formations	till silts	6	330	45	2 : 1	2
2. Peace River		silts till	18	4800	150	10 : 1	4.5
3. Tributary to Peace				1500	150	4 : 1	0.6
4. Little Smokey	Puskwaskau Formation	till	60	900	90	6 : 1	1
5. Claypit 14	Oldman Formation	till	6	180	21	0.5 : 1	0.4
6. South Saskatchewan (4) (5) (3) (1) (2)	Oldman Formation	till	6	1250 1250 1700 1250 1400	97 97 90 60 60	3 : 1	2.7 6.3 1.5 4 5
7. Cypress Hills	Ravenscrag Frenchman Whitemud	Tertiary gravels	15	1800	180		0.3
8. Michichi Creek	Edmonton Formation	till	9	400	75	1 : 1	1.2
9. University of Alberta trenches		till? (6 m) sand below		5	10	0.25 : 1	0.012
10. Athabasca-Oldman	Paskapoo Formation	silts till	15	700	120	1 : 1	2.5
11. Missouri 1	Pierre Shale			5000	110		12

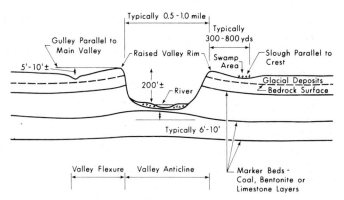

Fig. 2. Sketch illustrating the features resulting from valley rebound (after Matheson and Thomson, 1973).

Over the three-year period from 1970 to 1972, inclusive, Matheson (1970), Roggensack (1971), and Arvidson (1972) undertook air photo analyses of landslides along major portions of five selected rivers in central and southern Alberta. Landslides and slide areas were identified and areas of each were assessed from air photos. The resultant data were tabulated and plotted on topographic maps. Data from groundwater maps, bedrock topographic maps and geologic maps were superimposed on the landslide map. The rivers that were studied all were incised into the Tertiary and Upper Cretaceous bedrock. Parts of these river valleys were postglacial channels, whereas other sections were flowing in re-excavated preglacial channels. Conclusions arising from this study, which covered the entire southern half of Alberta, are outlined in the following paragraphs.

The stratigraphy and lithology of the bedrock formations forming the valley wall were a major factor in the number and areal extent of landslides along the valley walls (Fig. 3). Landslide activity was more pronounced within the marine Bearpaw Formation and the brackish to freshwater Edmonton

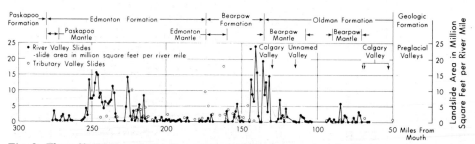

Fig. 3. The effect of geologic formation and buried preglacial valleys on the landslides along the Red Deer River, Alberta (after Arvidson, 1972).

Group than in the dominantly clastic sediments of the Tertiary (Paskapoo) and deltaic members of some Upper Cretaceous formations (Oldman).

However, the presence of strata prone to instability in the valley wall apparently is not in itself sufficient cause for landslides. The presence of groundwater exerts considerable influence on valley wall stability, both in bringing about new landslides and in maintaining slide activity or reactivating very old slide masses. In a rather indirect way, the overlying Pleistocene deposits influence groundwater availability to the bedrock. Topographic highs of the relatively impermeable glacial lake sediments tend to divert water, whereas surface depressions direct water into a given area. The presence of preglacial valleys, the large majority of which are floored by free draining clean sands and gravels, act as a deep, subsurface drain which effectively lowers the water table in its immediate vicinity. The result of this deep drain is that sections of a river valley otherwise characterized by frequent slides remain stable despite no change in lithology of the bedrock. The influence of these buried valleys is shown in Fig. 3. Recognition of this situation in the field is complicated by the fact that the preglacial channels very often have no surface expression, their presence being completely masked by Pleistocene deposits.

River erosion and deposition were also noted to be major factors, though their influence is universal and by no means unique to this study. At and immediately downstream of the point where meander bends impinge upon the base of the valley walls, landslides are active. Toe erosion removes load which allows lateral rebound and, in combination with the effects of interbed slip, enhance landslide activity. On the other hand terrace deposits provide a toe load and bring about a kind of stability, although of a somewhat precarious nature. A river that occupies a rather stable reach need only remove the toe debris of a landslide to reactivate that slide. The scarps of these landslides were observed to move farther into the top of the valley wall in a retrogressive manner.

These slide surveys over large areas indicated that the bedrock type was also apparent from the type of landslide movement, though a given type of landslide configuration was not unique to a particular formation. The continental Paskapoo Formation contained only minor rockfalls. Landslides in the Edmonton, Foremost and Oldman Formations were mainly translational or massive block movements, often with rotational components in upper slide areas associated with retrogressive action. Slides in the Bearpaw Formation are translational in some instances, but also contain surface viscous flow features where groundwater supplies are adequate.

Scott and Brooker (1968) reported a survey conducted on eight landslides in five areas, ranging geographically from western Manitoba to central Alberta. They discussed the stress history of the Bearpaw Formation and point out that as a result of the high overburden stresses imposed throughout the geologic history of the Bearpaw Formation, the value of the coefficient of earth pressure at rest is likely to be quite high. Peterson (1956) reported lateral stresses in the Bearpaw shale that were 150% of the vertical stresses imposed by the existing overburden. Stress relief, for example, due to rapid postglacial valley erosion, allows water uptake by the clay, which in turn results in a decrease in shear strength. This factor appears to contribute strongly to softening of the clay shale in the zone near surface.

In a large part of central Saskatchewan, the South Saskatchewan River is entrenched in the Bearpaw Formation. Extensive landslides have occurred and are considered to be the major cause of valley widening. The overall slopes of valley walls in the slide areas are in the range of 14—18°. Slide topography extends for many kilometres along the river.

Along a part of the St. Mary River in south-central Alberta, slope failures are not common and are confined primarily to steep undercut slopes. Adjacent valley slopes rise 30—60 m at inclinations of 30—40° with the horizontal. The Bearpaw Formation forms the major part of the valley walls and has a similar lithology to the outcrops in central Saskatchewan.

Along the valley of the Red Deer River, the Bearpaw Formation crops out for many kilometres along its southeasterly course in southern Alberta. However, in comparison to the St. Mary and South Saskatchewan River valleys, the Red Deer River valley is relatively lacking in landslide development although the general lithology seems similar to the other sites. In their report, Scott and Brooker (1968) sought geologic explanations for the different degrees of slide activity at these three sites; at each site, rock of similar lithology exists under similar climatic and apparently similar groundwater conditions, and has experienced similar loading histories. The Red Deer River valley contains extensive terraces in the area studied, which suggests that the downcutting in postglacial time took place in stages. This extensive terracing produced an overall flatter, and thus more stable, slope. In terms of valley rebound, as suggested by Matheson (1972), the Red Deer River terraces may distribute the interbed slip vertically throughout the valley wall such that the part of a flat-lying bed that was subjected to interbed slip underlies a terrace. The height of the terrace is low enough that local instability does not result. The St. Mary River site is the most westerly of the three sites, and thus closest to the source of sediments for Paleocene deposition. The overburden loads were probably somewhat higher here than farther east and Young's modulus is somewhat higher. Higher moduli result in less valley rebound and, in turn, less interbed slip. Therefore, though the lithology is similar to the other sites, the rock may be somewhat stronger. On the South Saskatchewan River, excavation was rapid and Young's modulus low. These combine to give rise to maximum upwarping of beds along the valley wall, maximum interbed slip and associated strength decrease along the slip zone. Augmented by the softening effect, landslides are prolific.

In summary, the prime geological factors affecting the engineering behavior of the highly overconsolidated shales and soft rocks are depositional environment, lithology and stratigraphy, stress history, structure, climate, geomorphology, and groundwater. For a two-dimensional analysis, the geometry of the failure surface appeared to be best approximated by a series of straight lines, the lowest of which often coincided with a weak layer such as a bentonite seam. These zones of weakness are often thin; unless particular care is taken, their presence can be missed in a subsurface drilling program.

In the following, some specific case histories of landslides that occurred in the flat-lying soft sedimentary rocks of the Interior Plain are discussed.

SPECIFIC CASE STUDIES

Analysis of older landslides

South Saskatchewan River Dam. Ringheim (1964), in a study of the Bearpaw shale on which the South Saskatchewan River Dam is founded, was able to classify the shale in the field into three consistency zones. The *upper soft zone* was disturbed and softened to a high degree by weathering and swelling. The natural moisture content of this zone was assigned values in excess of 29%. The *transition zone* between the upper soft layer and the deep, hard, or intact, shale had natural moisture contents between 25 and 31%; and the *hard zone* was characterized by moisture contents less than 27%. These divisions contributed to a better understanding of the physical variations within the formation and such mechanical defects as joints and slickensides.

During the twelve-year period prior to construction, extensive field investigations were carried out in the vicinity of the dam site. The geotechnical properties of the clay shale were relatively unknown because the proposed dam was the first major structure in the region on the Bearpaw Formation. Subsurface exploration consisted of churn drill holes and drive samples, 15-cm diameter core samples recovered by rotary drilling, and deep test pits 1.3 m in diameter from which block samples were extracted. A horizontal test drift 2.0 m wide by 2.5 m high was driven into the left bank 110 m. A pressure test section indicated that the horizontal pressure was 150% of the vertical pressure due to 20 m of overburden. Considerable lateral rebound of the clay shale was noted to have occurred.

From preliminary field observations and laboratory testing prior to construction, it became apparent that a large difference existed between field and laboratory shear strengths of the soft shale. In terms of effective stresses, the latter values were $c' = 40$ kN/m^2, $\phi' = 20°$. During construction, movements occurred as a result of construction activity which, without exception, extended through till or colluvium into the soft shale zone. The majority of these slides were the direct result of stress changes caused by excavation.

Though these slides were disturbing in many respects, they provided an excellent opportunity for "back analysis", from which field shear strengths of the soft shale zone could be determined. The field parameters were $c' = 0$, $\phi' = 9°$. The results of these analyses led to modifications in slope designs as the project progressed.

A large number of porous tip piezometers were installed in the shale, and observations indicated that the pore water was carrying a large percentage of the embankment load soon after its placement. Dissipation rates were very slow in the medium and hard shale zones and little relief was obtained during normal construction period. Some evidence was available that pore pressures in the soft shale dissipated sufficiently to achieve a measurable increase in strength.

Landslides in Edmonton, Alberta. Within the downtown area of the City of Edmonton, a large landslide was reported by Hardy (1957) and was re-analyzed by Pennell (1969). Recorded movements of this slide, known as the Grierson Hill slide, date back to 1887. In that year the average bank slope was about 22°. Following a series of very minor events, a major movement in 1905 resulted in a loss of about 23 m at the top of the bank. In the following decade the scarp retreated a further 17 m. From 1915 to the present only small movements have occurred in the top of the bank in the east area of the slide.

The precise reasons for the slide are not known with certainty, and the details of events were never recorded. However, several features are pertinent. The bentonitic clay shales, mudstones and coal strata of the Edmonton Formation comprise about three quarters of the material in the valley wall. The upper part is a capping of till and lake sediments of late Pleistocene age. Coal seams within the Edmonton Formation in the slide area were being actively mined prior to the turn of the century. Old timers reported a series of springs on the valley wall at the contact between the till and the clay shales. During a ten- to twelve-year period prior to 1900, the river was actively eroding the toe area and encroached into the bank about 30 m. In an attempt to arrest movement, the toe area was used as a dump for twenty years or so after 1900. Some 15 m of debris, which included brick rubble and clay fill, accumulated. This fill, following the 1905 movement, pushed the river out some 120 m and has resulted in an overall slope on the order of 12°.

Creep movements in recent years had to be arrested to sustain a roadway up the valley wall. A program of drainage was undertaken and a drainage gallery 1.5 m in diameter and a storm sewer were installed in 1960. Since that time only very minor movements have occurred as evidenced by observation on the paved road.

One might postulate that subsidence from the mining opened up cracks that allowed ingress of water to the slide mass, and that when combined

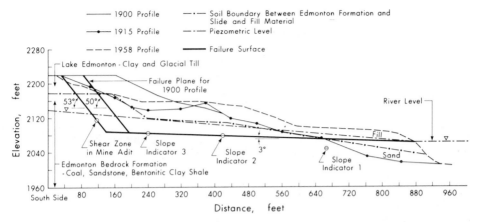

Fig. 4. The Grierson Hill landslide, Edmonton (after Pennell, 1969).

with toe erosion, the major movement occurred. Toe loading and drainage has essentially stopped movement, but the factor of safety is not much greater than unity. A stability analysis on the profile shown in Fig. 4, using peak parameters of $c' = 35$ kN/m^2 and $\phi' = 22°$ in the Pleistocene sediments and a residual angle of 10° in the clay shales along the lower part of the slip plane, yielded a factor of safety of 0.98 for the 1905 profile. A similar analysis of the present profile, using residual strength parameters throughout, resulted in a factor of safety of 1.06.

The area along the top of the south valley wall of the North Saskatchewan River, which forms the northern boundary of the University of Alberta campus, was considered for building sites during the expansion of university facilities. The river bank had a history of being unstable, as noted above, and it was necessary to establish the stability of the river bank within the campus prior to the design of any structures near or on the valley wall. The results of these studies have been reported by Thomson (1970).

Clay shales, bentonite, coal and bentonitic sandstone strata of the Edmonton Formation form more than half of the valley wall which is 54 m high. Overlying materials are Pleistocene sands, till and lake sediments. From geomorphological evidence augmented by old survey data, it is apparent that a terrace some 6 m above river level existed across the entire study area. On the western end, erosion has largely destroyed the terrace, and only vestiges remain. It is a situation geomorphically similar to the Lesueur landslide discussed in a subsequent paragraph. A large, buried preglacial channel underlies the northwestern side of the campus. This old channel is floored with a considerable thickness of clean, free draining sands but does not have surface expression. The geotechnical properties of the soils used in the analyses are listed in Table IV.

TABLE IV

Geotechnical properties of riverbank stability study (from Thomson, 1970).

Soil properties	Clay shale	Bentonitic shale	Bentonite	Bentonite, sandy-silty
Natural moisture content (%)	10—20	25—45	50—70	10—20
Liquid limit (%)	40—70	125—220	180—260	90—120
Plastic limit (%)	20--25	30—45	60	25—30
<2 μm (%)	30	37—52	92	20—25
Montmorillonite (%)		80—90	100	90
Shear strength parameters				
Peak c' (kN/m^2)	57.5	43		
ϕ' (degrees)	14	14		
Residual ϕ'_r (degrees)	12	8		

On the east side of the study area (Fig. 5), the valley slopes at about 10° and exhibits extensive though subdued slump topography. An "infinite slope" analysis was used to assess the stability of the slide area, which was thus shown to be marginally safe. On the west side (Fig. 6), the valley wall sloped directly to the river at an angle close to 30°. Analyses at the west end revealed that strength parameters near peak values were being mobilized. In view of other case histories this was a cause for some concern and dictated that any further river erosion along the toe area be arrested.

As a result of this intensive study it was recommended that (1) any future construction be set back 45 m from the valley wall, (2) armoring of the river banks be undertaken, (3) no encroachment on the present valley walls be permitted without the advice of a qualified geotechnical engineer, and (4) piezometer and tiltmeter observations be continued.

Little Smoky slide. Hayley (1968) investigated a large landslide in northern Alberta which has been affecting the foundations of a bridge across the

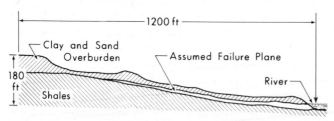

Fig. 5. East side of study area, riverbank stability study, Edmonton (after Thomson, 1970).

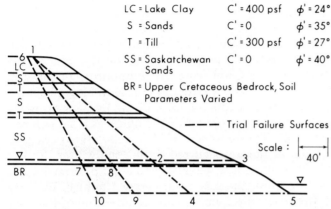

Fig. 6. West side of study area, riverbank stability study, Edmonton (after Thomson, 1970).

Little Smoky River. In the reach of the river involved in this study, the valley is preglacial. It was infilled by Pleistocene till deposits which the present river is re-excavating. An intensive drilling program was undertaken and ten tiltmeters in addition to three piezometers were installed. Direct shear tests to determine the residual angle of shearing resistance were carried out on undisturbed samples obtained by means of a Pitcher sampler (Terzaghi and Peck, 1967, pp. 310—312).

The location of the failure plane was deduced from the tiltmeter surveys and was found to consist of a downward projection of the scarp face and an essentially horizontal shear plane approximately parallel to bedding through the bedrock (Puskwaskau Formation), an Upper Cretaceous soft grey shale of marine origin. The residual angle of shearing resistance was found to be 14°; the soil had a liquid limit of 55%, plasticity index of 24%, and a natural moisture content of 21%. The water table was found to be well above the till-bedrock contact and to exhibit only minor variations over the summer period.

Of interest in this analysis is the finding that, once the general shear zone is located by normal tiltmeter readings at fairly large intervals of depth, it is necessary to take subsequent readings at intervals as small as 15 cm. This spacing is necessary to define the shear zone with adequate clarity. Although one might expect shear to occur at the till-shale interface, the shear zone was about a metre into the clay shale. The analysis of the failed slope indicated a sequential failure, from the riverbank to the scarp, with the residual angles of shearing resistance mobilized throughout along the failure zone. Movements of the slide continue to the present day; they are taken up by a system of rollers between the bridge superstructure and the piers.

Analysis of more recent landslides

Lesueur slide. A "first-time" slope failure in some circumstances can be initiated by relatively little disturbance. This appeared to be the case at the Lesueur landslide located some 6.5 km east of Edmonton (Thomson, 1971a). An outline map of the slide is presented in Fig. 7. In May of 1963 a narrow but well-defined crack arced across the lawn of the Lesueur property and seemed to stop about 3 m from the corner of the house as it sat perched on the top of the valley wall. The presence of this well-defined crack in early May suggested that initial failure occurred during the winter. Rather slowly over the summer period the crack widened and a scarp formed until, by midmorning September 3, 1963, the scarp was nearly 1 m high. By early evening of the same day the scarp was 2 m high and by morning of the next day, it was over 7 m high. The scarp was some 50 m wide but the slide fanned out to nearly 300 m in the toe area (Fig. 7).

At the time of the slide, the lawn was 31 m above river level and the immediate river bank about 3 m above water level. During the failure, the material at the toe was observed to move out over a practically horizontal plane and, after a small amount of cantilevering, chunks dropped off into the river. The next day the toe area was a jumble of blocks and fallen trees that completely masked the dominantly planar movement.

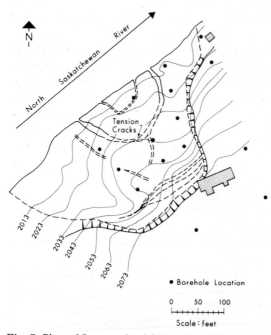

Fig. 7. Plan of Lesueur landslide (after Thomson, 1971a).

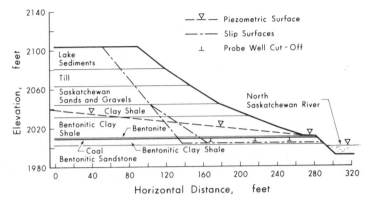

Fig. 8. Stratigraphic section of the Lesueur landslide (after Thomson, 1971a).

The subsurface stratigraphy and the groundwater configuration were determined by drill holes and piezometer installations in work carried out over the succeeding years. Pore water pressures were determined using a transducer piezometer (Brooker et al., 1968) and samples of the bedrock materials were successfully obtained using a Pitcher sampler. The stratigraphy and water table are shown in Fig. 8. Strength parameters were determined on representative samples obtained from the slide mass and are given in Table V.

The failure surface was defined, in large part, by the scarp face and closed-off probe wells. Analyses in terms of effective stresses were carried out using the method of Morgenstern and Price (1965). Preliminary analyses suggested that the strength parameters being mobilized at incipient failure were greater than residual on average, less than peak on at least some portion of the sliding surface; some cohesion must have been acting. It was reasoned that the original ground surface was an old slide and since the present slide scarp is some 10 m back of the crest of the slope (the old scarp), peak strength parameters were being mobilized in the soils lying above the Upper Cretaceous clay shales. The horizontal portion of the failure plane was mobilizing strength values less than peak. Subsequent analyses, in which a strength decrease was assumed along the horizontal shear plane toward the river bank, gave a factor of safety of unity. This progressive decrease in strength from within the mass to its outcrop on the valley wall seems consistent with valley rebound as discussed by Matheson (1972), though an explanation of why the slide occurred at that particular time remains to be given.

Westgate et al. (1969) showed that a terrace some 8 m above present river level is in the order of 6500 years old. In the past 6500 years the river has been widening its valley, rather than downcutting; that is, it has been alternately building and destroying terraces. In the past century or less the river has been moving laterally toward the toe of the Lesueur slide and destroying

TABLE V

Geotechnical properties of Lesueur slide (from Thomson, 1971a)

Soil properties	Clay shale	Bentonite shale	Bentonite	Coal
Liquid limit (%)	100	200		
Plasticity index (%)	75	170		
Sand (%)	20	15		
Silt (%)	35	30		
Clay (%)	45	55		
X-ray diffraction				
Montmorillonite (%)	80—100		100	
Illite (%)	10—0			
Kaolinite and chlorite (%)	10—0			
Shear strength parameters				
Peak c' (kN/m^2)	65	43	40	0
ϕ' (degrees)	24	14	14	32
Residual ϕ'_r (degrees)	17	10	8	32
γ Total (ton/m^3)	1.82	1.79	1.70	1.39

a fairly low but stabilizing terrace. By 1963 only vestiges of this terrace were visible, and some otherwise minor rise in the groundwater table might have been sufficient to provide the final trigger. Such a minor rise could be caused by groundwater build-up due to freezing of the ground surface at lower elevations. It is of interest to note that prevention of erosion of the low terrace would likely have prevented a major landslide. Although the seat of the landslide was in soft bedrock, its behavior and an understanding of the mechanism of the slide movement requires a consideration of the geomorphic processes of erosion in the area.

In March 1971, a stadia profile of the failed slope was obtained for a re-analysis. In the seven and a half years since the original slide, continuous but slow movements increased the scarp height to 10 m from the earlier value of 7 m. For the purpose of this analysis, it was assumed that stratigraphy, piezometric levels, and strength parameters were essentially unchanged from those used in the previous analysis. Results of this re-analysis suggests that the cohesion everywhere had decreased toward zero and that the angle of shearing resistance along the horizontal part of the slip plane had decreased toward a residual value. In considering the results of this re-analysis, the following comments are pertinent (Thomson, 1971b).

A shear zone parallel to bedding is likely to be thin since it is following an inherent line of weakness, whereas a shear zone oblique to bedding is likely to be somewhat jagged and therefore thicker. In the former case, relatively little deformation would bring about a major realignment of the clay particles, hence the residual angle of shearing resistance. The thicker shear zone oblique to bedding would require much larger deformations to smooth out the shear plane before the residual angle of shearing resistance could be approached. Downward displacement of the slide block may have resulted in two unlike soils being in contact, hence a residual angle of shearing resistance would be difficult to assess.

Devon slide. Eigenbrod and Morgenstern (1972) conducted a detailed examination of a slide in the Edmonton Formation which formed the backslope of a road cut. The slide is located near the town of Devon, about 20 km south and west of Edmonton, Alberta; the site is located on the North Saskatchewan River in a reach where the present river is flowing in a post-glacial channel. At this site, the bedrock consists of slightly indurated, interbedded claystone, mudstone, siltstone, and sandstone that have a regional dip of 3.7 m/km to the southwest. Coal seams a few centimetres to a metre thick and bentonite layers ranging from 2 cm to about 30 cm in thickness occur in the sequence. The upper portion of the bedrock exhibits distortions due to ice shove.

During the summer of 1965 rebuilding of the road down the valley wall required easing of the grade. At the slide site the backslope was trimmed and cut to a slope of 20° to the horizontal. From the ditch invert to the top of the slope, the vertical difference in elevation is about 20 m. In the fall of 1965, cracks appeared in the slope about 12 m vertically above the ditch, and a landslide rapidly developed. A detailed surface examination was made at that time, but subsurface investigations, consisting of drill holes and test pits, were not initiated until 1969.

The slip surface was located in the test pits in the toe area in a bentonite seam at its contact with an underlying coal seam. Samples were obtained for subsequent testing. The stratigraphy of the slide mass was determined from a series of boreholes augmented by the test pits and the outcrops on the slope. In addition to the bentonite layer containing the slip surface, another bentonite layer was found about 7.5 m above it. Both of these layers contained distinct, commonly oxidized, failure planes. All drill holes were outside of, but close to, the failed slide mass. The bentonite layers outside the slide zone were found to be sheared.

From piezometric observations, three distinct perched water tables were located, but in the coal underlying the lower bentonite layer the pore water pressure was, at most, less than 0.5 m of head.

From detailed consideration of the data and the site it became clear that the Devon slide, though a "first-time" slide, was actually located within a massive older slide block which presumably developed during valley formation, and whose scarp was partly obscured by erosion and by bush. A subsequent borehole in an area behind and not affected by the general slumping, revealed a harder, more intact bedrock.

The shear surface in the slide mass was clearly defined in the scarp area and ranged in dip from 55° to 60°. The shear surface was considered to consist of a downward projection of the scarp face to its intersection with an essentially horizontal projection of the failure surface visible in the toe area.

The residual parameters of the lower bentonite seam, which contained a part of the slip surface, were determined from laboratory tests to be $c' = 0$ and $\phi' = 8°$. The upper part of the shear plane in the scarp area passed through a highly fissured siltstone that varied in consistency; harder, drier material was contained in but less prevalent than, softened material of an otherwise similar nature. Triaxial tests were carried out on both materials. Results of laboratory tests on the softened materials were used in the stability analysis. Typical stress-strain curves for these materials are shown in Fig. 9.

The general non-circular limit equilibrium analysis described by Morgenstern and Price (1965) was used to analyze the section shown in Fig. 10. Zero water pressure was taken as acting along the shear plane in the lower bentonitic layer, and piezometric levels from the perched water tables were applied to the backslope portion of the slip surface. Using the strength properties shown in Table VI a factor of safety of 1.01 was determined. If c' of the backslope material was set equal to zero, the factor of safety dropped to 0.83. It would appear that some cohesion must have been acting oblique to bedding when the major movements occurred.

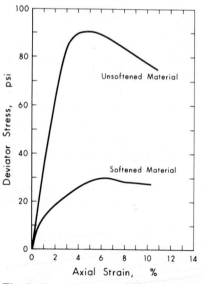

Fig. 9. Typical stress-strain curves for slide scarp material, Devon landslide (after Eigenbrod and Morgenstern, 1972).

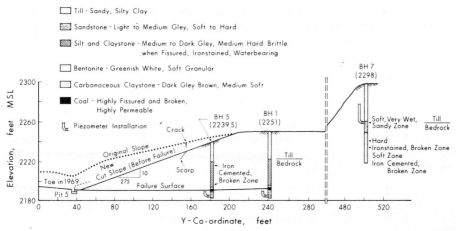

Fig. 10. Stratigraphic section of the Devon landslide, Alberta (after Eigenbrod and Morgenstern, 1972).

Of particular interest in this study were sheared surfaces in the bentonite beyond the slide mass, the effects of weathering in the siltstones and mudstones of the scarp area and the series of perched water tables. The latter demanded detailed study and a careful decision regarding the location of piezometer tips. The sheared surfaces in the bentonite may have arisen from one or a combination of the following processes: (1) bedding plane slip associated with valley rebound, (2) differential swelling of the bentonite constrained between adjacent non-swelling layers, (3) movements during the initial large slide, and (4) shearing associated with ice shoving of the bedrock during the Pleistocene.

The engineering significance of the sheared bentonite deserves emphasis even though its origin in this case may be ambiguous. Large-scale geomorphic

TABLE VI

Summary of index and strength properties (from Eigenbrod and Morgenstern, 1972)

Material	Liquid limit (%)	Plasticity index (%)	c' (kN/m^2)	ϕ' (degrees)	ϕ'_r (degrees)
Bentonite clay	100	60			8
Weathered backslope, clayey	85	65	9.5	33	
Weathered backslope, sandy-silty	45	24			
Unweathered backslope, sandy siltstone	43	23	185	19	

expression of locally sheared clay is often lacking and points out the special concern that minor geologic detail warrants, particularly in interbedded soft sedimentary rocks.

DISCUSSION

The geology and geomorphology of landslide areas is of major importance in understanding failure mechanisms and in the design of stabilizing works. Minor geologic details, such as the presence of thin bentonite seams or sheared zones resulting from old landslides, are difficult to detect either in drill holes or from geomorphic evidence; yet their influence on slope stability is profound. Non-homogeneous water pressure or perched water tables require careful field investigation and interpretation.

The soft, poorly indurated rocks of the Interior Plains are essentially flat-lying, but often comprise complex lithology due to lateral and vertical facies changes. Beds that are not flat-lying may reasonably be assumed not to be in place. Folding or brecciation to depths of 20 m may be locally present, due to ice shoving during the Pleistocene. The marine and brackish water deposits appear most prone to sliding.

During waning phases of glaciation, the erosion of incised valleys by meltwater caused many landslides. The topography of these old landslides has been subdued by subsequent erosion or masked by tree growth. Their stability, however, remains marginal, and remarkably little stress relief is required to reinitiate movement. Lateral river erosion or minor cut and fill associated with road building, for example, can precipitate a major slide. On the other hand, creation of a river terrace enhances stability although the overall factor of safety of the slope remains low. Valley uplift following rapid erosion may have caused interbed slip; such slip can be of sufficient magnitude to decrease shearing resistance parallel to bedding to low values. The subdued topography of old landslides and the valley lip formed by rebound may be revealed by a careful terrain analysis; both may be of geotechnical significance.

The presence of bentonite, either as a matrix constituent of clay shales and mudstones or as a thin pure stratum, results in low angles of shearing resistance. Except for the infinite slope analysis of long, low-angle slopes, a non-circular or wedge-shaped two-dimensional failure surface was found appropriate in stability studies. Observations of probe wells and tiltmeters provided strong evidence of a steeply dipping upper slide surface that typically merges rapidly with an essentially flat-lying lower slide surface. Many old slides have been analyzed, and there is abundant evidence from this region as well as from elsewhere that the residual strength is being mobilized along the slip surface. However, the composition of the slide material is generally not very uniform; hence the determination of the residual strength is not straightforward unless a dominant slip surface is identified.

In the case of first-time slides there is growing evidence that the most difficult factor to assess is the shearing resistance oblique to bedding. This strength value can range over wide limits; strong materials, if subjected to softening processes, can give rise to failure that is progressive in time.

Although much has been learned about the behavior of the soft sedimentary rocks of this region, they are still enigmatic; much remains to be investigated.

ACKNOWLEDGEMENTS

Most of the research from which this report has been drawn was carried out using funds from the National Research Council of Canada. Important contributions were also made by the Geological Survey of Canada, the Prairie Farm Rehabilitation Administration, and the Research Council of Alberta.

REFERENCES

Arvidson, W.D., 1972. *An Air Photo Study of Landslides along the Bow and Red Deer Rivers, Alberta.* Unpublished internal report, Dep. of Civil Engineering, Univ. of Alberta, Edmonton, Alta.

Barton, R.H., Christiansen, E.A., Kupsch, W.O., Mathews, W.H., Gravenor, C.P. and Bayrock, L.A., 1964, Quaternary. In: R.G. McCrossan and R.P. Glaister (Editors), *Geological History of Western Canada.* Alberta Society of Petroleum Geologists, Calgary, Alta., pp. 195—200.

Brooker, E.W., Scott, J.S., and Ali, P., 1968. A transducer piezometer for clay shales: *Can. Geotech. J.*, 5: 256—264.

Carrigy, M.A., 1970. Proposed revision of the boundaries of the Paskapoo Formation in the Alberta Plains. *Bull. Can. Pet. Geol.*, 18: 156—165.

Douglas, R.J.W. (Science Editor), 1970. *Geology and Economic Minerals of Canada.* Geological Survey of Canada, Ottawa, Ont., pp. 19—21.

Eigenbrod, K.D. and Morgenstern, N.R., 1972. A slide in Cretaceous bedrock, Devon, Alberta. In: C.O. Brawner and V. Milligan (Editors), *Geotechnical Practice for Stability in Open Pit Mining.* American Institution of Mining and Metallurgy, New York, N.Y., pp. 223—238.

Fleming, R.W., Spencer, G.S. and Banks, D.C., 1970. *Empirical Study of Behavior of Clay Shale Slopes. USAE NCG Tech. Rep.*, No. 15, 2 volumes, (Livermore, Calif.).

Gravenor, C.P. and Bayrock, L.A., 1961. Glacial deposits of Alberta. In: R.F. Legget (Editor), *Soils in Canada.* Univ. of Toronto Press, Toronto, Ont., pp. 33—50.

Irish, E.J.W., 1970. The Edmonton Group of south central Alberta. *Bull. Can. Pet. Geol.*, 18: 125—155.

Hardy, R.M., 1957. Engineering problems involving preconsolidated clay shales. *Trans. Eng. Inst. Can.*, 1: 5—14.

Hardy, R.M., 1963. The Peach River highway bridge — a failure in soft shales. *Highw. Res. Rec.*, 17: 29—39.

Hayley, D.W., 1968. *Progressive Failure of a Clay Shale Slope in Northern Alberta.* Unpublished M.Sc. Thesis, Dep. of Civil Engineering, Univ. of Alberta, Edmonton, Alta., 61 pp.

Locker, J.G., 1969. *The Petrographic and Engineering Properties of Fine-Grained Sedimentary Rocks of Central Alberta.* Unpublished Ph.D. Thesis, Dep. of Civil Engineering, Univ. of Alberta, Edmonton, Alta., 285 pp.

Locker, J.G., 1973. Petrographic and engineering properties of fine-grained rocks of Central Alberta. *Res. Counc. Alta. Bull.,* 30, 144 pp.

Matheson, D.S., 1970. *An Air Photo Study of Landslides along the Pembina River, Alberta.* Unpublished internal report, Dep. of Civil Engineering, Univ. of Alberta, Edmonton, Alta.

Matheson, D.S., 1972. *Geotechnical Implications of Valley Rebound.* Unpublished Ph.D. Thesis, Univ. of Alberta, Edmonton, Alta., 424 pp.

Matheson, D.S. and Thomson, S., 1973. Geological implications of valley rebound. *Can. J. Earth Sci.,* 10: 961—978.

Morgenstern, N.R. and Price, V.E., 1965. The analysis of the stability of general slip surfaces. *Geotechnique,* 15: 79—93.

Pennell, D.G., 1969. *Residual Strength Analysis of Five Landslides.* Unpublished Ph.D. Thesis, Dep. of Civil Engineering, Univ. of Alberta, Edmonton, Alta., 166 pp.

Peterson, R., 1954. Studies of Bearpaw shale at a damsite in Saskatchewan. *Proc. Am. Soc. Civ. Eng., J. Soil Mech. Found. Div.,* 80 (Sep. No. 476), 28 pp.

Peterson, R., 1956. Rebound in the Bearpaw shale, western Canada. *Geol. Soc. Am. Bull.,* 69: 1113—1124.

Ringheim, A.S., 1964. Experiences with the Bearpaw shale at the South Saskatchewan River Dam. *8th Int. Congr. on Large Dams, Edinburgh,* 1: 529—550.

Roggensack, W.D., 1971. *An Air Photo Study of Landslides along the Old Man and South Saskatchewan Rivers, Alberta.* Unpublished internal report, Dep. of Civil Engineering, Univ. of Alberta, Edmonton, Alta.

Rutherford, R.L., 1928. Geology of the area between North Saskatchewan and McLeod Rivers, Alberta. *Res. Counc. Alta. Rep.,* No. 19, 7 pp.

Scott, J.S. and Brooker, E.W., 1968. Geological and engineering aspects of Upper Cretaceous shales in western Canada. *Geol. Surv. Can. Paper,* 66-37, 75 pp.

Shepheard, W.W. and Hills, L.V., 1970. Depositional environments Bearpaw-Horseshoe Canyon (Upper Cretaceous) transition zone Drumheller "Badlands", Alberta. *Bull. Can. Pet. Geol.,* 18: 166—215.

Sinclair, S.R. and Brooker, E.W., 1967. The shear strength of Edmonton shale. *Proc. Geotech. Conf. Oslo,* 1: 295—299.

Taylor, R.S., Mathews, W.H. and Kupsch, W.O., 1964. Tertiary. In: R.G. McCrossan and R.P. Glaister (Editors), *Geological History of Western Canada.* Alberta Society of Petroleum Geologists, Calgary, Alta., pp. 190—194.

Terzaghi, K. and Peck, R.B., 1967. *Soil Mechanics in Engineering Practice.* Wiley, New York, N.Y., 2nd ed., 729 pp.

Thomson, S., 1970. Riverbank stability study at the University of Alberta, Edmonton. *Can. Geotech. J.,* 7: 157—168.

Thomson, S., 1971a. The Lesueur Landslide, a failure in Upper Cretaceous clay shale. *Proc. 9th Conf. on Engineering Geology and Soil Mechanics, Boise, Idaho,* pp. 257—287.

Thomson, S., 1971b. Analysis of a failed slope. *Can. Geotech. J.,* 8: 596—599.

Westgate, J.A., Smith, D.G.W. and Nichols, H., 1969. Late Quaternary pyroclastic layers in the Edmonton area, Alberta. *Proc. Symp. on Pedology and Quaternary Research.* Univ. of Alberta, Edmonton, Alta., pp. 31—55.

Williams, G.D. and Burke, C.F., 1964. Upper Cretaceous. In: R.G. McCrossan and R.P. Glaister (Editors), *Geological History of Western Canada.* Alberta Society of Petroleum Geologists, Calgary, Alta., pp. 169—189.

Chapter 15

ROCK SLOPE FAILURE AT HELL'S GATE, BRITISH COLUMBIA, CANADA

D.R. PITEAU, B.C. McLEOD, D.R. PARKES and J.K. LOU

ABSTRACT

During construction of the Trans-Canada Highway at Hell's Gate in the Fraser Canyon in British Columbia, extensive cracks and related slope movements were discovered upslope of the highway rock cut. The kinematics of failure involved outwards overturning of steeply dipping tabular fault blocks which were oriented parallel to the rock cut. Movement was cyclic, being a direct function of precipitation, and was retarded or accelerated depending on the presence or absence of low and/or high temperatures and low and/or high snowfall. Analyses of kinematically possible failure modes for the conditions involving variation of both the static head of water and excavation geometry indicated that an acceptable factor of safety against failure could be achieved by prevention of water pressure build-up, excavation and artificial support.

INTRODUCTION

During construction in 1964 of the Trans-Canada Highway in the Fraser Canyon, British Columbia, extensive cracks were discovered upslope of the highway rock cut in the area of the Hell's Gate Bluffs (Fig. 1). These cracks developed along a set of steeply dipping faults which were essentially parallel to the rock cut. Closer inspection of the area, which was largely covered with talus, revealed that one of these cracks, located along a feature referred to as the "Old Main Fault", was located some 90 m above the highway; the crack was about 100 m in length, with widths exceeding 1 m. In places, the cracks are more than 10 m in depth.

At the time it was not apparent whether the cracking resulted from slope readjustments due to the excavation or was due to natural movements which had taken place before excavation had commenced. Further highway excavation work was suspended in this area and in 1965-66 investigations and remedial measures were carried out. The remedial work consisted of rock scaling and removal of talus material up to 6 m depth, over 8500 m² of sur-

Fig. 1. Map showing the location of Hell's Gate.

face, and up to about 70 m above the highway. Overhangs were trimmed, and about 1800 rock bolts and 121 grouted 70-tonne anchors also were installed. Careful monitoring of the slope, however, indicated that movements continued to take place above the location where the scaling and other remedial work terminated. Further studies were conducted to assess the nature and extent of these movements with the objective that additional remedial measures could be devised.

The purpose of this paper is to describe the nature and results of these further studies. The kinematics of failure, which according to field evidence appears to involve toppling or overturning of fault blocks, are described. The significance of precipitation and other climatic factors which either enhance or detract from the stability of the slope are assessed, and comparative analyses of possible remedial solutions involving both preventing groundwater

Fig. 2. Oblique aerial view of the Hell's Gate Bluffs area showing the cut bench; upslope of this bench overturning movements have been recorded.

build-up and excavation are discussed.

The general physical setting of the Hell's Gate Bluffs area in question is indicated in Fig. 2. A close-up aerial oblique view of the cut bench and surrounding area is given in Fig. 3.

GENERAL GEOLOGY

The Fraser Canyon is the site of a pronounced zone of faulting. The Fraser River fault zone, and accordingly, a good portion of the river, follows

Fig. 3. Oblique aerial close-up view of the cut bench, upslope of which overturning movements have been recorded.

a geological contact zone between the Cascade and Coast Mountains. Rocks occurring within the range of influence of the Fraser River fault zone have been crushed, shattered and/or display intense faulting and jointing. The rocks at Hell's Gate Bluffs are no exception.

The badly broken nature of rocks in this area has resulted in a general weakening of the canyon wall rocks, and the area has been the site of concentrated glacial scouring and of excessive river downcutting before, during, and after glaciation. River erosion appears to have particularly concentrated in the weaker fault zones.

Regional uplift and elastic rebound of the land surface in postglacial times appear to have accentuated the downcutting action (Monger, 1969). In general this entire downcutting process appears to have taken place at a rate considerably faster than that required for adjustments in adjacent slopes. The result is that numerous slopes of varying degrees of instability exist; these are, geologically speaking, in an active state of failure. Postglacial landslides are common features; a major postglacial rockslide, for example, can be seen in Fig. 2 immediately to the north of the area discussed in this paper.

Lithology

The rock in the area is basically granodiorite, part of a 4.5-km-wide semicircular intrusive plug. A few narrow aplite and basic dykes are evident, but these are of little engineering significance. The rock is reasonably uniform in composition, consisting essentially of plagioclase feldspar and quartz with minor biotite and hornblende. The engineering properties of granodiorite have close affinities to that of granite. The intact, unweathered rock is tough and cohesive, and is expected to have a high compressive strength and high shear strength.

Structure

A well-developed set of northeasterly striking faults, which dip steeply either southeasterly or northwesterly (particularly the former), are abundant in the area. These faults are approximately parallel to the highway cut and accordingly they control the overturning failure. Fig. 4 shows the distribution of these faults in plan. Figs. 5—7 clearly show the nature of these pervasive fault structures. The faults are spaced approximately 3—12 m apart; the width of this whole zone of concentrated faulting is about 100 m. Individual fault zones seldom exceed 0.5 m in width and consist essentially of fine to coarse brecciated material. Plastic, clay-like alteration products are evident in some of these faults.

Several faults have been observed which strike approximately north-northeast to northeast and dip about 45° towards the highway. These faults are not as well developed as the fault set described above. They also display a more variable attitude and spatial distribution. Individual fault zones seldom exceed 10 cm in width. Two faults of this group, which can be seen cutting across the steeply dipping north-east-striking set of faults beneath the cut bench at highway level, are shown in Fig. 8.

Compared to fault structure, the commonly occurring tectonic joint appears to play a relatively less significant role in the overall stability of the Hell's Gate Bluffs. A large percentage of the joints in the immediate studied area, however, have developed sympathetic to faulting. The result has been that individually partitioned solid rock blocks are considerably smaller where

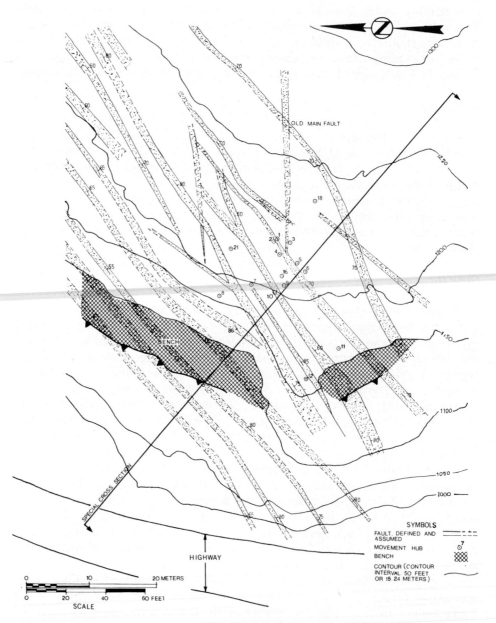

Fig. 4. Contour plan of the site showing the distribution and attitude of the major faults and the movement hubs in the area (refer to Fig. 17 for cross-section).

Fig. 5. Close-up oblique aerial view looking along the strike of the steeply inward-dipping to vertical well-developed set of faults which define the overturning tabular blocks.

faults are more closely spaced. In these highly jointed areas, weathering appears to be further advanced. These joints, presumably fault-induced, have developed mainly sub-parallel or at relatively small dihedral angles to the faults.

ASSESSMENT OF CLIMATIC CONDITIONS ON STABILITY

It was recognized that movement of the fault-controlled blocks above the cut bench on this slope at Hell's Gate Bluffs greatly accelerates in the fall

Fig. 6. View looking northerly from location immediately south of the cut bench; note the intensity of the steeply dipping northeasterly striking fault set and the associated breakage between faults.

and winter months. Attempts were made therefore to ascertain the relationship between the various climatic factors and movement. Climatic records taken at the Meteorological Station at Hell's Gate (about 210 m below the highway) were obtained from the Department of Transport. These climatic data were plotted against movement data recorded both above and below the cut bench. Movement information was obtained by plate tiltmeters, precise tape measurements between hubs, and triangulation surveys; measurements were generally taken on a two-week to monthly basis.

Cyclic movement pattern

Movement between hubs at nine different locations for the period 1965—1971 is shown in Fig. 9. These nine hubs were selected because they indicated the greatest and most significant movement of some 25 hubs that were

Fig. 7. Northeasterly striking, steeply inward-dipping faults viewed looking north on the cut bench in the area of overturning failure (for scale see man at lower right).

Fig. 8. View looking north from the south side of the cut bench; besides the steeply dipping faults, note the two northeasterly striking faults which dip to the west (indicated by arrows; for scale see man at lower right in photo).

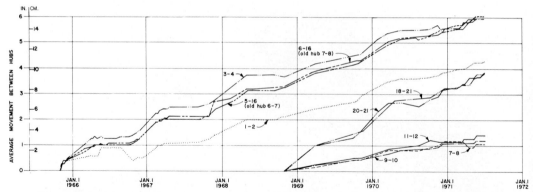

Fig. 9. Displacements between hubs located across cracks at nine different locations above the cut bench (see Fig. 4).

measured. The average displacement as a function of time between particular hubs is shown in Fig. 10.

Cyclic characteristics in the pattern of movement are clearly apparent in both Figs. 9 and 10. Movements in the summer months are small compared to the winter period. Because only movements between hubs at the nine locations mentioned above are considered significant, results from these locations only are considered in the following analyses.

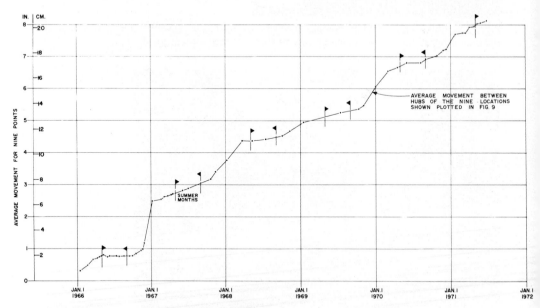

Fig. 10. Average movement between hubs at the nine locations plotted in Fig. 9. Note that the summer months occur between the "flags".

Interrelationship of movement and climatic conditions

The interrelationship of movement and climate is illustrated in Fig. 11. Plotted in Fig. 11a are the average monthly precipitation, average mean daily temperature and average monthly snowfall through the period 1966—1970. For correlative and comparative purposes, in Fig. 11b is a plot of the average daily rate of movement of the locations for the period October 1965 to June 1971. The average daily rate of movement was determined by making use of a segmented bar-type diagram, thus providing an estimate for movements on a daily basis.

Fig. 11 clearly indicates that there is a direct correlation between movement and precipitation. The greater the precipitation, the greater the movement, except during periods of freezing. It is interesting to note that instead of aggravating the problem, freezing and snowfall appear to retard movement. Although the average monthly precipitation peaks in January, the rate of movement starts to decrease after December. It is in December that freezing and snowfall conditions start to become significant.

The significance of climatic conditions is also illustrated in Fig. 12. Shown are plots from 1966 to 1970 of (a) total monthly snowfall, (b) total monthly precipitation, (c) number of days per month with freezing temperatures, (d) mean, maximum and minimum daily temperatures and (e) number of freezing cycles. Shown in (f) are plots of the weighted average daily displacement each year and the total displacement each year for the same nine locations discussed earlier. The movement data of Figs. 9 and 10 can also be compared to the climate data of Fig. 12.

Both Figs. 11 and 12f indicate that, if anything, the total annual movement is decreasing slightly over the period of study. However, as has been explained earlier, this appears to be highly dependent on the total annual precipitation and whether or not this precipitation has been able to infiltrate directly into the slope (i.e., whether ice conditions or a snow blanket exist). The irregular plot of the average weighted daily movement also suggests that anomalous rates of movement can arise with various combinations of extremely cold or warm winters and/or heavy or light snowfall. For example, in 1967 both the weighted daily movement and the precipitation were large, but both freezing and snowfall conditions were not severe. On the other hand, in 1968 the weighted daily movement was small even though the precipitation was the largest of all the years considered. The significant aspect in this case, however, is that both snowfall and freezing conditions were severe and accordingly appear to have restricted the infiltration of water and thus retarded slope movement.

Determination of time lag between precipitation and movement using cumulative sums technique

The cumulative sums technique was adopted to determine the amount of time, or time lag, that exists between precipitation and slope movement

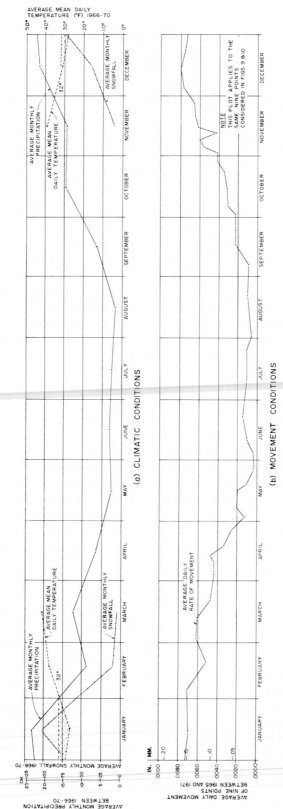

Fig. 11. Plots of average (a) monthly precipitation, monthly snowfall and monthly temperature, and (b) average daily movement, during the period 1966—1970.

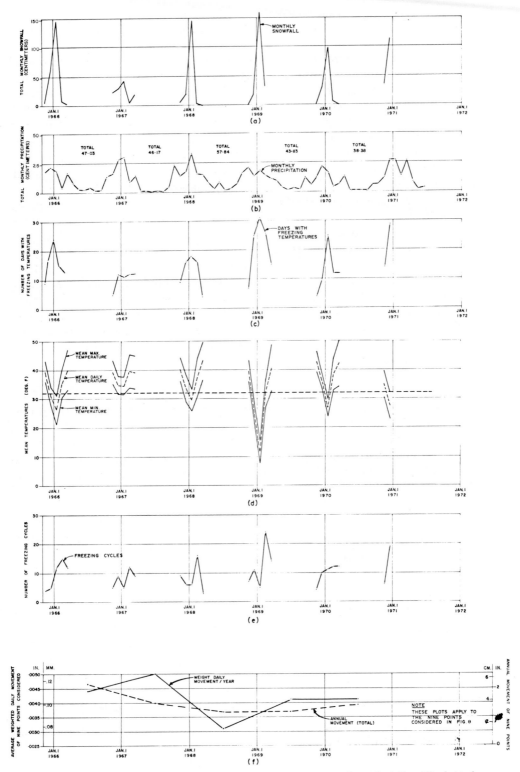

Fig. 12. Different climatic factors (a—e) and slope movement characteristics (f) plotted for comparative purposes (see text for details).

events. Cumulative sums, or "cusums" as they are also called, have been used extensively in industrial quality control (Woodward and Goldsmith, 1964). They have also been used for studying long-term trends in natural phenomena, such as river flows and silt deposition (Hurst et al., 1965) and the analysis of joints (Piteau and Russell, 1972). Experience has shown that there is a very close empirical relationship between the natural fluctuations of piezometric surfaces and the curve of the cumulative sums of precipitation. This relationship seems to apply to water surface fluctuations in bedrock as well as to water-bearing zones in unconsolidated material.

"Cusums" techniques have been applied mainly to series of events equally spaced in time. In this analysis, however, we use this technique to compare plots from observations which are both regularly (i.e., precipitation) and irregularly (i.e., movement) spaced in time.

The "cusums" technique provides a rapid and a relatively precise method of determining major trends above or below a selected reference value. Also, it can be used to ascertain the magnitude and location and/or period, whatever the case may be, of these variations. One important feature of this analysis method is that relatively small changes, say in the current mean value of precipitation, appear as distinctly different slopes. At the same time, erratic variation or "noise" in the data are smoothed out.

Basically, the approach is a simple one. It consists merely of subtracting a constant quantity (which in this instance is taken to be the average of the movement and precipitation values considered) from each value in the series, and accumulating the differences as each additional value is introduced. Successive accumulative differences are designated the "cumulative sums" of the original sequence of precipitation or movement values. Further details concerning the use of this technique can be obtained from the references cited earlier.

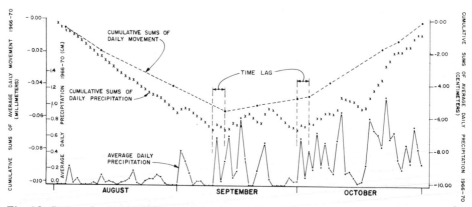

Fig. 13. Curve of average daily precipitation and cumulative sums curves of average daily precipitation and average daily movements.

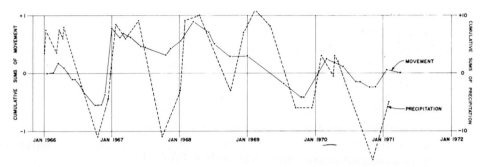

Fig. 14. Annual plots for comparative purposes of cumulative sums of movement and precipitation.

Since both precipitation and movement start to increase in the months of August, September and October, the "cusums" were first attempted for this period. The cumulative sum curves for both the movement and precipitation for August to October for the period October 1965 to June 1971 are shown in Fig. 13. The same movement information used in this analysis is shown in Fig. 11b. The two definite breaks in the slope of the curves indicate that the time lag between precipitation and subsequent movement is of the order of three days or less, as indicated by the pronounced change of slope of the curves. For comparative purposes to indicate how the "cusum" analyses brings out trends, a plot of the raw precipitation data is also shown.

In Fig. 14 movement and precipitation are compared on a yearly basis. The results show similar main trends but in considerably less detail. In this case, however, the movement information is not sufficient to make a reasonable estimate of the time lag between precipitation and movement.

Amount of precipitation required to generate movement

In Figs. 11b and 13 the plot of movement with time indicates that the movement typically starts to increase around mid-September. By smoothing out the curve of the average daily precipitation in Fig. 13, it appears that an average daily precipitation of about 3 mm was sufficient to cause this increased rate of movement.

MATHEMATICAL SIMULATION MODEL OF CLIMATE AND SLOPE MOVEMENT

Review of simulation procedure

Simulation may be defined as the development of a mathematical format to quantify events in a complex physical situation. The success or failure of this technique may be tested by application of simple linear regression where the dependent variable is a direct measure of the physical phenomenon and the simulated value of the same physical phenomenon is the independent variable. Since this analysis deals with the movement across cracks in a slope, the "physical phenomenon" of significance here is the average overall movement of the rock mass; the relationships of this movement to both daily precipitation and daily precipitation acting together with average daily temperature were investigated. The purpose of this analysis was to obtain a better understanding of the relationship of climate and movement, the ultimate objective being that practical remedial measures could be better applied once the cause and kinematics of movement were understood.

Basic assumptions and nature of the simulation input

The format and the coefficients selected should bear some reasonable relationship to the phenomenon being investigated. For example, drainage must occur in accordance with the basic laws of hydraulics. Based on the findings discussed earlier and other knowledge certain assumptions were made, as follows:

Flash flood. If the daily precipitation exceeds a certain amount, it is assumed that cracks in the slope will be unable to accommodate all of the available moisture. Hence, it is arbitrarily assumed that daily precipitation in excess of 5 cm will run off the slope.

Absorption. It is assumed that the quantity of moisture which enters the slope is probably related to the average daily temperature. For example, when ice forms in the cracks, the moisture will enter the slope at a reduced rate with the excess running off the slope. Precipitation which is likely to enter the slide will probably decrease gradually over a temperature range from 40 to 25°F. Above 40°F, the crack system is assumed to accept all daily precipitation of 5 cm or less. Below 25°F, precipitation will probably be in the form of snow; substantial quantities will either blow off the rock face or cause an equivalent excess-precipitation condition when temperatures rise above freezing. It should be noted that both the slope configuration and geometry of cracks in the slope are considered in mathematical terms to be constant quantities.

The following "absorption ratios" as per varying temperature were arbitrarily assumed:

Average daily temperature (°F):	40+	39—35	34—30	29—25	24—
Absorption ratio:	1.00	0.90	0.60	0.45	0.30

To illustrate the above, if 1.0 cm of daily precipitation occurs when the average daily temperature is at 32°F, it is assumed that 0.6 cm precipitation enters the slope and 0.4 cm runs off the slope.

Drainage. Based on the daily precipitation and absorption ratio, moisture will infiltrate into the rock mass which forms the slope. It is assumed that the rate of drainage through the system of cracks in the slope is linearly related to the accumulated "head". The following is a table of drainage rates which relate to average daily temperatures:

Average daily temperature (°F):	40+	39—30	29—
Rate of drainage (in cm) per centimetre accumulated:	0.50	0.38	0.25

Thus in cold weather the assumed drainage rate is one half of that assumed to occur in warm weather.

Time lag. Analyses discussed earlier suggest a time lapse between precipitation and subsequent movement of the slope. Intuitively one would expect the time lag not to be uniform throughout the year. When movement first begins in the fall, there is probably a greater time lag between precipitation and movement events than in the spring of the year, when the slope is more nearly saturated and movements are already well developed. However, based on the results in Fig. 13, and in the absence of a more precise evaluation, a constant two-day time lag was considered reasonable for simulation purposes.

To illustrate, if slope movements were recorded November 8 and November 25, the simulation procedure would consider precipitation and temperature data between November 6 and 8 and November 23 and 25, respectively.

Accumulation of moisture in the system of cracks in the slope. The following calculations illustrate the principle of accumulation, which is considered to be a result of interactions between precipitation, temperature, absorption and drainage.

If 1.0 cm of precipitation falls at 42°F, then at the end of the day the accumulation will be given by:

accumulation = 1.0 − 0.2(1.0) = 0.80 cm

If, on the following day 0.5 cm falls at 32°F, then the quantity absorbed will be given by:

absorption = 0.5 × 0.90 = 0.45 cm

This quantity is added to the previous residual (accumulation):

new accumulation = 0.8 + 0.45 = 1.25 cm

At a temperature of 32°F, the rate of drainage will be 0.15 cm per centimetre of accumulated "head":

drainage = 1.25 X 0.15 = 0.19 cm

The new residual will be given by:

new residual = 1.25 − 0.19 = 1.06 cm

Thus, 1.5 cm of precipitation occurred in two days; 1.06 cm remained trapped in the slope and the remainder drained away.

Relationship of movement and moisture accumulation. It is not possible to determine the porosity of the slope forming material and consequently the volume of moisture stored in the rock mass. It is assumed that the porosity of the slope forming material is such that if 2 cm of precipitation accumulates in the slope, the standing head will be twice as high as that associated with an accumulation of 1 cm of precipitation. Because hydraulic pressure acts equally in all directions, it is assumed that the rock face movement is directly proportional to the residual moisture at the end of each day. In the example shown, at the end of the first day the residual is 0.80 cm, rising to 1.06 cm at the end of the next day. If absorbed moisture is indirectly responsible for the movement of the rock face, then the degree of movement on day two will be about 1.3 times the amount recorded on day one. The accumulated movement for the two days will be proportional to the sum of the daily residual values, i.e., the "total accumulation":

movement "proportional to" $F(0.8 + 1.06)$

After the date and time associated with each movement observation, the accumulation of movement is re-set to zero while the residual moisture accumulation is retained. This procedure permits continuity in the analysis.

"Day zero" of the analysis. At the start of the analysis, the slope is assumed to be dry and movement is zero. The first movement observation was recorded November 3, 1965. Taking into account the time lag described above, November 1, 1965, becomes "day zero" for this analysis.

Sequence of the analyses

(1) *Listing of all input data.* This comprises a routine part of any computer analysis and is run to insure that the information has been correctly key-punched and is complete.

(2) *"Dating" the movement observations.* As described previously, this analysis is intended to assess the accumulation of rock face movement between consecutive movement observations while accepting a two-day time lag between the event of precipitation and the incidence of movement. The computer searched the input file and stored the day number associated with all non-zero movement records. The analysis was designed to cover some 2300 days of precipitation data and some 60 movement records; there were often periods of several months between the movement observations.

(3) *Simulation of the observed movements:*

(a) Test for excess precipitation: the daily precipitation was tested and the quantity in excess of 5.0 cm was deleted as previously described.

(b) Absorption ratio: the average daily temperature was stratified as described previously. The available moisture was computed as the product of the precipitation and the appropriate absorption ratio.

(c) Daily accumulation: the residual moisture from the previous day was added to the absorbed moisture giving a daily accumulation.

(d) Daily drainage: the drainage rate was computed by first taking into account the average daily temperature, then the flow was estimated as the product of the daily accumulation and the appropriate drainage rate.

(e) Daily residual: the daily residual was recorded as the difference between daily accumulation and the daily drainage.

(f) Total accumulation: daily residuals were accumulated for the time interval between consecutive recordings of the rock face movement. This accumulation is the output from the simulation procedure and is considered to relate directly to the daily accumulation of rock face movements.

The steps described above were repeated for the number of days between consecutive movement observations; the climatic data is offset by two days to account for the time lag. At the end of each period, the computer printed the relevant data comprising the observed movement, daily accumulation, daily residual, total accumulation and dates involved. The computer then reset the total accumulation to zero. The daily residual was retained to provide continuity.

The analysis was run twice, once with actual daily temperatures, and next using a fixed temperature of $40°F$; this procedure provided a "parameter test" concerning the influence of temperature.

Testing the effectiveness of the simulation procedure

If the simulation procedure is valid, total accumulation should plot linearly against rock face movement. It should also pass through the origin, so that a total accumulation of zero implies zero movement.

Initially, four sets of regressions were run (pertinent data are given in Table I):

TABLE I

Statistical data pertaining to simulation model

Test No.	Number of points	Correlation coefficient	Coefficient of determination *	Slope of line	Constant of regression
1	59	+0.795	0.633	+0.00662	+0.00018
2	49	+0.878	0.771	+0.00847	−0.00866
3	59	+0.762	0.536	+0.00674	+0.00268
4	49	+0.844	0.712	+0.00879	−0.00835

* Coefficient of determination is the proportion of the variance explained by the regression line.

(1) Includes both temperature and precipitation and all rock face movements.

(2) Includes both temperature and precipitation but excludes movement observations immediately following blasting, scaling or other works likely to induce unassignable movements.

(3) Includes precipitation and all rock face movements.

(4) Includes precipitation but excludes movement observations as described above.

The separation of movement data induced by climatic influences from that induced by blasting or scaling appears justified because these mechanisms were not considered in the simulation model.

From Table I the following observations were made:

(1) The constant of regression is relatively close to zero, thus complying with the fundamental requirement that zero accumulation implies zero movement.

(2) The exclusion of movement data which may have been induced by excavation, etc., appreciably improves the degree of linearity between simulation output and the movement observed.

(3) The degree of linearity is somewhat influenced by temperature.

Careful field examination of the rock slope indicated that the lower part of the slope was relatively dry, suggesting that most of the drainage was occurring at the base of the cracks behind the slope and possibly below the highway. The assumption that drainage was influenced by temperature was therefore revised, and a constant value of 0.5 cm per centimetre head was employed. Also, because the temperature measurements are assumed representative of conditions at the base of the slope at highway level, all temperatures were reduced by 2°F to account for the elevation difference at the top of the slope, where the precipitation enters via cracks.

The analysis was repeated, taking into account these two changes. Correlation coefficients were thus increased by +0.05 in each of the four cases, seemingly confirming the reasonableness of the revised assumptions.

Conclusions from the simulation analysis

The simulation analyses support the view that a deterministic mathematical relationship exists between slope face movement and climatic conditions. These results substantiate in a mathematical manner results of earlier analyses which indicated a direct relationship between climate and movement.

The accuracy of the simulation seems significantly enhanced by considering the interactions of both temperature and precipitation. This analysis is not presented, however, as an assessment of "cause and effect", and should not form a basis for predicting rock slope movement. However, these results could be used to establish a more rational monitoring program and to determine the degree of effectiveness of remedial measures.

MECHANICAL STABILITY ANALYSES

Various mechanical stability analyses were carried out to ascertain the quantitative significance of effects of groundwater and/or excavation on the stability of the slope. In this regard various kinematically possible failure modes were considered; varying static heads of water were used for evaluating both "excavated" and "unexcavated" conditions.

Mechanics of failure

Level plate and displacement data suggest that the failing rock blocks, at least those adjacent to the "Old Main Fault", appear to be hinging approximately where the faults defining the respective blocks intersect the "Old Main Fault". The level plate measurements indicate that the blocks in question are rotating outwards, away from the slope in an overturning manner. In support of this view, excessive fracturing and spalling of the rock, perhaps associated with high stress concentrations at the toe of the bench face but due to overturning, are apparent in Figs. 15 and 16. The displacement measurements are limited and confined to the slope surface only; they also can be interpreted as resulting from an overturning failure.

Based on the tiltmeter results the radii of the overturning movement and hinge points of the individual fault-controlled overturning blocks were approximated, as shown in Fig. 17. The section shown in Fig. 17 is a typical cross-section of the slope (see Fig. 4) in the area of greatest movement and was used in the analyses to follow.

Effect of water pressure on a typical overturning block

To assess the influence of water pressure on the slope, the effects of water pressure on the overturning moments of a typical wedge block occurring in

Fig. 15. Highly fractured rock at the toe of the cut bench; this fracturing is interpreted to be the result of the overturning action.

Fig. 16. Fractured rock and spalling at the toe of the cut bench, presumably caused by overturning of fault blocks towards the bench.

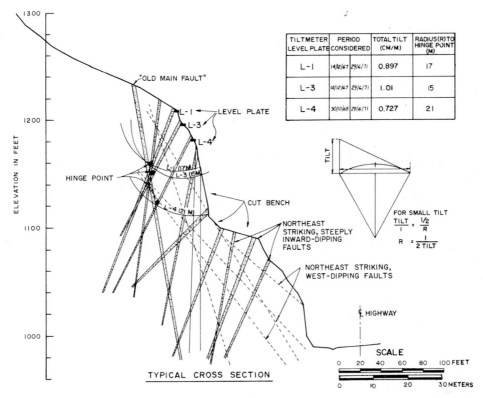

TILTMETER LEVEL PLATE	PERIOD CONSIDERED		TOTAL TILT (CM/M)	RADIUS(R) TO HINGE POINT (M)
L-1	14/12/67	29/6/71	0.897	17
L-3	14/12/67	29/6/71	1.01	15
L-4	30/10/68	29/6/71	0.727	21

Fig. 17. Estimated location of hinge points at "Old Main Fault" of overturning blocks (refer to Fig. 4 for location of section).

the slope was considered. It was assumed for the sake of simplicity that the block is an individual rotating unit unaffected by adjacent blocks and that the factor of safety F of the wedge is 1.0 when no water pressure is acting on it.

Water pressure is plotted against both the overturning movement and F in Fig. 18. This analysis clearly shows that the factor of safety of a typical rock wedge decreases appreciably with an increase in the static head of water.

Stability analyses for the case of overturning intact blocks in the slope

Five fault-controlled blocks, which are shown in Fig. 19 according to the fault geometry indicated in Fig. 18, were considered individually with respect to stability. The assumption was made that blocks 1—5 were intact and that the blocks overturned about a pivot point, A, located at the toe of the bench face. In the same manner as above, the effects of varying the static

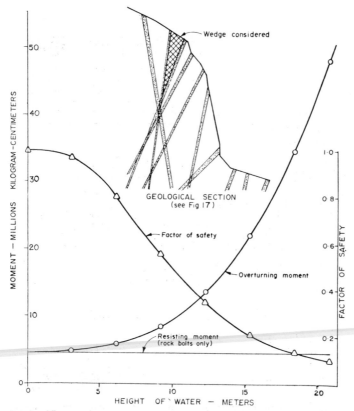

Fig. 18. Plots of overturning moment and factor of safety, indicating the effect of water.

head of water and the significance of the excavated and unexcavated conditions were examined. Two different excavation geometries were considered, but only results for the case of excavation 1 are given here. The results of the analyses are given in Table II.

Assuming that no external forces act, the moments of the normal and tangential components about the pivot point were treated as stabilizing and overturning moments, respectively, and their ratio was arbitrarily used as the factor of safety. That is, the factor of safety is the ratio of the resisting moment to the overturning moment.

The results in Table II indicate that although excavation considerably reduces the overturning forces of the blocks, the effect of the excavation is not nearly as significant when the full height of water is acting in the slope. For example, with respect to overturning, the factor of safety in the case of intact blocks 1 and 2 is essentially unchanged when water is at full height, regardless of whether excavation has taken place or not.

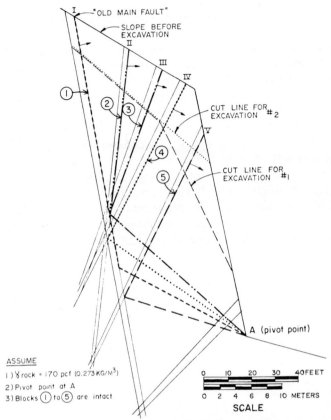

Fig. 19. Configuration of five individual overturning intact blocks (according to the geological section in Fig. 17) to be analyzed for stability.

TABLE II

Results of stability analyses for the case of intact blocks

Block No.	Factor of safety	
	before excavation	after excavation
No water		
1	2.65	3.92
2	2.68	3.40
3	2.74	4.68
4	2.90	4.03
5	3.73	5.01
Full height of water		
1	0.89	0.85
2	1.01	1.00
3	0.97	1.39
4	1.03	1.40
5	1.05	1.55

Stability analyses for the case of interacting overturning fault blocks

This analysis assumes that an upper block is overturning about a hinge point *A* onto a lower block; the lower block in turn rotates about a pivot point *B* located at the toe of the bench face. Various combinations of fault-controlled block configurations were analyzed. Various static heads of water were also considered, and two excavation geometries compared. Based on earlier findings, the results from these analyses appear to be more realistic.

One particular combination of blocks, where a wedge defined by faults II and IV overturns onto block 4, is depicted in Fig. 20. The results for this particular failure mode are shown graphically in Fig. 21; the slope seems stable except for extremely high groundwater conditions, regardless of whether or not excavation has taken place. This particular failure mode does not appear reasonable, since it suggests that slope movement will probably not occur unless the static water level is unreasonably high.

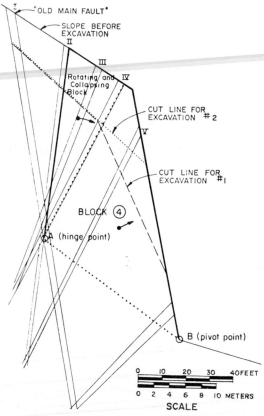

Fig. 20. Configuration of block 4 and an upper wedge for the case of interacting blocks.

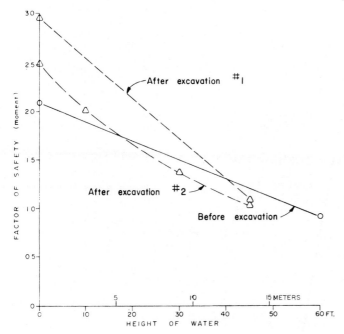

Fig. 21. Stability analysis results for block 4 interacting with an upper wedge (see Fig. 20), for cases of changing slope geometry and different groundwater conditions.

However, another combination of interacting blocks gives more realistic results, viz. for the case where a block defined by faults I, II and V overturns onto block 5. The configuration of the two blocks in the slope and the slope geometry for the three cases analyzed are shown in Fig. 22; the results are given in Fig. 23.

The curve in Fig. 23, which applies to the condition before excavation, indicates that when the height of water is 9 m above the hinge point A, the factor of safety is less than 1.0. It can be seen once again, that besides excavation, the prevention of water pressure buildup is most important. In this respect a maximum allowable height of 9 m of water above point A appears to be a reasonable design criterion for a groundwater drainage and/or groundwater infiltration prevention system.

GROUNDWATER GEOLOGY AND HYDROLOGY CONSIDERATIONS

Significant features affecting the groundwater geology and hydrology appear to be as follows:

(1) The major faults, as well as containing breccia, have abundant clayey material in some sections and secondary mineralization.

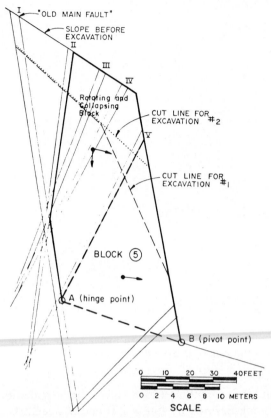

Fig. 22. Configuration of block 5 and an upper wedge, for analyses of the case of interacting blocks.

(2) Where movements are presently occurring, the surface traces of some
faults have opened to expose relatively large cracks.

(3) Several transverse faults and joints are present that form hydrologic
connections between the main fault blocks.

It is difficult to come to firm conclusions regarding the groundwater
configuration associated with slope movement. There may be no permanent
areally extensive water body in the fault and fissure system in the vicinity of
the cut bench. Instead, the water pressure buildup could come from a
"curtain" of water slowly moving downward through the fractures. That is,
little lateral groundwater flow due to normal percolation may exist, and vertical infiltration of precipitation may dominate entirely. When the "curtain"
of water extends higher than about 9 m above the hinge point (i.e., A in Fig.
22) of the rotating blocks it exerts sufficient pressure to initiate movement.

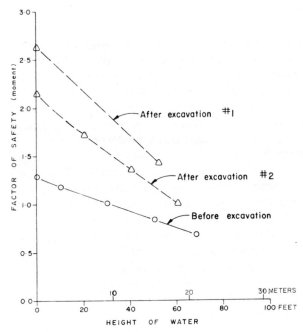

Fig. 23. Stability analysis results for block 5 interacting with an upper wedge (see Fig. 22), for the case of changing slope geometry and different groundwater conditions.

On the other hand, if there is a permanent water body, sufficient water must infiltrate into the ground to raise the water table to a level about 9 m above the hinge point. Because conclusive evidence one way or the other is not available, a groundwater prevention system in such instances should be designed to protect against both possibilities. It should be noted, however, that based on field results, it appears that the groundwater remedial measures should be directed more towards prevention of precipitation entering the slope than towards removing the groundwater once it has infiltrated the slope. An extensive system of drain holes had been attempted previously at about three different levels on the slope and proved to be entirely ineffective due to the abnormally cracked nature of the rock mass.

Mechanical analyses described in the previous section indicate that groundwater remedial measures should be designed such that the static head of water should not be allowed to build up to a height greater than about 9 m above the hinge point. Results of cumulative sums analyses of data for the last five years suggest that groundwater remedial measures should perhaps be such that water could be drained away in less than three days, to achieve the objective of preventing groundwater buildup.

SUMMARY AND CONCLUSIONS

The failure mechanism of the Hell's Gate Bluffs slope problem involves outward overturning of vertical to steeply dipping lenticular and tabular blocks. These blocks are basically defined by a well-developed set of parallel to sub-parallel faults which trend approximately parallel to the highway.

The slope is in a marginal condition of stability, and groundwater pressure is causing the slope to move. Besides indicating that movement is a direct function of precipitation, cumulative sums and other analyses indicate that the time lag or response between the causative precipitation event and subsequent movement of the slope is of the order of three days or so. These and other findings indicate that the approach to preventing water pressure development should be more a matter of preventing direct infiltration from precipitation on the slope, and since previous attempts at using drain holes were ineffective, not simply a matter of draining the slope once precipitation has entered the rock mass. These studies also show that the weighted daily movement from 1966 has not increased, but that significant variations can take place due to combinations of low and/or high temperatures and low and/or high snowfall. Combinations of these factors affect movement in that infiltration of precipitation can be retarded, in association with specific climatic controls. It appears, however, that the total annual movement has, if anything, decreased slightly since 1966.

The influence of groundwater pressure is as important, if not more important, than any other applicable remedial measures taken together. Mechanical stability analyses of kinematically possible failure modes for cases of varying static heads of water and two possible excavation geometries indicate that a factor of safety of not much over 1.0 can be achieved when water is at full height. For the same conditions but with no groundwater acting, the factor of safety after excavation is greater than 2.0. It was considered therefore that a combination of prevention of water pressure buildup, of excavation, and of artificial support, should provide an acceptable factor of safety against failure.

It should be noted that certain remedial measures involving provisions for excavation, in order to remove weight, and a shotcrete cover on the slope to prevent water infiltration, are being proposed to stabilize the slope. Other than to state that excavation will consist of about 18,000 m^3 of rock, and shotcrete will cover approximately 4000 m^2, further elaboration of these measures are not within the scope of this paper. (These remedial measures have now (1976) been carried out, and the slope appears to have stabilized.)

ACKNOWLEDGEMENTS

Grateful acknowledgement is extended to the British Columbia Department of Highways for permission to publish this paper. Thanks are extended to the many Department of Highways staff involved with rock slope problems in this area; the authors particularly acknowledge the cooperation of E.B. Wilkins and the assistance of J.W.G. Kerr and J.D. Austin, for providing records and their experience concerning the Hell's Gate problem; G. Miller, D. Lister, J. Horcoff, W.A. Richards, and E. Beswick for bringing important field information to the author's attention; N.T. Bain, for assistance with the simulation model; and T.G. Kirkbride, for the oblique aerial photographic work. Thanks are also extended to W.L. Brown for his assistance with the groundwater analyses, to R.A. Spence for suggestions and a general review of the analysis, and to D.C. Martin for assistance with the figures.

APPENDIX. FINITE ELEMENT ANALYSIS

A two-dimensional finite element analysis was subsequently carried out to investigate associated stress conditions (Kalkani and Piteau, 1976). The purpose of this analysis was to indicate zones of tension during the period when failure had been monitored; results were compared with the stability analyses and empirical movement data as presented in this chapter. A summary of the finite element study is presented in this Appendix; the summary was added in final proof by the volume editor in order to complete the case history.

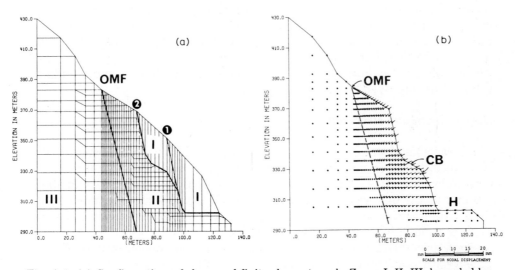

Fig. A-1. (a) Configuration of slope and finite element mesh. Zones I, II, III, bounded by heavy lines, refer to material property zones as discussed in text. Point *1*, boundary of excavation 1 prior to 1963. Point *2*, boundary of excavation 2, started but discontinued in 1963. Surface trace of Old Main Fault denoted by *OMF*. (b) Total displacement vectors due to excavations 1 and 2. *OMF* = Old Main Fault, *CB* = cut bench, and *H* = highway.

TABLE A-I

Physical properties assumed for finite element analysis

Granodiorite zone	Density (Mg/m^3)	Young's modulus (MN/m^2)	Poisson's ratio
I	2.68	10,400	0.37
II	2.84	48,100	0.20
III	2.98	82,100	0.20

Model assumptions

The slope profile analyzed is a section at right angles to the highway (Fig. A-1). The Old Main Fault is the only major structural discontinuity specifically included in the analysis. The slope was, however, considered heterogeneous, inasmuch as the granodiorite mass was assumed to increase in strength with depth. Three material zones were arbitrarily defined with assumed properties as indicated in Table A-I; Zone I includes rock excavated during or prior to 1963, Zone II extends from the slope or excavation boundary to the Old Main Fault, and Zone III is the rock mass behind the Old Main Fault.

The following properties were assumed for the Old Main Fault: cohesion = 98 kN/m^2, friction angle = 10°, normal stiffness = 49 MN/m^2 per metre, shear stiffness = 25 MN/m^2 per metre, stiffness after tensile failure = 98 kN/ m^2 per metre and stiffness after shear failure = 2.5 MN/m^2 per metre. The ratio of ambient horizontal to vertical stress was taken as 0.33. An earthquake "equivalent horizontal acceleration" was assumed as 0.1 g. A modified version of a static finite element code described by Duncan and Goodman (1968) was employed (see Kalkani, 1975).

Displacement and stress patterns

Displacement vectors for the "dry" slope condition are indicated in Fig. A-1. Magnitudes of upward and outward displacement are 1.0 mm at the crest above the cut bench, 1.3 mm at the berm of the cut bench, and about 0.6 mm at the highway. Similar analyses with high groundwater table conditions (water forces acting on the Old Main Fault) produce displacements two or three times greater than those illustrated.

Stress patterns for the dry slope condition reveal concentrations of compressive stress at the toe of the slope (Fig. A-2) and regions of tensile stress on the order of 6 m deep, above and just below the cut bench. Under conditions of high water table, the tensile stress region extends over an area about three times larger than indicated for the dry condition, to depths as great as 15 m (Fig. A-3; cf. Bukovansky and Piercy, 1965). The tensile zone thus

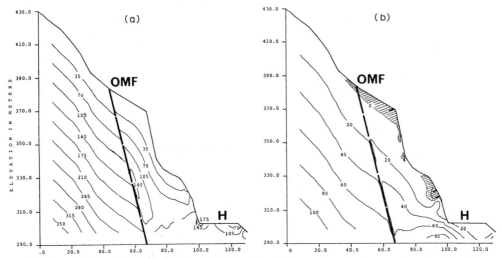

Fig. A-2. (a) Contours of maximum principal stress (maximum compression) due to excavations 1 and 2. Dry slope condition. (b) Contours of minimum principal stress due to excavations 1 and 2. Dry slope condition. Tensile region hachured. Old Main Fault and highway indicated by *OMF* and *H*, respectively. Stress contours in tonnes/m^2 (multiply by 9.8 to convert to kN/m^2).

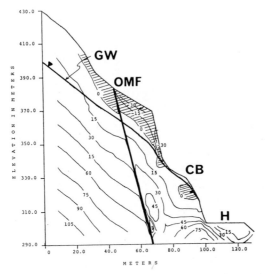

Fig. A-3. Contours of minimum principal stress due to excavations 1 and 2. High groundwater condition, with groundwater table indicated by *GW*. Compare extensive tensile region with Fig. A-2. Stress contours in tonnes/m^2 (about 10 kN/m^2). Old Main Fault, cut bench, and highway indicated by symbols *OMF*, *CB*, and *H*.

extends to within about 8 m of the upper hinge points about which toppling seems kinematically possible. This tensile zone, at least up to the Old Main Fault, closely approximates the rock excavation zone which contributed to the present stable condition of the slope.

REFERENCES

Bukovansky, M. and Piercy, N.H., 1975. High road cuts in a rock mass with horizontal bedding. Proc. 16th Symp. on Rock Mechanics, Univ. Minnesota, pp. 47—52.
Duncan, J.M. and Goodman, R.E., 1968. Finite element analyses of slopes in jointed rock. U.S. Army Eng. Waterways Exp. Stn., Rep. No. S-68-3, Corps of Engineers, Vicksburg, Miss., 271 pp.
Hurst, H.E., Black, R.P. and Simaika, Y.M., 1965. *Long Term Storage: An Experimental Study.* Constable, London, 160 pp.
Kalkani, E.C., 1975. Two dimensional finite element analysis for design of rock slopes. Proc. 16th Symp. on Rock Mechanics, Univ. of Minnesota, pp. 11—20.
Kalkani, E.C., and Piteau, D.R., 1976. Finite element analysis of toppling failure at Hell's Gate Bluffs, British Columbia. Bull. Assoc. Eng. Geol., 13: 315—327.
Monger, J.W.H., 1969. Hope map area, west half British Columbia. *Can. Geol. Surv. Paper,* 69-47.
Piteau, D.R. and Russell, L., 1972. Cumulative sums technique: a new approach to analyzing joints in rock. *13th Symp. on Stability of Rock Slopes. American Society of Civil Engineers, New York, N.Y.,* pp. 1—29.
Woodward, R.H. and Goldsmith, P.L., 1964. *Cumulative Sums Techniques.* Oliver and Boyd, Edinburgh, 74 pp.

Open Pit Mine Slopes

Chapter 16

ACOUSTIC EMISSION TECHNIQUES APPLIED TO SLOPE STABILITY PROBLEMS

RAYMOND M. STATEHAM and ROBERT H. MERRILL

ABSTRACT

Techniques and instruments used to detect instability in rock masses around underground mines have been refined and applied to the detection of instability in rock slopes. From the number and magnitude of the acoustic emissions in the rock mass, engineering estimates of slope instability can be made. The electronic apparatus has advanced from early tube-type circuitry to solid-state circuitry capable of amplifying the output of accelerometers, velocity gages and pressure gages used to detect acoustic emissions (rock noises). Some of the techniques, instruments, and results at four open pit mines are described. The results of these studies show that the technique can be used to predict failure, monitor the progress of the failure, and provide evidence that earthquakes can affect the stability of a pit wall.

INTRODUCTION

Acoustic emission techniques, under various names, have been applied for many years to the detection of instability in underground mines and for a lesser time in open pits. Early work in the use of acoustic emissions was done by Obert and Duvall (1957) who applied the term "microseismic method" to the techniques used and the term "microseismims" to the individual emissions. Other workers have used the terms microseismic techniques, rock noises, subaudible noises, and acoustic emissions, for the same purposes (Goodman and Blake, 1965; Broadbent and Armstrong, 1968; Wisecarver et al., 1969; Stateham and Vanderpool, 1971; Hardy, 1972). In the literature referenced in this article, these terms are now used synonymously.

Methods using acoustic emissions to detect instability in mines or rock slopes are based on the phenomenon that rock under stress normally emits noises and that sliding in the rock slope may also be expected to generate noises. These noises (usually subaudible) increase in number per unit time (noise rate) or magnitude (amplitude) as stresses in the rock approach the failure stress of the rock (Obert and Duvall, 1957). Consequently, techniques that detect and record these acoustic emissions or rock noises can be used as semiquantitative methods for predicting failure in the rock.

The first applications by Obert and Duvall (more than 30 years ago) involved the use of acoustic emissions to predict instability in the ribs, pillars, and backs of underground mines. Early studies in rock burst prediction were also initiated. Considerable progress has been made in the use of acoustic emissions during intervening years, both in underground mines and open pits. Techniques have been developed for accurate prediction of rock bursts (Blake and Leighton, 1969; Blake, 1971, 1972) and for determination of areas of first failure in block caving operations (Oudenhoven and Tipton, 1973). In open pits, techniques based on the detection of acoustic emissions have been used to monitor changes in slope stability during a full-scale test of pit wall steepening in Kennecott Copper Corporation's Kimbley pit, Ruth, Nevada, and to monitor failures in the Liberty and Tripp-Veteran pits (Wisecarver et al., 1969; Stateham and Vanderpool, 1971), and as a control technique in unloading an unstable pit wall (Paulsen et al., 1967).

GENERAL CONSIDERATIONS

The basic problem of detecting or monitoring slope instability with acoustic emission techniques can be separated into several components such as detecting and recording noises, isolating acoustic emissions from spurious noises, and correlating the resultant data to the stability or instability of the slope. These facets of the basic problem are influenced by one or more of the following:

Mine Environment. Broadbent's statement (Broadbent and Armstrong, 1968) that a mine presents a hostile environment to the microseismic system (acoustic emission techniques) seems especially true in open pit mines. Noise from mining equipment operating in the pit can be severe, although this noise can usually be avoided by recording between shifts. Adverse climatic conditions such as rain, wind, and electrical storms, may cause spurious signals to be superimposed on valid data. A skilled operator may recognize this type of noise so that it does not influence his analysis, but climatic noises, both sonic and electrical, often are so severe that the data are unusable for the duration of the disturbance.

The physical configuration of the pit and mining progress within the pit contribute to the adverse effect of the mine environment. Desirable instrument locations in the pit wall may be inaccessible, sloughing of rock from the pit wall may cover and/or damage cables and instruments placed on inactive benches, and mining progress may require the relocation of instrumentation.

Other seemingly minor problems related to the mine environment can be very irritating. For example, it is often necessary to use great lengths of instrument cable near open pits, and many types of rodents will chew through the insulation or protective covering on the cables, creating shorted or open circuitry.

Sonic transmissibility of the rock. The acoustical transmission through the pit wall varies greatly from one area to another due to changes in rock type, physical properties of the rock, and other geologic conditions (faults, fractures, jointing, etc.). For example, in the Kimbley pit near Ruth, Nevada, noises generated by striking the rock with a hammer were always detectable at distances of 30 m. At some locations, these noises could be detected at distances of 150—250 m. However, in the south wall of the Tripp-Veteran pit, about 3 km away, noises from hammer blows normally could not be detected at distances of more than 15 m with the same instrumentation. At Duval Corporation's Esparanza pit near Tucson, Arizona, (Fig. 1), the distances over which similar noises could be detected ranged from 8 to 225 m, depending on the presence of fractures in the slope wall. When failure occurs in a slope, fracturing can be severe, and nearby transducers, commonly called geophones, may be isolated from active failure in the rock; when this occurs, no further noises will be detected at that particular geophone location.

Acoustic emission characteristics. Blake and coworkers (1969, 1974) have shown that rock noise frequencies occur in a broad spectrum (50—10,000 Hz) and have complex wave forms. Although the studies by Blake and others were concerned primarily with noises generated by applied stress or fracturing, their statements are equally true of noises arising from rock sliding. Both of

Fig. 1. Location map of southwestern United States showing study areas.

these sources are present during the failure of pit slopes (Broadbent and Armstrong, 1968). A major cause of variation in rock noise frequency as recorded in the field is the fact that the higher frequency components of the wave are rapidly absorbed with increasing distance and travel time. In most slope stability studies of open pits, it is necessary to place instruments over large areas and, as a result, different geophones may receive different frequencies from the same rock noise. Obviously, this fact must be considered when selecting instrumentation for acoustic emission studies around open pits.

Operator requirements. One of the most important considerations in acoustic emission studies of pit slopes is the capability of the operator. While most of the conditions associated with the mine environment (such as geology, climate, or mine equipment) are outside the control of the investigator, the recorded noises may be influenced by any or all external conditions, and an operator trained in recognizing the presence of extraneous noise or the absence of valid noise (when such noise should be present) is essential to a successful microseismic program for detecting instability in a slope.

INSTRUMENTATION

A system of microseismic instrumentation used for pit wall instability studies consists of geophones, amplifiers, a listening circuit, and a recorder. The listening circuit makes possible real-time audio monitoring of signals entering the instrument system and is a valuable tool in the verification of true rock noises or the identification of spurious noise.

The geophone may be an accelerometer, velocity gage, or pressure gage, but essentially is a transducer that detects seismic (acoustic) energy and converts mechanical vibration to its electrical analog. An associated preamplifier (either housed in the geophone case or in line in the cable near the case) insures transmission of the signal through the long lengths of cable used for pit wall monitoring. The signal is then amplified and recorded, either on tape or an oscillograph.

Early models of the microseismic units used high output (20 v/g at 1000 Hz) rochelle salt crystals as transducers. A vacuum-tube preamplifier and/or a transformer housed in the geophone case amplified the signal so that electrical analogs of the rock noises were transmitted to tube-type amplifiers and were recorded on teledoltos paper by galvanometer-controlled pens.

As later models were developed, solid-state electronic components were used instead of the early vacuum tubes, and magnetic tapes were used for recordings reproduced on oscillographs for visual display. Ceramic transducers replaced the rochelle salt in geophones using this type of system. Other detectors, such as velocity gages or pressure gages were sometimes

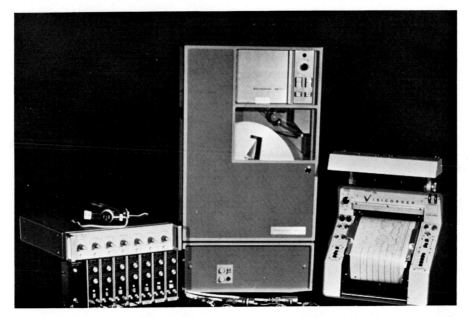

Fig. 2. Photograph of instruments for acoustic emission studies.

used as geophones because of their sensitivity at low frequencies.

The modern detectors normally have a wide band width (20—10,000 Hz). Although they are less sensitive than the resonant-type geophones (tuned to 1000 Hz) used in earlier investigations, they can be used more efficiently in slope stability studies because they respond equally well to seismic signals over a broad frequency range.

Use of the solid-state components with integrated circuitry has led to more instruments that are not only more portable and trouble-free, but also offer easier field maintenance through the use of replaceable printed circuit boards and integrated circuitry. Fig. 2 shows a modern system of instrumentation used for detection of acoustic emissions.

PROCEDURE

The number, spacing, and type of geophones required to effectively cover a study area, as well as the geometric configuration of the geophone array, are dictated by geologic structure, rock characteristics, pit wall configuration, and the size of the area monitored. Normally, planes of weakness in the rocks (faults, joints, fractures, bedding planes) are expected to provide failure surfaces for slides occurring in the slope. Consequently, geophones are placed as near as possible to known or suspected failure surfaces. Because the geophone spacing in the study area is also controlled by acoustical transmission, locations for the phones are usually selected after hammer-blow tests have been made to determine the distances over which rock noises will travel.

The geophones are normally placed in drill holes (either horizontal, inclined, or vertical) driven into the rock mass containing the suspected failure plane. This procedure not only aids in placing the geophone near to the failure, but also provides a means of isolating the geophone from many of the extraneous noises (especially airborne noise) around the pit or slope. Obviously, the coupling (area of surface contact) between the geophone and the wall of the drill hole may be poor. Several methods have been used to improve this geophone-to-rock coupling. The methods include the use of a spring-loading device to hold the geophone against the borehole wall, adhesives to bond the geophone to the wall, construction of matched surfaces for geophones to rest against, and the fitting of geophones into steel cylinders in the hole. Coupling of the detector to the rock by placing the instrument in water-filled holes normally should not be used. Much of the rock noise energy is in the shear wave and will not be transmitted through the fluid. If no other coupling method is practical, fluid coupling can be used, but signal strength will be reduced to the pressure wave (about 30—40% of the original signal).

After the instrumentation is in place and the collection of data has begun, recording periods are usually limited to seismically quiet times (between shifts, etc.) when the pit is not operating, because mining noises tend to obscure acoustic emissions. Initially, a background noise rate (noise rate for stable slope) is established for the rock mass being studied. Subsequent sampling is continued over long enough periods to establish representative rock noise rates. These rates are then displayed in graphic form; plots of noise rates against time on standard calendar paper have proven an easy, effective way of indicating changes in slope condition (Wisecarver et al., 1969), even though the number of emissions detected is dependent on other factors such as instrument sensitivity, rock transmissibility, etc. Studies thus far show the following correlations:

0—10 noises per hour: stable slope in an inactive mining area
10—50 noises per hour: stable slope in an active mining area
  >50 noises per hour: unstable slope

CASE HISTORIES

Brief descriptions of acoustic emission studies in four different pits are included in this report; each wall differs from the others in one or more of the following: height, slope, rock type, or planes of weakness. Because the geologic features of the pit walls are varied and complex, and because a discussion of the geology of each area is beyond the scope of this paper, only the acoustic emission studies will be described. Each of these studies was concerned with possible, probable or actual failure in a pit wall; therefore, it may be advantageous to consider definitions of failure before discussing the investigations. For purposes of this paper, failure is considered to have occurred when the first tension cracks are formed behind the failure block. The complete and final movement of material down the pit slope is called total failure or a slide. The term "intermediate failure" may then be applied to the interval between failure and the slide.

The location of the case history study areas is shown in Fig. 1.

Acoustic emission investigations, Boron pit

Acoustic emission studies were conducted by the U.S. Borax and Chemical Corporation and the U.S. Bureau of Mines in a cooperative study at the Boron pit, Boron, California (Paulsen et al., 1967; Wisecarver et al., 1969). An area of potential failure in the south wall of the pit was instrumented using microseismic equipment with the fundamental concepts developed by Obert and Duvall (1957).

Geophones were placed near potential failure surfaces in vertical drill holes, 10 cm in diameter, that had been cased to reduce noise caused by rock particles falling from the sides and collar of the drill holes.

Continuous recording (24 hours daily) provided data for graphs showing (1) the average hourly noise rate (acoustic emissions per hour) or (2) the daily noise rate (acoustic emissions per 24 hours). These plots over a period of time provided a graphic display of acoustic activity near the surface of potential failure.

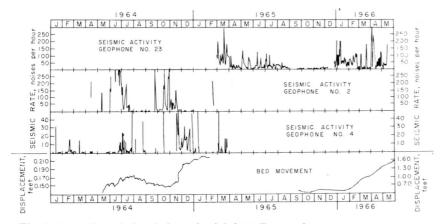

Fig. 3. Acoustic emission (microseismic) data, Boron pit.

Fig. 3 shows graphs of the hourly noise rate recorded at three geophone locations in the south wall of the Boron pit, together with a plot of measured displacements along the same surface of failure (Paulsen et al., 1967; Wisecarver et al., 1969). As shown in this figure, increases in the noise rate preceded the displacements (note the scale change in Fig. 3 to accommodate increased displacement rates); the use of acoustic emissions of microseismic techniques seems a usable technique for prediction of incipient failure because the first tension cracks were preceded by the very high noise rates recorded in September, 1964. Noise rates remained high with continued displacement, showing intermittent increases or decreases in number. Geophone 23 was added to the system in February, 1965, and immediately began to detect significant numbers of acoustic emissions (well in excess of background for the slope), indicating general instability in the slope.

While carefully observing the number of emissions recorded by this geophone, the company began removing approximately 1 million tons of rock from the surface of the problem area to correct the instability. This corrective excavation was completed in March 1965; a decrease in the number of acoustic emissions resulted, although the noise rates did not return to the

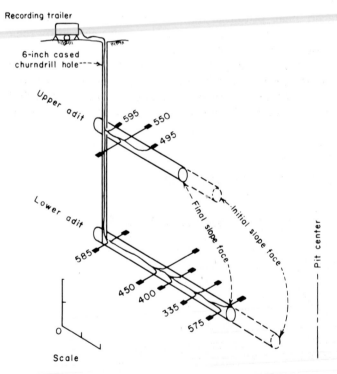

Fig. 4. Adit geophone locations, Kimbley pit (bar scale ca. 30 m).

originally low background count. At the same time, displacement slowed but did not completely stop. Both the rate of occurrence of acoustic emissions and rate of displacement indicated that the removal of weight from the failure block did not completely correct the instability of the slope. A significant increase in noise rates in December, 1965, followed by crack propagation necessitated the redesign of this part of the pit wall to prevent a slide in the finished slope. The slope was mined to a new, stable configuration.

Acoustic emission investigation, Kimbley pit

Studies involving the use of acoustic emissions were used to monitor the west wall of Kennecott Copper Corporation's Kimbley pit, near Ruth, Nevada. This technique was incorporated into a safety control program while the pit wall was being steepened from about 45° to about 60°; it was expected to provide early warning of any incipient failure in that portion of the pit wall. Two adits driven into the pit wall, and the pit wall itself, were instrumented with geophones placed in holes 5.2 cm in diameter and 1—15

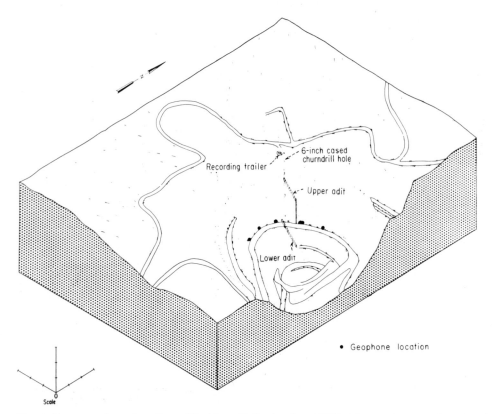

Fig. 5. Face geophone locations, Kimbley pit (bar scale ca. 100 m).

m deep. Figs. 4 and 5 show adit and pit wall locations, respectively. As
shown in Fig. 4, the geophones were connected by cables to amplification
and recording equipment in a mobile laboratory located on the surface
behind the pit slope.

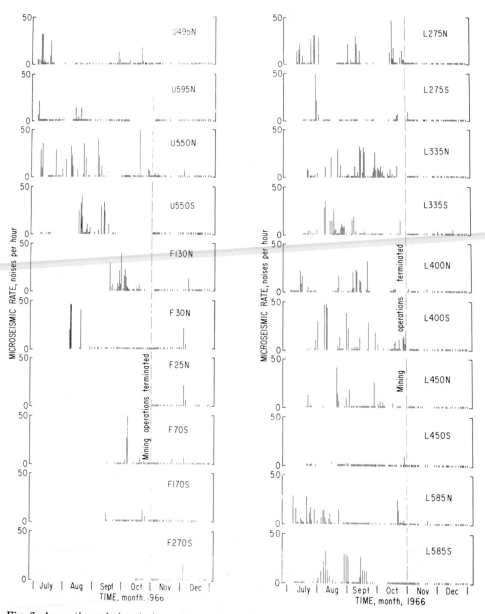

Fig. 6. Acoustic emission (microseismic) data, Kimbley pit.

The equipment used at the Kimbley pit was an improved, transistorized version of the equipment used at the Boron pit. Perhaps the most significant change in the design other than the use of transistors (instead of vacuum-tube electronics) was the use of a hollow cylinder of ceramic lead zirconate titanate as a detector to provide omnidirectional sensitivity.

Because mining operation in the pit generated a great amount of detectable noise, recording periods were restricted to periods of inactivity in the mine (between shifts and on weekends). The data obtained during these recording periods were converted to average hourly noise rates each day and graphed for ease of interpretation. Noise rates detected during the study are shown in Fig. 6.

Mining operations to steepen the Kimbley pit wall started before the acoustic emission study began. For this reason, no background noise count was available and correlations concerning pit wall stability are limited to a "mining to post-mining" relationship. Noise rates recorded during the time the pit wall was being steepened tend to be erratic. The erratic nature of these values is possibly related to variations in mining, locations in the pit, and/or variations in elevation along the bench created by mining. The bench at times varied in elevation by as much as 20 m and is believed to have created temporary stress concentrations, thereby increasing the number of acoustic emissions generated in that part of the rock mass.

Higher than normal noise rates were detected in the lower adit (denoted by "L" prefix in location number) during late July and the first half of August. During this time, the bench level approached and passed the lower adit; thus, the high noise rates appear to be related to the proximity of mining operations. Relatively high noise rates for pit face locations (denoted by "F" prefix in location number) were caused by raveling of the pit slope. The impact on the slope face of falling rock particles, loosened by weathering, generated noises that were detected by the geophones.

Relatively high noise rates were recorded prior to termination of mining operations even though monitoring occurred during mining "down" time. Evidently changing conditions created by pit wall steepening caused redistribution of stress in the rock mass. The low noise rates at the end of the study were in good agreement with stress and displacement data (from other phases of the full-scale test) that indicated the pit wall was stable (Wisecarver et al., 1969).

Acoustic emission investigations, Liberty pit

During the time of the Kimbley pit investigation, an additional small-scale acoustic emission study was made in the nearby Liberty pit. A geophone was placed in the pit wall below an area of suspected instability. The equipment used was identical to that used at the Kimbley pit except that noises were recorded on a portable, battery-powered tape recorder placed at the collar of

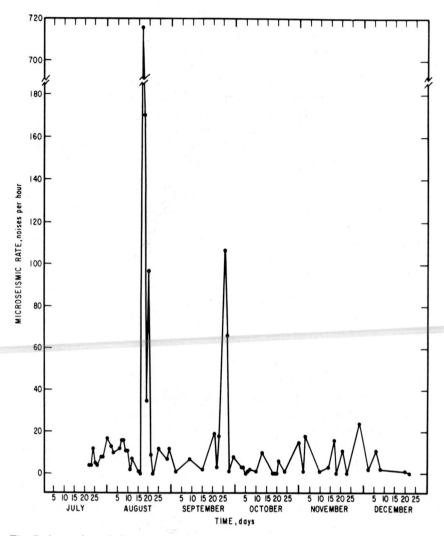

Fig. 7. Acoustic emission (microseismic) data, Liberty pit.

the geophone hole. A permanent paper record of the noises was then made by playing back the tapes through the amplifiers and an oscillograph.

Fig. 7 shows the hourly noise rates obtained in the Liberty pit wall. The most noticeable features of the data plot are extremely high noise rates in mid-August and late in September. These high noise rates appear to be related to earthquakes whose epicenters were approximately 250 km from the pit. Quakes occurred near Cedar City, Utah, on August 16 and 18 and on September 26, with magnitudes of 6.1, 5.1, and 3.9, respectively, on the

Richter scale. Apparently, the pit wall was affected by the shock waves from these earthquakes, and the number of acoustic emissions rose significantly. The reaction lasted for about two days after the earthquake.

The average noise rate, about 8 per hour, was less than expected and does not indicate an unstable area. However, in addition to the previously stated conditions that may reduce efficiency in acoustic emission monitoring, an additional problem was created by an 80-Hz low-frequency cutoff on the tape recorder. Many low-frequency noises may have been missed.

The Liberty pit study demonstrated the effectiveness of portable instrumentation for recording of acoustic emissions in remote or inaccessible locations.

Acoustic emission investigations, Tripp-Veteran pit

In a cooperative study by Kennecott Copper Corporation and the U.S. Bureau of Mines, a combined displacement-acoustic emission study was conducted in the north wall of the Tripp-Veteran pit, Ruth, Nevada. Because hairline tension cracks were noted on the surface above a section of the slope, displacement gages and acoustic emission recording instruments were installed (Fig. 8), and a study was initiated to (1) monitor the progress of the expected failure, and (2) determine possible correlations between displacement and acoustic emission data. The first of these objectives was

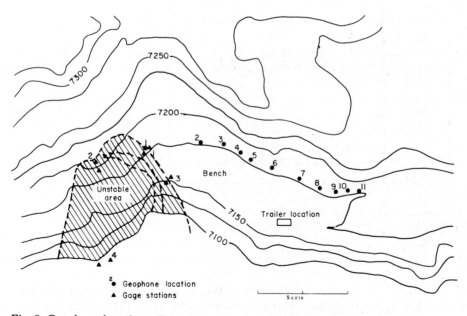

Fig. 8. Geophone locations, Tripp-Veteran pit (bar scale ca. 60 m).

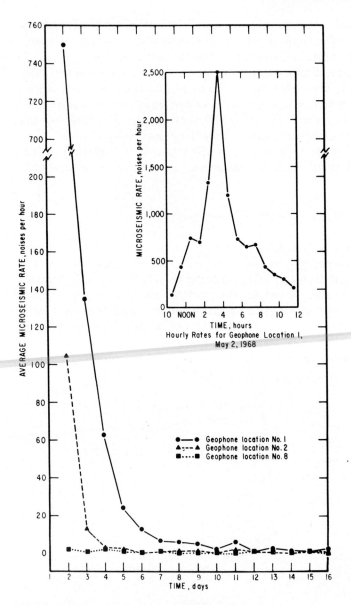

Fig. 9. Acoustic emission (microseismic) data, Tripp-Veteran pit.

especially important because mining operations were in progress at the toe of the slope in the failure area.

The solid-state instrumentation used in the acoustic emission study at this site was developed to reduce the electronic noise encountered in previous studies. In addition, a ceramic lead zirconate titanate disk was used as a unidirectional detector in the geophone; it was pre-stressed (in compression) for increased sensitivity.

Ten geophones were placed in 4-cm diameter holes, 2.5 m deep in the pit wall. They were located at intervals of 8—25 m and were from 50 to 200 m from the zone of expected failure. An additional geophone was buried in the pit slope near the expected failure surface.

A short-term acoustic emission study was made near the study area in January 1968 to monitor the effects of the Atomic Energy Commission's FAULTLESS event (an underground nuclear explosion at the Nevada Test Site). At that time, the average noise rate was about 3 per hour (Merrill, 1968). Consequently, 3 noises per hour was used as the background count in this study (Stateham and Vanderpool, 1971).

Fig. 9 shows the average daily noise rates for the three geophones, and the hourly totals for the most active location on the day of maximum movement. An average of 750 noises per hour were recorded over a 14-hour period from the location nearest the failure plane. During 1 hour of this period, a high of 2500 noises per hour was recorded. During this hour of maximum seismic activity, mining operations were terminated at the toe of the slope and both acoustic emission noise rates and displacement rates began to decrease. This decrease continued, and within ten days the noise rates detected at all geophones were within the background value of 3 noises per hour (Stateham and Vanderpool, 1971). Throughout the study, noise

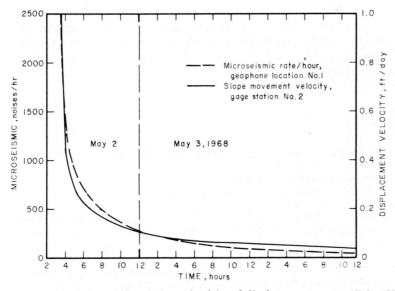

Fig. 10. Acoustic emission (microseismic) and displacement curves, Tripp-Veteran pit.

rates detected by geophones more than 90 m from the failure were too low to be used for failure detection.

Because correlation of noise rates to displacement rates was an integral part of the Tripp-Veteran study, best-fit curves were developed (using a non-linear optimization routine) from both the displacement and acoustic emission data (Stateham and Vanderpool, 1971). Fig. 10 shows these curves and their similarity, indicating (1) acoustic emission and displacement techniques can be correlated, and (2) acoustic emission techniques can be used to monitor the progress of pit wall failure.

CONCLUSIONS

Acoustic emission studies around slope walls are far more difficult to conduct and evaluate than around underground openings; however, these techniques provide an important method of monitoring open slopes and of assessing the question of their stability. As shown by the results of the studies at the Boron pit and the Kimbley pit, these techniques can be used to predict failure in a pit wall before the visual evidence of such failure (tension cracks) is formed. In addition, acoustic emission techniques (as shown from the studies at the Boron pit and the Liberty pit) can be used to monitor the effects of external events, such as mining, stripping, or earthquakes, on open slopes.

Factors on which the success or failure of an acoustic emission study is dependent are as follows: (1) accurate definition of the problem by evaluating the effects of the slope environment and rock characteristics; (2) proper installation of geophones in locations near the failure surface; (3) instrumentation; (4) suitable recording periods; (5) establishment of background noise rates for stable conditions; and (6) operator capability in data evaluation and interpretation.

REFERENCES

Blake, W., 1971. Rock burst research at the Galena Mine, Wallace, Idaho. *U.S. Bur. Mines, Tech. Progr. Rep.*, 39, 22 pp.

Blake, W., 1972. Rock burst mechanics. *Q. Colo. Sch. Mines*, 67(1), 64 pp.

Blake, W. and Duvall, W.I., 1969. Some fundamental properties of rock noises. *Trans. Soc. Min. Eng., Am. Inst. Min. Metall. Eng.*, 244: 288—290.

Blake, W. and Leighton, F., 1969. Recent developments and applications of the microseismic method in deep mines. In: W.H. Somerton (Editor), *Rock Mechanics — Theory and Practice. Proc. 11th Symp. on Rock Mechanics, Univ. of California, Berkeley, Calif., June 16—19*, pp. 429—443.

Blake, W., Leighton, F. and Duvall, W.I., 1974. Microseismic techniques for monitoring the behavior of rock structures. *U.S. Bur. Mines, Bull.*, 665, 65 pp.

Broadbent, C.D. and Armstrong, C.W., 1968. Design and application of microseismic devices. *Proc. 5th Can. Symp. on Rock Mechanics, December 6—7*, 20 pp.

Goodman, R.E. and Blake, W., 1965. An investigation of rock noise in landslides and cut slopes. *Felsmech. Ingenieurgeol.*, Suppl. 2, pp. 88—93.

Hardy, H.R., Jr., 1972. Application of acoustic emission techniques to rock mechanics research. *Am. Soc. Test. Mater., Spec. Tech. Publ.*, 505, 83 pp.

Merrill, R.H., 1968. *The Effect of the FAULTLESS upon Instrumented Areas of the Kimbley, Liberty, and Tripp-Veteran Pits, Ruth, Nevada.* Report prepared for the Safety Evaluation Division, U.S. Atomic Energy Commission, Nevada Operations Office, Las Vegas, Nev., 29 pp.

Obert, L. and Duvall, W.I., 1957. Microseismic method of determining the stability of underground openings. *U.S. Bur. Mines, Bull.*, 573, 18 pp.

Oudenhoven, M.S. and Tipton, R.E., 1973. Microseismic source locations around block caving at the Climax Molybdenum Mine. *U.S. Bur. Mines, Rep. Invest.*, 7798, 12 pp.

Paulsen, J., Kistler, R.B. and Thomas, L.L., 1967. Slope stability monitoring at Boron. *Min. Congr. J.*, 53(9): 28—32.

Stateham, R.M. and Vanderpool, J.S., 1971. Microseismic and displacement investigations in an unstable slope. *U.S. Bur. Mines, Rep. Invest.*, 7470, 22 pp.

Wisecarver, D.W., Merrill, R.H. and Stateham, R.M., 1969. The microseismic technique applied to slope stability. *Trans. Soc. Min. Eng., Am. Inst. Min. Metall. Eng.*, 244: 378—385.

Chapter 17

SLOPE FAILURE OF 1967—1969, CHUQUICAMATA MINE, CHILE

BARRY VOIGHT and B.A. KENNEDY

ABSTRACT

   Major pit slope movements on rock mass discontinuities were initiated
by an earthquake in December 1967 and sustained by excavation and blasting operations. Limiting equilibrium analysis suggests that average cohesion
mobilized in the slope movements was low (ca. 100 kN/m^2) and that friction
was also low (ϕ = 20—25°). Fluid pressures were apparently absent. The low
strength values suggest that failure occurred on smooth joint surfaces coated
with sheet silicate minerals. These surfaces are interpreted from analysis and
are apparently not reflected in available statistical analyses of rock fabric.
The portion of the slope which ultimately collapsed involved three relatively
distinct blocks separated by major discontinuities; sliding occurred on planes
dipping at perhaps 23—37°W. Displacement and acoustic emission data were
collected over a period of a year following initial observations of slope movement. Extrapolation of displacement data was used to predict slope collapse
over a month in advance of the date of actual collapse in 1969. Advance prediction permitted modification of mine transportation systems, and mine
production was stopped for only 65 hours. No injuries or equipment damage
occurred.

INTRODUCTION

   The Chuquicamata open pit mine is located high in the Andes in the arid,
mountainous Atacama Desert (Fig. 1). The pit (Figs. 2, 3) is carved in a
crystalline porphyry copper deposit; it is about 3.3 km long, 1.0 km wide,
and up to 400 m in depth. Over 1000 million metric tons of total material
had been removed by the end of August 1969 (see, e.g., Dunbar, 1952). With
the exception of a number of minor bench failures, no major slope problems
had occurred prior to 1967. In December 1967, however, pre-existing fractures on the south end of the east wall of the mine began to open. Various
instruments were then installed for continuous monitoring of slope behavior.
From the information obtained, the approximate "earliest possible" date of

Fig. 1. Location map of Chuquicamata Mine, Chile.

Fig. 2. Southeastern face of Chuquicamata open pit, late 1968. Area on crest of slope, to left of smelter stacks, was excavated in 1968 stripping program in an effort to control slope movements. Note sporadic small wedge failures on benches. See Fig. 7 for identification of benches and location of slide area.

slope collapse was predicted five weeks before actual collapse in February 1969. Mine truck and rail access routes were altered with a minimal mine shutdown of two and one-half days. No damage occurred.

The slide area was in the southeast corner of the pit (Figs. 3, 4). The original crest was at bench C-3 (bench elevations are given in Table I). A 3% mainline railroad ramp crossed the slide area from G-2 bench at 2742 m to the G-4 elevation of 2710 m, and a truck ramp descended at 8% from the H-1 bench at 2694 m to the H-3 bench at 2671 m. The overall slope angle varied from about 42° on the N3600 coordinate line to almost 47° at the N3440 coordinate line. The bottom of the mine at the time of failure was I-1, at elevation 2645 m, giving an overall slope height in the failure zone of 248 m.

Discussions of the Chuquicamata slope have been presented by Kennedy et al. (1971), and, in slightly modified form, by Kennedy and Niermeyer (1971). This account draws upon these papers as well as some additional sources; previously unpublished illustrations are included which more completely illustrate the slope collapse, and additional analyses are presented.

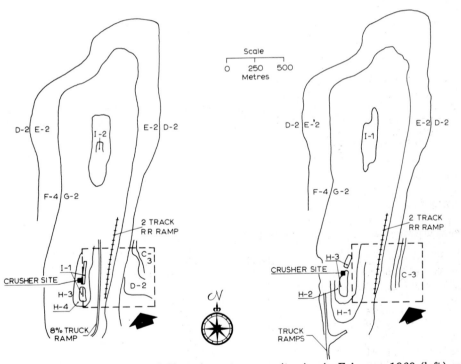

Fig. 3. Topographic maps of Chuquicamata open pit mine in February 1969 (left) and December 1967 (right) (after Kennedy et al., 1971). Slide area in southeast pit corner (arrows); see Figs. 4, 6 for map details.

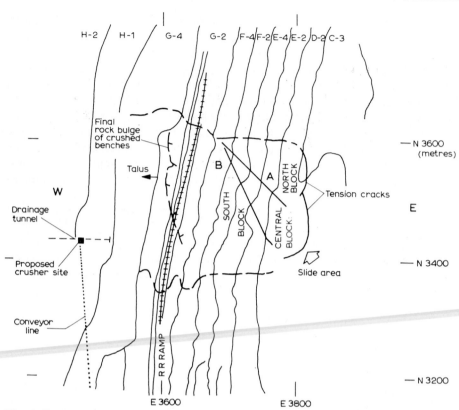

Fig. 4. Details of slide area (cf. Figs. 3, 7) showing major structures A, B and slide blocks in relation to benches (after Kennedy et al., 1971). See Table I for bench elevations. Profile of Fig. 9 indicated by E-W.

TABLE I

Bench elevations at Chuquicamata (after Kennedy et al., 1971)

Bench	Height above sea level (m)	Bench	Height above sea level (m)
C-3	2893	H-1	2694
D-2	2865	H-2	2684
E-2	2842	H-3	2671
E-4	2818	H-4	2658
F-2	2794	I-1	2645
F-4	2770	I-2	2632
G-2	2742		
G-4	2710		

GEOLOGIC SETTING OF SLIDE AREA

A complex contact between the Chuquicamata porphyry intrusives and the older metamorphic series of meta-sediments, volcanics and intrusives occurs in the southeast section of the mine (Lopez, 1939; Perry, 1952; Renzetti, 1957). The host rock in the slide area is granodiorite; this rock is virtually unaltered except adjacent to a few major structures which cross the slide area. Near these structures, plagioclase is most sensitive to alteration, followed in extreme cases by potash feldspar, with the main alteration products sericite, kaolinite, and chlorite. Little detailed information concerning the metamorphic rocks is available.

The degree of pre-slide cataclastic deformation (mechanical fracturing and crushing) within the area is exceptionally high in both igneous and metamorphic rocks. Diamond drill holes from the slide area indicated a zone of gypsum fracture filling, roughly concentric to the form of the pit; in this zone the rock mass is relatively strong (the gypsum may have been leached from overlying fractured rock, thus creating a relatively weak surficial mass about 50 m thick). Petrographic thin sections indicate that the gypsum is confined to fracture fillings and is not distributed interstitially. Leaching of the gypsum cement has also occurred along many of the larger structural zones, creating domains of weak rock within the stronger cemented mass.

The distribution of fractures in a spherical projection "pole" diagram appears to follow a relatively well-developed, broad "girdle" pattern with at least three reasonably distinctive maxima. These are identified here as sets 1, 2, and 3, respectively, indicating northwest, northeast, and east-west strikes (Fig. 5); all fracture sets are steeply dipping, generally with a westerly disposition. The northeast set is supposedly not well developed (Kennedy et al., 1971) but it seems statistically significant on the F-2 bench, at least in terms of quantity of observed fractures. The extent of individual fractures and their surface properties are separate questions that are of interest, but these cannot be illustrated or resolved by spherical projection techniques. A north-south-trending set with westerly dip has also been reported by Kennedy (1971, p. 328; cf., Kennedy et al., 1971, p. 57), although this set is not reflected by fracture data from bench F-2 (Fig. 5). This set (here called set 4) is apparently characterized by strong local variations and is interpreted as a face-parallel branching network of interconnected planes mainly dipping 50—90°W, with occasional shallower dips or eastward reversals (Kennedy et al., 1971, p. 57). Its approximate pole azimuth has been added to Fig. 5, where it effectively would fill in a "saddle" in the girdle pattern between pole maxima associated with the northwest- and northeast-striking fracture sets, as previously noted.

The statistical absence of set 4 probably reflects sampling bias, because a structural system parallel or nearly so to an exposed face will not intersect the surface of observation as frequently as highly oblique structures. The

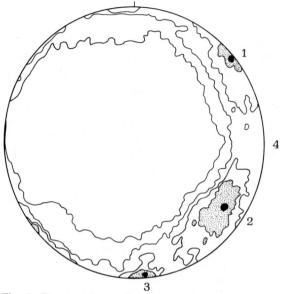

Fig. 5. Equatorial equal area lower-hemisphere stereonet summarizing pole orientations for 1270 fractures measured in the Chuquicamata slide area, bench F-2 (data from Kennedy, 1971, pp. 328—329). See Figs. 4, 7 for bench location. Contours at 0.6, 1.7, 2.9, 4.0, and 5.2% of 1.0% unit area. Fracture sets 1—3 defined by fabric maxima. Set 4 not evident in fabric diagram; see text discussion.

statistical problem has been generally discussed by Terzaghi (1965), Robertson (1971), and Broadbent and Rippere (1971). The well-developed northwest-trending set 1 appears to focus about a prominent structural zone [structure A of Kennedy et al. (1971), a major fault with a large number of associated, closely spaced minor fractures (Fig. 4)]. The maximum width of this zone is about 4 m; its strike is N45°W and its dip variable, in the range 60—90°SW. A second relatively major discontinuity zone (structure B) strikes N25°W and dips 80—90°SW (Fig. 4) and is thus also associated with fracture set 1; this is a single structure, in essence a splay from fault structure A; its maximum width is about 50 cm.

No water table could be detected in the slope. Water was not encountered in any of the vertical or horizontal holes drilled into the mass, and it is felt that groundwater considerations had little or no effect on the slope failure (Kennedy, 1971, p. 338).

The mine is located in an area of earthquake activity in which an average of about two tremors occur each day; the great majority of these are very low on the Richter scale, but magnitude-6 earthquakes have occurred in the region.

EARLY SLIDE HISTORY AND INSTRUMENTATION PROGRAM

In August 1966, tension cracks were first noticed on the C-3 bench between N3700 and N3800 and between N3470 and N3565. These cracks were mapped and monitored. Slope movements were small and eventually ceased, at which time monitoring was stopped. The next significant event was a strong earthquake, estimated at Richter magnitude 5, which occurred on 20 December 1967. A few days after the earthquake, concrete blocks at the southern end of the visitors overlook on C-3 bench were found to be tipped over. Tension cracks had formed at the crest area. Instruments to monitor rock movement were thus placed along the crest and over the entire slope face (Kennedy et al., 1971, pp. 57–58; Kennedy and Niermeyer, 1971, pp. 217–221).

The first measurement system installed was a series of "quadrilateral" stakes (Fig. 6). Displacement measurements gave the direction and magni-

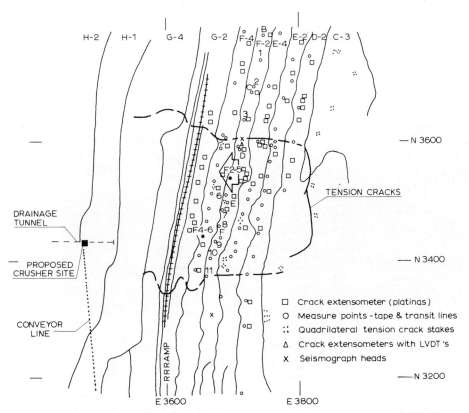

Fig. 6. Slide area instrumentation (cf. Figs. 3, 4; after Kennedy et al., 1971). Measurement system components indicated.

tude of movement. At least one stake was referenced to a known stable survey point at frequent intervals to provide a correction for large-scale movement. Lines of steel stakes were also installed along benches across the slide area and into stable ground on each side (Fig. 6). Horizontal and vertical offsets of each point were measured each day, and plots of horizontal, vertical and resultant total displacement, velocity, and acceleration were calculated and recorded. As the movements increased, velocity contours were drawn along the slope face (Fig. 7) to provide a graphical illustration of slide boundaries and of the distribution of active areas. Crack extensometers were installed; movements were ordinarily measured with dial gauges, but some extensometers were instrumented with LVDT units and strip-chart recorders. Finally, a three-channel short-period seismograph (D.C. to 12.5 Hz) was set up, and with experience it was possible to recognize the cultural noise generated by drills, trucks, etc., and earthquakes (Kennedy et al., 1971, p. 58). "Local movement" events in the slide area — with associated acoustic emission — were identified by characteristic envelope shape, a frequency of 6—9 Hz, and irregular amplitude. Records were kept of the number of "local movement" events occurring in given time spans, their frequency, duration, and associated earth motion.

Additional details of the instrument systems are given by Kennedy and Niermeyer (1971). In later stages of slide development, several rock blocks

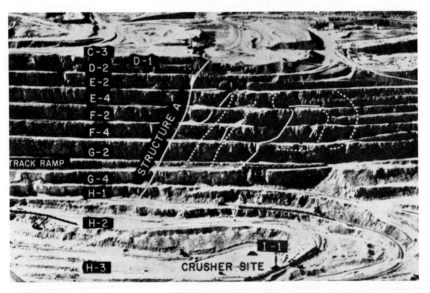

Fig. 7. Slide area, southeastern face of Chuquicamata mine (after Kennedy et al., 1971). Solid white lines are surface cracks; dotted lines are velocity contours. Structure *A* as indicated (cf. Figs. 2, 4). See Table I for bench elevations.

became unstable and minor slides developed. A system of electric contact monitors was then installed; rock movements activated a warning light in the dispatcher's office. When the alarm was observed, visual examination of the area was made and if necessary the troublesome rock mass was removed by blasting.

PROGRESSIVE FAILURE AND SLOPE COLLAPSE

When measurement was first started in early January 1968, movement rates of about 2—5 mm/day were recorded on crest tension cracks and 1—3 mm/day on the slope face. Limits of the slide were still imprecisely known but coordinate lines N3800 and N3300 appeared to define its approximate lateral boundaries. An estimated $11-14 \times 10^6$ metric tons of material appeared to be in motion (Kennedy et al., 1971, p. 58). Extensive monitoring was being carried out on benches D-2 to F-4, and rock mass discontinuities in the slide area were mapped in detail (Fig. 5). By September slope velocities had increased to about 7 mm/day; movements continued to progress at a more or less steady rate until 9 November 1968 (Fig. 8), when an abrupt increase was observed.

It should be noted that these slope movements were influenced by extensive mine excavations. During the latter part of 1967 and throughout 1968, excavation for a crusher site was being carried out at the slope toe (Figs. 4, 7, 9); the mine was deepened 39 m from H-2 to I-1 levels and existing lateral support between G-2 and H-2 was removed. In addition to these massive changes of slope geometry, the associated blasting undoubtedly contributed to deterioration of slope conditions. For example, blast 73 H-3, consisting of 211 holes loaded with 150 metric tons of explosive, was detonated close to the base of the slide area on 9 November 1968; the increase of slide movement rate mentioned above is specifically attributed to this event.

In addition, an unloading program commenced above D-2 at the crest of the slide area on 25 April 1968, and a total of 4×10^6 metric tons of material was removed. The effect of this unloading operation in delaying or minimizing the slide has been questioned (Hoek and Bray, 1974, p. 251); Kennedy et al. (1971, p. 57) remarked merely that the effect of unloading in delaying the slide was difficult to evaluate, but that the amount of material ultimately deposited in the bottom of the mine due to slope collapse was thereby much reduced. [1]

Velocity of slope movement sharply increased after the blast of 9 Novem-

---

[1] We were dealing with something that, at that time, very few people really understood from a practical standpoint. We felt, not knowing the mechanism of failure, that if we could remove some of the driving force, we might slow down the mass or even stop it. (B.A.K.)

ber; a few days afterward, slope velocities *decreased* suddenly to new steady
state values of as much as 3 cm/day (for measurement point 5; Fig. 8). By
this time the area north of the N3600 line seemed stabilized, whereas velocity contours suggested that distinctive zones of movement were being developed in the active slope; these are referred to as the "central block" [the

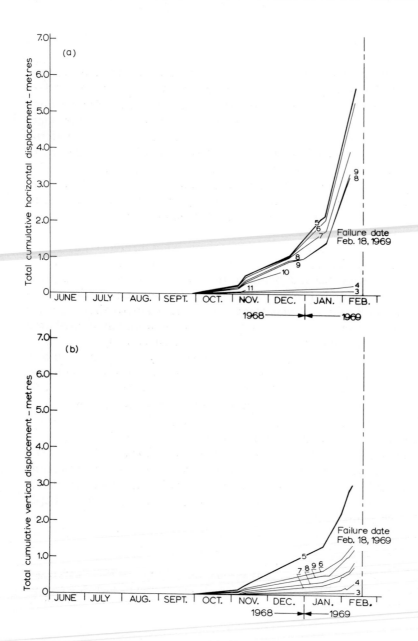

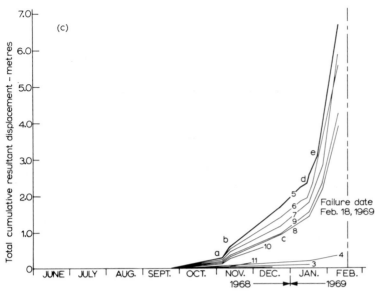

Fig. 8. Horizontal (a), vertical (b), and resultant displacement (c) versus time for F-2 bench measurement points, September 1968 to February 1969 (after Kennedy and Niermeyer, 1971). Data points associated with earlier displacement measurements were destroyed. Locations *a* to *e* on curve (Fig. 8c) indicate sharp breaks in displacement-time record; see text. Heavy line for point 5 data.

"central wedge" of Kennedy et al. (1971)], the "north block," and the "south block" (Fig. 4). The central block was bounded on the north by structure A, on the south by structure B, and at the crest by the southern set of tension cracks. The north block rested against the central block, with its approximate northern limit the N3600 coordinate line, and its crest the northern set of tension cracks (Kennedy et al., 1971). The south block was bounded by structure B and, approximately, the N3380 coordinate; its downslope boundary was not clearly defined. It was clear that shear movements were taking place within the slope, yet no definite toe had become apparent. Drill holes from solid ground behind the slope crest indicated an apparent failure plane dipping at about 70°W (precise location unspecified). Some drill holes penetrating the slide area were offset while drilling was in progress; one such hole closed off in one hour (Kennedy et al., 1971, p. 59).

In November the south and central blocks began to disaggregate into smaller joint-bounded blocks, especially in the area bounded by benches E-4 to G-2. Locally, wedges were formed with discontinuity intersections at a shallower angle than bench slopes, and a number of these small blocks (involving 300–1000 metric tons) became unstable.

By December the maximum resultant displacement of point 5 in the south

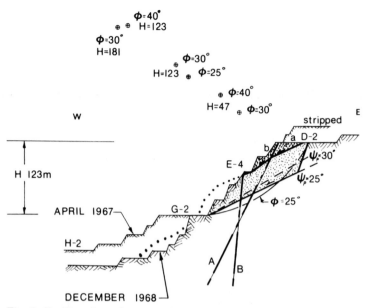

Fig. 9. Profile of southeastern face of Chuquicamata open pit at N3500 (modified after Kennedy et al., 1971). See Fig. 4 for location. Profiles for 1967 and 1968 shown; stripped bench crest above D-2 noted. Slide profile denoted by heavy dots. Structures A, B are heavy lines. Critical theoretical cylindrical failure surface for $\phi = 25°$ shown; also shown are critical planar surfaces, $\psi_p = 25°, 30°$, bounded at rear by steeply dipping fractures. Blocks a, b, with dotted patterns are parts of the north and central blocks, respectively, involved in the slope collapse. The deformed zone extends at least to $H = 123$ m. Centers of rotation for critical cylindrical failure surfaces are indicated for various H, ϕ.

block was well over a metre, and slope velocity was more or less steady (about 3 cm/day at measurement point 5).

A minor change in velocity occurred on or about 23 December 1968, and major changes developed in the periods 15—17 and 24—28 January, 1969 (see displacement-time curve locations a to e; Fig. 8c). These specific features of the data have not been discussed in the literature. Some may be due to deterioration of rock mass cohesion by specific blasting events, increased average shear stress-to-strength ratio associated with excavation-induced slope geometry change, and/or loss of cohesion due to progressive failure. Others are due to re-setting of survey points as sightlines became difficult, or as points were damaged. Examination of the blast and related excavation events as a function of time will be necessary for adequate interpretation of the displacement record. [2] It may be noted, however, that adjustment of

---

[2] Points 8 and 9 were located on a bench that failed at about time c. This bench failure may have released local slope constraints, leading to increased displacement rate as shown in Fig. 8c.

the slope to a new state of dynamic equilibrium appears to have required a period on the order of several days. Not all monitoring points appear to have been activated on the same day by a given trigger mechanism. For example, the cluster of significant rate changes noted for the approximate period 15–17 January suggest that the slope was first strongly affected in the vicinity of measurement points 5 and 6 (cf. Figs. 6, 8). About two days later the data suggest similar rate changes for point 4, near the north boundary of the slide area, and points 7–9, toward the south boundary. A similar trend can be noted for bench F-2 data for the period 24–28 January. Presumably similar evidence for progressive behavior also exists for points spread over the entire slope, but the data are not available for analysis. Following the 9 November blast, for example, the increase in displacement was first recorded at the toe and subsequently progressed with time up to the crest of slope (Kennedy, 1971, p. 328).

The effect of blasting on slope conditions is very likely a most significant factor. The sensitivity of the slope to vibration-induced damage is suggested by the effect of the December 1967 earthquake, and the effects of subsequent blasting events were probably of equivalent or greater severity in this respect, as the following remarks by Kennedy (1971, p. 328) suggest:

"During primary blasting it was customary to turn our seismographs off as the instruments were so sensitive that a seismic acceleration of the magnitude and proximity of a blast would put the unit out of range and possibly cause damage".

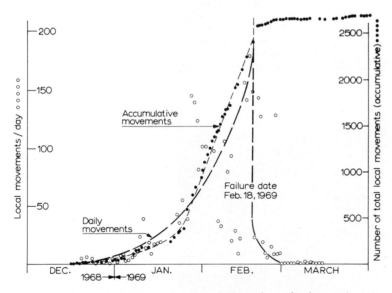

Fig. 10. Number of daily and accumulative microseismic events versus time (after Kennedy and Niermeyer, 1971). Range 6–9 Hz.

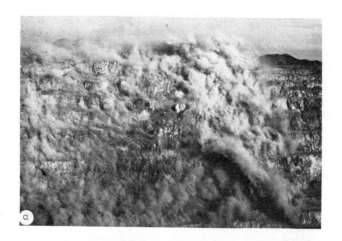

Fig. 11. Slope collapse at southeastern face of Chuquicamata open pit: (a) 6:59 p.m.; (b) 7:00 p.m.; (c) 7:01 p.m.; (d) 7:02 p.m.; (e) 7:13 p.m.; (f) post initial clean-up operations. Note positions of railway and haulroad; cf. Figs. 2, 7.

Records of blasting and excavation as a function of time, in the proximity of the moving slope, have not been made available.

By New Year's Day 1969 resultant displacements of 2 m had been obtained, and by another month the interior points of the south block had been displaced 4 m. The velocities had increased dramatically in two abrupt steps after mid-January, and by 11 February velocities between 19 and 46 cm/day were recorded for south block measurement points. Benches on the slope face were becoming dangerous and virtually inaccessible, and a measurement system was therefore set up on the crest of the slide (Kennedy et al., 1971, p. 60). Stadia rods were set up on two distinct blocks in the slide area. These were read hourly by transit with its telescope fixed at a set vertical angle. On 15 February the velocity of the mass increased once again and the slide area became extremely active with many small rock falls. At 3:00 p.m. on 16 February, the mine was closed down; falling boulders were becoming a hazard on the haul road. Displacement measurements were increased to 15-minute intervals. The velocity of the mass increased steadily. The frequency of acoustic emission had also increased significantly from an average of 3—5 events per day in October 1968 to a widely variable rate from late January to mid-February 1969, in which over 100 events per day were not uncommon (Fig. 10). At 6:45 p.m. on 18 February, the outer crest was moving at about 1.4 m/hr, and at 6:58—7:02 p.m. slope collapse occurred, involving rock from the crest of D-2 to bench F-2. The mass of debris according to Kennedy et al. (1971) involved about 1.4×10^6 metric tons and fell downslope as far as the H-3 bench (Fig. 11). The slide mass chiefly involved the central block; part of the north block slid out once the buttress formed by the central block had been removed. A small portion of the south block was sheared through in the toe of the slide mass, but most of the south block remained in place although it was permanently deformed by displacements as great as 8 m. The mass involved in the deformed zone was estimated at 4.1×10^6 metric tons (Kennedy et al., 1971).

AN ATTEMPT TO PREDICT THE DATE OF SLOPE COLLAPSE

Kennedy et al. (1971) and Kennedy and Niermeyer (1971) claimed that "the failure date was accurately predicted some five weeks before failure", [3] thus captivating the imagination of the geotechnical reader. Indeed, no aspect of the Chuquicamata slope failure received more attention than this prediction. Salamon (1971) probably spoke for many when he remarked that the accuracy was very impressive and the confidence put into these predictions was almost unbelievable. The prediction question is examined here.

Toward the end of 1968 and early 1969, displacement rates were suffi-

---

[3] "Poetic licence in the heat of the moment." (B.A.K.)

ciently high that some form of major slope failure seemed almost inevitable. In order to minimize interference with Chuquicamata operations, an attempt at prediction of the date of slope collapse seemed necessary. Management wanted an answer. Two important assumptions were made (Kennedy et al., 1971, p. 59): (1) if the fastest-moving point on the slope continued at the fastest rate, this would be the point most likely to fail first; (2) at any given point the rock mass could only absorb a finite amount of displacement before failure occurred; this critical displacement would presumably be maximum at the crest and minimum at the toe.

It was next estimated that the fastest-moving measurement point — point 5 on F-2 bench — could tolerate approximately 6 m of displacement prior to collapse. This "collapse criterion" was based on engineering judgment, following examination of the displacement data and study of the rock mass; two minor bench failures apparently yielded useful data on magnitudes of expected displacements (Kennedy et al., 1971, p. 59; Kennedy, 1971, p. 331). Attempts were then made to fit equations to the displacement-time curves. Least square, parabolic, and hyperbolic fits were among those tried (cf. Fig. 12). However, large differences in prediction date depended upon the curve-fitting technique used; furthermore, small variation in the constants of extrapolation equations caused enormous differences in the predicted failure date. Because of the great variance in these methods, the curves plotted from extrapolation of equations were used only as *rough guidelines* in drawing the final prediction curves (Kennedy et al., 1971, p. 59). The hand-drawn extrapolation of the cumulative displacement-time of the fastest-moving point, F-2-5 (Fig. 13), gave the *earliest* estimated date of possible failure as 18 February 1969; this estimate was made on 13 January 1969. This predicted date was not put forth as a definitive answer to the slope failure question; it was purely a "best estimate" from the available data at that time. [4] Field work continued, and discussions continued regarding the validity of the prediction date. Failure date predictions were not made prior to 13 January. Some revision in approach was subsequently necessary, because the collapse criterion was soon proven invalid: point F-2-5 had accumulated more than 6 m of displacement before the first week in February had ended.

Kennedy et al. (1971, p. 59; cf. Kennedy and Niermeyer, 1971, p. 222)

---

[4] The original plot contained the following qualifying statements:

(1) The data used to compile these prediction curves gives rather flat plots, resulting in difficulty in extending the curves to a prediction point.

(2) If a massive failure were to take place, the earliest possible date of failure would be 18 February 1969.

(3) The possible early date of 18 February is determined from plotting the data of one measure point only.

A range of possible failure dates was indicated by an arrow extending from 18 February to September.

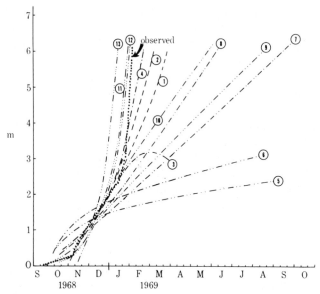

Fig. 12. Non-linear displacement-time regressions for Chuquicamata slope, time intervals 25 September to 13 January, or 1 February. Observed displacements noted by heavy dotted line. Numbers correspond to data of Table II.

suggest that the seismograph also yielded data suitable for predictive purposes. [5] This suggestion seems questionable. The data are presented in Fig. 10. Whereas in October 1968 an average of 3—5 acoustic events (interpreted as local movements) per day were recorded, the frequency of acoustic events increased (*not* steadily) to 182 recorded in the 12 hours prior to collapse. Plotting the number of daily events or accumulative events in the 6—9-Hz range against time, a similar curve to that of total displacement against time can be defined (Fig. 10). However, it seems unlikely that this curve could have been used for quantitative prediction of slope failure, because it is doubtful that a reasonable criterion for slope collapse based on acoustic emission could have been selected. Furthermore, insufficient data seem to have been available in early January for confident extrapolation of the curve.

The interpretative problem of slope collapse prediction hinges on resolution of several important questions:

(1) How can the movements which occur prior to slope collapse be distinguished from movements associated with slopes that stabilize?

(2) What *data* (displacement, velocity, seismic, etc.) are best suited to collapse prediction?

(3) What *extrapolation method* for available data is most suitable for a given set of conditions?

---

[5] Its use in this manner was not in fact attempted (Kennedy et al., 1971, p. 59).

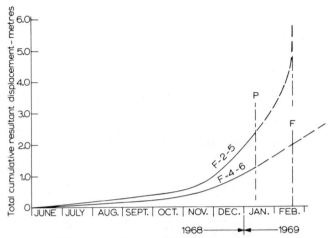

Fig. 13. Failure date prediction diagrams using total resultant displacement versus time (after Kennedy et al., 1971). Extrapolation made on data collected on and prior to 13 January 1969 (*P*). Predicted and actual failure is 18 February (*F*), based on fastest-moving point (F-2-5). Slow-moving point F-4-6 also shown.

(4) Given adequate data and a suitable extrapolation method, how does one select a reliable *criterion* for slope collapse?

There is no reliable and practicable set of answers to these questions available at the present moment, either from the point of view of theory or empiricism. It seems clear that more data than are currently available will be required in order to assess various extrapolation methods and collapse criteria.

To illustrate the problem, thirteen extrapolation curves are given based on Chuquicamata displacement-time data for, primarily, the periods 25 September to 13 January and 25 September to 1 February (Fig. 12). These curves are based on data for measurement point F-2-5 presented in Fig. 8 (not raw data). Regression equations, constants, and prediction dates (for the criterion, resultant displacement $d = 6$ m) are given in Table II.

Collapse dates thus predicted vary from January 1969 to infinity. Some plots (logarithmic) suggest deceleration of slope movements, whereas others (quadratic, exponential) suggest acceleration. The cubic extrapolation appears to be most sensitive to variations in constants; for the data to 13 January, stabilization is predicted after a displacement of about 3 m, whereas for the data to 1 February, slope collapse is predicted on 20 February. By guidance of hindsight one tends to favor curves which consistently predict a continuous increase of displacements with time. On these grounds the quadratic and exponential forms seem most suitable. Variation in predicted failure data for these two forms is in excess of two months! The exponential form appears to be the more conservative for the data employed

TABLE II

Non-linear displacement-time regressions for time intervals 25 September to 13 January and 25 September to 1 February

Curve number	Regression equation *	Regression constants	Approximate data interval	Predicted date for $d = 6$ m
1	$d = a + bt + ct^2$	$a = -0.023$ $b = 6.2 \times 10^{-3}$ $c = 1.4 \times 10^{-4}$	25 Sept.–13 Jan.	25 Mar. 1969
2	$d = a + bt + ct^2$	$a = 0.12$ $b = -3.4 \times 10^{-3}$ $c = 2.4 \times 10^{-4}$	25 Sept.–1 Feb.	8 Mar. 1969
3	$d = a + bt + ct^2 + ft^3$	$a = 0.097$ $b = -6.2 \times 10^{-3}$ $c = 4.2 \times 10^{-4}$ $f = 1.6 \times 10^{-6}$	25 Sept.–13 Jan.	value not reached $d < 3.2$ m
4	$d = a + bt + ct^2 + ft^3$	$a = -0.14$ $b = 2.1 \times 10^{-2}$ $c = 2.4 \times 10^{-4}$ $f = 2.5 \times 10^{-6}$	25 Sept.–1 Feb.	20 Feb. 1969
5	$d = a + b \ln t$	$a = -1.3$ $b = 0.91$	25 Sept.–13 Jan.	May 2070

6	$d = a + b \ln t$	$a = -1.8$ $b = 0.91$	25 Sept.—1 Feb.	Sept. 1982
7	$d = at^b$	$a = 0.014$ $b = 1.1$	25 Sept.—13 Jan.	13 Sept. 1969
8	$d = at^b$	$a = 0.010$ $b = 1.2$	25 Sept.—1 Feb.	8 June 1969
9	$d = at^2 + b$	$a = 3.6 \times 10^{-4}$ $b = -0.45$	25 Sept.—13 Jan.	16 Aug. 1969
10	$d = at^2 + b$	$a = 6.8 \times 10^{-4}$ $b = -1.6$	25 Sept.—1 Feb.	24 May 1969
11	$d = a\,e^{bt}$	$a = 0.12$ $b = 0.032$	25 Sept.—13 Jan.	31 Jan. 1969
12	$d = a\,e^{bt}$	$a = 0.13$ $b = 0.031$	25 Sept.—1 Feb.	2 Feb. 1969
13	$d = a\,e^{bt}$	$a = 0.092$ $b = 0.04$	25 Sept.—8 Nov.	13 Jan. 1969

* d = resultant displacement, m; t = elapsed time, days.

here, suggesting the earlier failure dates which span three weeks time.

These plots should put the question of accurate prediction into better perspective. Even if one knew collapse would in fact occur rather than stabilization, and assuming an absolutely reliable collapse criterion, a predicted failure date should not have been expected to be closer than a week to the actual collapse data.

In many discussions of slope behavior it has been tacitly assumed that slope failure was known to be impending even if the collapse date was uncertain. But why is this necessarily the case? Indeed, what criteria can be used to indicate the future failure, or future stability, of a slope?

Stability is achieved when velocities of points in the slope vanish. Considered in terms of a displacement-time diagram, a continuously increasing curve slope implies ultimate instability, whereas a continuously decreasing slope *may*, under certain conditions, imply stability. The proviso is that the maximum displacement achieved, as velocity approaches zero, must be less than some critical displacement associated with slope collapse. Thus, the concave-downward slopes of Fig. 12 (curves 5, 6) led to a prediction of failure in 1980 or beyond.

It is evident by inspection that many of these curves are not particularly good representations of the data on which they are based, for the ranges in which theoretical curve and observational data coincide. Thus there may be a tendency for the analyst to discard them without further consideration. Indeed, the original data cannot be described in all details by *any* continuous function. There are sharp breaks in velocity, as discussed previously, associated with specific blasting and rock excavation events. This fact suggests that for a given value of elapsed time, the incremental displacement-time data may be independently analyzed. Regression data for Chuquicamata are thus presented in Table III for the following time increments: 25 September to 8 November, 8 November to 13 January, 13—25 January, and 25 January to 10 February. Successive logarithmic curves (Fig. 14) are as accurate as any in terms of point-for-point representation of the data, for their range of coincidence with the data. On the basis of the 6-m displacement criterion, ultimate slope stability is predicted for the data to 8 November, whereas collapse is predicted in April 1970, March 1969, and February 1969, respectively, for the data to 13 January, 25 January, and 10 February, 1969. Linear approximations seem reasonable as an alternative form, yielding prediction dates somewhat more advanced than those cited above. In both general cases, the more recent approximations tend to converge upon the "correct" answer.

In terms of Chuquicamata, the incremental approach supports the contention that the displacement-time data falls into discrete regions bounded by positions *a* to *e* on curve 5, Fig. 8c (cf. Broadbent and Ko, 1972). According to this view the slope deforms continuously according to some regular pattern, until that pattern is disrupted by some internal disturbance (progres-

TABLE III

Displacement-time regressions for selected time increments

Curve number	Regression equation	Regression constants	Predicted date for $d = 6$ m	Data interval
1	$d = a + bt$	hand-fit	Mar. 1970	25 Sept.–8 Nov.
2	$d = a + bt + ct^2 + ft^3$	$a = 2.7 \times 10^{-3}$ $b = 1.1 \times 10^{-2}$ $c = -1.4 \times 10^{-4}$ $f = 1.5 \times 10^{-6}$	17 Mar. 1969	25 Sept.–8 Nov.
3	$d = a + b \ln t$	$a = -0.074$ $b = 0.092$ $(r^2 = 0.86)$	value not reached $d < 1.0$ m	25 Sept.–8 Nov.
4	$d = at^b$	$a = 0.05$ $b = 0.5$	after 1 year $d < 2.0$ m	25 Sept.–8 Nov.
5	$d = at^2 + b$	$a = 8.0 \times 10^{-5}$ $b = -1.9 \times 10^{-2}$	Aug. 1970	25 Sept.–8 Nov.
6	$d = a\,e^{bt}$	$a = 0.092$ $b = 0.04$	13 Jan. 1969	25 Sept.–8 Nov.
7	$d = a + bt$	hand-fit	23 May 1969	8 Nov.–13 Jan.
8	$d = a + bt + ct^2 + ft^3$	$a = -1.05$ $b = 4.1 \times 10^{-2}$ $c = -1.9 \times 10^{-4}$ $f = 9.1 \times 10^{-7}$	4 Apr. 1969	8 Nov.–13 Jan.
9	$d = a + b \ln t$	$a = -6.6$ $b = 2.2$ $(r^2 = 0.99)$	Apr. 1970	8 Nov.–13 Jan.
10	$d = at^b$	$a = 0.44$ $b = 0.37$	after 1 year $d < 4.0$ m	8 Nov.–13 Jan.

TABLE III (continued)

Curve number	Regression equation	Regression constants	Predicted date for $d = 6$ m	Data interval
11	$d = at^2 + b$	$a = 4.5 \times 10^{-4}$ $b = -1.3$	17 June 1969	8 Nov.–13 Jan.
12	$d = a\,e^{bt}$	$a = 0.20$ $b = 0.039$	2 Feb. 1969	8 Nov.–13 Jan.
13	$d = a + bt$	hand-fit	1 Mar. 1969	13 Jan.–25 Jan.
14	$d = a + bt + ct^2 + ft^3$	$a = 0.74$ $b = -4.7 \times 10^{-2}$ $c = 5.5 \times 10^{-4}$ $f = 2.7 \times 10^{-8}$	20 Feb. 1969	13 Jan.–25 Jan.
15	$d = a + b \ln t$	$a = -41$ $b = 10.4$ $(r^2 = 0.98)$	21 Mar. 1969	13 Jan.–25 Jan.
16	$d = at^b$	$a = 2.3$ $b = 0.12$	after 1 year $d < 5.0$ m	13 Jan.–25 Jan.
17	$d = at^2 + b$	$a = 1.9$ $b = -22$	15 Feb. 1969	13 Jan.–25 Jan.

18	$d = a\,e^{bt}$	$a = 0.058$ $b = 0.28$	27 Jan. 1969	13 Jan.–25 Jan.
19	$d = a + bt$	hand-fit	7 Feb. 1969 *	25 Jan.–10 Feb.
20	$d = a + bt + ct^2 +$ $\quad ft^3$	$a = 4.7$ $b = 9.8 \times 10^{-2}$ $c = -3.1 \times 10^{-3}$ $f = 1.8 \times 10^{-5}$	6 Feb. 1969	25 Jan.–10 Feb.
21	$d = a + b \ln t$	$a = -118$ $b = 29$ $(r^2 = 1.0)$	7 Feb. 1969	25 Jan.–10 Feb.
22	$d = at^b$	$a = 2.8$ $b = 0.26$	7 Feb. 1969	25 Jan.–10 Feb.
23	$d = at^2 + b$	$a = 5.8 \times 10^{-3}$ $b = -88$	7 Feb. 1969	25 Jan.–10 Feb.
24	$d = a\,e^{bt}$	$a = 0.26$ $b = 0.16$	7 Feb. 1969	25 Jan.–10 Feb.

* Note: observed date for $d = 6.0$ m is 7 February, *within* data interval.

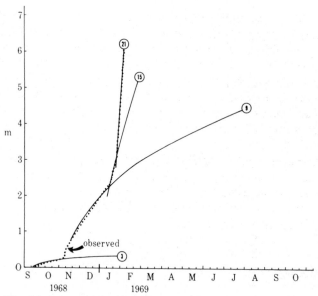

Fig. 14. Logarithmic displacement-time regressions for Chuquicamata slope, incremental analysis. Numbers correspond to data of Table III. Observed displacements denoted by heavy dotted line.

sive rupture of a major rock "bridge" separating discontinuities) commonly reflecting some external cause (blasting, excavation, earthquakes) or mass rock creep. Prediction of slope behavior requires knowledge of the nature of the existing dominant deformational pattern plus an effective means of estimating specific time-variable trends as influenced by external and internal factors. In the case of external factors, this seems potentially possible if excavation and blasting schedules are known in advance, but application would be difficult in detail. Internal factors (due to, say, mass rock creep) might be predicted on a statistical basis. A conservative alternative is to suspect that if the worst can happen, it will; this approach effectively leads to an exponential extrapolation or something equivalent to it. It may be noted that the exponentially based predictions of failure date based on the above time increments are conservative and fall within a relatively narrow range, viz., 13 January to 7 February 1969 (Table III). The success of this approach much depends on the local conditions; if the slope is relatively undisturbed by external factors, the method may greatly overestimate the rate of natural deformation.

Finally, it may be noted that each type of extrapolation curve displays its own characteristic sensitivity to the value of maximum displacement assumed for the limiting value criterion. Prediction dates taken from steeply sloping curves, such as exponential functions, are least sensitive. A range of

TABLE IV

Sensitivity of predicted failure date to assumed critical displacement, exponential curve

Data interval	Predicted date		
	$d = 6$ m	$d = 10$ m	$d = 15$ m
1 Oct.—9 Nov.	13 Jan.	23 Jan.	27 Jan.
9 Nov.—13 Jan.	3 Feb.	17 Feb.	27 Feb.
13 Jan.—25 Jan.	28 Jan.	30 Jan.	31 Jan.

dates associated with exponential extrapolations of incremental data from Chuquicamata, for assumed maximum displacements in the range 6—15 m, are given in Table IV. The range of predicted failure dates varies from a few weeks to only a few days, with the narrower range associated with higher velocities and lower net required displacements.

ANALYSIS OF MECHANICS OF SLOPE FAILURE

Rock fabric

Wedge failure can occur when the plunge of the line of discontinuity intersection exceeds the friction angle ϕ and yet is also less than the dip of the face of slope. This implies that intersections for an unstable wedge must fall, in a spherical projection, in the area between a great circle defining the slope face and a small circle defined by the angle of friction (see, e.g., Hoek and Bray, 1974, pp. 48—54). For Chuquicamata this region is small, since the slope face angle ψ_f is 42—46° and the estimated minimum friction angle is about 20°.

Examination of the fabric of discontinuities within a portion of the slope (Fig. 15) suggests that no statistically predominant set "daylights" at an angle appropriate to encourage slope failure. Four significant fracture sets exist, and in addition two major discontinuity zones (structures A and B) are present which can be considered individually as members of set 1; but the lines of intersections of these various fabric elements plunge more steeply than the slope face (Fig. 15), and the wedges thus formed may be regarded as stable.

Given this statistical display, a feasibility study would suggest few slope problems for pit slopes of moderate angle, say $\psi_f < 60°$. This conclusion is obviously based on a small sample, but it is supported by the complete absence of previous major slope failures at Chuquicamata. A number of minor bench failures have occurred (see, e.g., Figs. 2, 7); indeed, wedge fail-

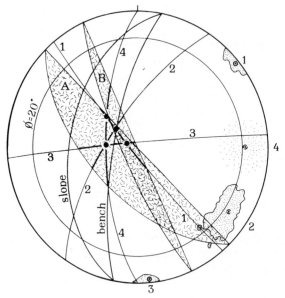

Fig. 15. Equal area stereonet summarizing important geometrical and geological relationships for Chuquicamata slope. Face and bench slopes indicated by great circles. Limiting
friction small circle denoted by $\phi = 20°$. Pole maxima for fracture sets indicated by numbers 1—4, as are great circles corresponding to mean orientations of fracture sets. Structures A and B each indicated by patterned areas bounded by two great circles indicating
range of observed dips. Heavy dots represent important intersections of planar discontinuities; these are located in the stable region with respect to pit slope face inclination.
Region of potential instability shaded. See text for details.

ures are predicted on intersections involving all significant fabric elements.
Many could be eliminated by restricting bench face angles to 60° or less, but
there are limits to the practicality of this suggestion. With respect to major
slope failure, discontinuity set 3 is likely to have been of significance in providing low-friction lateral boundaries of release parallel to the direction of
slope displacements. The rear boundary of the slide area and associated tension cracks were likely influenced by sets 1, 2, and 4.

Analysis of mobilized strength suggests that pre-existing weakness planes
were important in slope movements. Basal shear must have occurred predominantly on discontinuities dipping (at angles ψ_p) toward the pit; since
$\phi < \psi_p < \psi_f$, $\psi_p = 21-45°$. Now this orientation is completely lacking on
the fabric diagrams of Figs. 4 and 14 [6], although the existence of such discontinuities can be inferred from the fact of the slope failure and considera-

[6] However, it must be emphasized that these data reflect fabric measurements from one
bench only; additional fabric information was obtained but the data have not been
released.

tion of elementary mechanical principles. The Chuquicamata problem may thus serve as an excellent example of the significance of Karl Terzaghi's "minor geologic details", i.e., features that are not discovered from the results of careful investigations (cf. Terzaghi, 1929, 1962; Chapter 2, this volume). The purely statistical (discontinuity orientation) approach is simply inadequate for complete resolution of the question of stability when only a single, well-defined discontinuity of appropriate orientation can control the kinematics of deformation.

Next we consider analysis of limiting equilibrium. Two different idealizations are first examined and the data given; the results are then interpreted in relation to Chuquicamata slope behavior.

Planar failure surface idealization

The two-dimensional idealization may not be greatly in error. [7] For computation with regard to the planar mode, it is assumed that failure occurs on a single plane which strikes approximately parallel (say $\pm15°$) to the slope face, and with dip ψ_p such that $\phi < \psi_p < \psi_f$. A slice of unit thickness is considered, subjected to the following assumptions:

(1) The tension crack as well as slide surface strikes parallel to the slope face.

(2) Groundwater is not a factor and therefore fluid pressure can be ignored in analysis.

(3) Shear strength is governed by the Coulomb law involving cohesion (c) and friction (ϕ) terms.

(4) Shear forces at lateral slide boundaries are negligible.

(5) Because they were not necessary to sustain slope movements, dynamic forces associated with earthquakes and blasting can be ignored.

From a simple force balance:

$$c = \frac{W \sin \psi_p}{A} \ [F - \cot \psi_p \tan \phi] \tag{1}$$

where the weight of slide block, W, is given by:

$$W = \tfrac{1}{2}\gamma H[(1 - (z/H)^2] \cot \psi_p - \cot \psi_f \tag{2}$$

the area of the slide surface, A, is given by:

$$A = (H - z) \operatorname{cosec} \psi_p \tag{3}$$

and F is factor of safety (the ratio of total resisting force to total shear

---

[7] An attempt to model the stress distribution in the Chuquicamata slope is given in Chapter 22, this volume (see section "Open pit mine B").

force), γ is unit weight, z is depth of tension crack and H is height of failed portion of slope. Depth of the critical tension crack is given by:

$$z_c = H[1 - (\cot \psi_f \tan \psi_p)^{1/2}] \tag{4}$$

and its associated position (distance b from the crest of slope) is:

$$b_c = H[(\cot \psi_f \cot \psi_p)^{1/2} - \cot \psi_f] \tag{5}$$

The critical orientation of failure plane, ψ_{pc}, is given by:

$$\psi_{pc} = \tfrac{1}{2}(\psi_f + \phi) . \tag{6}$$

All terms correspond to those employed by Hoek and Bray (1974, pp. 133—136, 141, 143—147). Limiting equilibrium is then assumed for back analyses. By setting $F = 1$, and defining H, equation (1) gives a solution for c required for equilibrium as a function of ϕ; in all calculations, $\gamma = 2.6$ Mg/ m^3 and $\psi_f = 44°$.

Data from minor bench failures perhaps would be informative in regard to selection of ϕ values, but these have not been published. The value $\phi = 20°$ is judged to be a practical minimum value. [8] The maximum ϕ angle is limited by the slope face angle of 42—46°; thus a practical range $\phi = 20—40°$ seems indicated.

Data based on calculations for $\phi = 20°$, 30° and 40°, and for $H = 123$ m slope height (from the crest at bench D-2 to the base of the bulged zone at bench G-2) and $H = 61.5$ m (exactly half of the previously assumed height, but also approximately the average height of slope from the crest to the base of slope collapse between benches E-4 and F-2), are presented in Table V and Fig. 16.

---

[8] The apparent (therefore minimum) dip of the slide surface of the unstable portion of the north block is about 23°W (Kennedy et al., 1971), but this may be diagrammatic. The range $\phi = 35—45°$ seems suitable for partially weathered granitic rock (cf. Rocha, 1964; Vargas, 1967; Hamel, 1971; Hoek and Bray, 1974, pp. 152, 169), and heavily kaolinized granite (Ley, 1972). On the other hand, the possibility of somewhat lower values cannot be excluded inasmuch as fracture coatings may contain sericite (fine-grained mica), chlorite, kaolinite. A residual friction angle of 15° was determined for kaolinite (Kenney, 1967, table 1). The residual friction angle for mica and chlorite is a function of humidity; friction values as low as 10—15° have been determined for the condition of water saturation, whereas the angle 18—28° applies to an oven-dry state (Horn and Deere, 1962; Kenney, 1967; cf. Voight, 1973). For the oven-dry, air-equi-librated condition the range is 14—20°, which conceivably could correspond to condi-tions at Chuquicamata. However, because of roughness of the failure surface, imperfect parallelism of sheet silicate minerals, and mixtures of massive minerals, shear resistance is necessarily higher than these minima and may more closely approach that of granular minerals.

TABLE V

Two dimensional failure idealizations, Chuquicamata slope

	H = 61.5 m			H = 123 m		
	ϕ = 20°	30°	40°	ϕ = 20°	30°	40°
Planar mode						
ψ_p (deg.)	32	37	42	32	37	42
z (m)	12	7.2	2.1	24	15	4.2
b (m)	16	10	2.2	31	21	4.5
A (m²)	93	90	81	187	180	178
W (t */m)	2490	1360	362	9960	5460	1450
c (kN/m²)	59	21	2.0	118	43	3.8
Cylindrical mode						
c (kN/m²)	110	50	10	220	90	20

* Metric tons.

Cylindrical failure surface idealization

   In this idealization it is assumed that the distribution and orientation of rock mass discontinuities are sufficiently widespread and diverse such that

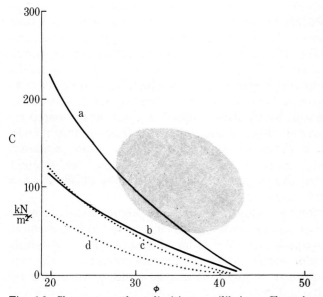

Fig. 16. Shear strength at limiting equilibrium, Chuquicamata slope. Cylindrical failure mode, *a* and *b*; planar failure mode, *c* and *d*. For curves *a*, *c*, slope height is 123 m; for *b*, *d*, height is 61.5 m (see Table V). Typical range for partially weathered granite according to Hoek and Bray (1974, p. 169) indicated by shaded pattern.

development of a failure surface along a path of least resistance is possible. In a statistically homogeneous material such a path is approximately cylindrical. The following additional assumptions are made:

(1) Shear strength is governed by a Coulomb law.

(2) Fluid pressures are negligible.

(3) Shear forces at lateral boundaries are negligible.

(4) A tension crack exists behind the slope crest.

(5) The position of the center of curvature for the failure surface and the tension crack location are both such as to minimize the slope factor of safety.

The lower-bound solution for static equilibrium, obtained by the "friction-circle method" of Taylor (1948; cf. Janbu, 1945; Lambe and Whitman, 1969, pp. 357—363), modified to include the influence of a tension crack (Hoek and Bray, 1974, p. 212), is used here. The approximate locations of the rotation centers for critical cylindrical surfaces are given in Fig. 9, determined from the rotation center method of Hoek (1970, fig. 20; cf. Taylor, 1948; Janbu, 1954). Accordingly, the dimensionless ratio $c/\gamma HF$ is determined as a function of tan ϕ/F and the slope angle ψ_f (Hoek and Bray, 1974, p. 218). Setting $F = 1$ yields the limiting value of c mobilized at limiting equilibrium. Values are summarized in Table V and Fig. 16.

Interpretation

Irrespective of idealization, the "average" cohesion required for stability was typically less than 100 kN/m² (1 bar); such values are even lower than those considered "reasonable" by Hoek and Bray (1974, p. 169) for partially weathered granite (Fig. 16). This result seems to underscore the dominant role of rock mass discontinuities in Chuquicamata slope movements. The total shear resistance to slope movement must have been quite small to permit either slope collapse or large-scale displacements. Sliding must have occurred on fairly continuous joint planes. Possibly these were coated by sheet silicate minerals, thus reducing the angle of sliding friction; but in addition such surfaces must have been smooth, with only minor apparent strength due to undulation, asperities, or intact rock "bridges". Subsequent stabilization of a portion of the slope could reflect decrease of shear stress (due to stripping) or strength increase, e.g., dilation.

The values established for the cylindrical mode are higher than those for plane failure, simply because the geometry of the associated failure surfaces are on the whole more favorable. It is the order of magnitude of the results that is significant. The reader should not be overly concerned about the precision of the stated values; any of the c solutions could be increased by assuming progressive failure and restricting cohesive resistance to portions of the failure surface.

A cylindrical failure surface for the $H = 123$ m slope failure was depicted by Kennedy et al. (1971, fig. 9). This was purely a diagrammatic representation (Kennedy, 1971, p. 328), and it was recognized that the actual failure surface could have been irregular, following various joint surfaces. However, the present analyses suggest that the depth of the failure zone may have been overemphasized. Both cylindrical and planar analyses suggest relatively shallow failure zones. Planar failure is limited by $\psi_p > \phi$ with the critical value $\psi_{pc} = \frac{1}{2} (\psi_f + \phi) > 32°$, assuming minimum $\phi = 20°$. For cylindrical failure the minimum "critical" radius of curvature is nearly $2H$, significantly greater than shown by Kennedy et al. (1971). Volumes associated with the two idealizations are comparable. On the other hand, the actual positions of the tension cracks were about 40—70 m behind the slope crest (60 m at section N3500); if these extended to F-2 level on discontinuities dipping at about 70°W (set 4), as implied by Kennedy et al. (1971, p. 60), then the angle of the basal shear plane (ψ_p) must be no greater than 25°. The critical surface for cylindrical failure, assuming $\phi = 25°$, intersects the ground *at* the surface location of the tensile cracks; this, then, also suggests the possibility that $\phi \leqslant 25°$. An alternative interpretation associates these cracks with deformation involving a greater height of slope, say $H = 181$ m; however, no firm evidence has yet been presented to support this view.

In this analysis the best estimate of the slide mass is given by the planar case for $\phi = 20°$, namely $W = 9960$ metric tons for a section of unit width, or about 2×10^6 metric tons for a width of 200 m. This is about half of the 4.1×10^6 metric tons estimated by Kennedy et al. (1971) from pit surveys and assumed failure plane from tension cracks to the toe.

In regard to the slope collapse from levels D-2 to F-2, two-dimensional analysis suggests very low values of average cohesion (Table V). Again, nearly continuous, relatively smooth rock mass discontinuities seem to have been required. This is borne out by field observations which note collapse in two stages, i.e., collapse of the central block through a small toe buttress in the south block, followed by slip of the north block. At the N3500 section the critical failure plane for $\phi = 30°$ nearly coincides with the major shear plane, and structure A intersects this plane at the approximate location where, in the theoretical analysis, a tension crack is assumed. Therefore the planar analysis seems equivalent to the first stage of collapse, at least to a first approximation in which the lateral push of the north block is ignored. [9] Cohesion along the failure surface had been virtually destroyed when collapse occurred.

The north block then slipped at a low angle into the void created by the first stage of movement. The N3500 section suggests a very low angle for the slide surface (although it is an apparent dip), and on this basis ϕ could be 20° or even less. Cohesion must have been very small, perhaps on the order

---

[9] A more correct solution would have to account for bending moments induced in the central block by thrust of the north block; moreover, a three-dimensional analysis would be desirable. However, the available data do not permit such refinements.

of 10 kN/m² or less. These values suggest a through-going, smooth joint surface coated with sheet silicate minerals.

Effect of the slope unloading program

What of the effect of slide area unloading? This question cannot be completely resolved by any theoretical analysis because the distribution and
extent of discontinuities in the removed section are not known, the geometry in three dimensions of the failed blocks has not been specified, and their
surface properties are unknown. However, Fig. 9 suggests that the lateral
force applied by the north block on the central block may have been materially reduced by stripping. A very rough approximation can be given for
the two dimensional idealization by comparison of required cohesion values
for $H = 61.5$ (post-stripping) and 90 m (pre-stripping). The actual value of ϕ
is not important, and it is assumed here that constant $\phi = 30°$; accordingly,
required cohesion is about 47 and 68 kN/m², respectively (the cylindrical
mode is used, for convenience). The larger value is the average cohesive resistance available in a stable slope prior to stripping; the smaller value is the
cohesive resistance required to support the slope after stripping. The ratio of
the larger to the smaller value thus suggests a partial factor of safety with
respect to cohesion (F_c) of about 1.5, due to the excavation. This assumes,
however, that the material could be excavated by blasting without affecting
the average shear strength (ϕ, c) of the immediately subjacent rock mass. It
is conceivable that excavation blasting could worsen the situation rather than
improve it. In any case, average cohesion must have decreased in order to
permit slope collapse, either as an effect of heterogeneously distributed cohesion, progressive failure, blasting, or some combination of factors. The
mass that ultimately fell was only a fraction of the rock excavated.

The solution for partial factor of safety with respect to cohesion can be
written in more general terms, because for any given constant values of ϕ
and ψ_p, the dimensionless term $c/\gamma HF$ is a constant. For constant γ and
$F = 1$, cohesion is simply proportional to height of slope. Thus:

$$F_c = H \text{ (pre-stripping)}/H \text{ (post-stripping)}$$

subject to the implications discussed above.

For 123 m height, the effect of 28 m slope unloading was to provide a
partial factor of safety:

$$F_c = 151 \text{ m}/123 \text{ m} = 1.2$$

assuming that cohesion was uniformly distributed and that the existing
slope was not damaged further in the excavation process. Consideration of
planar failure idealization leads to a similar result with somewhat more
computational effort.

Hoek and Bray (1974, p. 251) have stated that it was "unlikely that the unloading of the slopes had any significant influence on the slide" at Chuquicamata. In a hypothetical example problem involving somewhat similar geometry, Hoek and Bray (1974, pp. 169—171) showed that height reduction only began to show "significant" benefits (presumably in reference to a design $F = 1.5$) once the height reduction approached 40%. They correctly point out that once this has been accomplished, most of the unstable mass will have been removed and it would then be worth removing the rest of the potentially unstable block and the remains of the problem. This point seems relevant to the Chuquicamata slope, where the amount of rock stripped from the slope crest greatly exceeded the collapsed mass. On the other hand the size of the slope collapse was not known by foresight. It should also be recalled that the 123-m slope did stabilize after at least 8 m of resultant displacement. While the direct influence of stripping on slope displacement rate cannot be assessed from data thus far published, it seems conceivable that the small margin of diminished shear stress produced by the stripping program could have provided the margin for ultimate stabilization of the slope rather than its collapse. In this sense, "significant influence" as previously referred to may consist in the difference between stability and collapse, irrespective of the magnitude of the artificial factors of safety attached to it.

Slope movements ceased following the relatively small failure of the slope crest, which itself can be considered as "stripping" by natural causes. Analytically it can be approximated by a reduction in slope height from 123 m to roughly 100 m, thus providing a $F_c = 1.2$ associated with slope unloading. While small, the margin was apparently sufficient to bring slope movements to a halt.

CONCLUSIONS

In summary, slope movements at Chuquicamata appeared to have been reactivated by the earthquake of 20 December 1967. Movements reportedly progressed at a more or less steady rate (the actual data have not been published) until the fall of 1968, when several abrupt changes in displacement-time behavior were noted that were probably associated with the excavation history of the pit and blasting events. These subsequent events were instrumental in aggravating slope conditions and led to large (>8 m) displacements in a deformed zone from the slope crest to (at least) 123 m below the crest, and to slope collapse for a portion of this deformed zone in February 1969. There is reasonable doubt that failure of this magnitude would have occurred were it not for disturbances by excavations in 1968. Collapse might have been delayed for at least a year; at best, the slope would have stabilized. Interpretations must be based on extrapolation of displacement-time data to 8 November 1968. Because these data are quite incomplete and the appropriate regression form is not obvious, firm conclusions are not justifiable;

still, the data suggest that the December 1967 earthquake should not be considered as the major cause of the slope collapse.

In January 1968, an enormous mass was in motion between N3300 and N3800 coordinate lines; subsequently zones north of N3600 and south of N3400 became stabilized. The kinematics of the slope movements are not completely clear, but seemed to involve load transfer from the north block to the central and south blocks, across dominant structures A and B. Slope movements were almost completely dominated by rock discontinuities; the amount of cohesion required to prevent movements in the quasi-static loading state (neglecting dynamic effects) was quite small, and it seems clear that little intact rock support was present. Local breakage and slip was inferred from microseismic events which increased in frequency to the period of slope collapse. Fabric studies indicate three sets of steeply dipping fractures that could have defined the rear of the deforming mass, and an east-west set that minimized lateral shear resistance. The basal slide planes are, on the other hand, wholly unrepresented by published fabric statistics. Their existence and orientations can be inferred; analysis based on known position of tension cracks and mineralogy of the rock mass suggests a dominant failure surface dip of approximately $25°W$ ($\pm5°$), with strike approximately that of the slope face ($\pm20°$). The dominant angle of friction ϕ is perhaps about $20-25°$; this value seems low for weathered granodiorite, but is compatible with relatively smooth fractures coated with mica or chlorite.

Collapse occurred only in the upper part of the slope; the toe of the collapsed zone involved a small portion of the south block. When this failed, the central block slid out along a fracture dipping at perhaps $37°W$; this, in turn, permitted the north block to slide out on a fracture dipping at perhaps $23°W$ (see Fig. 9). An extensively crushed zone involving a height of about 123 m stabilized soon after collapse of the slope crest; acoustic emission from the slope greatly diminished. Ultimate stabilization may have been in part aided by a stripping program carried out in 1968, which perhaps provided a small margin favoring stability.

It should be emphasized that the date of actual slope collapse at Chuquicamata was simply the *earliest* of various predicted dates. The 6-m criterion had already been "met" during the first week of February; the slope was moving both farther and faster than anticipated. Under slightly different circumstances an engineering staff might be taken by surprise. Yet the work at Chuquicamata has emphasized that displacement measurements are probably the most valuable predictive tool for the geotechnical engineer. The acoustic approach is potentially useful for qualitative information on slope deformation (cf. Chapter 16, this volume) and clearly can outline the existence and perhaps the location of slope movements. However, an acoustic criterion for collapse prediction has not yet been developed. Even in regard to displacements the matter of extrapolation functions and collapse criteria are incompletely resolved.

It is of interest to compare the 6-m criterion of maximum displacement employed at Chuquicamata with the only 2—4 m of apparent displacement which occurred prior to the Vaiont reactivation, involving an enormously greater mass. A simple relationship between maximum displacement and slide volume may not be forthcoming, and a reliable failure criterion may be developed only with difficulty.

Finally, too much emphasis to date may have been placed on the apparent absolute accuracy of the Chuquicamata prediction by the casual reader. The overall result might not have been greatly different if the prediction had been one or a few weeks off (indeed, it *was*, if one considers that the collapse criterion had been satisfied prior to collapse). Perhaps the main lesson of this example is that "by knowing what to look for and by making full use of available data, a set of sound engineering decisions could be made and the serious consequences which could have resulted from this failure were avoided" (Hoek and Bray, 1974, p. 251).

ACKNOWLEDGEMENTS

The regression data were examined as part of a classroom exercise in slope stability at the Pennsylvania State University in which the contributions of Stan Suboleski and M. Pagano were notable. The manuscript was reviewed by W.G. Pariseau.

REFERENCES

Broadbent, C.D. and Ko, K.C., 1972, Rheological aspects of rock slope failures. *Proc., 13th Symp. on Rock Mechanics, Urbana, Ill., 1971.* American Society of Civil Engineers, New York, N.Y., pp. 573—593.

Broadbent, C.D. and Riperre, K.H., 1971. Fracture studies at the Kimberly Pit. *Proc. Symp. on Planning Open Pit Mines, Johannesburg, 1970.* Balkema, Amsterdam, pp. 171—179.

Dunbar, D.M., 1952. History of Chuquicamata copper. *Min. Eng.,* 4: 1164—1165.

Hamel, J.V., 1971. Kimbley pit slope failure. *Proc., 4th Pan Am. Conf. Soil Mech. Found. Eng., San Juan,* 2: 117—127.

Hoek, E., 1970. Estimating the stability of excavated slopes in opencast mines. *Trans. Inst. Min. Metall.,* pp. A109—A132.

Hoek, E. and Bray, J.W., 1974. *Rock Slope Engineering.* Institution of Mining and Metallurgy, London, 309 pp.

Horn, H.M. and Deere, D.U., 1962. Frictional characteristics of minerals. *Geotechnique,* 12: 319—335.

Janbu, N., 1954, Stability analysis of slopes with dimensionless parameters. *Harvard Soil Mech. Ser.,* 46, 81 pp.

Kennedy, B.A., 1971. Reply to discussion, Session 7. *Proc., Symp. on Planning Open Pit Mines, Johannesburg, 1970.* Balkema, Amsterdam, pp. 328—329, 332, 337.

Kennedy, B.A. and Niermeyer, K.E., 1971. Slope monitoring systems used in the prediction of a major slope failure at the Chuquicamata Mine, Chile. *Proc., Symp. on Planning Open Pit Mines, Johannesburg, 1970.* Balkema, Amsterdam, pp. 215—225.

Kennedy, B.A., Niermeyer, K.E., Fahm, B.A. and Bratt, J.A., 1971. A case study of slope stability at the Chuquicamata Mine, Chile. *Trans. Soc. Min. Eng., Am. Inst. Min. Metall. Eng.,* 250: 55—61.

Kenney, T.C., 1967. The influence of mineral composition on the residual strength of natural soils. *Proc., Geotech. Conf., Oslo,* 1: 123—129.

Lambe, T.W. and Whitman, R.V., 1969. *Soil Mechanics.* Wiley, New York, N.Y., 553 pp.

Ley, G.M.M., 1972. *The Properties of Hydrothermally Altered Granite and Their Application to Slope Stability in Opencast Mining.* M.Sc. Thesis, Emperial College, Univ. of London, London.

Lopez, V.M., 1939. The primary mineralization at Chuquicamata, Chile. *Econ. Geol.,* 34: 674—711.

Perry, V.D., 1952. Geology of the Chuquicamata ore body. *Min. Eng.,* 4: 1166—1168.

Renzetti, B.L., 1957. *Geology and Petrogenesis at Chuquicamata, Chile.* Ph.D. Thesis, Indiana Univ., Bloomington, Ind.

Robertson, A.M., 1971. The interpretation of geological factors for use in slope theory. *Proc. Symp. on Planning Open Pit Mines, Johannesburg, 1970.* Balkema, Amsterdam, pp. 55—71.

Rocha, M., 1964. Mechanical behavior of rock foundations in concrete dams. *Proc., 8th Int. Conf. on Large Dams,* 1: 785—831.

Salamon, M.D.G., 1971. Discussion, Session 7. *Proc., Symp. on Planning Open Pit Mines, Johannesburg, 1970.* Balkema, Amsterdam, p. 331.

Taylor, D.W., 1948. *Fundamentals of Soil Mechanics.* Wiley, New York, N.Y., 700 pp.

Terzaghi, K., 1929. Effect of minor geologic details on the safety of dams. *Am. Inst. Min. Metall. Eng., Tech. Publ.,* 215: 31—44.

Terzaghi, K., 1962. Does foundation technology really lag? *Eng. News-Rec.,* February 15, pp. 58—59.

Terzaghi, R., 1965. Sources of error in joint surveys. *Geotechnique,* 15: 287—304.

Vargas, M., 1967. Design and construction of large cuttings in residual soils. *Proc., 3rd Pan Am. Conf. Soil Mech. Found. Eng.,* 2: 243—254.

Voight, B., 1973. Correlation between Atterberg plasticity limits and residual shear strength of natural soils. *Geotechnique,* 23: 265—267.

Chapter 18

PIMA MINE SLOPE FAILURE, ARIZONA, U.S.A.

JAMES V. HAMEL

ABSTRACT

Between 1964 and 1967, a slope failure occurred in the north wall of the open pit Pima Mine, which is located near Tucson, Arizona. The mine wall was 210 m high with an average inclination of 37°. The upper 67 m of the mine wall consisted of gravelly desert alluvium deposited unconformably over the underlying rock. The rock consisted of a complex of carbonate hornfels and quartzitic clastics with intrusions of rhyolite, syenite, and quartz monzonite porphyry. This rock was extensively jointed and faulted and contained a major thrust fault parallel to and generally underlying the mine wall.

The cross-section of the failure mass was determined from inclinometer measurements and topographic surveys made by the U.S. Bureau of Mines. Sliding occurred on several surfaces in a zone about 30 m thick and generally parallel to the mine wall. This zone of sliding is believed to have coincided with the shear zone of the thrust fault in the mine wall. Average Coulomb shear strength parameters required for limiting equilibrium of the failure mass were calculated with the Morgenstern-Price method of slope stability analysis.

It was concluded that the average in-situ strength parameters of the alluvium and of the rock were on the order of $c' = 0$, $\phi' = 28-33°$. These friction angle values are typical for the residual (large displacement) shear strength of both granular soils and hard rocks. Mobilization of residual shear strength in the rock is consistent with its previous faulting. Development of residual shear strength in the alluvium is attributed to loosening and progressive failure, due to movement in the underlying rock.

INTRODUCTION

The Pima Mine is an open pit copper mine located about 32 km southwest of Tucson, Arizona (Fig. 1). The orebody was discovered in 1950 by geophysical methods and some underground development was done initially

Fig. 1. Location map.

(Thurmond et al., 1954, 1958; Thurmond and Storms, 1958). Open pit mining began in 1956 and ore production was 2700 tonnes per day in 1957. Four expansions in open pit operations (Komadina, 1965, 1967; Martin, 1969; Beall, 1973) have subsequently increased ore production to 48,000 tonnes per day.

A slope failure occurred in the north wall of the Pima Mine during the period from approximately 1964 to 1967. This slope failure was investigated by the writer in 1968 and 1969 as part of a research project on open pit mine slope stability. Coulomb shear strength parameters required for limiting equilibrium of the failure mass were calculated and the most probable in-situ strength parameters and failure surfaces were selected (Hamel, 1969, 1970). The slope failure and stability analyses are described herein. Conclusions are drawn regarding the probable in-situ rock strength and the mechanism of slope failure.

DESCRIPTION OF SLOPE FAILURE

Location and pit geometry

A 1966 aerial photograph of the Pima Mine shows the slope failure to be located in the inward bulge near the center of the north wall of the mine (Fig. 2). A plan sketch map of the failure area is given in Fig. 3, and Fig. 4 shows a cross-section through the slope failure. At this location the mine wall had a total height of about 210 m and an average inclination of 37°.

Geology

The Pima Mine is situated in the very active Pima or Twin Buttes mining district of southern Arizona (cf. Seegmiller, this volume). Cooper (1962) and Lacy and Titley (1962) described the complex geology of the area. Cooper gave a preliminary geologic map of the eastern part of the Pima mining district, which includes the Pima Mine, and presented an extensive

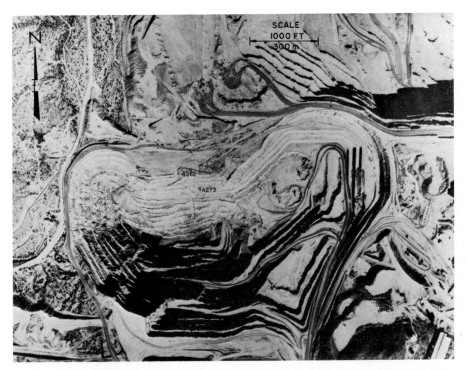

Fig. 2. Aerial photograph of Pima Mine, 1966. Slope failure is located on north pit slope in vicinity of inclinometer wells A-262 and A-273.

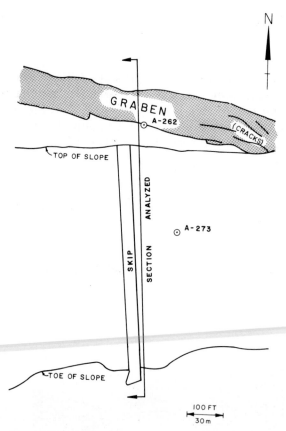

Fig. 3. Sketch map of failure area (cf. Fig. 2).

list of references on regional and areal geology. The geology of the Pima Mine and orebody has also been described in numerous publications (Thurmond, 1955; Thurmond et al., 1954, 1958; Thurmond and Storms, 1958).

The Pima or Twin Buttes mining district is located on the eastern edge of the Sierrita Mountains. These mountains are composed of a generally granitic core with metamorphosed sedimentary rocks on the west slope and less altered sedimentary rocks on the east slope. Formations on the east range from Cambrian age Bolsa Quartzite and Abrigo Limestone through Devonian Martin Limestone, Mississippian Escabrosa Limestone, Pennsylvanian Naco Limestone, Permian shales, limestones, quartzites, and gypsum to Upper Cretaceous volcanics and sediments. Coarse-grained intrusives including granite, quartz monzonite, and granodiorite underlie recent alluvial fan deposits in the eastern piedmont area where the Pima Mine is located. The regional geologic structure is complex due to extensive folding and faulting (Thurmond and Storms, 1958).

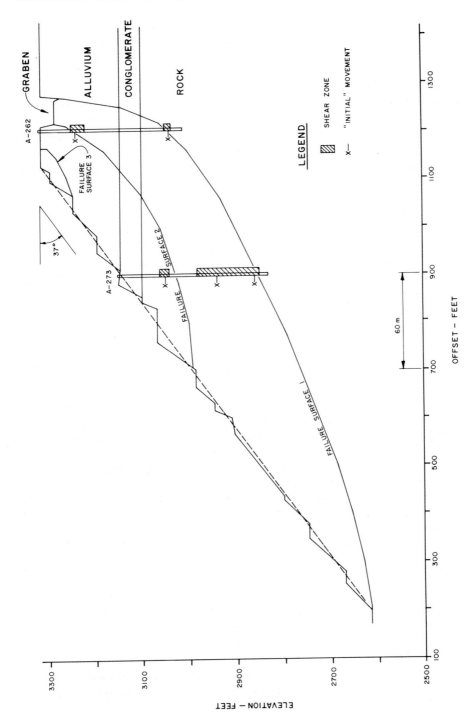

Fig. 4. Slope cross-section with failure surfaces 1, 2, and 3. For location of section refer to Fig. 3.

The north wall of the Pima Mine contains three main zones of geologic material. These zones are designated *Alluvium*, *Conglomerate*, and *Rock* in Fig. 4. The upper 67 m of the slope consists of gravelly desert alluvium which was deposited unconformably on the underlying rock. The upper 52 m of unconsolidated material is designated Alluvium. The lower 15 m of desert alluvium has been cemented by siliceous and calcareous materials and is designated Conglomerate. The base of the Conglomerate was essentially level in the failure area (Fig. 4).

The underlying Rock consists of a complex of carbonate hornfels with some quartzitic clastic rocks and intrusions of rhyolite, syenite, and quartz monzonite porphyry (Thurmond et al., 1958). The sedimentary rocks in the Pima Mine area are thought to be of Permian age (Thurmond, 1955). The Pima orebody is a pyrometasomatic (contact metamorphic) deposit, and carbonate hornfels constitutes the main high-grade ore zone with chalcopyrite the main ore mineral. Additional low-grade ore occurs as widely disseminated chalcopyrite in the clastic rocks. The igneous intrusives are generally not mineralized (Thurmond et al., 1958). Bedding, where discernible in the altered sedimentary rocks, strikes approximately N80°E and dips 30—50° south (Thurmond, 1955; Thurmond and Storms, 1958).

The Rock is extensively faulted and jointed but very little information is available on the locations, attitudes, and physical characteristics of the discontinuities. It is known, however, that a major thrust fault zone formed a substantial part of the north wall of the mine in the failure area. This thrust fault has a strike of approximately N80°E and a dip of about 45—50° south (Thurmond and Storms, 1958). The brecciated fault zone in the north wall of the mine was on the order of 30 m thick and was generally composed of angular particles of limestone and quartzite in a matrix of highly altered limestone (Thurmond, 1955).

According to information furnished by the Denver Mining Research Center, U.S. Bureau of Mines, the groundwater level had been drawn down throughout the district and was at the level of the bottom of the pit (elevation 2615 ± ft (797 m)) at the time of the slope failure. The average annual precipitation in this arid portion of southern Arizona is only about 300 mm (Thurmond and Storms, 1958).

Movement history

The Denver Mining Research Center reported that failure indications were first noted in 1964 when tension cracks were observed behind the crest of the slope. Most of the movement associated with the slope failure apparently occurred in 1966 and much of this movement was monitored by the Bureau of Mines and the Pima Mining Company. Inclinometer measurements of slope movement were obtained during 1966 at inclinometer wells A-262 and A-273 (Figs. 2—4). Some additional movement data were obtained by surveying surface reference points. The relationship of slope movement to mining activities was not established but the writer infers that mining activities did cause some short-term and localized movement accelerations.

Inclinometer well A-262 at the top of the slope had a surface elevation of about 3315 [1] and a depth of about 90 m. Movement measurements taken at A-262 from May 11, 1966, to January 10, 1967, were furnished to the writer by the Denver Mining Research Center. These measurements indicated two definite zones of sliding (Fig. 4). The initial measured movement between elevations 3040 and 3050 was apparent by mid-May 1966. The data showed a net movement of 18 cm south (downslope) and 64 cm east by November 29, 1966, the date of the last movement reading at this elevation. The November 29, 1966, profiles of measured horizontal movement versus elevation were essentially linear from elevation 3315 to elevation 3040. A second definite zone of movement appeared at elevation 3230—3240 in June 1966. The rock at this elevation had moved a total of 18 cm south and 46 cm east (10 cm south and 25 cm east from its November 29, 1966, location) by January 10, 1967, the date of the last readings. The January 10, 1967, movement profiles from the surface to elevation 3240 were also essentially linear.

Well A-273, located on a bench at the top of the Conglomerate, had a surface elevation of about 3150 and a depth of about 90 m. Movement measurements taken from June 3, 1966, to November 30, 1966, were furnished by the Denver Mining Research Center. These measurements indicated several zones of movement (Fig. 4). The first movement was apparent by July 1966 at about elevation 2870. The data showed a net movement of 3.0 cm south and 2.5 cm east by August 8, 1966, the date of the last reading taken at this elevation. At another definite zone of movement at about elevation 2940, the data showed a net movement of 3.8 cm south and 1.0 cm east by August 8, 1966. A third definite zone of sliding at elevation 3040—3050 had a net movement of 14 cm south and 18 cm east by November 30, 1966, the date of the last reading.

Movement profiles for well A-273 were generally much more irregular with depth and showed much wider zones of movement than did the profiles for well A-262. The A-273 profiles showed shear zones from elevation 2850 to 2980 and from elevation 3040 to 3060. This indicates that the failure mass was breaking up more, i.e., ceasing to move as a rigid body, in this lower part of the slope. It should be remembered, however, that measurements began at A-273 about three weeks after they began at A-262 and measurements began at A-262 more than a year after tension cracks were first observed at the crest of the slope. The amount of movement which occurred before measurements began is, of course, unknown.

---

[1] All elevations given in this paper are in feet because of standard practice at the Pima Mine.

Movement measurements were also taken at a surface reference point at well A-262 for the period from June 16 to November 29, 1966. Movement components of 1.2 m south, 1.4 m west, and 1.7 m vertical (subsidence) were measured during that period. These components give a resultant downslope movement to the south of 2.1 m on an angle of 54° with the horizontal and a resultant downslope movement to the west of 2.2 m on an angle of 50° with the horizontal; the total resultant movement was 2.5 m downslope and southwesterly. Most of this measured surface movement occurred from September 6 to November 29, 1966.

A graben formed on the north side of well A-262, as shown in Figs. 3 and 4. The graben block was 21—30 m wide (north-south) by about 180 m long (east-west). A relatively small slide occurred on December 8, 1966, below the southeast end of the graben. The graben block moved down about 1.2 m from November 29 to December 9, 1966. Exact measurements of this movement could not be taken because of mining activity in the area and subsidiary cracks and movements.

Failure surfaces

The locations of the three failure surfaces in Fig. 4 were inferred by U.S. Bureau of Mines personnel on the basis of the inclinometer measurements and other field observations, e.g., graben and crack formation. Failure surface 1 extends from the toe of the slope, through points of initial movement observed in the inclinometer wells, to the back of the graben block. It has a total length of about 420 m; 86% of this length is in the Rock zone. The portion of this failure surface passing through the Alluvium (upper 43 m) is almost vertical and probably represents a tension crack.

Failure surface 2 extends from about the middle of the slope through points of movement in the inclinometer wells to the front of the graben block. It has a total length of about 200 m, 59% of which is in the Rock zone. The upper 8 m of failure surface 2 is nearly vertical and probably represents a tension crack. The inclinometer data indicated that movement began in the vicinity of failure surface 2 only slightly after it began in the vicinity of failure surface 1.

Failure surface 3 extends from the back of the second bench to a 5-m tension crack about 12 m back from the edge of the pit. It has a total length of about 41 m (including the 5-m tension crack) and passes entirely through the Alluvium. No movement data were available for failure surface 3 as it was between the two inclinometer wells.

No information on the lateral extent of failure was available to the writer. It is inferred, however, that the main failure mass had a width on the order of the length of the graben block, i.e., about 180 m.

STABILITY ANALYSES

General remarks

A soil or rock slope failure can be viewed as a large-scale natural shear test. This in-situ test includes all the heterogeneity, anisotropy, and environmental effects that cannot be modeled in laboratory or theoretical investigations. If the boundary conditions of a slope failure are known or can be evaluated, it is possible to gain an understanding of the mechanism of failure and to calculate the average shear strength of the soil or rock at the time of failure.

It is impossible, of course, to calculate unique values of the effective Coulomb cohesion intercept c' and angle of internal friction ϕ' for a failure surface in a slope. A *set* of c' and ϕ' values corresponding to limiting equilibrium of the failure mass can, however, be calculated. A question then arises concerning which pair of c'-ϕ' values existed in the slope at the time of failure. If there are laboratory or field strength test data available for the slope-forming materials, these data generally provide guidance in the choice of a pair (or range) of c'-ϕ' values at failure. The material involved in the Pima Mine slope failure was removed after the failure, so it was impossible to obtain samples for testing. Field values of c' and ϕ' at failure must therefore be estimated from calculated c'-ϕ' data on the basis of descriptions and geologic history of the slope-forming materials, general shear strength behavior principles, and geotechnical judgment.

The Morgenstern-Price (1965, 1967) two-dimensional limiting equilibrium method of slope stability analysis was used for calculation of strength parameters and for the other stability analyses described herein. The Morgenstern-Price method is well-suited to this work because it treats a failure surface of general shape and requires the failure mass to be in complete static equilibrium. Calculations were performed with an IBM 7090 computer at the University of Pittsburgh Computer Center. Values of c' were determined for various values of ϕ' such that the factor of safety against sliding along the failure surface was equal to or greater than 1.0. The factor of safety is here defined as the ratio of shear strength available to shear strength required for limiting equilibrium.

In all analyses performed, each zone of geologic material — Alluvium, Conglomerate, and Rock — was assumed homogeneous and isotropic with respect to shear strength. Groundwater level was assumed to be at or below the pit bottom in all stability calculations. The presence of benches was included in the analysis of failure surface 3 as this failure mass was small compared to the sizes of the upper benches. A uniform $37°$ slope was used in analyses of all other failure surfaces.

The following unit weights were considered reasonable for the materials in the slope.

Alluvium: 2.08 Mg/m³
Conglomerate: 2.40 Mg/m³
Rock: 2.56 Mg/m³

They were used in all the analyses described herein.

Given failure surfaces

The three failure surfaces shown in Fig. 4 were analyzed initially. Failure surface 3, the shallow failure surface passing entirely through the Alluvium, was analyzed first in order to obtain strength parameters for Alluvium for use in later analyses. The Alluvium is basically gravel, so it probably had little cohesion. With $c' = 0$, a value of $\phi' = 28°$ was determined for equilibrium along failure surface 3. The strength parameters $c' = 0$, $\phi = 28°$ were then used for Alluvium in analyses of failure surfaces 1 and 2. Relatively small percentages of failure surfaces 1 and 2 pass through Conglomerate, which is also composed of predominantly gravelly material. The strength parameters $c' = 0$, $\phi' = 28°$ were therefore also used for Conglomerate in the analyses of failure surfaces 1 and 2.

Coulomb strength parameters calculated for limiting equilibrium in Rock along failure surface 1 and 2 are given in Table I along with strength parameters calculated for limiting equilibrium in Alluvium along failure surface 3. Values of average shear and normal stresses on the Alluvium of failure surface 3 and on the Rock portions of failure surfaces 1 and 2 are also given in Table I. The average shear stresses are numerically equal to cohesion values

TABLE I

Strength parameters and stress values calculated for limiting equilibrium

Failure surface	Friction angle, ϕ' (degrees) with $c' = 0$	Cohesion intercept, c' (kN/m²) with $\phi' = 0$	Average normal stress, σ' (kN/m²)	Average shear stress, τ (kN/m²)
1	28	647	1245	647
2	28	335	637	335
3	28	67	129	67
1-A	33	479	742	479
2-A	32	407	671	410

Notes:
(1) Strength parameters and stress values for failure surfaces 1 and 2 are for Rock with $c' = 0$, $\phi' = 28°$ in Alluvium and Conglomerate.
(2) Strength parameters and stress values for failure surface 3 are for Alluvium only.
(3) Strength parameters and stress values for failure surfaces 1-A and 2-A are for Rock with $c' = 0$, $\phi' = 30°$ assumed for Alluvium and Conglomerate.

required for equilibrium with $\phi' = 0$. The average effective normal stresses were obtained by dividing each average shear stress by the tangent of the friction angle calculated for equilibrium with $c' = 0$.

Additional failure surfaces

As the zone of shearing indicated by the inclinometer measurements at well A-273 was more than 30 m thick, eighteen additional failure surfaces passing through this zone were analyzed assuming strength parameters $c' = 0$, $\phi' = 28°$ for all materials. Several of these additional failure surfaces and their computed factors of safety are shown in Fig. 5.

The two most critical of these additional failure surfaces, denoted 1-A and 2-A, are shown in Fig. 6. Strength parameters and average shear and normal stress values calculated for limiting equilibrium in Rock along failure surfaces 1-A and 2-A are given in Table I. In these latter strength calculations, Alluvium and Conglomerate were both assumed to have had $c' = 0$, $\phi' = 30°$ as analyses of the Alluvium and Conglomerate failure surfaces in Fig. 5 indicated that a minimum ϕ' value of 30° (with $c' = 0$) was required along these failure surfaces for equilibrium.

Discussion

Values of average shear and effective normal stress calculated for each failure surface and given in Table I can be used to construct in-situ Coulomb strength envelopes for the materials involved in the slope failure. Such strength envelopes are useful in interpretation of calculated strength data and in comparison of calculated and measured strength data in cases where the latter are available.

Average shear and normal stress values for each failure surface in Table I are plotted in Fig. 7 with upper and lower-bound strength envelopes. The lower-bound strength envelope corresponds to the Rock strengths calculated for failure surfaces 1 and 2 and to the Alluvium strength calculated for failure surface 3. This lower-bound strength envelope is characterized by $c' = 0$, $\phi' = 28°$. The upper-bound strength envelope, which corresponds to the Rock strength calculated for failure surface 1-A, is characterized by $c' = 0$, $\phi' = 33°$.

Table I and Fig. 7 indicate that the effective in-situ strength of the Rock at the time of the slope failure can be characterized by $c' = 0$, $\phi' = 28-33°$, for effective normal stresses on the order of 620–1250 kN/m². These ϕ' values are in the range of residual (large displacement) friction angles commonly cited for hard rock materials (Deere et al., 1967; Krsmanovic, 1967; Coulson, 1972). Mobilization of a near-residual level strength is consistent with the geologic history of the Rock. The Rock material in the failure area was extensively fractured, brecciated, and altered; it would probably have

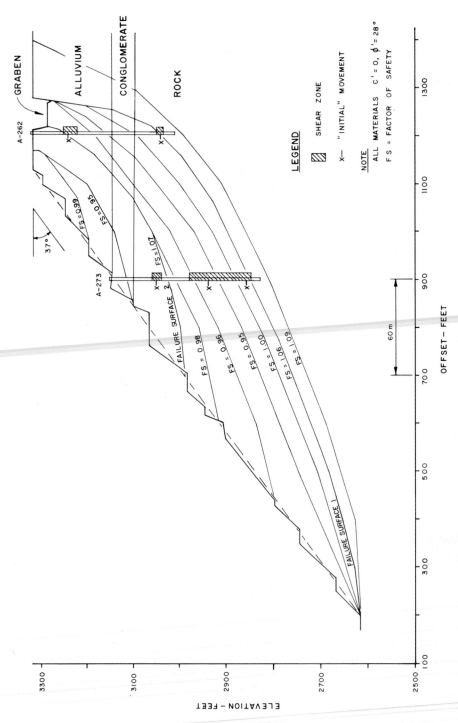

Fig. 5. Slope cross-section with additional (hypothetical) failure surfaces analyzed.

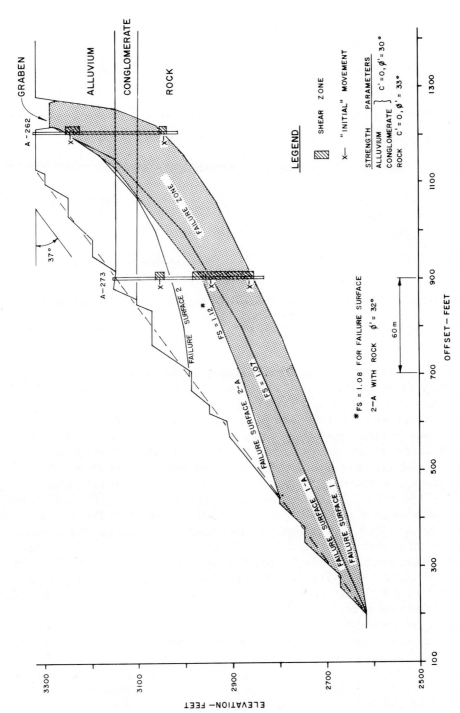

Fig. 6. Slope cross-section with critical failure surfaces.

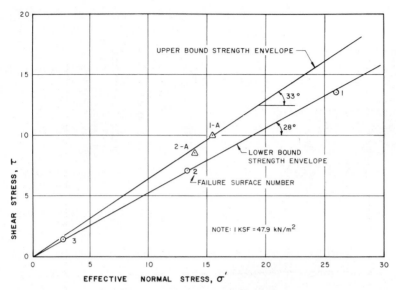

Fig. 7. Inferred Coulomb shear strength envelopes based on slope failure analyses. Units given refer to 10^3 lbf/ft²; to convert to kN/m² multiply by 47.9.

behaved more like a cohesionless soil aggregate than an intact rock material. The large displacements associated with tectonic activity would have reduced the available Rock strength along a potential failure surface from a peak (small displacement) to a residual (large displacement) value.

Analysis of failure surface 3 suggests that the effective in-situ strength of the Alluvium at the time of slope failure can be characterized by $c' = 0$, $\phi' = 28°$, for effective normal stresses on the order of 140 kN/m². Typical peak strength parameters for desert alluvium in this low normal stress range are $c' = 0$, $\phi' = 40-50°$ (Strohm et al., 1964; Hamel, 1969, 1975). The limited available data on residual strength of desert alluvium (Strohm et al., 1964; Hamel, 1969, 1975) plus soil mechanics concepts regarding the shear strength of granular soils (Lambe and Whitman, 1969) suggest that typical residual strength parameters for desert alluvium at low normal stresses might be $c' = 0$, $\phi' = 30-35°$. The calculated ϕ' of 28° for failure surface 3 is significantly below the range of measured peak ϕ' values for desert alluvium and even slightly below the range of probable residual ϕ' values.

As noted previously, a minimum ϕ' value of 30° (with $c' = 0$) was required for equilibrium along the Alluvium and Conglomerate failure surfaces in Fig. 5. It can be shown, using the stability chart of Hoek (1970) for circular failure surfaces, that a minimum ϕ' value of about 32.5° (with $c' = 0$) was required in the Alluvium and Conglomerate for equilibrium of the upper portion of the 37°-average mine slope. These minimum ϕ' values of 30—

32.5° fall in the range of probable residual friction angles given above for desert alluvium. Friction angles even larger than 32.5° were required, however, for equilibrium of the steeper bench slopes in the Alluvium.

The low calculated ϕ' value of 28° for failure surface 3 through the Alluvium probably reflects a residual shearing condition developed during or after progressive failure of the larger sliding mass below. As the large rock mass moved both downslope and laterally, stresses redistributed within it, cracks opened, and the graben block subsided. All these factors removed lateral support from the Alluvium and weakened it. No information on the history of movement along failure surface 3 was available but it is hypothesized that no appreciable movement occurred along failure surface 3 until well after movement had begun in the Rock below.

CONCLUSIONS

The Alluvium, Conglomerate, and Rock are believed to have had essentially zero cohesion and effective friction angles of 28—33° (Fig. 7) at the time of the 1964—1967 slope failure in the north wall of the Pima Mine. The strengths mobilized during the slope failure were field residual strengths [2] for the respective materials. Mobilization of a residual strength in the Rock is consistent with its previous shear displacement due to faulting. Mobilization of a residual strength in the Alluvium and Conglomerate is attributed to loosening and progressive failure due to movement in the underlying Rock.

Available movement data were insufficient to establish the surface(s) of initial movement. Inclinometer measurements at wells A-262 and A-273 showed that movement began in the vicinity of failure surface 1 but the amounts and locations of movement before measurements began are unknown. Inclinometer measurements also showed that substantial movements occurred at several levels in the slope, especially in the later stages of slope failure.

Measured movements and the results of stability analyses (Fig. 5) suggest that sliding occurred along several surfaces in a wide zone rather than along an individual failure surface. The failure zone was about 30 m thick and was roughly bounded by failure surfaces 1 and 2-A (Fig. 6). Much of the failure zone probably coincided with the brecciated zone of the thrust fault in the mine wall, and the graben marked the intersection of the failure zone with the ground surface. The material in this failure zone is believed to have had average strength parameters $c' = 0$, $\phi' = 28$—33° at the time of the slope failure.

---

[2] The concept of field residual strength is discussed in more detail by Hamel (1976); see also Chapters 8 and 14, this volume.

ACKNOWLEDGEMENTS

   This paper is drawn largely from a dissertation presented by the writer to the University of Pittsburgh as partial fulfillment of the requirements for the degree of Doctor of Philosophy (Hamel, 1969). Analysis of the Pima Mine slope failure was supported by the U.S. Department of the Interior, Bureau of Mines, under Grant No. G0180476 (MIN-11) to the University of Pittsburgh. Information on the slope failure was furnished by the Denver Mining Research Center, U.S. Bureau of Mines, and by the Pima Mining Company. Permission to publish this paper was granted by both of these organizations. R.E. Gray and G.R. Thiers of GAI Consultants, Inc., and L.P. Kettren of Dames and Moore reviewed the manuscript of this paper and suggested several improvements. The writer takes full responsibility, however, for the opinions and conclusions expressed herein.

REFERENCES

Beall, J.V., 1973. Copper in the U.S. — a position survey. *Min. Eng.*, 25(4): 35—47.

Cooper, J.R., 1962. Some geologic features of the Pima mining district, Pima County, Arizona. In: *Contributions to Economic Geology, 1959. U.S. Geol. Surv. Bull.*,1112: 63—103.

Coulson, J.H., 1972. Shear strength of flat surfaces in rock. *Proc. 13th Symp. on Rock Mechanics, Univ. of Illinois, 1971.* American Society of Civil Engineers, New York, N.Y., pp. 77—105.

Deere, D.U. et al., 1967. Design of surface and near-surface construction in rock. *Proc. 8th Symp. on Rock Mechanics, Univ. of Minnesota, 1966.* American Institute of Mining and Metallurgical Engineers, New York, N.Y., pp. 237—302.

Hamel, J.V., 1969. *Stability of Slopes in Soft, Altered Rocks.* Ph.D. Dissertation, Univ. of Pittsburgh, Pittsburgh, Pa. (Order No. 70-23, 232-University Microfilms, Ann Arbor, Mich.), Chapter 3 and Appendix C.

Hamel, J.V., 1970. The Pima Mine slide, Pima County, Arizona. *Geol. Soc. Am., Abstracts with Programs*, 2 (2): 335.

Hamel, J.V., 1975. Large-scale laboratory direct shear tests on desert alluvium. *Proc. 15th Symp. on Rock Mechanics, South Dakota School of Mines and Technology, 1973.* American Society of Civil Engineers, New York, N.Y., pp. 385—414.

Hamel, J.V., 1976. Libby Dam left abutment rock wedge stability. In: *Rock Engineering for Foundations and Slopes.* American Society of Civil Engineers, New York, N.Y., Vol. 1, pp. 361—385.

Hoek, E., 1970. Estimating the stability of excavated slopes in opencast mines. *Trans. Inst. Min. Metall.*, 79: A109—A132.

Komadina, G.A., 1965. Pima Mining Company — a progress report. *Min. Eng.*, 17(6): 48—53.

Komadina, G.A., 1967. Two-stage program boosts Pima to 30,000 TPD. *Min. Eng.*, 19(11): 68—72.

Krsmanovic, D., 1967. Initial and residual shear strength of hard rocks. *Geotechnique*, 15: 145—160.

Lacy, W.C. and Titley, S.R., 1962. Geological developments in the Twin Buttes District. *Min. Congr. J.*, 48(4): 62—64, 76.

Lambe, T.W. and Whitman, R.V., 1969. *Soil Mechanics*. Wiley, New York, N.Y., Chapters 10 and 11.

Martin, M.D., 1969. Pima Mining Company — a further major expansion. In: H.L. Hartman (Editor), *Case Studies of Surface Mining*. American Institute of Mining and Metallurgical Engineers, New York, N.Y., pp. 179—194.

Morgenstern, N.R. and Price, V.E., 1965. The analysis of the stability of general slip surfaces. *Geotechnique*, 15: 79—93.

Morgenstern, N.R. and Price, V.E., 1967. A numerical method for solving the equations of stability of general slip surfaces. *Comput. J.*, 9: 388—393.

Strohm, W.E. et al., 1964. Stability of crater slopes. *U.S. Army Eng. Waterw. Exp. Stn.*, Paper No. PNE-243F, 128 pp.

Thurmond, R.E., 1955. A description of the Pima ore body. *Min. Congr. J.*, 41(1): 27—30, 64.

Thurmond, R.E. and Storms, W.R., 1958. Discovery and development of the Pima copper deposit, Pima Mining Co., Pima County, Arizona. *U.S. Bur. Mines Inf. Circ.*, 7822, 19 pp.

Thurmond, R.E. et al., 1954. Geophysical discovery and development of the Pima Mine, Pima County, Arizona. *Trans. Am. Inst. Min. Metall. Eng.*, 199: 197—202.

Thurmond, R.E. et al., 1958. Pima: a three-part story — geology, open pit, milling. *Min. Eng.*, 10(4): 453—462.

Chapter 19

TWIN BUTTES PIT SLOPE FAILURE, ARIZONA, U.S.A.

BEN L. SEEGMILLER

ABSTRACT

A major pit slope failure occurred at the Twin Buttes Mine, Arizona during 1970—1971. Significant mining problems were caused by the failure and consequently a program of failure analysis and remedial action was undertaken. The program began in mid-1971 and was completed in late 1973. Geologic structure, mechanical properties and hydrology parameters were examined, and various analyses were performed. The cause of failure was believed to be weak, adversely oriented, geologic structures which under-cut the pit slope. Remedial action to prevent future similar failures was investigated and it was determined that rock anchors offered a possible solu-tion. A test section of pit slope was supported with 40 rock anchors of up to 234 tonnes working force. Subsequent mining operations through December 1973 demonstrated the field effectiveness of rock anchors for this type of failure prevention. Cost analysis indicated rock anchors to be a viable eco-nomic solution.

INTRODUCTION

A major pit slope failure occurred at the Twin Buttes Mine, Sahuarita, Arizona, during 1970—1971. This slope failure caused significant planning and operational problems, particularly during the period September through November, 1970. An examination of the failure problem and its causes and consideration of future remedial action began in December 1970; an initial study report was completed in March 1971. In June 1971, an in-depth two-year study (Seegmiller, 1972, 1973; Stewart and Seegmiller, 1972) was begun to further analyze the failure and examine various courses of remedial action. In June 1972 a program of slope support using rock anchors was undertaken. By December 1973 the program had demonstrated that rock anchors could economically prevent similar future slope failures (Seegmiller, 1974; West, 1974). The purpose of this paper is to present the details of the failure problem and describe the remedial action program that was under-taken.

The Twin Buttes Mine is located some 40 km south of Tucson, Arizona. It is a pyrometasomatic deposit from which some 32,000 tonnes/day of copper-bearing ores are produced. Mining in the pit is carried out using a conventional pushback system. Alluvial gravels, which form a 135 m thick blanket over the bedrock, are removed by a scraper-conveyor-wagon system. Ore and waste rock are removed by a shovel-truck-conveyor system. Depth of mining had reached 300 m by 1973 in the deepest portion of the pit. Initial pit planning called for 37.5° slopes in the alluvium and 45° slopes in the rock.

This paper begins with a detailed description of the failure problem. The basic parameters governing stability are then examined including the geologic structure, mechanical properties and hydrology. The cause of the failure is then discussed, and is followed by a complete description of the rock anchor support system including its design, installation, cost and effectiveness.

FAILURE DESCRIPTION

Historical aspects

Minor stability problems in the south slope of the Twin Buttes pit have existed since early 1969. Many of the problems were associated with two major geologic structures, the Main Pit Fault and the East-West Fault, which cut through the south slope as shown in Fig. 1. The stability problems initially amounted to only a nuisance. However, as mining continued in late 1969 and early 1970, subsequent slope failures became larger and presented more serious problems. By December 1970, sections measuring up to 100 m

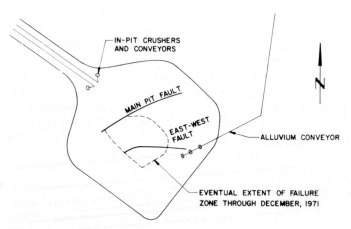

Fig. 1. Twin Butte pit plan (late 1969—early 1970).

Fig. 2. Classic wedge failures in south slope.

Fig. 3. South slope failure late in 1970.

Fig. 4. South slope failure in March 1971.

in length and consisting of two and three slope benches had displaced up to 3 m in some zones, as shown in Figs. 2 and 3. Many new tension cracks began to appear on most of the south slope benches. New areas of slope displacement and greater magnitudes of movement continued to occur throughout the period from December 1970 through March 1971. At that time essentially the entire south slope had failed and was in continuous movement (Fig. 4). Crude surface extensometers and survey nets were used to monitor surface displacement and velocity. Standard displacement charts (Seegmiller, 1973) were plotted and used to predict future slope displacement. As of July 1971, more than 15 m of vertical and 20 m of horizontal displacement had been measured, and displacement was continuing. By December 1971, the failure involved vertically more than 150 m of rock and 30 m of alluvium (Fig. 5).

Mining at or near the toe of the slide continued throughout the period of slope failure. The material removal rate was such that one 15-m bench was mined out every three weeks. This activity aggravated the instability condition and a direct correlation between displacement magnitude and mining was noted. This situation, however, could not be avoided because a major zone of ore at the slide toe had to be mined. Displacement monitors successfully warned of any possible catastrophic movement; no injuries or equipment damage resulted.

Fig. 5. South slope failure in December 1971.

Failure observation details

The minor failure problems which initially occurred adjacent to the Main Pit and East-West Faults consisted of small zones up to 15 m in length. These zones progressively disintegrated into rubble slopes. Generally, only one bench was involved and never more than three. The larger zones which appeared in late 1970 resembled classic wedges, with sliding occurring down the structural intersection line. The wedges would begin on one bench and gradually enlarge to involve two and three benches. By March 1971, numerous wedges had failed and enlarged to the extent that essentially the entire south slope had become a massive block or slab failure. Very little rotation was observed, indicating that the failure zone was probably quite shallow and a relatively near-surface phenomenon. Detailed examination of the failure area revealed that the failure plane was composed of a large number of relatively planar and parallel structures. These structures were found to strike in the same direction as the slope and dip at approximately $38°$ into the pit. They appeared to limit the failure in the vertical direction to less than 25 m of depth. Some undercutting of the alluvial gravels took place causing ramp maintenance problems in the higher south slope areas. Total slide size including the undercut alluvium was estimated to be approximately 3×10^6 tonnes.

Failure mode analysis

Field observations of the actual failure planes indicate that while the failure began in the classical wedge manner, it proceeded to completion in a slabbing fashion. In other words, failure initially took place on steeply dipping structures known as set B and moderate dipping structures known as set A, the two intersecting sets forming wedges of rock. It then continued by slabbing parallel to set A structures which were striking approximately parallel to the pit slope. The faults of set B acted as detachment planes for the sides of the slabs. In plan these wedge and slab failures would have appeared as depicted in Fig. 6. Several low-angle (10°) shear planes were noted, particularly in the lower portions of the failure. These shears were thought to be caused by the excessive driving forces of the slabs. Such driving forces were calculated in some cases to have possibly exceeded more than 300 tonnes per linear metre of slope length during slab failure. They not only caused the low-angle shears but, in addition, caused toe heaving through solid, but well-fractured, rock. In section the failure was thought to appear as shown in Fig. 7. This figure depicts failure of a 45° slope along structures dipping into the pit at 38°. Overall slope angle was lessened to approximately 28° after failure.

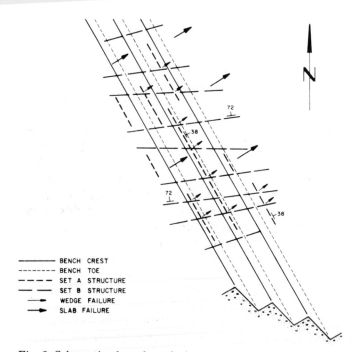

BENCH CREST
BENCH TOE
SET A STRUCTURE
SET B STRUCTURE
WEDGE FAILURE
SLAB FAILURE

Fig. 6. Schematic plan of south slope failure zone.

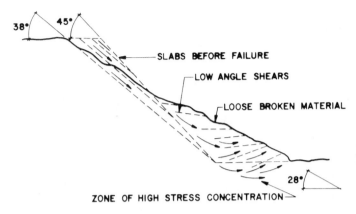

Fig. 7. South slope failure section.

BASIC STABILITY PARAMETERS

Geology and geologic structure

The rock type comprising the south slope area is predominantly a mixture of feldspathic quartzite, conglomerates, tuffs, andesites, and dacites; the upper 40% of the entire south slope is an alluvial gravel, some 120—135 m thick. Numerous geologic structures in the form of faults and tectonic joints are found throughout the rock mass. Data on the faults and joints had been collected by pit geologists since rock was first exposed in January 1968. These data were recorded in the field on 50-scale sheets and then transferred onto 100-scale geologic plan maps. By November 1970, more than 5000 structures had been plotted. Information, including structural type, strike, dip, coordinate location, bench elevation, and rock type, was subsequently digitized and punched onto computer cards for processing and analysis. A computer program was used to produce equal area nets for each structural type in the south slope. These were compared to nets produced in other parts of the pit and in other rock types. Variation in the strike and dip of major structure sets was found to be minor when various rock types and pit areas were compared. A typical equal area net for the two major structure sets is shown in Fig. 8. A summary of the characteristics of these two sets is given in Table I.

Mechanical properties

Unconfined uniaxial strength of typical solid rock exceeds 140 MN/m^2, in contrast to the shear strength of the much weaker geologic structures. Failure of the south slope occurred predominantly along clay-filled structures.

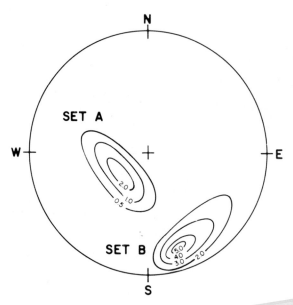

Fig. 8. Idealized equal area net for geologic structures in failure zone (lower hemisphere projection, 1% contour interval).

Field observations indicated that the actual shearing occurred in the clays themselves rather than along the clay-rock interface. Field and laboratory testing with direct shear apparatus was used to establish shear strength parameters for the clays. In addition, back-analysis of a series of failed wedges was used to further estimate design parameters for future work. A summary of approximate strength characteristics of the geologic structures in the south slope area is shown in Table II.

The strength of the rock mass as a function of distance into the pit face appeared to change quite drastically over the first 20 m. Rock mass deformation characteristics were tested using both Goodman Jack and seismic

TABLE I

Characteristics of major structure sets in the south slope area

	Set A	Set B
Predominant structural type	joints	faults
Average strike	N50° W	N75° E
Average dip	37° NE	80° NW
Relative occurrence	moderate	major
Exposed continuity	3—4 m	30—50 m
Approximate spacing	2 m	1 m

TABLE II

Approximate shear strength characteristics of geologic structures in the south slope area

	Cohesion (kN/m²)	Friction angle (degrees)
Joints		
peak	120 ± 50	34 ± 5
residual	50 ± 24	32 ± 5
Faults		
peak	52 ± 34	32 ± 9
residual	20 ± 17	31 ± 6

Values given are mean ± 1 standard deviation.

refraction techniques. Indications were that the rock mass progressively tightened inward into the bench face. These results are suggestive of the effects of production blasting and differential rebounding across geologic structures.

Hydrology

Water studies carried out during the period 1969—1971 indicated that the water table in the vicinity of the pit had been substantially lowered. In the immediate vicinity of the south slope, the water table appeared to have been lowered 75—100 m below the failure zone. A series of three piezometers was subsequently placed about the slide area. They showed the water table to be some 150 m below the upper portion of the failure, or approximately on the same horizon as the lowermost portion of the failure. Intense desert thundershowers caused considerable runoff in and around the slide area whenever they occurred. Some small zones in the south slope are known to have had their stability adversely affected by these thundershowers. However, the extent of the major zone of failure and the cause of its failure were believed due to factors other than hydrology.

CAUSE OF FAILURE

Following a two-year study of the failure problem, including its mode and the basic stability parameters, conclusions were drawn as to the basic cause. In effect, the failure came about as a result of a series of weak geologic structures being undercut by mining. That is, two sets of weak geologic structures, one mainly faults, the other joints, existed in such a combination that wedges and slabs of rock slid into the pit when these structures were undercut. Whether or not production blasting and differential rebounding produced or merely aggravated a pre-existing condition was not determined.

REMEDIAL ACTION: SLOPE STABILIZATION WITH ROCK ANCHORS

General

Because additional mining was desired in the south slope area, particular concern was focused on preventing similar failures in the future. Examination of drill core in the slide area indicated that the adversely oriented low-strength structures would again be found as mining was undertaken. It was therefore believed that unless some type of remedial action was taken, similar future failures should be expected. As a consequence, a slope stabilization program (Seegmiller, 1974) was undertaken in June 1972 to investigate the costs and effectiveness of rock anchors to prevent future slope failures. A test section adjacent to the failure area was selected as shown in Fig. 9. The site was approximately 60 m in width and covered four successive 15-m benches. The basic stability parameters of the test section were

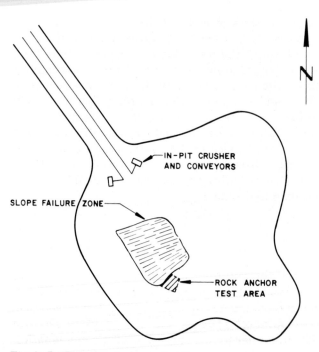

Fig. 9. Rock anchor test area location.

similar to those of the adjacent failure area, but failure had not yet developed. The reason for this was that the test section was essentially in a corner of the pit and was not undercut. Future mining would push back the adjacent east wall, undercutting the rock anchor test zone. The effectiveness of the system would then be demonstrated.

Support system design

Several basic assumptions were necessary in arriving at a suitable design for a rock anchor system of slope support. These assumptions include:

(1) A potential plane of shear failure exists which passes through the toe of the slope.

(2) Rock situated in the section between the failure plane and the slope face is potentially unstable, but may be regarded as a rigid body.

(3) Rock beyond the failure plane is stable and suitable for anchorage purposes.

(4) Rock anchors create compressive forces which act through a 90° cone and are uniformly distributed over the potential failure plane.

(5) Slope failure will only take place when the shear stresses exceed the shear strength along the potential failure plane.

(6) Shear strength along the potential failure plane is a function of the angle of friction, cohesion and rock anchor force.

Using the above assumptions as a basis, the total required rock anchor force could be calculated for any particular safety factor. A safety factor (defined as the ratio of shear strength to shear stress) of only 1.10 was selected because the test was being conducted for research purposes and a higher degree of safety was not required. Rock anchors of the various sizes, lengths, and with differing spacings were selected to reflect the research nature of the test. The lengths were chosen to adequately ensure that the anchorage section would be located in firm rock. Anchorage length was calculated using rock-to-grout bonding strengths applicable to the rock type. Anchor orientation was selected to maximize the forces across the major geologic structure sets. The optimum inclination of the anchors was deter-

TABLE III

Test area characteristics

	Bench A	Bench B	Bench C	Bench D
Number of anchors	13	12	8	7
Spacing (m)	4.6	6	7.6	7.6
Hole bearing	due south	due south	due south	due south
Hole inclination	−5°	−5°	−5°	−5°
Hole length (m)	45	45	30	30

TABLE IV

Rock anchor data

	A-1, A-3, A-4, A-8, A-9, A-10, A-11, A-12, B-1, B-2, B-4, B-5, B-6, B-8, B-9, B-11	C-2, C-3, C-5, C-6, C-7, D-3, D-5, D-6	A-5, A-6, D-1, D-2	B-10	A-2, A-7, A-13, B-3, B-7, B-12	C-1, C-4, C-8, D-7, D-4
Anchorage length (m)	10.7	10.7	13.7	13.7	10.7	10.7
Hole diameter (cm)	8.9	8.9	8.9	12.7	8.9	8.9
Initial force (tonnes)	117	88	176	234	100	76
Number of 1.27-cm diameter cables	8	6	12	16	8	6
Load cell used	no	no	no	no	yes	yes

mined to be +7° from horizontal, but for practical purposes was set at −5° from horizontal. A summary of the rock anchor sizes and spacings is given in Tables III and IV.

Installation

All holes were drilled with a track mounted percussion drill which averaged about 45 m per eight-hour shift. Each hole was cleaned and water-tested for leaks. Major leaks were grouted and redrilled. The anchor tendons were placed into the cleaned holes by hand or with the aid of a tugger. The tendon consisted of strands of 1.27-cm diameter cable and a grout tube. The lower end of each tendon had three spreader rings on 1.5-m centers about which the strands were taped. The spreader rings gave the tendon a "sausage effect" and insured that grout would flow onto all sides of the strands. Following emplacement, the anchorage section of the tendon was grouted and then let cure for fourteen days. A steel reinforced concrete "blockout" was constructed about each hole collar. The purpose of the blockout was to transmit the rock anchor load to the surrounding rock. Following anchorage and blockout curing the tendons were tensioned using a center hole jack. A VSL-type head and wedge system was used to hold the forces on the tendons. Secondary grouting was used for corrosion protection on all units except the ones containing load cells. On these units the stressing portion of

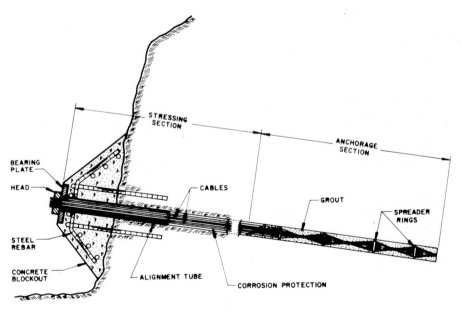

Fig. 10. Section through typical rock anchor.

Fig. 11. Completed rock anchor.

Fig. 12. Completed rock anchors on uppermost test bench.

the cables were greased and wrapped with plastic to insure load transmission from the anchorage section to the head. A typical section through a completed rock anchor is shown in Fig. 10. Photographs of the completed units in the test area are shown in Fig. 11 and 12. Cost and economic considerations have been discussed by Seegmiller (1974).

Effectiveness

Approximately six weeks after the project began heavy rains caused approximately 10 cm of displacement along a major geologic structure which bisected the test area. The displacement took place before any rock anchors were completed. Subsequent tensioning was successful in arresting the displacement and stabilizing the displaced portion of the test section.

Waste stripping adjacent to the test area in the east wall took place during the last six months of 1973. Material down to the lowermost test bench was removed for a distance of 183 m adjacent to the test area. Slope displacements on the order of 3 m were noted in portions of the newly stripped area as well as bench sloughing and spalling. However, the test area remained in place and was in essence the only stable zone along the entire south slope as shown in Fig. 13. By late 1973 the research project had demonstrated the effectiveness of the support system and all project objectives were consid-

Fig. 13. Stable rock anchor test section (circled) and surrounding slope failures.

ered fully met. In other words, a method had been found to economically stabilize future pushbacks in the Twin Buttes south slope.

REFERENCES

Seegmiller, B.L., 1972. Rock stability analysis at Twin Buttes. *Proc. 13th Symp. on Rock Mechanics, University of Illinois, 1971.* American Society for Civil Engineers, New York, N.Y., pp. 511—536.

Seegmiller, B.L., 1973. Slope stability research: its payoff in mining. *Min. Congr. J.*, 7: 32—39.

Seegmiller, B.L., 1974. How cable bolt stabilization may benefit open pit operations. *Min. Eng.*, 26: 29—34.

Stewart, R.M. and Seegmiller, B.L., 1972. Requirements for stability in open pit mining. In: *Geotechnical Practice for Stability in Open Pit Mining.* Society of Mining Engineers, American Institute of Mining and Metallurgical Engineering, New York, N.Y., pp. 1—7.

West, L.J., 1974. Rock mechanics annual review. *Min. Eng.*, 26: 44—46.

Chapter 20

PIT SLOPE PERFORMANCE IN SHALE, WYOMING, U.S.A.

G. WAYNE CLOUGH, LAWRENCE J. WEST and LARRY T. MURDOCK

ABSTRACT

Slope stability studies at three open pit mine sites in Wyoming were performed over the period 1970 through 1973. At each of the sites substantial
strata of shale were encountered. In each case, the groundwater table is at
a level such that the shale strata in the slopes are saturated. The shale at site 1
is of the Wind River Formation while that at sites 2 and 3 is of the Wasatch
Formation. Results of undrained strength tests and classification tests suggest that the Wind River and Wasatch shales have properties similar to those
reported for shale of the Fort Union Formation.

The authors were asked to undertake studies of the Wind River shale at
site 1 after severe slope failures were encountered during early mine excavation. Slopes of 32—55° were observed to fail over periods of several weeks to
one year after excavation in either massive rotational slippages or progressive
failures of benches which worked their way back up a slope. Stability analyses of selected rotational failures suggested that the undrained field strength
was about half way between that found for intact and fissured samples in
undrained triaxial tests. Finite element analyses showed that zones of local
yielding existed under steep benches on the slopes, and it was concluded this
phenomenon led to progressive bench failures. In order to minimize the
slope problems, it was recommended that slopes be flattened to 45° or less,
benches be eliminated, and sandstone layers be drained using lateral drains.
Behavior as of March 1975 suggests that the recommendations have improved stability of the slopes although shallow slides in the shales continue
to be a problem because adequate drainage of the thick shale beds (up to
26 m) cannot be effected.

The studies of the Wasatch shale at site 2 were undertaken before excavation was begun. Slopes recommended for site 2 ranged from 45° to 60°
depending upon the thickness of the shale layers and sandstone interbeds in
the slope. Excavation at this site, begun in 1972, was accompanied by a
thorough program of lateral drainage of the sandstone layers which has been
very effective in removing free water from the slopes. Also, ripping procedures have been utilized to remove the rock and, as a result, the slopes are

smooth without the sharp breaks caused by bench and bank excavation. Slope performance at site 2 has been satisfactory with nothing more than minor shallow slippages occurring.

Excavation at site 3 has not yet been undertaken. Slope recommendations were made in the form of design charts which allow adjustment of the excavation slopes to fit field conditions for slope height and groundwater table.

INTRODUCTION

Prediction of slope performance in clay shale represents one of the most challenging tasks to face the geotechnical engineer. Analytic methods have been shown to be at best only a guide to slope performance in shale because of problems involving unusual initial stresses (Smith and Redlinger, 1948; Mencl et al., 1965; Duncan and Dunlop, 1969), widely varying properties (Peterson et al., 1960) and susceptibility to weathering and swell (Peterson et al., 1960; Bjerrum, 1967). These difficulties have led to a strong reliance on records of past experience in design of slopes in shale. In this paper, three cases of slope design and two cases of slope performance for deep open pit mining projects in Wyoming are examined so as to provide useful experience records. The projects involved one investigation of a stability problem and two design studies. The studies included detailed geologic investigations, numerous physical property and strength tests, limit analyses of pit stability and finite element analyses of stress distribution in the pit slopes.

PROJECT DESCRIPTIONS

The locations of the three mines are shown in Fig. 1. Site 1, owned by Utah International Inc., is located in the Shirley Basin (Mud Springs Quadrangle) about 60 km southwest of Casper, Wyoming; site 2, owned by

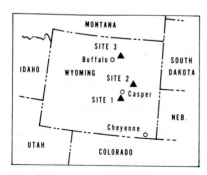

Fig. 1. Locations of mine sites.

Humble Oil and Refining Company, is located in the Powder River Basin (Highlands Flats Quadrangle) about 139 km east of Casper; and, site 3, owned by Texaco Corporation, lies on the western part of the Powder River Basin (Lake DeSmet Quadrangle) about 15 km north of Buffalo, Wyoming. Sites 1 and 2 are uranium mines and site 3 is a coal mine.

The sites are located in semi-arid land with gently rolling topography. Maximum mine depths are 115 m for site 1, 82 m for site 2, and 165 m for site 3. Mining is in progress at sites 1 and 2; site 3 has been designed but mining has not been implemented.

GEOLOGIC AND GROUNDWATER CONDITIONS

Geologic and groundwater conditions at the sites are in some ways similar. Subsurface materials at all the sites are horizontally layered beds of alternating shale and sandstone of early Tertiary age. The water table is within 50 m of the surface at all sites. Differences occur in the subsurface conditions because of variations of bed thicknesses and location of the water table relative to major shale beds.

Site 1. A typical geologic cross-section through a slope at site 1 (Shirley Basin) is shown in Fig. 2a. The upper 20-26 m consist of moderately cemented sandstones of the White River Formation. Beneath it are alternating beds of shale and sandstone of the Wind River Formation (Eocene). The shale beds predominate and range in thickness from 7 to 26 m. Highly fractured lignitic beds up to 3 m thick occur in the shale. The water table is about 26 m below the ground surface, thus the shale beds in the Wind River Formation that are exposed in the mine slopes are fully saturated and are subjected to seepage flow. Extreme problems of slope stability were encountered in the shale of the Wind River Formation.

Site 2. A typical cross-section through a slope at site 2 (Powder River Basin) is shown in Fig. 2b. Slopes at this site are excavated in the Tertiary sandstones and shales of the Wasatch Formation which conformably overlies the Fort Union Formation. Shale and sandstone beds of approximately equal thickness alternate in the typical section.

The shale beds at this site range in thickness from less than 3 m to about 16 m. Lignite is found in the shale but only as lenses generally not more than 15 cm thick. The water table is from 23 to 40 m below the ground surface.

Site 3. Subsurface conditions at site 3 were more heterogeneous than those of sites 1 and 2. In the areas where shale was prominent, three different sections could be defined:

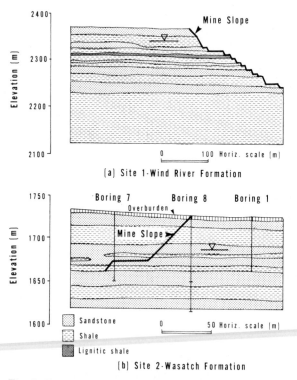

Fig. 2. Typical geologic cross-sections for (a) site 1, Wind River Formation, and (b) site 2, Wasatch Formation.

(1) Essentially homogeneous shale beds up to 100 m thick, overlying coal.

(2) Alternating beds of sandstone and shale with bed thicknesses ranging from 10 to 20 m thick, overlying coal.

(3) Sandstone beds up to 30 m thick, overlying shale beds up to 60 m thick, which in turn were underlain by coal.

Typical sections for each of the conditions are shown in Fig. 3. The water table location varied for the sections; in the case of the very thick shale beds in the section of Fig. 3a, the water table was at or near the surface. For the other sections the water table was located from 0 to 35 m below the ground surface.

A number of other materials occurred at this site in addition to shale, sandstone, and coal. One of the most interesting of these was clinker, a material formed by the baking of shale and sandstone by subterranean coal fires. The clinker usually consisted of chunks or bricklets of fire-hardened shale or sandstone and was found at the surface in beds up to 50 m thick.

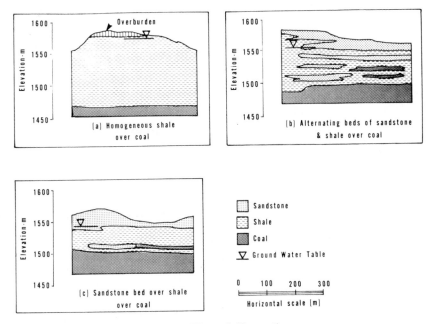

Fig. 3. Geologic sections at site 3, Wasatch Formation.

GEOTECHNICAL PROPERTIES OF TERTIARY SHALES

Classification, consolidation, and undrained triaxial tests were performed on all of the shales; drained direct shear tests were performed on the shale at site 2. Strength and deformation tests were conducted on undisturbed samples; the classification tests were conducted on disturbed samples collected at the natural moisture content.

Classification data

The results of the Atterberg limit classification tests from all three sites are shown in Fig. 4, a plot of plasticity index versus liquid limit. Also shown on the plot are the classification boundaries defined by the Unified Soil Classification System. A large percentage of the points fall above the A line and to the right of the B line, indicating that the shales are inorganic and have a relatively high plasticity. Samples whose values fell below the B line were classed as low-plasticity materials and contained higher percentages of silt and sand. The data show that the shales at all the sites yielded generally similar results, with those from site 3 (Powder River Basin) being slightly less plastic than those from site 1 (Shirley Basin) and site 2 (Powder River Basin). For purposes of comparison, the range of values reported for Fort

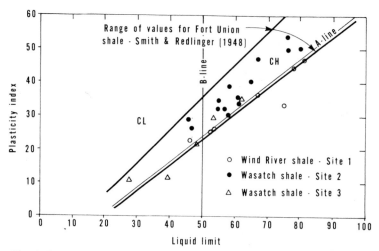

Fig. 4. Results of Atterberg limit tests on shale samples from sites 1, 2 and 3.

Union shale by Smith and Redlinger (1948) are also shown in Fig. 4. Most of the data from sites 1, 2, and 3 fall within the range of results reported for Fort Union shale.

A summary of average values of the Atterberg limits, natural water contents and dry densities for the shales at the three sites is given in Table I. Also shown are similar data reported in the literature for other well-studied shales; in several instances, where this figure was available, percent montmorillonite is also indicated. In terms of the Atterberg index data the shale samples from sites 1, 2, and 3 appear to have similar properties to shale samples from the Pepper, Cucaracha and Fort Union Formations. The Bearpaw shale has a much higher plasticity index.

As to dry densities and water contents, the Wind River shale of site 1 gave the highest average water content and lowest average density of all samples tested in a "fresh" condition. Weathered Bearpaw shale samples yielded a higher water content and lower density than any of the intact samples.

Consolidation parameters

A few consolidation tests were performed at each site to gain some insight as to possible preconsolidation load and to determine expected values of rebound in the shales. Unfortunately, the equipment employed in the tests could not load the samples past the preconsolidation load, and the preconsolidation load could not in fact be determined. However, the loading and rebound patterns were instructive. Typical results for samples from sites 1 and 2 are shown in Fig. 5, plotted as void ratio versus effective vertical stress. Two curves each are plotted for the Wind River shale of site 1 and the

TABLE I

Average geotechnical data for shales

Shale formation	Geologic age	Condition of samples	Natural water content (%)	LL (%)	PL (%)	PI (%)	Dry density (Mg/m³)	Montmorillonite (%)	Reference
Wind River (site 1)	Eocene	fresh	23	67	33	34	1.60	n.a.	present investigation
Wasatch (site 2)	Paleocene	fresh	17	65	35	30	1.80	2—15	present investigation
Wasatch (site 3)	Paleocene	fresh	18	50	25	25	1.68	n.a.	present investigation
Bearpaw (intact)	Late Cretaceous	fresh	19	120	25	95	1.80	n.a.	Peterson et al. (1960, p. 780)
Bearpaw (weathered)	Late Cretaceous	weathered	35	120	25	95	1.36	n.a.	Peterson et al. (1960, p. 780)
Pepper	Late Cretaceous	fresh	22	58	26	32	1.67	13	Wright (1969, p. 255)
Cucaracha **	Miocene	fresh	17	65	28	37	1.76	n.a.	Lutton and Banks (1970, table B-1)
Fort Union	Paleocene	n.a.	20	*	*	*	1.67	0—15	Smith and Redlinger (1948, p. 63)

n.a. = information not available

* Range only given, see Fig. 4.

** See Chapter 4, this volume.

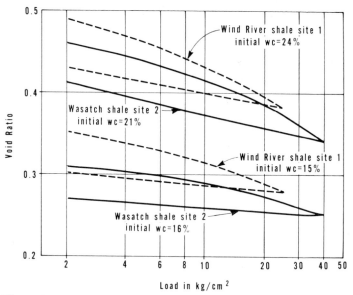

Fig. 5. Results of consolidation tests on shales from Wind River and Wasatch Formations.

Wasatch shale of site 2; one of the curves is for a higher water content sample (~22%) and the other for a lower water content sample (~15%). Consolidation results for tests on Wasatch shale from site 3 showed behavior similar to that obtained for the samples from site 2 and thus are not shown.

The data show that none of the curves ever reached a straight line virgin curve behavior even though vertical test pressure reached as high as about 4 MN/m². Thus, preconsolidation pressures are apparently well above 4 MN/m². Also, data for either shale demonstrate that the sample with the higher water content is more compressible and shows more rebound and water intake during unloading. It is also interesting that the samples of the Wind River and Wasatch shales show similar behavior; in terms of compressibility, the two shales appear quite similar when samples with equivalent water content are compared. However, it is important to recognize that the average water content of the Wind River shale (23%) is higher than that of the Wasatch shale (17%). Thus, on average, the Wind River shale is likely the most compressible.

Strength test results

Undrained tests. Undrained-unconsolidated triaxial tests were performed on about 70 undisturbed shale samples obtained from exploratory borings from sites 1, 2, and 3. No attempts were made to measure pore pressures in these tests; confining pressures ranged from 0 to about 3 MN/m². The results of

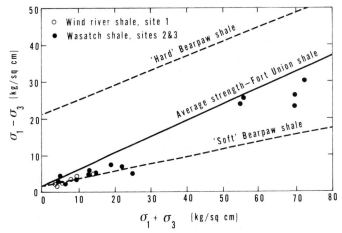

Fig. 6. Results of undrained triaxial tests on fractured and fissured samples of shale.

the tests are summarized in Figs. 6 and 7, plotted as values of principal stress difference at failure, $(\sigma_1 - \sigma_3)_f$, versus the sum of the principal stresses at failure, $(\sigma_1 + \sigma_3)_f$. This type of plot is related to the traditional Mohr diagram except that failure conditions are represented by points and not by Mohr circles.

The results are plotted on two different figures so that data obtained on fractured and fissured samples (Fig. 6) can be interpreted separately from those of intact samples (Fig. 7). The distinction between fractured and intact samples was made on the basis of the condition of the sample at the begin-

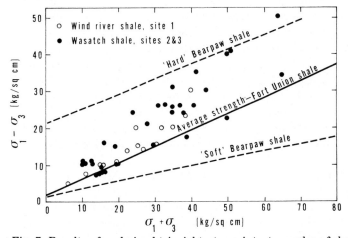

Fig. 7. Results of undrained triaxial tests on intact samples of shale.

ning of the test, and whether failure occurred on a pre-existing fracture or slickenside or through intact material. For purposes of reference, limiting strength values of undrained triaxial test results for "soft" and "hard" Bearpaw shale (Peterson et al., 1960, fig. 8) and average strengths of the Fort Union shale (Smith and Redlinger, 1948, fig. 3) are also shown in Figs. 6 and 7.

The data in Fig. 6 for the fractured and fissured samples suggest that the shale from sites 1, 2 and 3 sustains undrained strengths somewhat below those of the average results for Fort Union shale and slightly above those for "soft" Bearpaw shale. The test results for the intact shale from sites 1, 2 and 3 in Fig. 7 show that the strength of the samples of intact shale is clearly higher than those for "soft" Bearpaw shale, generally higher than the average strength of the Fort Union shale, and generally lower than those of "hard" Bearpaw shale. Samples of the Wind River Formation from site 1 seem to possess lower strength values than Wasatch samples from site 2 and 3.

In comparing the data of Figs. 6 and 7, it is clear that the fractured and fissured shale samples possess substantially lower strength values than the intact shale. For all of the sites, however, only 10—20% of the shale samples recovered were found to be fractured. Thus, these data present the designer with a problem: how much weight should be given to the lower strength values of the fractured and fissured samples? This quandary can perhaps be answered through studies of field slope performance, such as those reported in a subsequent section of this paper.

Drained tests. Twelve drained direct shear tests were conducted on the Wasatch shale samples from site 2. The tests were conducted at a very low strain rate to insure full drainage of excess pore pressures in the sample. Shearing of the sample took place along horizontal planes, planes corresponding to the bedding planes of the shale samples. In order to establish residual strength values, the sample was sheared to failure, then separated and set back to the original position and sheared again; this process was repeated until the shear strength showed no tendency to decrease.

Results of the tests are shown in Fig. 8 as values of peak and residual strengths plotted versus normal pressure applied in the test. Data for the peak strengths are scattered; an average failure envelope for the peak strengths yields a drained cohesion of 100 kN/m^2 and a drained friction angle of $32°$. Data for the residual strengths show less scatter and the drained cohesion and friction angle of the failure envelope are 40 kN/m^2 and $22°$, respectively.

Table II shows a comparison of the peak and residual strength parameters of the Wasatch shale to those reported for other well-studied shales. The peak strengths of the Wasatch shale are intermediary between the highest and the lowest values reported, falling nearer the higher values. The residual strengths of the Wasatch shale are in general agreement with other reported

TABLE II

Comparison of results of drained direct shear strength values

Shale formation	Average natural water content (%)	Peak strength parameters		Residual strength parameters		Specimen condition	Reference
		ϕ' (degrees)	c' (kN/m²)	ϕ' (degrees)	c' (kN/m²)		
Wasatch (site 2)	17	32	100	22	40	fresh specimens, not precut	present investigation
Bearpaw (intact)	19	35	70	16	10	fresh specimens, not precut	Peterson et al. (1960, p. 782)
Bearpaw (weathered)	35	28	40	22	10	weathered specimens not precut	Peterson et al. (1960, p. 782)
Pepper	22	14	40	9	0	precut specimens for residual tests	Beene (1967, p. 41)
Cucaracha *	17	18	500	7.5	0	precut specimens for residual tests	Lutton and Banks (1970, p. 196)
Fort Union	20	20	0	n.a.	n.a.	n.a.	Smith and Redlinger (1948, p. 63)

n.a. = not available.

* See also Chapter 4, this volume.

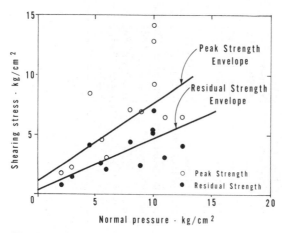

Fig. 8 Results of drained direct shear tests on shale from Wasatch Formation from site 2.

values except for those where the specimens were precut; these values are lower than those for the Wasatch shale.

CASE HISTORY DISCUSSIONS

Slope stability at site 1

Slope studies at site 1 were undertaken after substantial slope stability problems (subsequently described) had occurred in the newly opened pit. The pit excavation was made using a bench and bank excavation technique; the benches were about 9 m wide and 20—25 m high and sloped at between 45° and 65°. Initial overall slopes ranged from 55° along the west wall to 32° along the east wall. Slope failures occurred on all sides of the pit irrespective of the differences in overall slope. General descriptions of the failures are given below; Atkins and Pasha (1973) provide additional detail.

Excavation was begun in January 1969. The early excavation took place in sandstones of the upper White River Formation and no stability problems were experienced. Further excavation was carried through the upper mixed sandstone and shale units without problems until the pit bottom reached a depth of 62 m in January of 1970. This excavation depth exposed a relatively thick seam of saturated lignitic shale (see Fig. 2) and a failure in the west wall involving 500,000 m³ of the overburden which occurred along the seam. From visual observation, the sliding surface appeared to follow the lignitic seam closely until it veered upwards through the overlying shales and sandstones. As the excavation proceeded and more of the shale was exposed, additional slides occurred; these subsequent slides differed from the first

TABLE III

Summary of slide events — site 1

Slide	Location	Date	Volume of slide mass (m³)	Type of slide
1	west wall	Jan. 1970	500,000	movement along lignite seam
2	south wall	Feb. 1970	400,000	rotational slump
3	north wall	Aug. 1970	400,000	progressive slump-earthflow
4	south wall	Apr. 1971	1,400,000	rotational slump
5	east wall	Sep. 1971	200,000	rotational slump
6	west wall	1972	300,000	progressive slump-earthflow
7	west wall	Oct. 1972	1,600,000	rotational slump

slide in that they were largely confined to the shale and were either a rotational or progressive slump-earthflow type. The major slope failures are documented in Table III as to the type, date, and mass involved in the slide. In addition to these slides numerous small slides occurred, generally of the slump-earthflow type and largely confined to one or two benches at the most. All of the slides occurred relatively quickly, within several months to one year after the shales were exposed.

Photographs of one of the massive rotational slides and of a progressive slump-earthflow slide are shown in Fig. 9a and b, respectively. The rotational slide has passed behind two benches and the top of the slide mass has dropped about 15 m. The progressive failure began with the failure of the lowest bench in the photograph; following the first bench failure, a second and then a third bench collapsed as each successive bench was undercut by the previous failure.

The authors were engaged in early 1972 to investigate the slope problems and propose solutions. Exploratory holes were drilled around the site, undisturbed samples obtained and twelve piezometers installed and sealed off in a number of the sandstone lenses and layers at different elevations. The piezometer observations showed that perched groundwater tables existed in many of the sandstone lenses and that the intervening shale layers were essentially acting as aquicludes. In many cases initial excavation into the sand lenses produced a sudden burst of flow from the lenses which gradually diminished with time. The early high flow rate was apparently due to artesian pressures in these sandstone lenses.

The undisturbed samples of the sandstone and shale obtained from the exploratory holes were laboratory tested for classification and strength

Fig. 9. Types of stability problems at site 1: (a) massive rotational slide, (b) progressive bench failure.

purposes. In order to supplement the laboratory data available on shear strength of shale of the Wind River Formation, back-calculations of two of the best documented rotational slump failures were undertaken. No attempts were made to analyze the progressive slump-earthflow slides for the purpose of determining shear strength because these failures seemed to be strongly affected by surficial weathering, a factor which would lead to an erroneous estimate of shear strength of intact shale.

In the analyses the pervious sandstone was assumed to be sheared slowly enough to justify usage of drained strength parameters. The drained strength parameters were obtained from laboratory tests as $\phi' = 45°$ and $c' = 200$ kN/ m^2. Because the tests yielded very consistent values, these parameters were considered reliable.

The choice of the type of shear strength to be used for the relatively impervious shale presented a problem since the slope failures had occurred only several months to one year after excavation and it was not clear whether a drained or an undrained strength was applicable. Because the shale obviously had a low permeability and the slope failure surfaces were deep into the shale where drainage would be difficult, it was decided to assume that the shale was shearing in essentially an undrained state. The shale and sandstone were assumed saturated with a phreatic surface parallel to the slope face. All artesian pressures were assumed to be relieved from the sandstone layer since drainage of excess pressures from these layers occurred during exposure by excavation. The analyses were performed using the "modified Bishop" stability analysis (Bishop, 1955).

The undrained strength values employed for the shale were varied within limits dictated by the undrained laboratory strength test results shown in Figs. 6 and 7. After several trials it was found that the values of $c = 70$ kN/ m^2 and $\phi = 20°$ yielded factors of safety of one for the back-analyzed slope failures. These parameters may be reasonably interpreted as representative of an "operational field shear strength" for relatively short-term, deep slump-type failures in Wind River shale. Strength values which could be back-calculated from the progressive slump-earthflow failures would be lower. It is interesting that the $c = 70$ kN/m^2 and $\phi = 20°$ parameters agree exactly with those reported by Smith and Redlinger (1948) and Lane (1961) for similar back-calculations performed on known failures in Fort Union shale. This result appears reasonable in view of the fact that the undrained shear test results (Figs. 6 and 7) demonstrated that the average undrained strength of Fort Union shale seemed to correspond with much of the strength data from site 1. It is also interesting to note that strength parameters $c = 70$ kN/m^2 and $\phi = 20°$ yield an envelope which agrees with that shown in Figs. 6 and 7 as the average for the undrained laboratory shear test results for Fort Union shale. This envelope seems to represent an upper bound for the site 1 data obtained from the fractured and fissured samples and a lower bound for the site 1 data obtained from the intact samples. Such

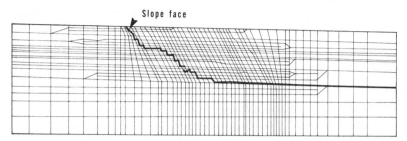

Fig. 10. Finite element mesh for south slope site 1.

a correlation may, however, be somewhat fortuitous, because the back-calculated field shear strength may not represent a true undrained shear strength.

The above parameters were subsequently employed in design stability studies of potential deep failures. The studies consisted of limit analyses using the "modified Bishop" method for assumed circular failure surfaces and Wright's (1969) method for noncircular surfaces. Finite element analyses were employed to examine stress conditions within the slopes which might lead to progressive failure.

The finite element mesh employed for the south slope is shown in Fig. 10; over 1000 elements were included in the mesh. The configuration of the mesh allowed for an accurate representation of the different rock layers and various stages of excavation. Supports for the left and right mesh boundaries were used so as to permit no lateral movement but free vertical movement. The right boundary was at the centerline of the excavation while the left boundary was located at a sufficient distance from the crest of the slope so as to have little effect on the slope behavior. At the bottom of the mesh, nodal points were assumed to be fixed against lateral or vertical movement.

In the finite element analysis the mine construction sequence, including water level changes as a result of seepage, was modeled as closely as possible; the actual sequence was broken into a series of stages, each of which was simulated by a loading step. Details concerning the modeling techniques are given by Clough and Duncan (1969). The stresses and displacements from each loading step were calculated and superimposed onto those which existed at the beginning of that loading step. At the beginning of the analyses the initial vertical stresses were calculated from gravity loading while the horizontal stresses were calculated as:

$$\sigma_{xi} = k_0 \sigma_{yi}$$

in which σ_{xi} is the initial lateral stress, σ_{yi} is the initial vertical stress and k_0 is the coefficient of earth pressure at rest. Choice of the k_0 values was based on evidence from published literature, since at the time of the analyses no testing procedures existed to accurately determine k_0. Based upon data from Peterson et al. (1960) and Brooker and Ireland (1965), a k_0 value of 0.5 was employed for the sandstone and 1.5 for the shale. Higher values of k_0 could be easily justified for the shale. However, use of a higher value would simply exaggerate the trends discussed in the following paragraphs for the finite element analysis.

Stress-strain behavior for the shale and sandstone was modeled by a non-linear elastic procedure described by Duncan and Chang (1970). Shale behavior was assumed undrained and sandstone behavior drained. Corresponding triaxial test results were used to determine the model parameters. In both types of tests the samples were loaded about half way to failure, unloaded, and then reloaded to failure. This allowed a determination of modulus values for both first loading and unloading-reloading. Average parameters were determined from the tests on each material. Strengths for each material were the same as those employed in the limit analyses.

In the analysis, the elastic modulus value for each element was calculated in accordance with the stress value in the element at each stage. Thus, the stress-strain behavior was adjusted as the stresses changed during each stage of loading. If an element was found to be unloading or reloading (shear stresses reduced from some previous stage), the model employed an unloading-reloading modulus value; if, however, the shear stresses consistently increase in an element, the model calculated a modulus value appropriate for the particular level on the first loading stress-strain curve. If failure was reached, the modulus in the failed element was reduced to a nominally small value. Strain softening stress-strain behavior cannot be modeled by this technique.

The results of primary interest from the finite element analyses were the stress levels in the shale. Stress level is defined as the ratio of the maximum shearing stress to the maximum shearing stress at failure; a stress level of 1 indicates failure. It was desired to ascertain if zones of failure might exist in the shale even though limit analyses might indicate that the slope had an adequate factor of safety. This was of concern because the shale demonstrated a significant loss of strength after failure, a characteristic which could lead to progressive failure as a result of locally overstressed zones. On the average, the ratio of the post-failure strength to the peak strength was 0.5 for the shale from undrained triaxial tests.

Overstressed or failure zones predicted in the pit slopes by one of the finite element analyses of the south wall of the pit is shown in Fig. 11 along with the results of several limit analyses performed using the "modified Bishop" method. All of the critical failure surfaces analyzed by the "modified Bishop" method yielded factors of safety of 1.10 or higher. However, according to the finite element model, certain zones within the shales had reached their allowable strength and thus plastically yielded. Such zones occurred at the bottom of the highest bench and near the toe of the slope;

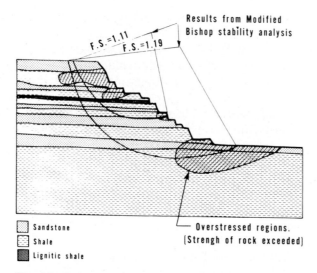

Fig. 11. Zones of overstressed shale and sandstone predicted by finite element analyses for pit slope at site 1.

they are located approximately at the lower ends of the critical failure surfaces as predicted by the "modified Bishop" method. Given the post-failure loss of strength observed in triaxial tests on the shales it seems reasonable to postulate that such yielding could indeed lead to progressive failure.

Based on the results of these analyses it was recommended that the use of bench and bank excavation techniques be minimized and eliminated where possible, since it was apparent that high shear stresses and zones of local overstress were induced at the base of the banks. It was further recommended that design slope angles be maintained in the 35—45° range, depending upon the height of shale in the slope. Finally, recognizing that a guarantee of slope stability could not be made and that some slope problems would probably still occur, it was proposed that a temporary catchment berm be left for as long as possible, immediately above the ore-bearing sandstone, in order to retain smaller failures on the berm rather than allowing them to cover the ore.

Groundwater control measures at site 1

It was clear from previous slope performance that groundwater flow into the mine was having an adverse effect on slope stability. Dewatering of the shale was not considered feasible, however, because of its low permeability. Instead, it was decided to drain the interbedded sandstone layers and lenses using horizontal drain techniques. If this effort were successful it was reasoned that the groundwater in the shale would be induced to flow vertically into the sandstone layers and thence out the lateral drains. This would result in a favorable gradient in the flow, reducing that along the direction of potential failure surfaces. Also the water available for surficial weathering should be reduced. Lane (1961) described the beneficial effects of this flow pattern, which he termed a "favorable seepage condition".

Some 150 horizontal PVC drain pipes were installed in the sandstone layers and lenses, with lengths varying from 80 to 150 m. Flows from these drains varied from 0.08 to 0.75 m^3 per minute, with flow generally diminishing with time. Drains into some sandstone lenses ran completely dry, and piezometers set into the lenses have verified that they have been fully drained. Total inflow from the drains and natural seepage from the sandstone layers has amounted to as much as 13,000 m^3 per day. This inflow is collected at the pit bottom and pumped to the surface.

Slope studies at site 2

At site 2, studies were undertaken for the design of the pit and were used to determine allowable slope angles and to develop groundwater control techniques. Both limit analyses and finite element analyses were conducted. Material parameters for the analyses were largely obtained from the laboratory test data because there was no previous slope performance experience which could be used for back-calculation of rock mass strength parameters. For the shale, an operational undrained strength of $\phi = 23°$ and $c = 100$ kN/m^2 was selected; drained parameters of $\phi' = 45°$ and $c' = 200$ kN/m^2 were used for the sandstone. At-rest pressure coefficients of 0.5 and 1.5 were employed for the sandstone and shale, respectively. Rationale for selecting these values was the same as that for site 1.

Limit analyses were performed using the "modified Bishop" method. These analyses were used to give preliminary slope designs, which were subsequently examined further by finite element analysis. Procedures followed in the finite element analyses were the same as those used at site 1. The results of the analyses yielded similar information to that obtained at site 1; namely, the slopes which, according to the limit analyses, had factors of safety greater than one contained significant zones of local yielding. On the basis of these results, the design slopes were lowered from those considered to be safe using only limit theory. Recommended slope inclinations varied with the nature of the materials in the slope and the depth of the pit. The final design slope varied from 45° to 60°, with 45° being applied to the slopes with the greatest proportion of shale (see, e.g., Fig. 2). Slope angles of 60° were applied to slopes predominantly of sandstone. These recommendations were made with the assumption that a thorough program of groundwater control was to be instituted.

Groundwater conditions were studied with data obtained from seventeen piezometers installed at the site. Because these data showed that the water table was to be encountered at about 30 m below the ground surface, it was considered imperative to provide some means of groundwater control. Deep wells were initially recommended but later rejected as being too costly. Instead, horizontal drains were eventually employed, using the same procedures as those used at site 1. These drains were installed in sandstone layers as the excavation proceeded and were found to perform satisfactorily. The drains generally ran full upon installation, showed progressively diminishing flows with time, and in most cases actually ran dry within several months of installation.

Slope stability at site 3

The investigation for site 3 was also for initial design; the studies consisted of groundwater observations, pumping tests, field studies of nearby slope performance, and stability analyses using limit techniques. The slope studies were performed having the advantage of knowledge of the results of the slope studies for sites 1 and 2. However, the slopes at site 3 are to be higher, involve thicker shale beds, and initially have more difficult groundwater conditions. In addition, some of the slopes are expected to remain stable for periods up to 30 years.

The field studies involved documentation of shale slope performance from nearby mines and natural slopes; no man-made or natural slopes in shale were in the mine area itself. A wide variety of behavior was observed, depending upon the age of the slopes and local groundwater conditions. Where no groundwater problems occurred, satisfactory slopes in shale were observed up to 55°. However, in natural slopes with obvious groundwater problems, slope angles greater than 30° were rarely observed. In fact, several kilometres north of the site some natural slopes at 30° were clearly unstable.

Analytical studies of the slopes were confined to limit analyses, again using the "modified Bishop" method, with occasional employment of the method developed by Wright (1969) for noncircular surfaces. Both long- and short-term stability of the slopes was considered. In the short-term analyses, it was assumed that an effective groundwater control program would be instituted in the form of lateral drains, of the type used at sites 1 and 2. Thus, favorable groundwater conditions were modeled in these analyses. For long-term stability, both favorable and unfavorable groundwater conditions were considered. The effects of a range of shale strengths were investigated for both short- and long-term analyses.

Design slope recommendations were presented in the form of charts (Fig. 12) where curves of slope height versus slope angle are plotted for given values of factor of safety for temporary and long-term slopes. This particular chart is for those regions of the mine where the overburden above the coal is essentially homogeneous shale. Other charts were prepared for regions of the

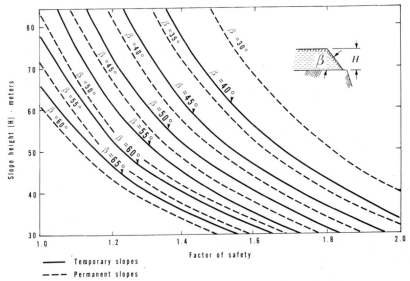

Fig. 12. Design slope chart for short- and long-term slopes for site 3 in regions of homogeneous shale slopes.

mine where different overburden conditions would be encountered. This type of chart allows mining personnel in the field to select an allowable slope for a given slope height and a desired factor of safety against failure. It was recommended that a minimum factor of safety of 1.5 be used where possible. As the mine is developed, the design charts can be updated on the basis of observed slope performance, particularly in areas where local faulting might lead to special slope problems.

SUBSEQUENT SLOPE PERFORMANCE

As of December 1974, slopes at sites 1 and 2, using the design recommendations, have been open for several years. At both sites the lateral drainage program has been very effective in draining the sandstone layers. The results of this drainage have been particularly beneficial in cases where the sandstone and shale layers alternate, where the layers are not too thick, and where the layers are about of equal thickness. In such cases the slope face is relatively dry and stable. Where the sandstone layers are very thin and the shale layers very thick, not as much benefit has apparently been derived; in this situation, which is confined to site 1, surficial saturation of the shale can still be observed.

Slope performance at the sites in general reflects the design recommendations. Only minimal slope failures have occurred at site 2. At site 1 some additional slope failures have occurred; the degree of this problem has, however, been reduced on three counts: (1) slope failures have been shallow, not deep-seated, (2) the slopes remain effectively stable for a longer period of time than previously, (3) the temporary catchment berms at the base of the slopes have prevented slope failures from interfering with mining operations until such time as the failure mass has to be removed.

SUMMARY AND CONCLUSIONS

Shale of the Wind River and Wasatch Formations in Wyoming pose serious slope stability problems in open pit mines, particularly in areas with a high groundwater table. Both shale units are relatively weak and deteriorate when subjected to wetting and drying cycles. At the locations studied, properties of the shale appear to be similar to those of the Fort Union Formation as described by Smith and Redlinger (1948) and Lane (1961). The average dry densities of all the shales are within ±10% of that of shale of the Fort Union Formation and the average natural water contents are within ±3%. Atterberg limits for all of the shales fall well within the range of values reported for Fort Union shale.

Of the samples examined, some distinctions can be made between the Wind River and Wasatch shales. The Wind River shale appears to be the weaker of the two units; its water content is higher, dry density lower and undrained strength generally lower. At the one pit location where slope performance in Wind River shale was studied, site 1, substantial failures were occurring. Shale beds ranged in thickness up to 26 m and groundwater conditions were such that the beds were fully saturated. Failures in the shale occurred anywhere from several weeks to one year after the slopes were exposed. Studies of the slope problems, begun in 1972, determined that the shear strength of the shale back-calculated from slope failures was approximately midway between that found in undrained laboratory tests for intact and fractured and fissured samples. The field shear strength was also found to be the same as that determined by Smith and Redlinger (1948) in their studies of slope failures in Fort Union shale.

Subsequent analyses of new design slopes were performed using limit analysis and finite element techniques. The finite element analyses demonstrated that even though a slope had an adequate factor of safety according to limit analysis procedures, local zones of yielding existed, particularly at the base of steep benches. Because the shale showed a significant drop in strength after yielding in laboratory tests, the local yielding zones were considered to be undesirable since they could lead to progressive failure.

Based on the results of the analytical and field investigations, the following recommendations were made:

(1) Eliminate, where possible, bench and bank excavation.

(2) Use lateral drains to remove free water from sandstone layers and lenses.

(3) Flatten the slopes to 45° or less, the exact amount depending upon slope height.

Subsequent slope performance during 1973 and 1974 has generally been satisfactory, although small, surficial sliding continues to be a problem, and one large slide of 400,000 m³ of shale has occurred.

Of the sites investigated involving Wasatch shale, sites 2 and 3, only site 2 has been mined as of March 1975. The studies conducted at site 2, begun in 1972, were oriented towards design of a uranium pit. Geologic conditions at this site are more favorable than those at site 1 since the shale beds are not as thick (16 m maximum) and are interbedded with relatively thick sandstone beds. Studies consisted of a field exploratory program, a laboratory testing program, and limit and finite element analyses of the slopes. Recommended slopes ranged from 45° to 60° depending upon the amount of shale in the slope face. Construction at site 2 began in 1972. Excavation is being performed using ripping techniques; the resulting slopes are smooth (few benches) and the stress concentrations caused by the bench and bank procedure are avoided. Lateral drains are used to remove water from the sandstone layers. The drains have been very effective in eliminating free water in the sandstone layers; the resulting seepage regime appears to be largely favorable for slope stability. Sliding in the shale at this site has been minimal.

Slope performance at sites 1 and 2 suggest the following conclusions:

(1) Groundwater conditions play a major role in stability of Wasatch and Wind River shale slopes.

(2) Removal of free water from sandstone layers interbedded between shale units is an effective means to improve slope stability, particularly where the sandstone units are nearly equal in thickness to the shale units.

(3) Bench and bank excavation techniques result in stability problems by creating zones of local overstressing which, in shales, usually leads to progressive failure.

(4) Slopes over 30 m in height in Wasatch or Wind River shale will not generally be stable at angles greater than 55°; unfavorable seepage conditions will produce failures in such slopes at even flatter angles.

REFERENCES

Atkins, J.T. and Pasha, M., 1973. Controlling open pit slope failures at Shirley Basin. Proc., Soc. Min. Eng., Am. Soc. Civ. Eng., June, pp. 38—42.

Beene, R.W., 1967. Waco Dam slide. Proc. Am. Soc. Civ. Eng., Soil Mech. Found. Div., 93: pp. 35—44, July.

Bishop, A.W., 1955. The use of the slip circle in the stability analysis of earth slopes. Geotechnique, 5: 7-17.

Bjerrum, L., 1967. Progressive failure in slopes of overconsolidated plastic clay and clay shales. *Proc. Am. Soc. Civ. Eng., Soil Mech. Found. Div.*, 93: 1—49, September.

Brooker, E.W. and Ireland, H.O., 1965. Earth pressure at-rest related to stress history. *Can. Geotech. J.*, 2: 1—15.

Clough, G.W. and Duncan, J.M., 1969. Finite element analyses of Port Allen and Old River locks. *U.S. Army Eng. Waterw. Exp. Stn., Contract Rep.*, No. S-69-6 (Corps of Engineers, Vicksburg, Miss.).

Duncan, J.M. and Chang C.Y., 1970. Nonlinear analysis of stress and strain in soils. *Proc. Am. Soc. Civ. Eng., Soil Mech. Found. Div.*, 96: 1629—1653, September.

Duncan, J.M. and Dunlop, P., 1969. Slopes in stiff-fissured clays and shales. *Proc. Am. Soc. Civ. Eng., Soil Mech. Found. Div.*, 95: 467—492, March.

Lane, K.S., 1961. Field slope charts for stability studies. *Proc., 5th Int. Conf. Soil Mech. Found. Eng.*, 2: 651—655.

Lutton, R.S. and Banks, D.C., 1970. Study of clay shale slopes along the Panama Canal. *U.S. Army Eng. Waterw. Exp. Stn., Tech. Rep.*, No. S-70-9, Rep. 1, 285 pp. (Corps of Engineers, Vicksburg, Miss.).

Mencl, V., Peter, P., Jesenak, J. and Skopek, J., 1965. Three questions on the stability of slopes. *Proc., 6th Int. Conf. Soil Mech. Found. Eng.*, 2: 512—516.

Peterson, R., Jaspar, J.L., Rivard, P.J. and Iverson, N.L., 1960. Limitations of laboratory shear strength in evaluating stability of highly plastic clays. *Proc., Am. Soc. Civ. Eng. Res. Conf. on Shear Strength of Cohesive Soils, Boulder, Colo.*, pp. 765—791.

Skempton, A.W. and Hutchinson, J., 1969. Stability of natural earth slopes and embankment foundations. *Proc., 7th Int. Conf. Soil Mech. Found. Eng.*, State of the Art Volume, pp. 291—340.

Smith, C.K. and Redlinger, J.F., 1948. Soil properties of Fort Union clay shale. *Proc., 3rd Int. Conf. Soil Mech. Found. Eng.*, 1: 62—66.

Wright, S.G. , 1969. *A Study of Slope Stability and the Undrained Shear Strength of Clay Shales.* Thesis submitted in partial fulfillment of the requirements for the Ph.D. degree, University of California, Berkeley, Calif.

Chapter 21

HOGARTH PIT SLOPE FAILURE, ONTARIO, CANADA

C.O. BRAWNER and P.F. STACEY

ABSTRACT

The northwest wall of the Hogarth Pit at the Steep Rock Mine near Ati-kokan, Ontario, Canada, developed signs of instability in August 1974. After evaluating the alternatives of attempting to fail the affected rock mass or of monitoring the movement while mining, Steep Rock Mines Ltd. decided to adopt a "mine and monitor" program. The slope displacement monitoring systems adopted included triangulation, electronic distance measuring, and extensometers. A seismograph was also installed.

Mining of recoverable ore reserves was successfully completed in March 1975. Movements showed rapid acceleration commencing in late April, 1975, and failure finally occurred on 23 June. An estimated 200,000 m³ of rock were involved in the toppling failure.

INTRODUCTION

This paper describes the monitoring program developed to assess movements in the highwall of the Hogarth No. 1 zone of the Steep Rock Mine at Atikokan, Ontario (Fig. 1). It constitutes a case example of a successful monitoring program which permitted removal of the recoverable ore reserves from that area of the mine, while at the same time maintaining safety for the operating mine personnel.

The success of the program was enhanced by effective communication between the stability consultant and the mining company, the company and its employees, and the involvement of all levels of company personnel in the program. The emphasis throughout was to provide a practical solution to a practical problem.

GEOLOGIC SETTING

The Steep Rock ore zone, composed of soft iron ore, dips steeply to the west; it is overlain by pre-Cambrian volcanics, and underlain by conformable

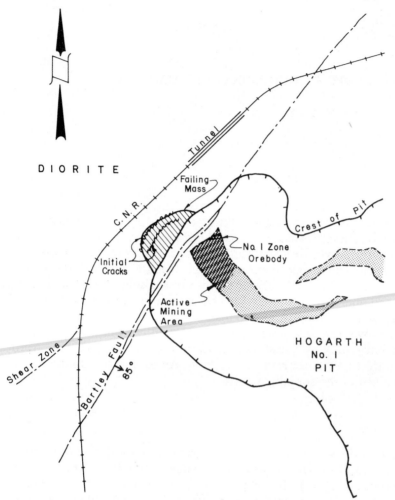

Fig. 1. Plan of Hogarth No. 1 zone, Steep Rock Mine.

sediments. The total Steep Rock complex is divided by regional faults into three major sections, each of which originally outcropped under an arm of the Steep Rock Lake (Jolliffe, 1955, 1966).

The section of the orebody being mined by Steep Rock Iron Mines Ltd. is termed the Middle Arm Orebody, referring to its respective arm of the lake. This orebody is separated into several sections by faulting and folding. The major divisions from south to north are termed the Errington zone, the Roberts zone, and the Hogarth zone. At the north end of the Hogarth zone the ore terminates against the Bartley fault, which strikes northeast and dips to the southeast at 85°. Northwest of the fault the original lakeshore was

Fig. 2. Airphoto of the Hogarth No. 1 zone highwall where movement occurred. Photo taken September 1974. The dashed line shows the general outline of movement zone with the railway behind.

formed by a steep, 90-m highwall comprised of a hornblende-biotite meta-diorite, which is termed "diorite" in this paper.

The diorite contains sheared basic dykes which parallel the fault. It also contains two well-developed vertical joint sets which strike respectively parallel to and perpendicular to the major fault direction. These joint sets controlled the development of the pre-mining lakeshore, and also controlled the failure to be described. The entire failure was restricted to the diorite highwall, although the monitoring program also covered the possibility of the failure extending to the iron formation below.

The crest area of the pit in the area of movement was covered by loose blocks from the excavation of a cut for the railway line which runs to the pellet plant. This line is located approximately 45 m behind the rear of the zone of movement (Figs. 1, 2). These loose blocks caused initial problems in locating cracks, and subsequently added to the difficulty of traversing the area.

ONSET OF FAILURE

A crack behind the crest of the diorite highwall was initially observed in October 1973. Movement observations commenced weekly. Further signs of instability developed in April 1974. A small blast on the hanging wall resulted in the toppling failure of a small wedge of diorite bounded by near-vertical joints. Talus spilled over the ore berm and down the iron formation face below.

The crack on the crest began to open and extend laterally. As a result the company installed a simple monitoring pin system, which was read on a daily basis.

In mid-August 1974, a second crack was noticed approximately 45 m behind the crest, when craters appeared in debris from the rock cut. At this point the pin monitors were supplemented by triangulation stations and three elementary wire extensometers constructed of drill steel and clothes line. On 20 August movements increased after a period of heavy rain. At this point the advice of Golder Brawner and Associates was sought.

REVIEW OF MINING ALTERNATIVES

It was apparent that a serious slope instability situation was developing. Accordingly, a temporary halt was called to mining operations in the Hogarth No. 1 zone while the slope movements were reviewed. To reduce vibration effects due to train traffic an 8 km/hr "slow order" was instituted on the track behind the highwall.

The data review indicated movement involving up to 200,000 m³ of diorite. The safest procedure would have been to discontinue mining in the area. However, since the Hogarth No. 1 zone was the only major source of ore available at the time, two alternatives which would allow ore production were considered. Attempts could be made to intentionally fail the slope by flooding the cracks, or by blasting, after which the ore could be mined following a delay for clean-up of the debris. Alternatively a comprehensive monitoring system could be installed to provide sufficient warning of potential failure, such that pit crews could be removed prior to major failure. The former course carried with it the risk of partial failure, possibly resulting in a less stable slope which would terminate all mining activities. The latter course was felt to be a safe approach, since geologic evidence, existing movement data and the author's past experience (Brawner, 1968, 1970, 1971, 1974; Brawner and Gilchrist, 1970) indicated that a "toppling" mode of movement would be expected; this mode would provide ample warning of failure in the form of an increasing rate of movement. This factor, combined with the requirement to maintain committed pellet shipments, influenced the company's decision.

To assess whether water pressure was contributing to instability, a piezometer was installed in a hole drilled from behind the crest to below the toe of the diorite slope at the 975-ft (352 m) level (Fig. 3). The piezometer remained dry until the standpipe was finally sheared by movement along the back crack.

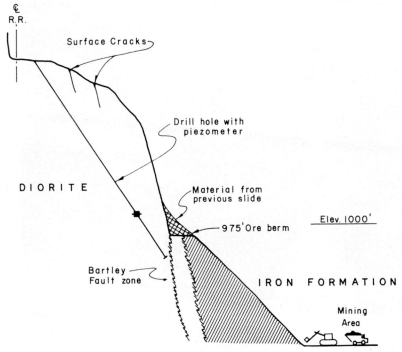

Fig. 3. Section through highwall showing generalized geology and the piezometer hole (see Figs. 6 and 9 for plan location of section).

MONITORING SYSTEM

When the decision was made to proceed with the "mine and monitor" approach, the monitoring system established by the Steep Rock staff was expanded to include an additional seven pin monitors (Fig. 4), seven further wire extensometers mounted on tripods (Fig. 5), and an increased number of triangulation stations. Fig. 6 shows the layout of the complete monitoring system.

The wire extensometers were tensioned with 25-kg weights so that wind and contact with bushes would not influence readings.

The results of the monitoring of the fourteen triangulation stations confirmed that a "toppling" mode of failure was occurring, with essentially horizontal movement of the crest towards the east-northeast, and with negligible movement at the toe of the exposed diorite slope. Daily triangulation by the survey crew proved to be both slow and impractical as a long-term program. Therefore, a laser electronic distance measuring (EDM) device was introduced to sight across the pit in the approximate direction of movement. (It should be noted that even if the line of sight is off-line of the direction of

Fig. 4. Pin monitor points used to measure horizontal and vertical components of movement across the cracks.

displacement by as much as 45°, the scale of the time-movement graph is only reduced by about 30%; warning of impending instability is still readily available.) Initially ten EDM reflector prisms were installed on the crest, on the 975-level (352 m) berm and on the iron formation face (Fig. 6). This general system remained in operation until completion of mining, with three further prisms being added on the 712-level (257 m) bench. The EDM monitoring was supplemented by periodic triangular surveys to obtain three-dimensional vectors of movement.

In addition to the "remote" EDM—triangulation monitoring system, movement of major individual blocks within the mass was monitored on a daily basis by the network of extensometers (Figs. 5, 6). Limit switches connected to a warning system involving sirens and flashing lights were attached to strategic lines. A light on the tripod indicated that power was on and that the switch had not been operated. Throughout the mining operation the limit switches were set for a predetermined movement and were reset daily. Prior to the onset of winter conditions this tolerance was 1.3 cm.

During the initial stages of the monitoring program the movements across the two major cracks were monitored at seven individual points using a three-pin system. However, by mid-winter the movement at several points made this operation too dangerous to continue (Fig. 4).

Fig. 5. Extensometer tripod. The wires were connected to tripods beyond the cracks. Limit switches were incorporated in the system and set to trigger flashing lights and sirens if the movement exceeded pre-set limits. The wires were tensioned with weights.

An extremely important element of the monitoring system was a continual visual guard by pit crew members. A heated observation shack was established across the pit from the zone of movement. Two guards per shift had a comprehensive view of the face of the highwall as well as the warning lights and sirens. Radio and visual contact was maintained with the crews working in the pit below. All events noted by the guards were recorded in a logbook which was initialled daily by the Mine Superintendent and by the Golder Brawner representative. Based on past experience, it was envisioned that any major failure would be preceded by ravelling of rock blocks too small to be detected by the monitoring system. The rate of ravelling of a

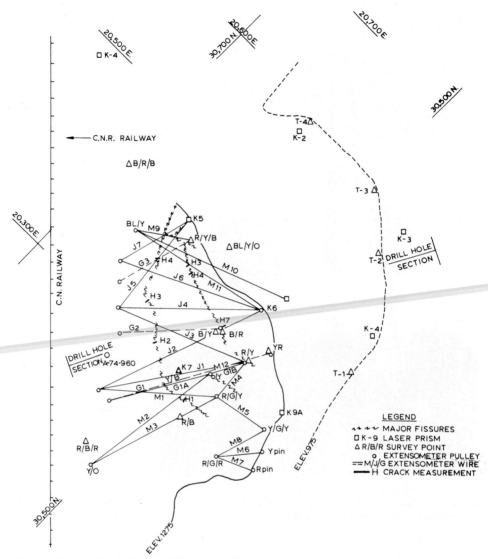

Fig. 6. Map of monitoring system at crest of slope (see section Fig. 3; data summarized in Figs. 8 and 9).

slope usually increases as failure approaches. The guards were instructed to record all ravelling events.

A Kinemetrics VR1/SS1 seismograph system was ordered installed by the Ontario Department of Mines. This was installed in December. The seismograph head was installed in the centre of the failure area, and the remote drum recorder was located in the guard shack.

The daily monitor reading and compilation were performed by a Golder Brawner field engineer, who interpreted and reported the results to the Steep Rock Senior Engineer. Movement plots were posted daily in the Mine Dry Room for the information of the operating crews.

MINING SEQUENCE

The monitoring system was complemented by a planned mining sequence. At the toe of the highwall a 12-m-wide berm with rock pile was left on the 712 level (257 m). This berm was intended to catch ravelling rock. A more desirable, wider berm would have greatly reduced ore reserves in the zone. Below the 750 level (271 m), overall slopes in the iron formation were designed at 40° instead of the originally planned 42.5°.

Prior to the mining of each bench, an approximately 30-m-wide slot was drilled, blasted and mined along the foot- or hanging-wall contact of the ore up to the proposed highwall toe (Fig. 7). At the same time a 30-m-wide zone

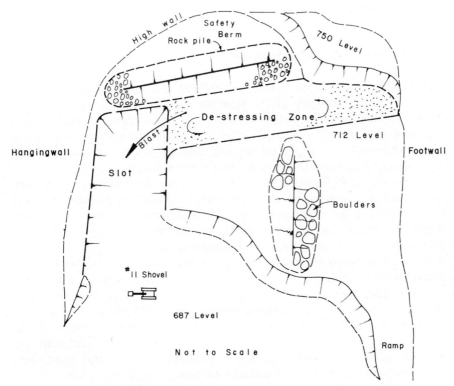

Fig. 7. Mining sequence in toe area.

was drilled off along the base of the highwall, perpendicular to the slot. The holes in the zone had a maximum loading of 540 kg of explosive per delay and were fired into the slot. This approach was aimed at reducing the exposure of men and machines by initiating potential failure while the operating equipment was still well back from the highwall.

At the same time a "Green Line", based on the dimensions of the moving block, was defined 90 m out from the highwall toe on each bench. This safety line, which was indicated in the field, became highly revered by the mining crews. It formed an inner boundary for all night-time and bad-weather operations. Further, no operations were permitted within the Green Line for 48 hours after a toe blast and for 12 hours after a production blast. It is interesting to note that when the slope failed, little, if any, material passed this line.

It was recognized that time was an important factor. Accordingly, a series of six high-powered lights were installed to illuminate the entire face and mining area, thereby providing conditions under which the initially specified daytime operations could be extended to a 24-hour basis. This procedure was not accepted by the union, even though the union was advised that the longer mining period which resulted from single-shift operation increased the danger and reduced the possibility of recovering all the ore.

MOVEMENTS — SEPTEMBER TO JANUARY

The "toppling" movement pattern established in early September 1974 continued until early January 1975. Movement was confined to the diorite above the 975 level (352 m) berm, with the two sets of vertical cracks opening up and propagating downwards. A small "wedge" on the east side of the main mass showed more rapid movement, and on several occasions loose material was removed from this area by Steep Rock scaling crews.

The extensometer ("J" Monitors) and EDM data showed that there were periods of constant movement rates separated by periods of more accelerated movement (Fig. 8). The acceleration of movement appeared to correlate with heavy rainfall, there being a time lag on the order of one day. No distinct relationship was evident between movements and blasting.

The material between the face and the original front crack showed a horizontal/vertical movement ratio in excess of 3 : 1 with no appreciable movement at the toe, indicating toppling of this block (Fig. 9). The block between the front and back cracks showed a horizontal/vertical movement ratio of approximately 1 : 1, indicating slumping behind the front block. Monitoring of survey stations on the 975 level (352 m) berm and iron formation face below indicated only minor movement. All major movement continued to be on an east-southeasterly azimuth, i.e., in the general direction of the EDM unit.

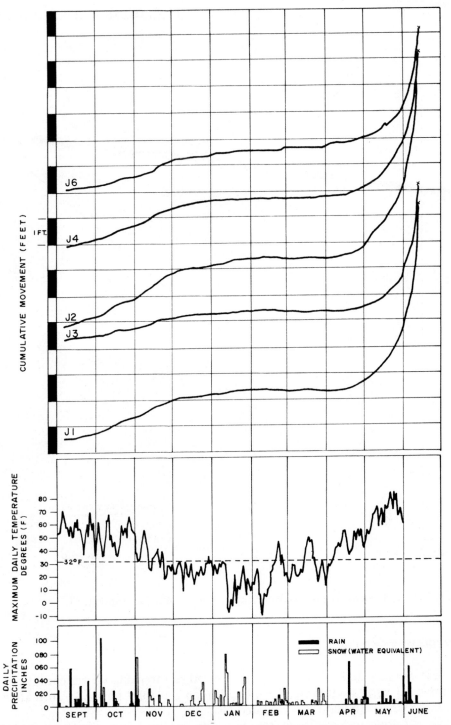

Fig. 8. Hogarth No. 1 zone. "J" extensometer monitors, total displacement history to 10 June, when most extensometers were destroyed. Data beyond this point in time acquired by triangulation (Fig. 9; see Fig. 6 for monitor locations).

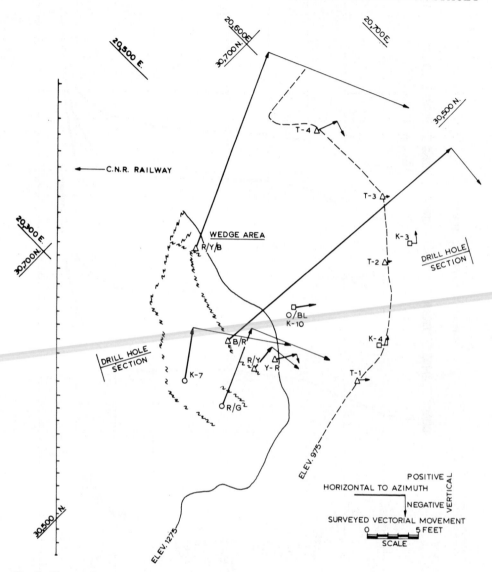

Fig. 9. Vector diagram showing displacements just prior to failure (cf. Fig. 8).

WINTER MOVEMENTS

On 9 January 1975, a period of prolonged sub-freezing winter conditions commenced with a blizzard. These conditions generally persisted until after 10 March when mining was completed (Fig. 8).

The sub-freezing conditions required changes in monitoring techniques.

Rapid temperature fluctuations from +30° to −20° F, experienced over a few hours, would result in rapid contraction of the extensometer wires by as much as 2 cm, with resultant tripping of the alarm system. The maximum daily permissible movement on an extensometer limit switch was therefore increased to 2.5 cm to accommodate this phenomenon.

During extremely cold conditions the EDM readings across the pit were affected by the temperature gradients developed in the air. As a result there could be a difference of 2 cm between a reading made in the early morning, and one taken in the late afternoon, particularly on still, clear days. This problem was overcome partially by referencing measurements to a stable station on the crest of the railway cut.

Blizzards or poor visibility resulted in the temporary cessation of operations, while the onset of thawing conditions resulted in a mandatory withdrawal behind the Green Line. Snow made access on top of the movement area a treacherous operation. Accordingly the cross-crack pin measurements were halted; by the time the snow had melted, many of the cracks were too wide to permit safe operation of this method.

Contemporaneously with the onset of permanent sub-freezing conditions, major movement essentially ceased (Fig. 8). By this time failure was confined to the area in front of the original front crack, and the main mass had been divided into a series of slabs by the opening of vertical fractures.

Although daily movement was not detectable by the extensometer and EDM systems, and variations in the triangulation survey data were generally less than the limits of accuracy, there was still evidence that the slope was active. In cold weather, water vapour was emitted from the major cracks. During the night shift when no equipment was operating in the zone, occasional small movement signatures appeared on the seismograph trace. In the daytime however, seismic indications of this type were masked by mine equipment signatures so that warning was not available from the seismic instrumentation. This confirms the experience of Kennedy (1972) on the practical limitations of the seismic monitoring technique.

Much of the movement in this period was restricted to a slight opening of the forward cracks on the face, and later to the occasional loosening of debris on the slope by freeze-thaw action.

COMPLETION OF MINING

The planned ultimate depth of the No. 1 zone at the 600 level (216 m) was reached in late February. Movement on the highwall during February was negligible. Limit equilibrium stability analyses were performed on the iron formation to assess the effect of taking an additional lift. The analyses showed only a minor influence on the factor of safety. Accordingly an additional 7-m bench was mined. At the completion of this level there was just sufficient width of ore exposed in the floor to make an additional 7-m-deep scram, or rob cut feasible.

With continuous monitoring the scram was drilled up to the toe of the previous bench and fired in one blast using a maximum loading of 544 kg per delay. An intensified monitoring program involving daily triangulation of four critical stations, two daily extensometer readings and three daily EDM surveys were instituted; after a delay of 48 hours during which no movement was detected, mining recommenced.

Mining of the scram cut to the 560 level (204 m) was completed on 10 March 1975. In total, 985,000 tonnes of ore was recovered from the Hogarth No. 1 zone compared with an original estimate of 950,000 tonnes. Part of this gain resulted from the mining of the two additional 7-m benches.

MOVEMENTS FOLLOWING COMPLETION OF MINING

Two monitoring programs of the highwall continued after completion of mining. The purpose of the first was to protect ore reserve drilling in the No. 1 zone; the second monitored movements which could affect the safety of the railway.

Drilling started immediately after completion of mining operations. Shortly after the onset of spring thaw conditions in early April, ravelling and small failures in the hanging wall volcanics led to the temporary removal of the drill to a safer location. By late April 1975 weekly reading of the extensometers and triangulation of selected survey points by Steep Rock personnel indicated a marked increase in movement on the highwall. This coincided with heavy rainfall. Increased activity was also noted on the seismograph. The movements followed the original pattern, with vertical fractures opening from the top to impart a toppling rotation on the front block, and the back block slumping behind the front block.

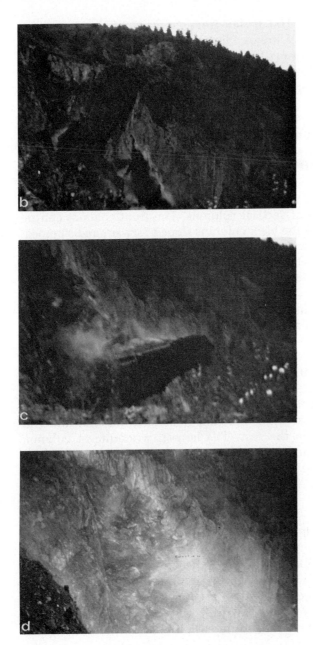

Fig. 10. Sequence of failure, 23—24 June 1975. (a) 9:00 p.m. — block has commenced toppling with over 11 m of horizontal movement and 1.5 m of vertical movement (cf. Fig. 9); (b) 9:15 p.m. — horizontal movement has reached about 27 m; (c) Seconds after photo (b) was taken the block toppled; (d) immediately following collapse.

On 21 May a slab of about 100 m^3 fell from the front of the main mass. Most of the material from this minor failure was caught on the 975 level (352 m) berm, with none reaching the pit floor. This event was preceded by minor ravelling and was followed by accelerated movement accompanied by increased seismic activity and constant ravelling, particularly from the "wedge" area.

On 10 June the "wedge" failed, involving several hundred cubic metres. This was followed on 11 June by an additional fall from the front of the main mass. By this time both the 975 level (352 m) berm and the catch berm at the 712 level (257 m) were choked with debris, and large amounts of rubble started to reach the pit floor.

As movement continued, sets of northwest-southeast joints opened to divide the mass into a series of vertical columns. The front columns continued to topple forward, while the back columns tipped forward and slumped. Most of the extensometers and EDM targets were destroyed in the period 10—11 June, but immediately prior to this date extensometers and laser targets in the centre of the mass were recording movements in excess of 0.3 m/day (Fig. 8).

Heavy rainfall on 20 June induced renewed heavy activity, and by 23 June the face was showing signs of major distress. On 23 June, 5 hours before failure, a triangulation point on the front of the old front crack was computed to have cumulative movement of 11.1 m horizontally and 1.7 m vertically since August 1974 (see Fig. 9). The remaining mass was keyed in by a block on the west side. This was seen to be cracking badly, and ravelling was general over the whole face. At 9:00 p.m. the back mass slumped, pushing out the front columns. Finally, at 9:15 p.m. the remaining material failed in three portions commencing in the east. The largest, westernmost column, failed in a typical toppling mode (Fig. 10) whereas the remainder broke up and ravelled. The column shown in Fig. 10 had approximate dimensions of 20 m × 20 m × 55 m. [1]

DISCUSSION

Almost certainly the slumping of the rear main block added driving force to the toppling of the front blocks. However, the exact mechanism of failure of the rear blocks is uncertain, since up to the actual failure there was no evidence of the toe being pushed out; in the last few days the toe area of the diorite did show some signs of distress. With regards to the base of the failure, it would appear to have resulted from tension failure. The resulting scar

---

[1] The dimensions given for the photographed column are "best estimates", since it was judged too dangerous to go out on the slope crest behind the failing mass in the last few days. It was not known how far back the failure would extend beneath the loose blocks.

contains only rubble, but there was never any evidence of near-horizontal jointing in the face.

Data obtained during the final days prior to failure indicated that even if failure had occurred while the mining operations were still in progress, the monitoring systems would have given ample warning. The decision to proceed with the mining under a controlled "mine and monitor" system was therefore justified. The instrumentation initially installed during August 1974 proved totally adequate, and the time delay and mine shut-down period required for additional instrumentation installation proved to be unnecessary. The choice of a 90-m Green Line was confirmed, since little material, if any, fell beyond this line.

REFERENCES

Brawner, C.O., 1968. Three major problems in rock slope stability in Canada. *2nd Int. Conf. on Surface Mining, Minneapolis, Minn.*
Brawner, C.O., 1970. Stability investigations of rock slopes in Canadian mining projects. *Proc., 2nd Int. Conf. on Rock Mechanics, Belgrade, 3,* Paper 7-8.
Brawner, C.O., 1971. Case studies of stability on mining projects. In: *Stability in Open Pit Mining.* American Institute of Mining Engineers, New York, N.Y., pp. 205—226.
Brawner, C.O., 1974. Rock mechanics in open pit mining. *Proc., 3rd Int. Conf. on Rock Mechanics, Denver, Colo.,* Suppl. Rep., Theme Three — Surface Workings, 1-A: 755—773.
Brawner, C.O. and Gilchrist, H.G., 1970. Case studies of rock slope stability on mining projects. *8th Annu. Symp. on Engineering Geology and Soil Engineering, Pocatello, Idaho.*
Jolliffe, A.W., 1955. Geology and iron ores of Steep Rock Lake. *Econ. Geol.,* 50: 373—398.
Jolliffe, A.W., 1966. Stratigraphy of the Steep Rock Group, Steep Rock Lake, Ontario. *Proc., Precambrian Symp., Geol. Assoc. Can., Spec. Paper,* 3: 75—98.
Kennedy, B.A., 1972. Methods of monitoring open pit slopes. *Proc., 13th Symp. on Rock Mechanics, Urbana, Ill., 1971,* pp. 537—572.

Chapter 22

CANADIAN EXPERIENCE IN SIMULATING PIT SLOPES BY THE FINITE ELEMENT METHOD

Y.S. YU and D.F. COATES

ABSTRACT

This paper discusses some experiences encountered in the application of the finite element method to rock slope analysis. Parametric studies of stress and deformation patterns around typical slopes indicate that a knowledge of the pre-mining stress field in geological materials is important for any meaningful rock slope simulation. If a rock slope is cut in an inhomogeneous rock mass, the stiffness of different formations will affect the resulting stresses considerably; large tensile zones could be expected around the toe areas, which are unfavorable conditions for slope stability. Therefore, in order to obtain accurate results from a finite element analysis, representative and detailed deformation moduli for the in-situ geological materials would be essential. Excavation displacements at the crest and along the slope face should be detectable with an appropriate type of instrument; such displacements may be a valuable guide in examining the ground reaction to excavations.

To obtain information on the mechanism of a slope slide, a plane strain model with joint elements was constructed. A brecciated rock mass in the slope wall, bounded by fault faces parallel to the slope face, was simulated by sets of joint elements bounding the quadrilateral elements. The simulation, however, was not satisfactory, possibly due to the nature of the three-dimensional geometry of the slide. The lack of realistic material properties and field stress conditions, and the coarseness of the mesh within the brecciated zone (even though the total mesh was quite large) may also contribute to the less-than-satisfactory simulation.

INTRODUCTION

Rock slope research is primarily motivated by the need for more information on the stability of walls in large open pit mines. With the advent of the high-speed digital computer, the finite element (FE) technique has become a useful tool for stress analysis of rock structures (Zienkiewicz, 1971; Desai and Abel, 1971). Although the FE method has proven to be a flexible tool for engineers to deal with complex boundary conditions, knowledge of actual field conditions remains a problem. However, natural geological discontinuities such as layering, joints and faults, which are usually prevalent in a rock mass, have been modelled by Duncan and Goodman (1968), Goodman et al. (1968), Wang and Voight (1970) and Ghaboussi et al. (1973). Other modifications, such as elasto-plastic approaches, have been suggested by various researchers to account for the possible yielding of rock; time-dependent approaches are available for non-linear material properties. Some rather more extensive and critical reviews on the FE analysis of elasto-plastic problems in the mechanics of geologic media have been given by Pariseau et al. (1970), and by Voight and Dahl (1970).

In this paper, some results of parametric studies on stress distribution around typical, elastic rock slopes are briefly described; and some difficulties experienced in simulating real-life slopes are presented and discussed. Where possible the results are compared with field measurements. Discrepancies between the computed and observed deformations seem due to the inadequacy of the input data, such as the physical properties of the in-situ rock mass, geological boundaries, and initial field stress conditions. In addition, the geometries of many open pit mines do not lend themselves well to two-dimensional idealizations, which require essentially that the variations in geometry and geology occur in one plane with the configuration in the third dimension being constant. A static analysis program for three-dimensional solid structures has been made available by Wilson (1972), but cost makes it impractical at the present time as far as the general simulation of rock slopes is concerned. These statements, however, are not aimed at belittling the important applications of the FE method to rock slope analysis; on the contrary, the results obtained, despite all of these difficulties, provide useful information on the behavior of rock in slopes due to excavations, and these results can be used for comparative purposes in both research and preliminary design work.

PARAMETRIC STUDIES ON STRESS DISTRIBUTION AROUND TYPICAL ELASTIC SLOPES

Finite representation of infinite rock mass

Since the FE method involves a physical approximation of the actual system, the location of the finite boundaries from the area of interest will influence the results of an analysis. This effect must be minimized to ensure realistic results. This is feasible because the effect of loading or disturbance decreases with increasing distance from the area of disturbance. No theoret-

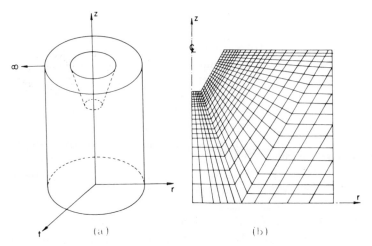

Fig. 1. The geometry of the slope studies: (a) definition of axes, (b) finite element model.

ical solutions are available for comparison even for a simple, typical slope; it is possible, however, to determine the significant extent of the medium to be modelled by a trial and error procedure which varies the extent and computes the resulting effect upon the solutions. Yu and Coates (1970) have judged that finite boundaries placed at a distance approximately four times the size of opening should be adequate, while others (e.g., Duncan and Goodman, 1968) have the boundary located at a distance of three times the total depth to bed rock. The first criterion was adopted for all subsequent studies.

Yu and Coates (1970) conducted a series of studies for which a total of 45 slope conditions were examined taking into account slope angles, slope shapes and various loading conditions. Fifteen cases were axisymmetric slopes with arbitrary loadings; others were two-dimensional plane strain models. Only a few highlights of their findings are extracted here for discussion. An example of a 60° slope is illustrated in Fig. 1. Fig. 1a shows the coordinate system and Fig. 1b represents the two-dimensional FE model that was used to determine the stresses and displacements. The investigated model slopes all had a slope height of 300 m and the same width of excavation except for the vertical slope, which was wider. The rock formations were assumed to be perfectly elastic with modulus of deformation of $E = 7.03 \times 10^5$ kg/cm^2 (ca. 7×10^4 MN/m^2) [1] and Poisson's ratio of 0.25.

---

[1] In this paper all figures reference stress in units of kg/cm^2, hence this will serve as the standard unit of stress herein. 1 kg/cm$^2 \simeq 0.098$ MN/m^2.

Stress distribution in slopes subjected to various field stresses

Field stress conditions in rock formation are known to have many patterns. Gravity loading alone results in the major stress being vertical with the horizontal stress being some fraction, such as one-third of the major stress. Stress measurements from various areas of the earth crust, as reported and compiled by Herget (1973), indicate that the horizontal stresses can be much greater than the vertical. The higher horizontal stress is, perhaps, caused by tectonic action. Thus the effect of horizontal tectonic force on stress distribution in slopes was also studied. The magnitude of the applied tectonic stress is expressed by the factor K, which is the ratio of the horizontal stress to the vertical at the toe level prior to excavation. For gravity loading only, K is the ratio of $v/(1-v)$ where v is Poisson's ratio.

Fig. 2a and b shows the stress distribution trajectories for a typical 60° slope with $K = \frac{1}{3}$ and 3, respectively. The mining of a 60° slope in an elastic and homogeneous rock mass under gravity load, i.e., $K = \frac{1}{3}$, did not create

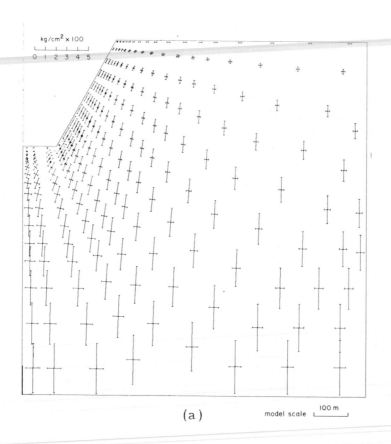

(a)   model scale

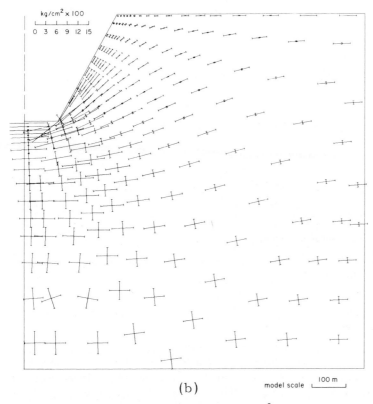

(b)

model scale |100 m|

Fig. 2. Trajectories of principal stresses for a $60°$ slope model: (a) $K = 0.33$, (b) $K = 3$.

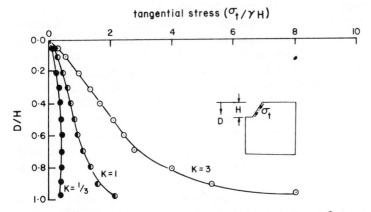

Fig. 3. Tangential stresses parallel to the slope face of a $45°$ slope subjected to various loading conditions. D is depth measured from ground surface.

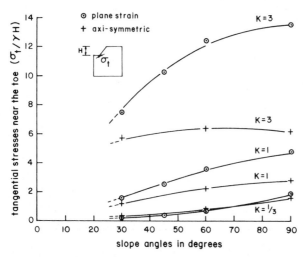

Fig. 4. Tangential stresses near the toe (approximately 5 m behind toe) as a function of slope angles and loading conditions.

any high stresses around the toe, as shown in Fig. 2a. However, with the introduction of tectonic horizontal forces, the stresses in the slope have been altered considerably, particularly around the toe area.

Fig. 3 shows a plot of the tangential stresses parallel to the slope face (approximately 5 m in from the face) of a 45° slope under three loading conditions, i.e., with $K = \frac{1}{3}$, 1 and 3. The tangential stresses, with the sign convention of compression positive, are normalized by dividing the stresses by γH where γ and H are the unit weight of rock and the slope height, respectively. As can be seen (Fig. 3), the stress distribution around rock slopes will be greatly influenced by the initial state of stress; the tangential stresses vary approximately linearly with the factor K.

Fig. 4 is a composite plot of tangential stresses near the toe (approximately 5 m behind the toe) for various slope angles and various loading conditions. The stresses did not increase significantly with an increase of slope angle under gravity loading for the plane strain case. With the presence of tectonic horizontal stresses, they increased more rapidly, and this tendency seems to be more pronounced with a higher K, say 3. However, for the axisymmetric case with $K = 3$, the tangential stresses behind the toe did not change significantly as the slope angle increased. Comparing results between the plane strain solution and the axisymmetric analysis, it is interesting to note that the difference in stress, for the same slope angle, becomes larger as K increases. For example, the tangential stress behind the toe of a 60° slope obtained from the plane strain solution with $K = 3$, can be as much as twice as high as the three-dimensional axisymmetric analysis. For gravity loading, both solutions yield approximately the same results. Thus, good

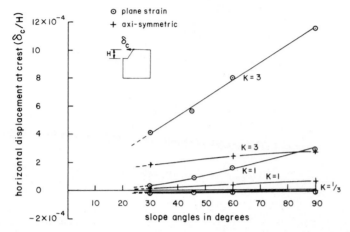

Fig. 5. Excavation displacements at crest of a slope as a function of slope angles and loading conditions. Negative signs indicate that movement is directed towards the pit wall.

engineering judgement is required to select either a plane strain or axisymmetric solution to simulate the actual geometry.

Excavation displacements

Excavation displacements are defined as the movement of rock mass induced by the excavations and it is an important aspect when examining the ground reaction to the removal of pit material. To obtain excavation displacements a procedure suggested by Coates (1970) was adopted. Excavation displacements can be calculated by first performing a FE analysis to obtain displacements from the FE model without excavation, followed by calculations of displacements with excavation and subtraction of the results. If the initial stresses, i.e., stress field prior to mining, are known, then, these can be incorporated into the FE analysis to obtain excavation displacements directly.

The general displacement patterns of a slope vary considerably with loading conditions, particularly when higher horizontal stresses are present. Fig. 5 shows the horizontal displacement at the crest in relation to loading conditions and slope angles. The excavation displacements are normalized by dividing the displacement by the pit depth H; the negative number indicates the movement is directed into the wall. Under gravity loading, the displacements are generally directed into the walls and they are not very sensitive to the increase of slope angle. With the introduction of horizontal stresses, the displacements are directed more towards the opening (Fig. 6); they increase almost linearly with an increase of slope angle.

Similarly, the plane strain solution yields much higher displacements at

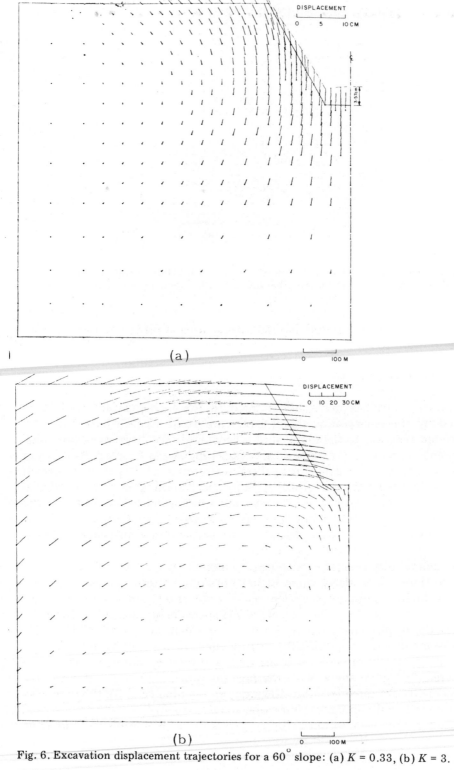

Fig. 6. Excavation displacement trajectories for a 60° slope: (a) $K = 0.33$, (b) $K = 3$.

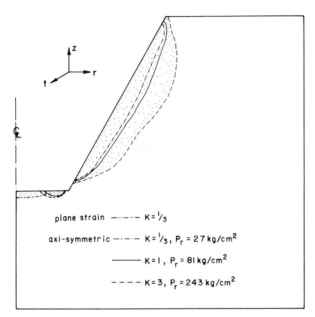

plane strain —··—··— K = ¹/₃

axi-symmetric —·—·— K = ¹/₃, P_r = 27 kg/cm²

—————— K = 1, P_r = 81 kg/cm²

- - - - K = 3, P_r = 243 kg/cm²

Fig. 7. Development of tensile zone (in *rz* plane) around a 60° slope under various loading conditions. *P* is the applied horizontal stress and parallel to *rz* plane. Dotted areas indicate tensile zone.

the crest as compared to those of axisymmetric analysis, under tectonic horizontal stress field. With gravity loading alone, the two solutions provide approximately the same results.

Tensile zone development

Only minor tensile stresses have been observed around the pit bottom from the plane strain models. For the axisymmetric case of a 60° slope, the tensile zone seems to be developed near the face of the pit walls in the section parallel to the direction of the applied horizontal stress (Fig. 7). This zone was shown to penetrate into the wall about 70 m at approximately the middle portion of the slope wall for $K = 3$; for $K = 1$, the tensile zone penetrates approximately 30 m into the wall. It is supposed that a typical rock mass contains numerous fissures and behaves as a material incapable of transmitting tension; hence, a tensile zone around excavations is important when considering slope stability.

Potential benefits from changing slope angles with depth

To investigate whether benefits can be derived on a purely geometric basis, six different slope configurations, as shown in Fig. 8, were examined;

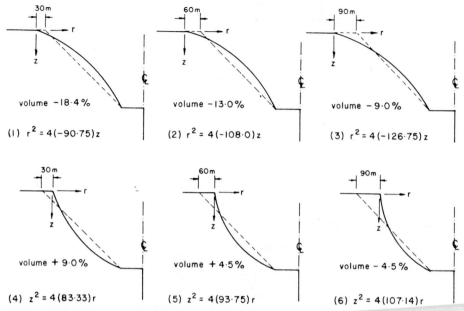

Fig. 8. Definition of the modified slopes. The difference in excavation volume is expressed as a percentage of a straight $45°$ slope.

stresses were compared with that of a typical $45°$ slope. The first three cases involve convex-shaped slopes, where the slope angle increases with increasing depth (maximum angle of $56°$); and the other three are concave-shaped. The difference in volume of excavation is expressed as a percentage of a typical $45°$ slope and is given in Fig. 8.

Examining the tangential stresses parallel to the slope face for various

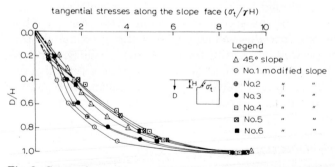

Fig. 9. Comparison of tangential stresses along the slope walls between a typical $45°$ slope and the modified slopes under plane strain condition and with $K = 3$. D is depth measured from ground surface.

configurations, the stresses are generally lower for the first three cases, i.e., the convex-shaped slopes, along a large portion of the excavation face (Fig. 9). Taking into consideration the benefits of a lesser volume of waste material from convex-shaped slopes, it seems that configurations 1 to 3, which would not be penalized by stresses, are worth trying in practice and where conditions permit.

SIMULATION OF OPEN PIT MINES

It has been realized that the idealization of a geological structure for a FE analysis, particularly for open pit mines, is not an easy task because a large number of parameters — such as in-situ material properties and field stress conditions — cannot be predicted or measured with certainty in this early stage of the science of rock mechanics. Therefore, idealization errors resulting from geometry, rock type distribution and their properties, and the initial state of field stresses, could be large and thus will affect the accuracy of the solution to a problem. However, in spite of all of these difficulties and uncertainties, comparative analyses with regard to stresses for different conditions of a slope can still be very valuable and should provide some additional information for research and design work. It was with this in mind that the following analyses were carried out.

Two-material slope models

The stresses in a typical slope, which is cut in a homogeneous geological medium, will not be affected by the modulus of deformation of rock materials. For most open pit mines, however, the rock mass cannot be assumed as homogeneous, isotropic and elastic. Therefore, the homogeneity of rock formations may have a significant bearing on the stress distribution around excavations of a rock slope. In order to examine this concept, investigations were conducted on simple slope models which have slope angles of 60° and 45° and consisted of two types of rock only, i.e., the orebody and wall rocks.

In this simple slope model, the orebody was assumed to be vertical and extended to the bottom of the model; the pit bottom had the same width as the orebody. Two slope angles (60° and 45°) with the same pit depth of 300 m were examined individually. Figs. 10—15 show the trajectories and contours of principal stresses for a 60° slope with the ratios E_o/E_r being 10, 2, and 0.1 (where E_o and E_r stand for deformation moduli for the ore and wall rock, respectively). When the ore is much stiffer than the wall rocks, i.e., $E_o/E_r = 10$, a large tensile zone developed around the pit floor and extended into the walls (Figs. 10 and 11). Subsequent work indicated that the extent and magnitude of these tensile stresses was greatly dependent on

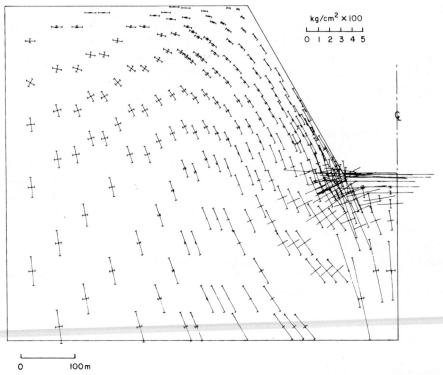

Fig. 10. Trajectories of principal stresses for the two-material slope model (60°) with $E_o/E_r = 10$.

the ratio of E_o/E_r and was less dependent on slope angles within the range of 45—60°. The larger the ratio of E_o/E_r, the larger the extent and magnitude of the tensile stresses. For the case of $E_o/E_r = 0.1$ (i.e., the ore was much softer than the wall rock), the pattern of tensile stresses changed considerably. A large volume of wall rock above the pit floor level, extending up to surface, was in a state of tension (Figs. 14 and 15). Clearly, the stresses are very sensitive to the deformation moduli if homogeneity does not apply. In other words, in order to obtain accurate results from a FE analysis, a thorough knowledge of rock mass deformability would be essential.

Open pit mine A

General description. The mine dimensions will be approximately 900 m × 600 m and up to 300 m deep. The orebody is mainly hematite (iron) ore as shown in Fig. 16. The hanging wall, in relation to the orebody, consists of blocky, very strong, elastic, banded hematite-quartzite (BHQ); the footwall consists of alternate beds of very strong elastic quartzite, low-strength elastic

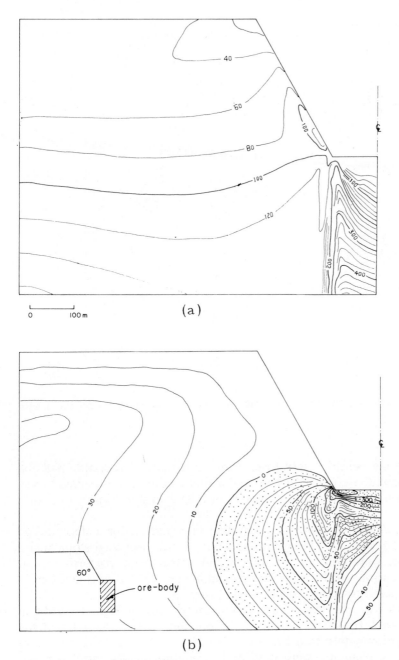

Fig. 11. Contours of equal-principal stresses (compression positive) for the two-material slope model (60°) with $E_o/E_r = 10$. Dotted area indicates tensile zone. Contour intervals are 20 and 10 kg/cm^2, respectively, for (a) major principal stress, and (b) minor principal stress.

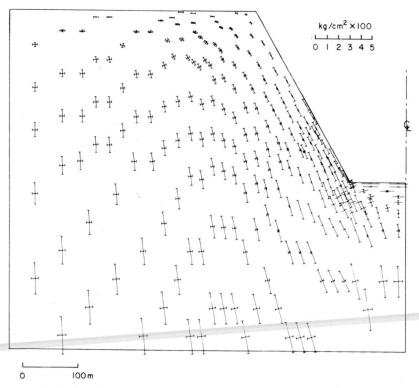

Fig. 12. Trajectories of principal stresses for the two-material slope model (60°) with $E_o/E_r = 2$.

schist and strong, elastic, schistose, banded hematite-quartzite. Bedding planes dip approximately 63°, and no major faults have been observed in this area. In addition, groundwater presents no problems in this mine.

Geometric simulation. To provide additional information for a slope design program, the FE models, taking into account several proposed slope angles and pit depths, were constructed. Stress and deformation patterns around several configurations of the slopes were examined and the resulting information was considered for the final pit design program.

Dimensions of the models were approximately 1200 m high and 2500 m wide; the distance from the centre of the opening to the boundary of the model was approximately four times the size of the opening. Both triangular and quadrilateral elements were used and a length-to-width ratio of 5 was maintained wherever possible. Although the FE technique is capable of handling very complex structures and geometry, cost makes it essential to simplify the geology. The general geology for the section of the model is

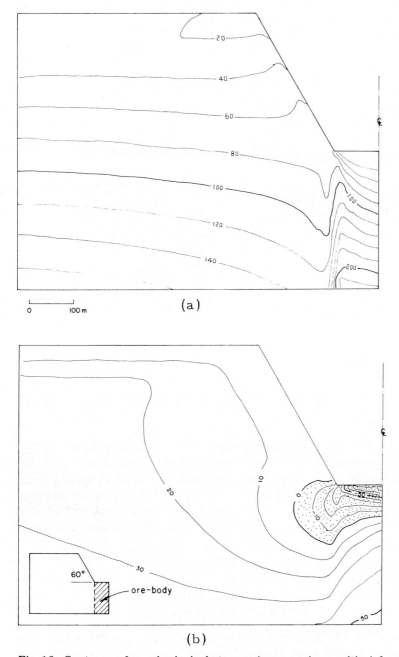

(a)

0   100 m

(b)

Fig. 13. Contours of equal-principal stresses (compression positive) for the two-material slope model (60°) with $E_0/E_r = 2$. Dotted area indicates tensile zone. Contour intervals are 20 and 10 kg/cm², respectively, for (a) major principal stress, and (b) minor principal stress.

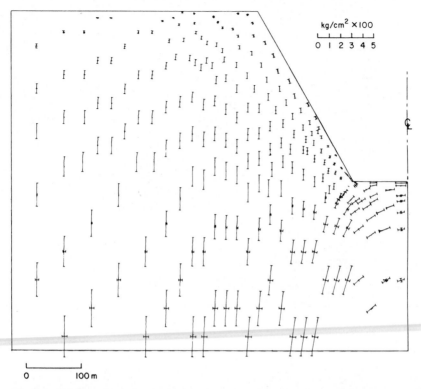

Fig. 14. Trajectories of principal stresses for the two-material slope model (60°) with $E_o/E_r = 0.1$.

shown in Fig. 16 and the simplified geology of the section is shown in Fig. 17. The model consists of six different types of rock numbered 1 to 7 for material identification with 5 as a dummy number. Several proposed slope angles for the footwall (FW) and hanging wall (HW) in relation to the orebody, as shown in Figs. 16 and 17, were examined. The first model (Fig. 18), with pit floor elevation at 346 m level, has four combinations of slope designs, i.e.,

Case 1: slope angle of 35° on FW and 45° on HW
Case 2: slope angle of 45° on FW and 56° on HW
Case 3: slope angle of 50° on FW and 65° on HW
Case 4: slope angle of 35° on FW and 65° on HW.

The fifth case (Fig. 17) is a little different from the conventional slope design. Two slope angles were designed for each side of the walls, the pit floor being located at 334 m level with a narrow bottom width. The footwall and hanging wall have an angle of 30° and 50° respectively from the floor, up to 400 m elevation; from here they were modified to 50° and 60° for the

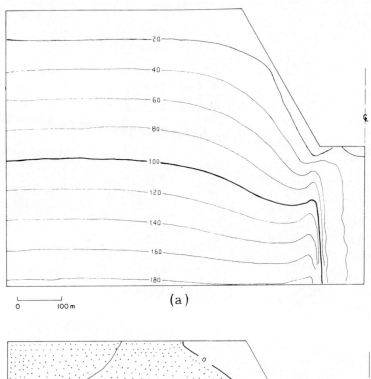

(a)

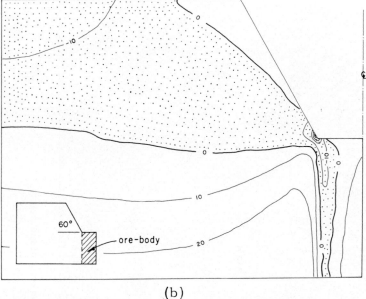

(b)

Fig. 15. Contours of equal-principal stresses (compression positive) for the two-material slope model (60°) with $E_o/E_r = 0.1$. Dotted area indicates tensile zone. Contour intervals are 20 and 10 kg/cm², respectively, for (a) major principal stress, and (b) minor principal stress.

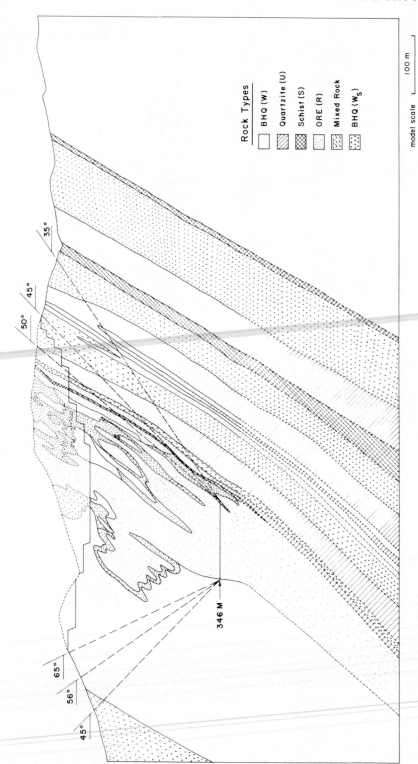

Fig. 16. Geological features of mine A.

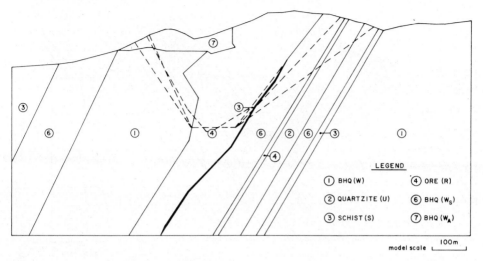

Fig. 17. Simplified geology for the finite element analysis. Dotted lines are the pit outline.

footwall and hanging wall. The FE mesh is shown in Fig. 19. It should be noted that Figs. 18 and 19 are the central portions of the models and that the full model extends 600 m further on the bottom and both sides.

Material properties and boundary stress conditions. The modulus of deformation, E, Poisson's ratio, ν, and the unit weight, γ, for the mine rocks were

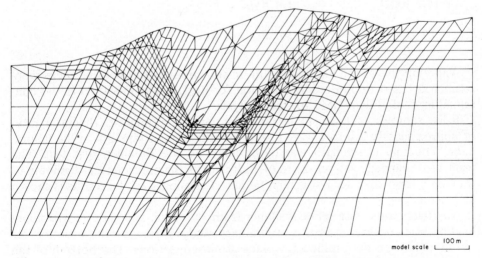

Fig. 18. Finite element mesh with pit elevation at 346 m. Shaded element, as indicated by an arrow, is the location where the largest tension occurred.

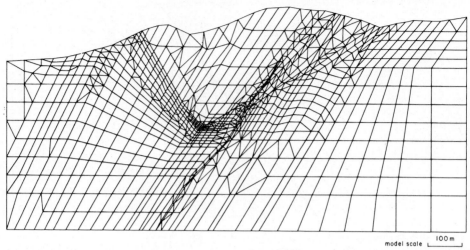

model scale $\underset{\text{100 m}}{\underline{\quad\quad\quad}}$

Fig. 19. Finite element mesh for the kinked slope with pit elevation at 334 m.

determined in the laboratory (see Table I). It is generally recognized that the modulus of deformation of rock substance is higher than that of the rock mass. Reduction factors based on a modified core recovery, designated as RQD (Rock Quality Designation), have been proposed by Deere (1964). An empirical equation has been deduced to estimate the modulus of deformation of rock mass (E_{rm}) from that of the intact rock substance (E_s):

$$E_{rm} = (4.5 \text{ RQD} - 3.05)\, E_s \quad \text{for RQD} > 70\%$$

$$E_{rm} = 0.1\, E_s \quad\quad\quad\quad\quad \text{for RQD} < 70\%$$

Applying the above equations, the in-situ moduli for the mine rocks were estimated based on limited information for RQD's. The modified values contained in Table I were used for the analysis.

Rock samples of 6.35 cm in diameter, having a height-to-diameter ratio of approximately 2.5 to 1 were tested for compressive strength in the laboratory. The tensile strength was determined by the Brazilian disc test. The tests were carried out on discs of 6.35 cm in diameter, and a diameter to thickness ratio of 4 to 1 was maintained. A summary of the results is shown in Table II.

No field stress data were available from this area, and only gravitational loading was considered in this preliminary study. Stress and deformation for the models were determined for plane strain conditions. The bottom of the model was fixed in the z-direction (vertical) and the sides were fixed in the r-direction (horizontal).

TABLE I

Physical properties of the mine rock for mine A

Code No.	Rock type	RQD (%)	Unit weight (Mg/m³)	Young's modulus (kg/cm² × 10⁵)		Poisson's ratio	
				laboratory	estimated	laboratory	assumed *
1	BHQ (W)	50	3.4	9.32 (9.15) **	0.91 (0.89) **	0.196	0.20
2	quartzite (U)	80	2.6	8.76 (8.60)	4.57 (4.48)	0.132	0.13
3	schist (S)	40	3.0	5.93 (5.82)	0.56 (0.55)	0.130	0.13
4	ore (R)	80	4.0	6.10 (5.98)	3.31 (3.25)	0.236	0.24
6	BHQ (A)	50	3.4	9.32 (9.15)	0.91 (0.89)	0.196	0.20
7	BHQ (W_A)	50	3.4	9.32 (9.15)	0.91 (0.89)	0.196	0.20

BHQ = banded hematite-quartzite.
* Smoothed from laboratory values.
** Numbers in brackets are expressed as MN/m² × 10⁴.

TABLE II

Average strength data of the mine rock for mine A

Code No.	Rock type	Uniaxial compressive strength (kg/cm²)	Standard deviation (kg/cm²)	Tensile strength (kg/cm²)	Standard deviation (kg/cm²)
1	BHQ (W)	2290 (8)	980	170 (7) *	30
2	quartzite (U)	2770 (4)	1050	210 (6)	15
3	schist (S)	1050 (5)	460	—	—
4	ore (R)	770 (5)	160	90 (8)	15
6	BHQ (W$_s$)	1690 (4)	1010	90 (11) **	13
				15 (12) *	1

Numbers in brackets are the number of samples tested.
 * By loading parallel to bedding.
** By loading at right angles to bedding.

Summary of results. The results for the several slope configurations were displayed by a computerized plotter which plotted the principal stress trajectories, and contour maps of principal stresses and displacement vectors.

The stress distribution patterns, under gravity loading, for the different cases examined, did not show very significant differences although local variations were noticed near the boundaries of excavation. Figs. 20—22 are examples of the principal stress trajectories for cases 1, 2, and 3, respectively. The principal stress directions are indicated by the orientation of the two line segments, mutually perpendicular and bisecting each other at the centroids of elements. The longest of the two lines indicates the direction of the major principal stress (compression positive) provided the principal stresses are all compressive; one-half length of a vector represents the scaled stress value. The line with a bar at its end indicates compression and a line without a bar indicates tension.

Figs. 23—28 show the contours of equal principal stresses. These plots permit a quick examination of the patterns and intensities of stress distribution in slopes. These plots clearly indicate that the quartzite bed in the footwall, which has a higher stiffness than the adjacent layers, carries a large proportion of the load and stresses are higher here than in the adjacent layers. It is interesting to note that large areas around the excavations are in a state of tension with high intensity concentrated at the corners of the toe for all the cases examined; both major and minor principal stresses around the toe in the hanging wall are tensile. However, these tensile stress patterns did not change significantly from the different combinations of slope angles on both sides of the pit wall except for case 5. Case 5 is a kinked slope with a narrow pit bottom, which produced much higher stress concentrations around the toe. This was probably due to the pit geometry, and this seems

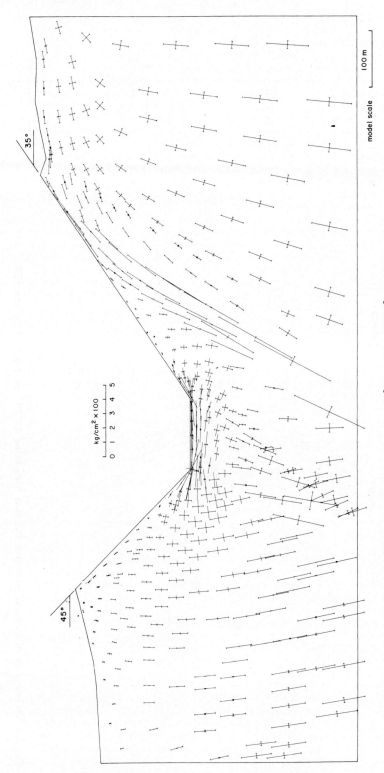

Fig. 20. Trajectories of principal stresses for case 1 of mine A (35° on FW and 45° on HW).

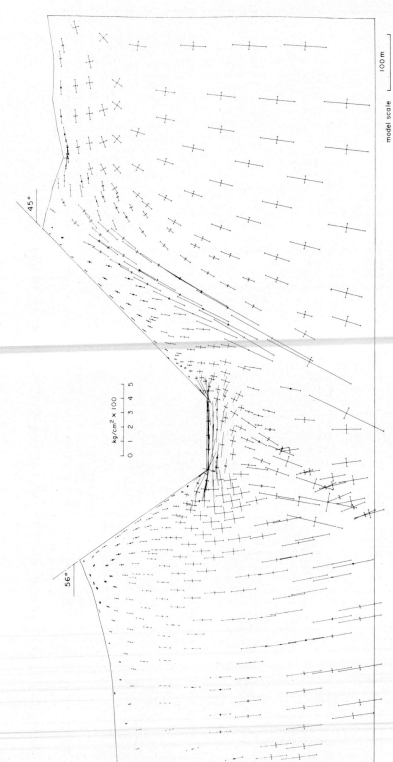

Fig. 21. Trajectories of principal stresses for case 2 of mine A (45° on FW and 56° on HW).

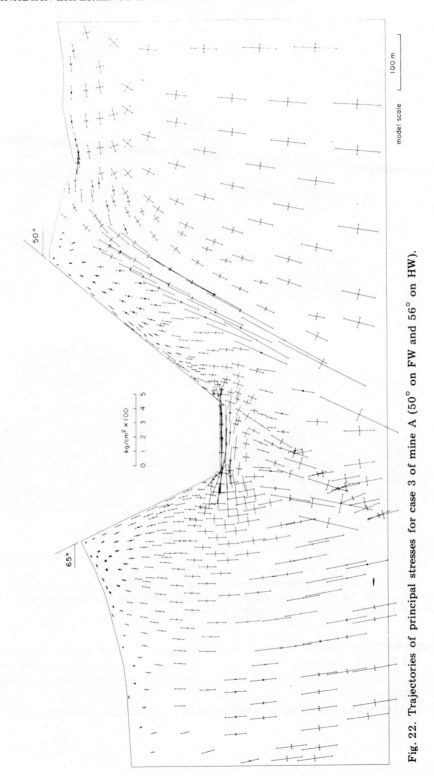

Fig. 22. Trajectories of principal stresses for case 3 of mine A (50° on FW and 56° on HW).

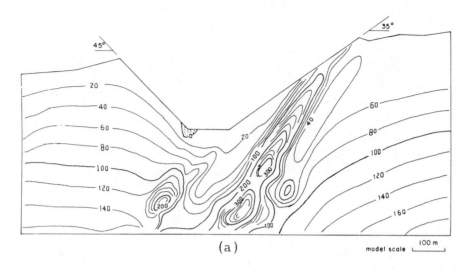

(a)

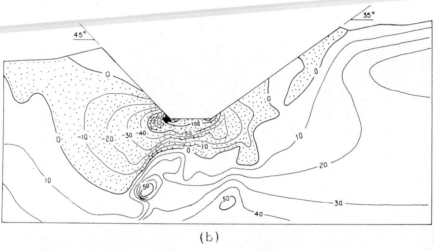

(b)

Fig. 23. Contours of equal-principal stresses (compression positive) for case 1 of mine A. Dotted area indicates tensile zone. Contour intervals are 20 and 10 kg/cm^2, respectively, for (a) major principal stress, and (b) minor principal stress.

to agree with the results of parametric studies that higher stresses would be expected from a concave-shaped slope. Owing to the unfavorable stress conditions resulting from the kinked slope (case 5), it was ruled out for design considerations.

It has been learned from the simple slope model that the differences in

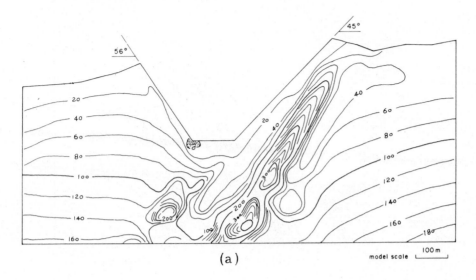

(a)

model scale |——| 100 m

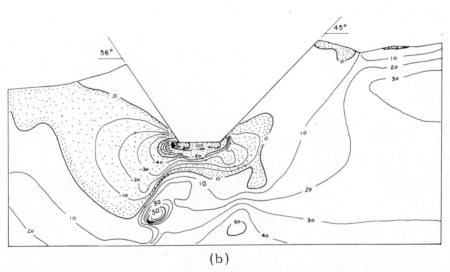

(b)

Fig. 24. Contours of equal-principal stresses (compression positive) for case 2 of mine A. Dotted area indicates tensile zone. Contour intervals are 20 and 10 kg/cm^2, respectively, for (a) major principal stress, and (b) minor principal stress.

stiffness between the different beds is an important factor in the development of these tensile zones. In the case of the model of mine A, E_o/E_r was 3.6, which seems to account for the high tensile stresses developed at the toe along the floor of the pit.

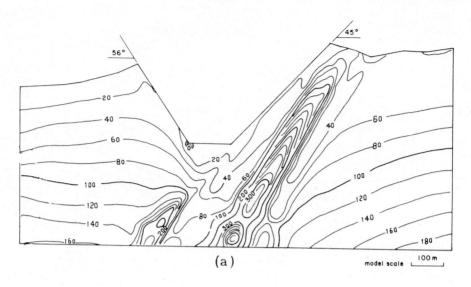

(a)

model scale

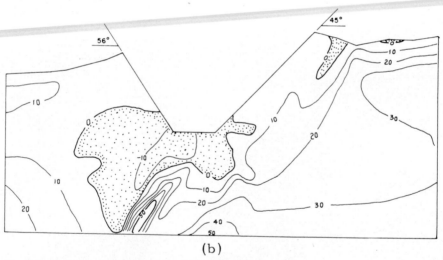

(b)

Fig. 25. Contours of equal-principal stresses (compression positive) for case 2 of mine A from "no-tension" analysis. Dotted area indicates tensile zone. Contour intervals are 20 and 10 kg/cm^2, respectively, for (a) major principal stress, and (b) minor principal stress.

The directions of these tensile stresses in the footwall of the model of mine A are approximately perpendicular to the direction of bedding. This would be an unfavorable condition as far as stability is concerned because the bedding planes are usually the weaker plane; when the tensile stress

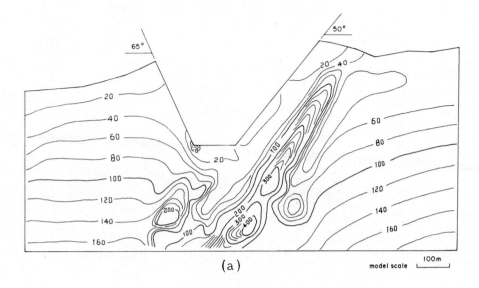

(a)

model scale 100m

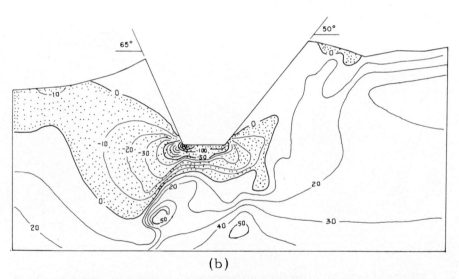

(b)

Fig. 26. Contours of equal-principal stresses (compression positive) for case 3 of mine A. Dotted area indicates tensile zone. Contour intervals are 20 and 10 kg/cm², respectively, for (a) major principal stress, and (b) minor principal stress.

exceeds the tensile strength of the rock mass, separation would occur along bedding planes. However, due to the reality of the presence of the numerous joints and fissures, most rock masses are incapable of sustaining tensile stresses; hence, the tensile stresses developed in the elastic model would have to be large-

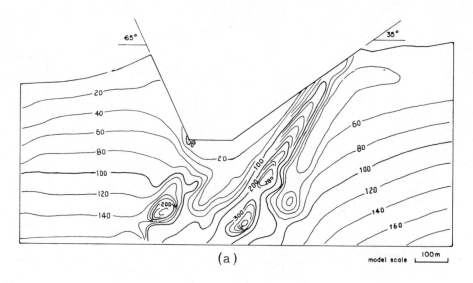

(a)

model scale └─────┤ 100m

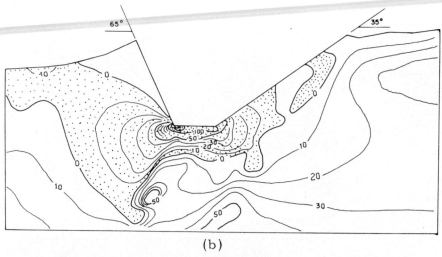

(b)

Fig. 27. Contours of equal-principal stresses (compression positive) for case 4 of mine A (35° on FW and 65° on HW). Dotted area indicates tensile zone. Contour intervals are 20 and 10 kg/cm², respectively, for (a) major principal stress, and (b) minor principal stress.

ly dissipated through a redistribution of the compressive stresses during mining. Therefore, an important non-linearity in practical applications in competent rocks is included in the "no-tension" idealization. A "no-tension" condition was simulated by an iterative procedure previously described in detail by

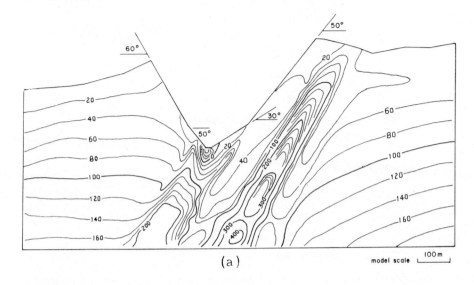

(a)

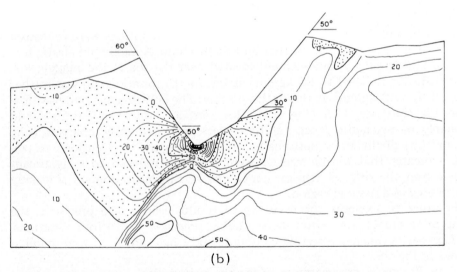

(b)

Fig. 28. Contours of equal-principal stresses (compression positive) for case 5 — the kinked slope of mine A. Dotted area indicates tensile zone. Contour intervals are 20 and 10 kg/cm², respectively, for (a) major principal stress, and (b) minor principal stress.

Zienkiewicz et al. (1968). In practice, the process converges to small tensile stresses after about 10—30 iterations.

Only one case of the mine model was analyzed for a "no-tension" condition. Due to the size of model (approximately 1500 elements and 1300

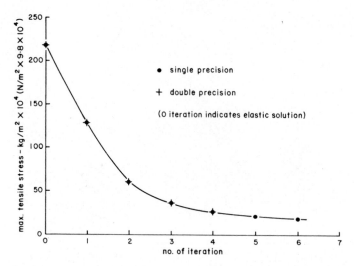

Fig. 29. Comparison between the results of double precision and that of single precision from "no-tension" analysis for one of the elements located at the toe as shown in Fig. 18.

nodes with a half bandwidth of about 40), only six iterations were processed using a single precision on an IBM 360/85 machine. A maximum elastic tension of 220 kg/cm² in one element located near the toe in the hanging wall (Fig. 18), was reduced to 26 kg/cm² at the fourth iteration and was further relaxed to 17.8 kg/cm² at the sixth iteration. The convergence rate was slow after the third iteration (Fig. 29). However, a "no-tension" state would probably be reached if more iterations were processed. A comparison (Fig. 24 and 25) of the stress patterns before and after the "no-tension" relaxation indicates that tensile stresses were generally reduced to a significant degree over the original tension zone. The small tensile domain is usually interpreted as a fissured region.

In order to examine the accuracy of results using single precision, the same mine model was re-run using double precision up to four iterations. Differences between the results of double and single precision were small as shown in Fig. 29.

Figs. 30 and 31 are examples of the excavation displacements for cases 1, 2, 3 and 4. The displacement patterns did not change significantly from one case to another; all showed that the pit floor heaved up and the crest moved into the pit wall.

In order to examine the effect of the relaxation of these tensile stresses on the displacement patterns, differences in displacements between elastic and "no-tension" cases are plotted in Fig. 32a. It might logically be expected that with the release of these tensile stresses the elements might expand, and hence, the pit bottom would heave up. However, the model showed that the

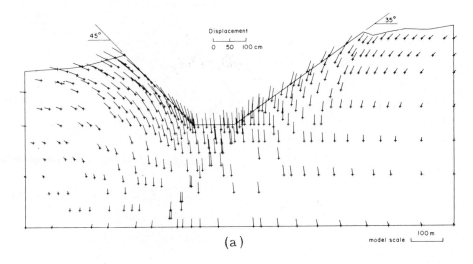

(a)

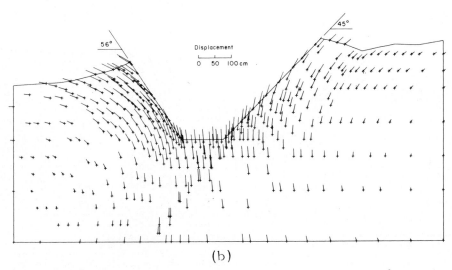

(b)

Fig. 30. Excavation displacement trajectories for mine A: (a) case 1 (35° on FW and 45° on HW), and (b) case 2 (45° on FW and 56° on HW).

relaxation displacements along the pit floor and its adjacent area were directed downward. This perhaps can be explained by the load transfer to adjacent elements which would cause a decrease in volume of these elements, and hence, downward displacement of the pit bottom. A decrease in volume of these elements would tend to close up numerous cracks in the rock mass;

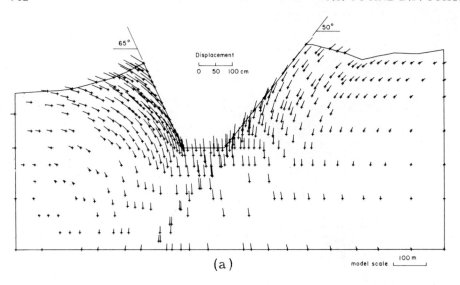

(a)

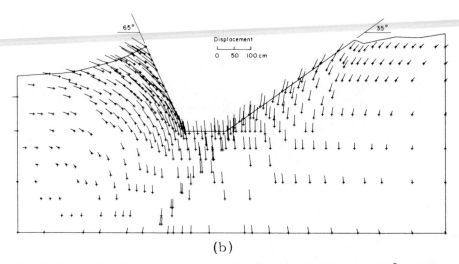

(b)

Fig. 31. Excavation displacement trajectories for mine A: (a) case 3 (50° on FW and 65° on HW), and (b) case 4 (35° on FW and 65° on HW).

consequently, rock mass quality would be improved, and hence, slope stability improved.

Differential displacement patterns were also examined for the simple two-material slope models. The modulus ratio between the orebody and the wall rock E_o/E_r was kept the same as that of the mine model, i.e., 3.6. The differ-

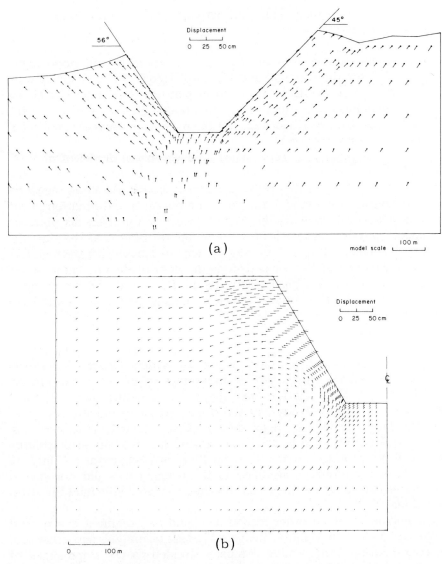

Fig. 32. Relaxation displacement due to "no-tension" analysis for: (a) case 2 of mine A, and (b) two-material slope model.

ential displacements calculated between the elastic solution and that of the "no-tension" analysis (six iterations) are plotted in Fig. 32b. A similar pattern was observed as a result of stress redistribution due to the relaxation of tension for a large portion of the slope; however, along the crest and one-third of the slope from the crest down, the displacements are directed inward and into the opening. This was, perhaps, caused by the steeper angle in the simple two-material model.

Discussion. Model studies show clearly that changes in the pit slope angles from 35° to 50° for the footwall and from 45° to 65° for the hanging wall, had very little effect on stress distribution patterns within the pit slopes except for some local variations along the boundaries. However, for the kinked slope (case 5), the stresses around the pit bottom increased considerably as compared to other cases.

Due to the complexity of rock types involved in the model of mine A, there should be no surprise that tensile zones were developed even at the lowest slope angle (35° on the footwall of the orebody). Based on the study of the two-material slope model, the ratio of modulus of deformation of the ore to that of the wall rock has the greatest effect on the stress distribution. The estimated physical properties for the model of mine A at the time the model was run, had a ratio of 3.6, which seems to account for the high tensile stresses developed at the toe and along the floor of the pit. These tensile stresses around the toe had been greatly reduced as a result of "no-tension" relaxation, and the resulting smaller tensile domain is usually interpreted as a "fractured region".

The footwall rocks consist of alternating beds of layered, strong elastic quartzite, layered low-strength elastic schist and layered strong elastic, schistose, banded hematite-quartzite. The most critical planes of weakness are the bedding planes which dip at an average of 63° into the pit. By making the overall slope flatter than the dip of bedding, say 50°, these planes of weakness will not daylight the slope. Although the stresses developed in the quartzite bed are high, they are below their uniaxial compressive strength (Table II). The rock of the hanging wall is a very strong elastic material, and it was judged that rotational shear or plane shear are unlikely to occur. Also, the stresses developed in the hanging wall did not exceed their strengths (Table II) and hence the hanging wall will withstand the stress of a 65° slope.

Taking into account all other factors as would be examined in a normal pit design program such as fabric analysis, blasting techniques and other economic considerations, the mining company decided to use slope angles of 65° for the hanging wall and 50° for the footwall. Excavation is now being carried out based on these decisions. However, no field deformation measurements are available at present.

Open pit mine B

General description. A slide took place on the northeast wall of mine B involving the displacement of 12—15 × 10⁶ tonnes of rock. The open pit mine where the slide occurred is approximately 3.3 km long, 1.0 km wide and up to 400 m in depth. The long axis of the pit trends N10°E. The overall slope angle varied from 42° to 46.5°. At the slide section the slope height was approximately 248 m. Details are given in Chapter 17.

Geologically, the mine is characterized by a complex contact between porphyry intrusives and an older metamorphic series of metasediments, volcanics and intrusives. The host rock in the slide area is granodiorite.

Several structural systems exist in the mine area. Three sets are present within the slide area, trending generally 000°, 045° and 135°. The well-developed 135° set includes a prominent fracture (labelled structure A) which strikes 135° and dips 60—90° southwest, as shown in Fig. 33. Structure A was the northern limit of the slide. A second large fracture (structure B) in the set formed the southwest margin of the slide.

Fortunately, the monitoring system which consists of a series of quadrilateral tension stakes, tape and transit lines, extensometers and seismograph,

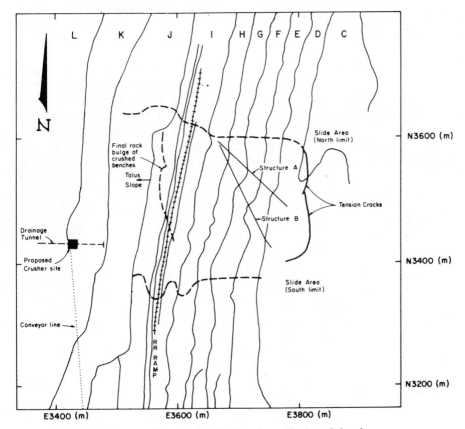

Fig. 33. Slide area in east side of mine B. C to L are the bench levels.

was established before the slide and served its purpose in predicting the occurrence. Consequently, this slide caused only a minimal interruption in production at the mine.

Tension cracks were first noticed on the top benches (see Fig. 33) at about two and one-half years before the slide occurred. However, the movement was small and eventually ceased. These tension cracks opened again in the same area after a strong earthquake which occurred just about fourteen months before the slide. Instrumentation to monitor rock movement was then established along the crest and over the entire slope face. The rate of movement was steady until a blast consisting of 211 holes loaded with 165 tonnes of explosives and breaking about 400,000 tonnes of material, was detonated at the base of the slide area just about three months before sliding took place. A typical plot of displacement versus time for the slide mass is shown in Fig. 34. The blast was made for the excavation of a crusher site, which required deepening of the pit by 39 m. Owing to the increased rate of movement, an unloading program, which involved the removal of 4.5 × 10⁶ tonnes of material, was started on top of the potential slide area. However, this unloading program did not achieve any stabilization and the slide took place about ten months later (Chapter 17).

Geometric simulation. It was not easy to select an adequate two-dimensional model to simulate a slope slide of this kind inasmuch as the phenomenon is really a three-dimensional one. A three-dimensional wedge analysis was, however, considered inappropriate because the dips of these two prominent faults do not produce a daylighted intersection. Examination of the in-situ displacement measurements showed that the horizontal displacements were greater than those of the vertical, particularly in bench *H*, as shown in Fig.

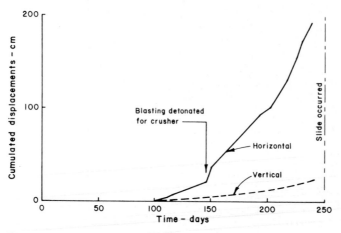

Fig. 34. A typical plot of displacement versus time. The reference date is arbitrary.

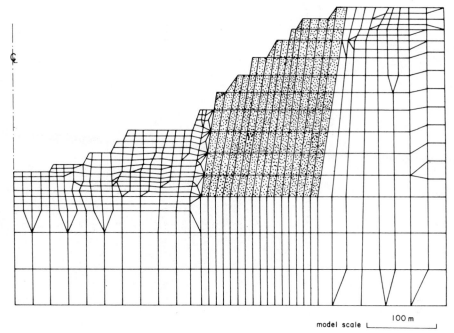

model scale |_____| 100 m

Fig. 35. Finite element model with toe elevation at 2684 m for the 1st-cut geometry. Dotted area represents the assumed fractured zone.

34. No toe movement was detected before the slide. Consequently, failure seemed to be a general break-up of the rock mass rather than a plane or rotational shear failure.

It was hoped that a plane strain analysis could be used to obtain some information concerning this slope slide. It was assumed that the loosening of this zone produced the measured deformations and that the broken mass exerted pressure against the rock toe-buttress. The fractured zone was simulated by sets of joints bounding conventional quadrilateral elements. The section N3500 looking north was selected for analysis. One of the FE meshes is shown in Fig. 35, and the dotted elements represent the assumed fractured zone. Each solid block is represented by one quadrilateral element, and each joint surface by one joint element per block face. This represents a joint spacing of approximately 10 m. Three stages of mining were modelled with the toe at elevations of 2710 m, 2684 m and 2645 m, the last one being the geometry prior to the slide. Elements were removed progressively to simulate each mining stage. The three configurations referred to later as the 1st cut, 2nd cut and 3rd cut correspond to the geometries for these stages (total height of slope being 183, 209 and 248 m, respectively). The third cut includes excavations at the toe for the crusher site and unloading of the top benches.

TABLE III

Physical properties of mine rocks for mine B

Physical properties	Case 1: joint model ($K_n/K_s = 100$)		Case 2: joint model ($K_n/K_s = 1.0$)		Case 3: elastic continuum model	
Modulus of deformation, E (kg/cm^2)	3×10^5	$(2.9 \times 10^4 \text{ MN/m}^2)$	3×10^5	$(2.94 \times 10^4 \text{ MN/m}^2)$	3×10^5	$(2.94 \times 10^4 \text{ MN/m}^2)$
Poisson's ratio, ν	0.20		0.20		0.20	
Unit weight (Mg/m^3)	2.7		2.7		2.7	
Shear joint stiffness, K_s (kg/cm^2/cm)	10^2	(9.8 MN/m^2)	10^3	$(9.8 \times 10 \text{ MN/m}^2)$	—	
Normal joint stiffness, K_n (kg/cm^2/cm)	10^4	$(9.8 \times 10^2 \text{ MN/m}^2)$	10^3	$(9.8 \times 10 \text{ MN/m}^2)$	—	
Tension limit for two-dimensional element (kg/cm^2)	1	$(9.8 \times 10^{-2} \text{ MN/m}^2)$	1	$(9.8 \times 10^{-2} \text{ MN/m}^2)$	1	$(9.8 \times 10^{-2} \text{ MN/m}^2)$
Tension limit for joint element (kg/cm^2)	0.1	$(9.8 \times 10^{-3} \text{ MN/m}^2)$	1	$(9.8 \times 10^{-2} \text{ MN/m}^2)$	—	

Physical properties. One of the major difficulties for modelling this slope was the inadequate knowledge of the physical properties of the rock mass, e.g., modulus of deformation, density, and joint stiffness. Therefore, estimates were made based on engineering judgements. The properties used in this study are shown in Table III. K_n and K_s, as shown in Table III, are the normal and shear stiffness of the joints, respectively. Since the concept of joint stiffness is relatively new, very little information regarding their values can be found in recent literature. However, De Rouvray and Goodman (1972) have reported that a reasonable variation range might be 1×10^2 to 1×10^7 and 1×10 to 1×10^3 kg/cm^2 per centimetre for K_n and K_s, respectively.

3.5×10^5 kg/cm^2 and 0.20 for modulus of deformation and Poisson's ratio, respectively, should be reasonable values for the type of rock involved. It was also assumed that the tensile strength for two-dimensional elements and joint elements was 1 and 0.1 kg/cm^2, respectively.

The model had 1647 nodes and 1516 elements (including 269 joint elements) with a half bandwidth of 59 for the jointed model. It was run with two sets of joint stiffness, referred to as case 1 ($K_n/K_s = 100$) and case 2 ($K_n/K_s = 1.0$). An elastic model (i.e., excluding joint elements) referred to as case 3 was also examined.

Computer program. The computer program was written at the mining Research Laboratories, in FORTRAN IV for an IBM 360/85 computer, and was substantially based on the previous work of Duncan and Goodman (1968). The program is developed for plane strain structure and is capable of handling joint perturbation and "no-tension" analysis for both two-dimensional elements and joint elements. The joint element is one-dimensional in the sense that it is defined by four nodal points like a two-dimensional element but a pair of nodes have identical coordinates in the state of rest.

Summary of results. For economic reasons, only five iterations were used in the process of analysis. The results described here were from the fifth iteration of each case..

The change of geometry from the 1st cut to the 2nd cut did not alter the stress patterns significantly because only a small volume of material was removed from the toe, and therefore, only stress trajectories and contours for each case of the 1st and 3rd cuts were presented as shown in Figs. 36—43.

The stress patterns for case 1 and case 2 are slightly different as shown in Fig. 36 and 37, respectively. As expected, a stress discontinuity is noticeable, particularly for case 2, when joint slip occurred at the boundary between the elastic continuum and the modelled fractured rock mass. Away from the slope face and within the fractured zone, the maximum compressive stress for a jointed model, cases 1 and 2, were both directed more vertically than for the elastic model (case 3). This indicates less load transfer through shear.

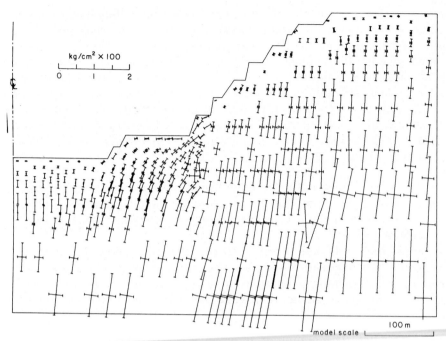

Fig. 36. Trajectories of principal stresses for the 1st-cut geometry — case 1 (K_n/K_s = 100).

Tensions on the order of 0—7 kg/cm² were observed around the pit bottom and along the slope face for all three cases. However, for case 2 of the jointed model, almost the entire face seemed to be in a state of tension. When the open pit was deepened at the toe for the crusher site at the same time as the unloading program was carried out at upper benches (3rd cut), the stress patterns were altered to some extent. For case 1 of the jointed model, the stresses around the area of lower benches were slightly increased as shown in Fig. 39 and the tensile zone on the surface and behind the crest had disappeared. This decrease in tension along the slope face was particularly noticeable for case 2 (compare Fig. 37 and Fig. 40). For the elastic model, the patterns of stress remained approximately the same for all three cuts although the magnitudes were slightly increased along some boundaries. The compressive stresses generated from all the cases did not seem to create any problem for slope stability.

The fractured rock mass is incapable of sustaining tension even though these tensions are small; consequently, local failure could first be initiated as a result of breakdown at elements where their stresses exceed their tensile strengths. As individual elements rupture, the released load will be redistributed to the adjacent elements; this increased load would cause further breakdown in that area if the increased stresses exceed their strengths. Conse-

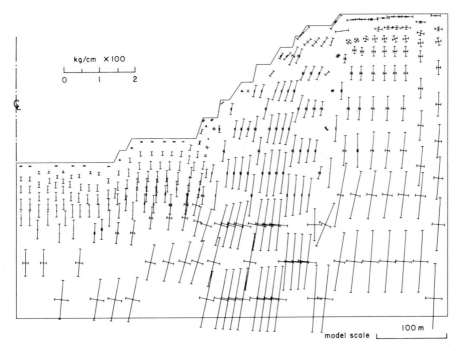

Fig. 37. Trajectories of principal stresses for the 1st-cut geometry — case 2 ($K_n/K_s = 1.0$).

quently a large scale of failure would be expected if this action or working of the ground continues.

The displacement patterns did not reveal any great differences among all the cases examined. They are all directed more or less upward. Figs. 42 and 43 show the displacement patterns for each case of the cuts 1 and 3. The magnitude of displacements along the upper benches were larger for case 2 and more separations were observed as joint slip occurred. This seems to be consistent with the fact that larger tensile zones have also developed for case 2.

Magnitudes of displacements in models are an order of magnitude less than those measured in the field. By the time deformation measurements were taken in the field, loose surface rock was undoubtedly contributing to a large portion of the displacement.

Discussion. Joint elements have successfully demonstrated their usefulness in simulating joint characteristics in small-scale models by Heuzé et al. (1971), Goodman and Dubois (1972), and De Rouvray and Goodman (1972). The two-dimension models with joint elements on such a large scale as mine B did not satisfactorily simulate the creation of a brecciated rock within the major structural features in the wall of a rock slope; this was possibly due to the geometrical nature of the slide and the inadequate represen-

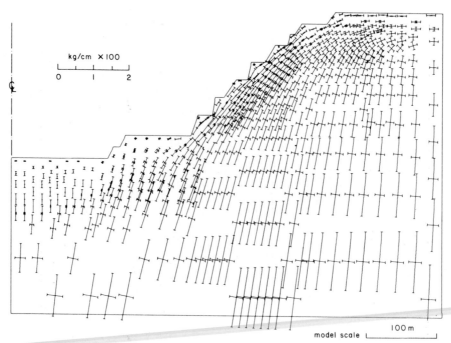

Fig. 38. Trajectories of principal stresses for the 1st-cut geometry — case 3 (elastic continuum).

Fig. 39. Trajectories of principal stresses for the 3rd-cut geometry — case 1 ($K_n/K_s = 100$).

Fig. 40. Trajectories of principal stresses for the 3rd-cut geometry — case 2 (K_n/K_s = 1.0).

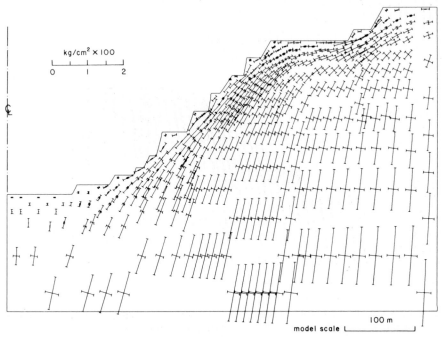

Fig. 41. Trajectories of principal stresses for the 3rd-cut geometry — case 3 (elastic continuum).

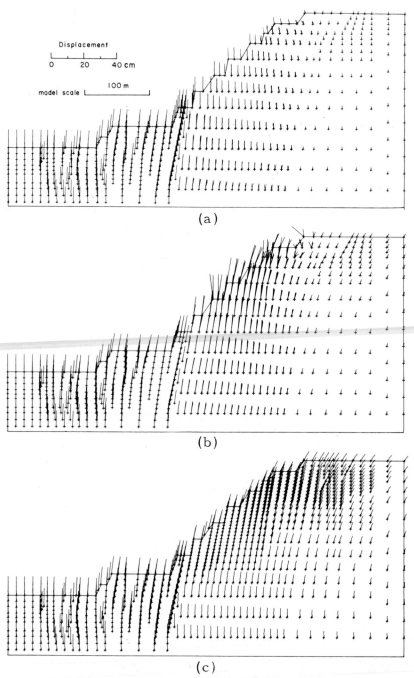

Fig. 42. Excavation displacement trajectories from the 1st-cut geometry for: (a) case 1 $(K_n/K_s = 100)$, (b) case 2 $(K_n/K_s = 1.0)$, and (c) case 3 (elastic continuum).

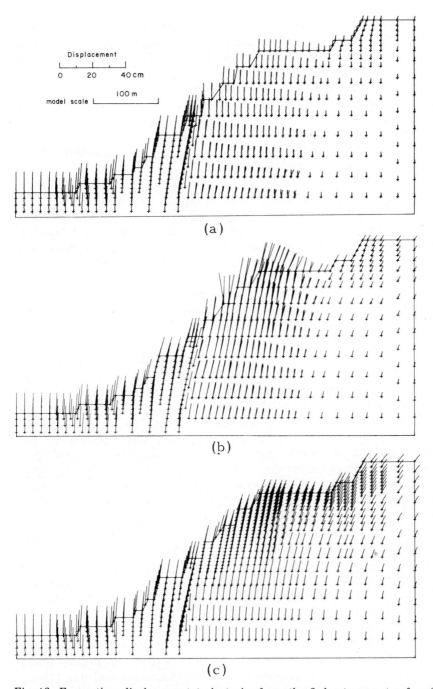

Fig. 43. Excavation displacement trajectories from the 3rd-cut geometry for: (a) case 1 $(K_n/K_s = 100)$, (b) case 2 $(K_n/K_s = 1.0)$, and (c) case 3 (elastic continuum).

tation of FE mesh for the geological structures. Furthermore, the physical properties of the mine rocks, particularly the normal and shear stiffness of joints, and the initial state of stress might be other factors contributing to the less-than-satisfactory simulation.

Despite all of these difficulties, however, the stress distributions obtained from the simulations provided some very useful information for analyzing rock slope stability.

CONCLUDING REMARKS

Parametric studies on stress distribution in typical slopes indicate that the mining of an open pit mine, under gravity loading, will not create high stress around critical areas such as the toe; but stresses in that area will be greatly increased as the ratio of the horizontal stress to the vertical increases, and hence, a knowledge of the pre-mining stress field is important for any meaningful rock slope simulation.

It has been shown that if a rock slope is cut in an inhomogeneous rock mass, the stiffness of the different formations will affect the resulting stresses greatly. Tensions occur not only at the surface some distance outward from the crest of an open pit, as reported by Blake (1967), but also can be expected around the toe areas; the extent and intensity of this tensile zone will depend on the stiffness ratio between the orebody and the wall rocks. Therefore, in order to obtain accurate results from a finite element analysis, representative and detailed deformation moduli for the in-situ rock would be essential. In other words, geological factors will have great impact on slope stability analysis. Tension zones were also observed around the toe of benches, and thus, the development of tension around excavations of an open pit would be also geometry-dependent.

The convergence of the stress transfer "no-tension" solution is slow for a large model such as open pit mine A and the cost of this analysis could be great. Thus a real "no-tension" condition cannot be easily obtained practically, i.e., a liberal tolerance must be accepted. In addition, the stress re-distribution pattern due to the relaxation of tension or the smaller tensile domain as a result of the stress relaxation in the larger domain, could be approximately predicted based on an elastic solution. Hence, for comparative purposes, an elastic solution can provide most of the useful information regarding the slope reactions to excavations.

Although joint characteristics have been successfully modelled with joint elements on a small scale, the capability of a one-dimensional joint element to simulate a brecciated rock mass on a large scale such as for mine B seems limited, possibly because: (1) it provides only a continuum solution, (2) the lack of realistic joint stiffness properties will usually compromise the results, and (3) sufficient refinement of the mesh within the brecciated rock mass will result in a large number of elements and will increase the bandwidth considerably. Furthermore such refinement is likely to be unacceptably expensive in computing time.

Excavation displacement or the movement of rock mass induced by excavation might be a valuable guide in examining the ground reaction to the removal of pit material. With an appropriate type of instrument, the movements at the crest and along the slope face should be detectable.

It has been demonstrated that the stresses around the toe of a slope obtained from plane strain solution, under higher tectonic stresses, can be as much as twice as high as the three-dimensional axisymmetric analysis; therefore, good engineering judgement should be exercised in applying the finite element results to slope stability analysis.

ACKNOWLEDGEMENTS

The authors wish to thank the mining companies who supplied the data for analysis. Messrs. N. Toews and D. Dugnore were very helpful for their assistance in programming and preparing the illustrations, respectively. Professor B. Voight read the manuscript and his comments and suggestions were most helpful.

REFERENCES

Blake, W., 1967. Stresses and displacements surrounding an open pit in a gravity-loaded rock. *U.S. Bur. Mines, Rep. Invest.*, 7002, 20 pp.

Coates, D.F., 1970. *Rock Mechanics Principles. Mines Branch Monograph 874.* Department of Energy, Mines and Resources, Ottawa, Ont., Chapter 2.

Deere, D.U., 1964. Technical description of rock cores for engineering purposes. *Rock Mech. Eng. Geol.*, 1: 17—22.

De Rouvray, A.L. and Goodman, R.E., 1972. Finite element analysis of crack initiation in a block model experiment. *Rock Mech.*, 4: 203—223.

Desai, C.S. and Abel, J.F., 1971. *Introduction to the Finite Element Method.* Van Nostrand-Reinhold, New York, N.Y., 477 pp.

Duncan, J.M. and Goodman, R.E., 1968. Finite element analysis of slope in jointed rock. *U.S. Army Corps Eng. Contr. Rep.*, No. S-68-3.

Goodman, R.E. and Dubois, J., 1972. Duplication of dilatancy in analysis of jointed rock. *Proc. Am. Soc. Civ. Eng., J. Soil Mech. Found. Div.*, 98 (SM4): 399—422.

Goodman, R.E., Taylor, R.L. and Brekke, T.L., 1968. A model for the mechanics of jointed rock. *Proc. Am. Soc. Civ. Eng., J. Soil Mech. Found. Div.*, 94 (SM3): 637—649.

Ghaboussi, J., Wilson, E.L. and Isenberg, 1973. Finite element for rock joints and interfaces. *Proc. Am. Soc. Civ. Eng., J. Soil Mech. Found. Div.*, 99 (SM 10): 833—848.

Herget, H., 1973. Variation of rock stress with depth at a Canadian mine. *Int. J. Rock Mech. Min. Sci.*, 10: 37—51.

Heuzé, F.E., Goodman, R.E. and Bornstein, A., 1971. Numerical analysis of deformability tests in jointed rock — "joint perturbation" and "no-tension" finite element solutions. *Rock Mech.*, 3: 13—24.

Pariseau, W.G., Voight, B. and Dahl, H.D., 1970. Finite element analysis of elastic-plastic problems in mechanics of geologic media — an overview. *2nd Int. Congr. on Rock Mechanics, Belgrade,* Paper 3-45: 311—323.

Voight, B. and Dahl, H.D., 1970. Numerical continuum approaches to analysis of non-linear rock deformation. *Can. J. Earth Sci.,* 7: 814—830.

Wang, Y.J. and Voight, B., 1970. A discrete element stress analysis model for discontinuous material. In: T.L. Brekke and F.A. Jørstad (Editors), *Int. Symp. on Large Permanent Underground Openings, Oslo, 1969.* Universitetetsforlaget, Oslo-Bergen-Tromsö, pp. 111—115.

Wilson, E.L., 1972. SOLID SAP — a static analysis program for three-dimensional solid structures. *Struct. Eng. Lab. Rep.,* No. UC SESM 71-9, (Univ. of California, Berkeley, Calif.).

Yu, Y.S. and Coates, D.F., 1970. An analysis of rock slopes using the finite element method. *Res. Rep.,* No. R 229 (Mines Branch, Department of Energy, Mines and Resources, Ottawa, Ont.).

Zienkiewicz, O.C., 1971. *The Finite Element Method in Engineering Science.* McGraw-Hill, London, 2nd ed., 521 pp.

Zienkiewicz, O.C., Valliappan, B.E. and King, I.P., 1968. Stress analysis of rock as a 'no-tension' material. *Géotechnique.* 18: 56—66.

REFERENCES INDEX, VOLUMES 1 AND 2

Coulson, J.H., (2) 643, 648
Coulter, H.W., (1) 161
Court, J.E., (1) 580, 601
Cowin, S.C., (1) 769, 791
Crandell, D.R., (1) 155, 161, 181, 183, 184, 194, 196, 482, 504
Crary, A.P., (1) 684, 688
Crosby, G., (1) 508, 515, 532, 548, 556
Cross, W., (2) 49, 51, 86, 192, 201, 223
Cruden, D.M., (1) 22, 23, 24, 63, 97, 111, 112, 675, 688
Cummans, J.E., (1) 118, 165
Cundall, R., (2) 30, 86
Curray, J.R., (1) 568, 570, 573, 574, 575, 576, 577, 578, 579, 580, 581, 585, 586, 587, 590, 591, 594, 596, 597, 598, 600, 601, 602, 603

Da Costa Nunes, A.J., (2) 421, 422, 423, 424, 427, 429, 431, 432, 434, 436, 438, 441, 443, 444, 446
Da Crus, P.T., (2) 444
Dahl, H.D., (1) 679, 688, 690, 691; (2) 29, 89, 452, 469, 710, 758
Dahlin, B.B., (1) 813, 829
Dahlstrom, C.D.A., (1) 102, 111, 112
Dake, C.L., (1) 39, 63, 424, 436
Dale, T.N., (1) 516, 556
Dallmeyer, R.D., (1) 522, 556
Daly, R.A., (1) 98, 99, 112
Dansgaard, W., (1) 794, 828
Dantas, H.S., (2) 444
D'Appolonia, D.J., (2) 9, 87, 454, 455, 463, 469
D'Appolonia, E., (2) 9, 87, 454, 455, 463, 469
Da Rocha Filho, I.P., (2) 444
Darton, N.H., (2) 250, 265
Da Silveira, I., (2) 444
Davidson, D.D., (2) 282, 283, 284, 287, 288, 298, 301, 304, 308, 309, 310, 311, 312, 313, 314, 322, 333, 340, 341, 342, 343, 344, 345, 348, 349, 350, 392, 395
Davidson, G., (1) 60, 63
Daviess, S.N., (1) 579, 590, 601
Davis, E.H., (1) 678, 690
Davis, H.E., (1) 687, 690
Dawson, G.M., (1) 151, 161, 265, 274
De Angulo, M., (2) 178, 180, 213, 223, 289, 301, 302, 304, 308, 309, 310, 311, 313, 314, 392, 395

Deere, D.U., (2) 62, 87, 301, 314, 434, 444, 453, 455, 469, 470, 511, 624, 631, 643, 648, 728, 757
Defensa Civil, (1) 321, 322, 336, 360, 361
De Jong, K.A., (1) 41, 64, 65, 66, 67, 165, 556, 557, 558, 560, 656, 726
Delach, M.N., (1) 541, 557
DeLory, F.A., (1) 350, 361
Denny, C.S., (1) 122, 161; (2) 450, 454, 469
Denton, P.E., (2) 85, 87
De Quervain, M., (1) 762, 785, 791, 804, 829
De Rouvray, A.L., (2) 749, 751, 757
Desai, C.S., (1) 670, 679, 688; (2) 87, 89, 710, 757
De Sitter, L.U., (1) 601
Dewey, J.F., (1) 4, 64, 510, 515, 518, 532, 535, 555, 556, 579, 601, 651, 653
Dibblee, T.W., (1) 626
Dickson, G.O., (1) 712, 725
Dietz, R.S., (1) 711, 725
Dill, R.F., (1) 571, 603
Dishaw, H.E., (1) 11, 112
Dodds, R.K., (1) 609, 622, 646, 647, 652, 653
Dodge, R.E., (1) 67
Doeringsfeld, Amuedo and Ivey, (1) 123, 161
Dolgushin, D.L., (1) 824, 828
Doll, C.G., (1) 508, 556
Donath, F.A., (1) 549, 556
Donovan, T.D., (2) 453, 457, 470
Dorf, E., (2) 257, 265
Dosch, E.F., (2) 62, 65, 90, 453, 471
Dott Jr., R.H., (1) 564, 601, 602, 603
Douglas, R.J.W., (2) 516, 539
Douglass, P.M., (2) 77, 87
Drake, C.L., (1) 561, 579, 581, 582, 583, 584 585 601, 603
Drake, L.D., (1) 815, 828
Drake Jr., A.A., (1) 530, 531, 549, 556
Drew, I.M., (2) 58, 88
Dronkers, J.J., (2) 322, 395
Dubois, J., (2) 751, 757
Duke, C.M., (1) 161
Dunbar, D.M., (2) 595, 631
Duncan, D.C., (1) 119, 120, 123, 163
Duncan, J.M., (2) 29, 87, 212, 223, 572, 574, 668, 682, 683, 690, 710, 711, 749, 757

Prucha, J.J., (1) 120, 164; (2) 250, 266
Pyrogovsky, N., (2) 414, 417

Radbruch-Hall, D.H., (2) 55, 56, 57, 89
Radhakrishnan, N., (1) 670, 679, 690
Radley-Squier, L., (2) 21, 89
Radok, U., (1) 795, 797, 812, 827
Ragle, R.H., (1) 198, 203, 205, 215, 248, 257, 258
Raisz, E., (1) 8, 66
Raleigh, C.B., (1) 41, 66, 535, 538, 549, 558, 725, 830
Ramsey, J.R., (1) 102, 112
Raney, D.C., (2) 331, 348, 351, 386, 387, 388, 389, 390, 397
Rapp, A., (1) 19, 66, 149, 156, 164; (2) 24, 68, 69, 89, 450, 454, 471
Rasmussen, L.A., (1) 822, 827, 832
Ratcliffe, N.M., (1) 506, 508, 510, 513, 514, 515, 516, 518, 532, 535, 544, 545, 546, 549, 553, 558, 559
Ray, L.L., (1) 457, 479
Raymond, C.F., (1) 61, 793, 798, 807, 808, 809, 811, 819, 824, 827, 832
Reade, T.M., (1) 44
Ready, D.W., (1) 832
Redlich, K.A., (2) 96, 107, 113, 132
Redlinger, J.F., (2) 71, 90, 668, 672, 673, 676, 677, 681, 688, 690
Reed, J.C., (1) 120, 121, 123, 125, 126, 163; (2) 257, 266
Reed Jr., J.C., (1) 556
Reese, L.C., (1) 670, 679, 690
Reeves, F., (1) 394, 395, 396, 397, 398, 399, 400, 407, 408, 420
Reiche, P., (1) 616, 656
Reid, H.F., (2) 49, 90
Reid, J., (1) 198, 203, 258
Reimnitz, E., (1) 579, 591, 592, 594, 602
Rendulic, L., (2) 114, 133
Rengers, N., (1) 90, 93
Renzetti, B.L., (2) 599, 632
Reusch, H., (2) 357, 397
Reyer, E., (1) 543, 559
Reynaud, L., (1) 807, 832
Reynolds, H.R., (1) 98, 112
Reyzvikh, V.N., (1) 356, 361
Rezende, S.H., (2) 443
Rice, G.S., (1) 98, 99, 112
Richards, A.F., (1) 541, 558, 559, 602
Richards, D.B., (2) 475, 514
Richards, P.W., (2) 250, 253, 257, 266

Richmond, G.M., (1) 120, 164
Richter, C.F., (2) 10
Rickard, L.V., (1) 508, 557
Riehl, N., (1) 802, 829, 832
Rigby, J.K., (1) 617, 654
Rigsby, G.P., (1) 804, 805, 832
Ringheim, A.S., (2) 527, 540
Rippere, K.H., (2) 600, 631
Rissler, P., (2) 34, 91
Rivard, P.J., (2) 668, 673, 676, 677, 683, 690
Roberson, M.I., (1) 581, 602
Roberts, A., (1) 650, 656
Roberts, D.G., (1) 579, 586, 603
Roberts, J., (1) 715, 726
Roberts, J.L., (1) 533, 559
Robertson, A.M., (2) 600, 632
Robertson, E.C., (1) 549, 559, 725, 726
Robin, G. de Q., (1) 229, 258, 795, 808, 809, 824, 825, 832
Robinson, A.H., (1) 6, 17, 64; (2) 2, 87
Robinson, C.S., (1) 631, 656; (2) 474, 475, 476, 478, 479, 482, 486, 488, 489, 491, 494, 496, 507, 508, 509, 513, 514
Roch, A., (1) 738, 741, 747, 752, 777, 785, 792, 797, 798, 806, 828
Rocha, M., (2) 624, 632
Rodgers, J., (1) 506, 507, 519, 553, 559; (2) 452, 471
Rodine, J.D., (1) 156, 164
Rodrigues, L.F.V.C., (2) 445
Röthlisberger, H., (1) 62, 66, 794, 802, 819, 820, 821, 826, 832
Roggensack, W.D., (2) 524, 540
Rona, P.A., (1) 579, 581, 594, 603, 651
Root, S.I., (1) 530, 559
Rosengren, K.J., (1) 497, 504
Rosenkrantz, A., (1) 20, 21, 66
Ross, C.P., (1) 124, 125, 164
Ross, D.A., (1) 541, 559, 564, 579, 587, 603
Rothpletz, A., (1) 72, 93
Rouse, G.E., (1) 266, 274
Rouse, J.T., (1) 424, 427, 437
Rubey, W.W., (1) 41, 44, 66, 123, 164, 398, 403, 420, 532, 533, 537, 538, 557, 705; (2) 115, 132, 257, 266
Ruedemann, R., 506, 559
Ruiz, M.D., (2) 422, 445
Runcorn, S.K., (1) 716, 727

Shor Jr., G.G., (1) 579, 590, 604
Shreve, R.L., (1) 27, 28, 29, 30, 31, 32,
    34, 75, 77, 79, 80, 88, 89, 90, 91, 93,
    126, 153, 165, 178, 180, 196, 198,
    203, 205, 212, 214, 215, 216, 219,
    220, 222, 223, 224, 225, 226, 228,
    230, 231, 238, 243, 249, 250, 258,
    304, 312, 314, 481, 482, 483, 484,
    486, 489, 490, 494, 497, 498, 500,
    501, 504, 702, 705, 798, 806, 808,
    811, 815, 819, 828, 832
Shumaker, R.C., (1) 512, 513, 559
Shumskiy, P.A., (1) 794, 808, 832
Silliman, B., (1) 155, 165
Silva, J.X., (2) 445
Silva Filho, B.C., (2) 439, 446
Silver, E.A., (1) 576, 603
Silverberg, N., (1) 581, 603
Simaika, Y.m., (2) 554, 574
Simpson, A., (1) 287
Simpson, E.S.W., (1) 579, 586, 601
Sims, P.K., (2) 481, 514
Sinclair, S.R., (2) 521, 540
Singh, A., (1) 609, 656, 670, 671, 672,
    673, 674, 682, 684, 689, 690
Skempton, A.W., (1) 133, 165, 350, 361,
    402, 415, 418, 421, 542, 559, 610,
    637, 652, 656; (2) 9, 51, 71, 78, 90,
    113, 132, 211, 212, 224, 457, 471,
    690
Skibitsky, A., (1) 671, 691
Skopek, J., (2) 668, 690
Slingerland, R.L., (2) 317
Smedes, H.W., (1) 425, 433, 437, 619
Smith, B.M.E., (1) 795, 832
Smith, C.K., (2) 71, 90, 175, 224, 668,
    672, 673, 676, 677, 681, 688, 690
Smith, D.G.W., (2) 522, 533, 540
Smith, J.H., (1) 677, 689
Smith, R.B., (1) 124, 126, 165
Smith, S.W., (1) 667, 668
Smoluchowski, M.S., (1) 44, 67
Smyth, W.R., (1) 519, 522, 523, 560,
    561
Snead, D., (1) 670, 672, 684, 688, 690
Snead, R.E., (1) 9, 67; (2) 63, 90
Snow, D.T., (1) 322, 361; (2) 34, 90
Sobral, H.S., (2) 436, 446
Sokolov, N.I., (1) 615, 621, 656
Solberg, G., (2) 383, 395
Sollas, W.J., (1) 44, 67
Somerton, W.H., (2) 592

Sommerfeld, R.A., (1) 737, 748, 752
Sommerhalder, E., (1) 774, 787, 792
Spencer, E., (1) 398, 421
Spencer, G.S., (1) 402, 409, 419, 421,
    541, 557; (2) 52, 53, 54, 71, 76, 87,
    539
Sproule, D.O., (1) 684, 687
Stacy, J.R., (1) 120, 121, 123, 125, 126,
    163, 675, 690
Stagg, K.G., (1) 654, 656, 691
Stanley, D.J., (1) 557, 581, 603
Stanley, R.S., (1) 515, 559
Stateham, R.M., (2) 577, 578, 582, 583,
    584, 587, 591, 592, 593
Stearns, D.W., (1) 120, 165; (2) 250, 257,
    266
Steinemann, S., (1) 799, 803, 804, 833
Stemberg, H.O., (2) 446
Stenborg, T., (1) 819, 833
Stephens, G.C., (1) 531, 558
Stevens, E.H., (1) 424, 437
Stevens, R.K., (1) 507, 508, 519, 520,
    521, 524, 532, 534, 547, 550, 553,
    559, 560, 561
Stewart, R.H., (2) 204, 223
Stewart, R.M., (2) 651, 666
Stini, J., (1) 610, 622, 650, 656; (2) 95,
    96, 97, 98, 99, 100, 101, 102, 103,
    104, 105, 106, 107, 108, 109, 121,
    126, 132
St. Julien, P., (1) 524, 525, 526, 527,
    528, 529, 534, 535, 537, 551, 559
St. Lawrence, W.F., (1) 735, 737, 748,
    751, 752
Storms, W.R., (2) 634, 636, 638, 649
Stose, G.W., (1) 530, 531, 560
Stout, M.L., (1) 484, 504
Strahler, A., (1) 98, 112, 117, 165, 616,
    656
Strakhov, N.M., (1) 7, 9, 67
Stride, A.H., (1) 579, 586, 587, 594, 603
Strøm, H., (2) 357, 397
Strömquist, L., (2) 68, 89
Strohbach, K., (1) 713, 727
Strohm Jr., W.E., (2) 151, 178, 180, 213,
    223, 282, 288, 292, 296, 301, 302,
    304, 307, 308, 309, 310, 313, 314,
    340, 393, 395, 646, 649
Stroud, R.A., (2) 58, 88
Studer, B., (1) 59
Subitzky, S., (1) 556
SUDESUL, (2) 446

SUBJECT INDEX, VOLUMES 1 AND 2

---

* Compiled by B. Lester, R. Sweigard and B. Voight, Department of Geoscience, Pennsylvania State University.